Cybernetic Revelation

Deconstructing Artificial Intelligence

JD Casten

ISBN: 978-0-9854802-0-2

Post Egoism Media
Eugene, Oregon, USA
www.postegoism.net

Contents

Part I
Introducing Cybernetic Revelation: Deconstructing Artificial Intelligence

Part II
Major Advances in the Philosophical History of Cognitive Conceptualism

Part I

Introducing Cybernetic Revelation Deconstructing Artificial Intelligence

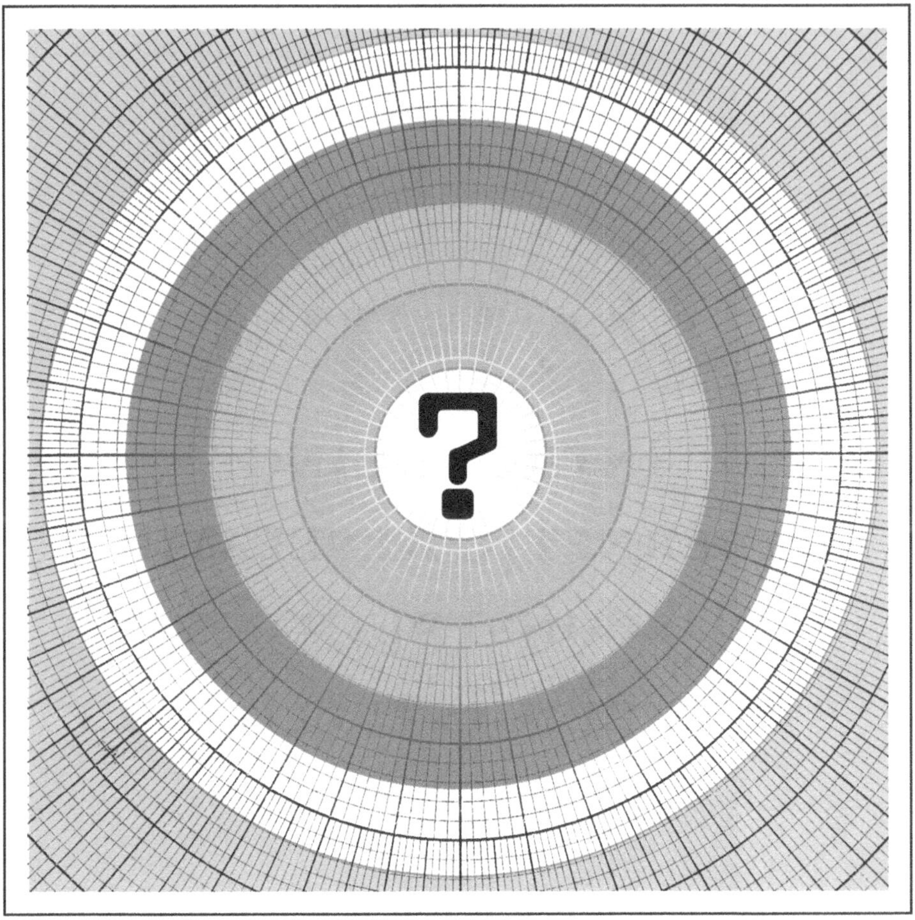

Chapter 1

Aim, Scope & Methodology

INTRODUCTION: I/O = ∞

> "For the genetics that we have attached to the machine to be the basis of a kind of evolution through natural selection, we must account for it by variation and the inheritance of variations. However, the type of machine genetics which we suppose has room for both. Variation occurs in the inaccuracy of the realization of the copying process we have discussed, while the copied machine exemplified in our white box is itself available as an archetype for further copying. Indeed, whereas in the original one-stage copying the copy resembles the original in operative image, but not in appearance, in the next stage of copying the spatial structure is preserved and the replica is a replica in that as well" (Norbert Wiener, *God & Golem, Inc.: A Comment on Certain Points where Cybernetics Impinges on Religion*, The MIT Press: Cambridge, MA (1964), p. 48).

When I was a high school student in the mid 1980's, and first started to think about programming artificial intelligence into a machine, I thought about writing a novel—a novel from a stream-of-consciousness perspective, like a James Joyce novel, but in this novel the self-narrator would slowly discover that they were an android. Later in college, I recalled having a realization, partly inspired by my studies of deconstruction, that I too, as a human being, was at least partly a machine. These two trains of thought are connected: of course, I did not discover in my own stream-of-consciousness that I was an android... but the realization of being "constructed"—both by culture and DNA—was a bit humbling.

Despite the above quote from Norbert Wiener's book, *God & Golem, Inc.*, and even with the title, *Cybernetic Revelation*, this book is not primarily focused on issues of religion. However, it might be related to religious issues in a manner of "negative theology." To my thinking, although there is much to understand about human nature, the world, universe, etc. in a mechanical and scientific way—there are some questions which science and mechanical understanding, technological understanding, cannot touch. Technology is more about the *form* of *information*, rather than the information content itself. The laws of physics tell us little about history, for example—and it is my contention that artificial intelligence tells us little about qualitative consciousness, other than by delimiting it in the negative. And seeing that some aspect can be defined by its "other" is a deep insight illustrated by many deconstructionists.

In 2011, IBM's artificial intelligence system, Watson, competed and won on the game show *Jeopardy!*—winning against two other top human contestants. Of course, Watson had access to much of the information contained on the internet (a vast knowledge base)—but the ability to understand the answer-questions and pose question-answers in natural language is an astounding feat, far beyond IBM's Deep Blue beating world chess champion Garry Kasparov in a 1997 tournament—for it is a leap from a logical game with quite limited rules in a fixed domain (a chessboard), to the flux of real world knowledge that we live in. Watson required more types of know-how than Deep Blue—and this was also applied with the programming of Watson, as team leader David A. Ferrucci wrote in a 2012 piece in the New York Times:

> "We had to keep the team's collective intelligence from being overcome by egos" (David A. Ferrucci, "Building the Team That Built Watson," *New York Times*, January 7th, 2012).

And:

> "Likewise, the scientists would have to reject an ego-driven perspective and embrace the distributed intelligence that the project demanded [....] We learned to depend on a philosophy that embraced multiple tracks, each contributing, relatively small increments to the success of the project" (ibid).

Such quotes demonstrate the complex decentralization of intelligence for both machines and humans; and indeed we find that much of what we know comes from without, and in many different ways—as well as our own brains having differing "sub-organs" that perform different types of tasks. We will most likely find no general theory of everything, like "1/0 =∞" that will be a holy grail for figuring out intelligence. And despite the complexity of deconstructive theory and artificial intelligence studies, I believe we will find much that eludes these contemporary paradigms of thought.

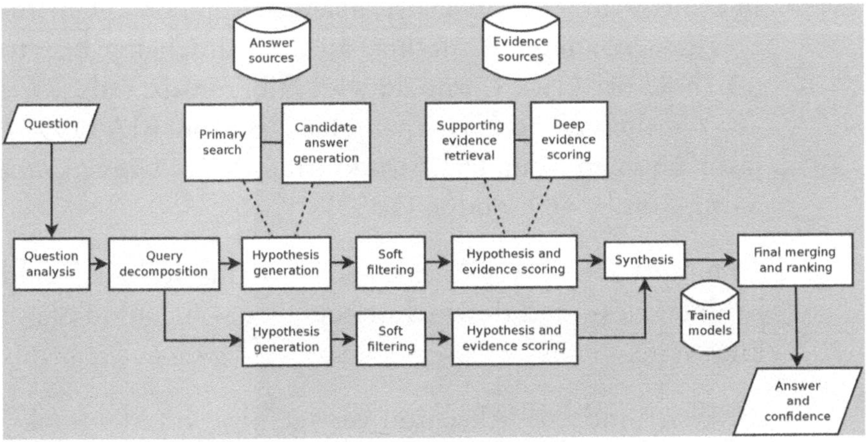

Flow chart for Watson's processing

WISDOM & WIT

> "You can find in a text whatever you bring, if you will stand between it and the mirror of your imagination. You may

not see your ears, but they are there" (Mark Twain, *The Wit and Wisdom of Mark Twain*, Running Press: Philadelphia, PA (1990), cited from "A Fable").

"This triple world resembles a net, or water in a mirage that is agitated; it is like a dream, *maya*; and by thus regarding it one is emancipated.

Like a mirage in the springtime, the mind is found bewildered; animals imagine water but there is no reality to it.

There is nothing but thought construction, it is like an image in the air; when they thus understand all, there is nothing to know.

Eternity and non-eternity; oneness, too, bothness and not-bothness as well: these are discriminated by the ignorant who are confused in mind and bound up by errors since beginningless time.

In a mirror, in water, in an eye, in a vessel, and on a gem, images are seen; but in them there are no realities anywhere to take hold of" (Jack Kornfeild with Gil Fronsdal (eds.), *The Teachings of t he Buddha*, Shambhala: Boston, MA (1993), p. 136, quoted from, D.T. Suzuki (trans.), *The Lankavatara Sutra*, Routledge: London (1932)).

"[Humpty Dumpty:] 'You've been listening at doors—and behind trees—and down chimneys—or you couldn't have known it!'

'I haven't, indeed!' Alice said very gently. 'It's in a book'" (Lewis Carroll, *Alice's Adventures in Wonderland and Through the Looking-Glass*, J.J. Little & Ives Company: New York (undated), p. 185).

"My father began to receive a series of anonymous and threatening letters. They said that, although no legal action would be taken against him at present, he was being

> watched (as if he didn't know that!) and that if he were to visit a certain psychotherapist (name and address supplied) he would find any eventual penalty for his immorality much reduced. His first instinct was to ignore this pressure, but when the letters eventually spelled out the alternative outcomes—a warning from the police versus a faked suicide in prison—he decided it would do no harm to comply.
>
> And so he began twice-weekly sessions [...]" (Momus (a.k.a Nicholas Currie), *The Book of Jokes*, Dalkey Archive Press: London (2009), p. 113).
>
> "It is true that it is common to hear one say 'I *made* a joke,' but one feels that one behaves differently during the process than when one pronounces a judgment or offers an objection. Wit shows in a most pronounced manner the character of an involuntary 'inspiration' or a sudden flash of thought. A moment before one cannot tell what kind of joke one is going to make, though it lacks only the words to clothe it. One usually experiences something indefinable which I should like most to compare to an absence, or sudden drop of intellectual tension; then all of a sudden the witticism appears" (Sigmund Freud, A.A. Brill (trans.), *The Basic Writings of Sigmund Freud*, The Modern Library: New York (undated), p. 752, from *Wit and Its Relation to the Unconscious*).

These wisdom and wit quotations may at first reading seem irrelevant to a book on the intersecting histories of artificial intelligence and deconstruction. But on closer inspection, one finds them packed with relevant issues: Twain's idea that we interpret a text through the *mirror of imagination*; the Buddhist notion that "there is nothing but thought construction;" the "para-logical" and paranoid speculation on sources of knowledge with Humpty Dumpty and Momus; and of course, that which lies on the verge of consciousness discussed by Freud. These themes are tied together by implying different kinds of epistemological theories of how we can know what we think we know.

Intelligence and reason as well as philosophy and theory may be contrasted with jokes and wisdom too—where wisdom can be seen as more poetic and possibly divine in origin—or at least developed from life experience rather than mere theoretical reflection. Although philosophers may seem wise, wisdom seen as guides to living life, rather than speculations on mind-body metaphysics and epistemology, might be better sought in Aesop's fables, or in Zen meditation. But our jokes and wisdom—like everything entwined with our world views—reflect our understandings of the world—and even philosophical *systems* are often implied in the simplest of witticisms. The *form* of intelligence and wisdom as *content* may be inextricably intertwined in we humans... but might an artificially intelligent robot have a sense of humor, or wisdom to share?

FENDER / BENDER

In a way, cartoons convey "digested" information. Usually the stories told are simple, and the visualizations are much less complex than those found in real life. There is a possible danger in predigested information—much like candy-cereal, chicken-nuggets or junk food having possibly deceptive appearances as to what they contain. Newspaper headlines may grab your attention, and advertisements may evoke positive feelings for a product, but the over-simplification of the issues found in such can detour one from the complexities of the issues involved, often so with some cloaked agenda in the background. Consider the following jokes from two cartoon robots, Fender from the movie "Robots," and Bender from the television show "Futurama":

> "Fender: I know that sounds bad, but I'm just doing musical arm farts. You know how to do those? They're hard to do because we're made of metal, but that's where the skill comes in" (*Robots*, Dir. Chris Wedge and Carlos Saldanha, Perf. Robin Williams, Blue Sky Studios, 2005, film).

> "Fry: If your programming told you to jump off a bridge, would you do it?

Bender: I dunno, I'd have to check my programming... yup."

"Bender: You really want a robot for a friend?

Fry: Yeah, ever since I was six.

Bender: Well, okay. But I don't want people thinking we're robo-sexuals, so if anyone asks, you're my debugger" ("Space Pilot 3000," *Futurama*, Writs. David X. Cohen and Matt Groening, Dir. Rich Moore and Gregg Vanzo, Perf. John Di Maggio and Billy West, Fox, 28 Mar. 1999, television).

Despite these jokes, some knowledge about robotics is implied—Fender references "skill," Bender talks of being locked in to his "programming," and the issue of robot-human friendship (even sexuality) is raised. But most clearly, these cartoon robots demonstrate *personality*.

Personality. Wit. Wisdom. These are themes that will largely be absent from this book! Yet, I bring them up because we find so many ideas, I believe, defined in relation to their "other." Again, this is an insight found with structuralism, post-structuralism, postmodernism, and deconstruction (as well as with Buddhist teachings on Pratītyasamutpāda, or dependent origination). It is my hope that by covering a history of what might be called "conceptual engineering"—the mechanics of thought or the technology of knowledge—we will elucidate not only this topic, but by taking this course to its limits, we may better see what lies on the other side of this border. But moreover, if we are to truly replicate human intelligence to its full extent, our mechanized friends would necessarily be funny and witty, along with a host of other personality traits.

WHAT THIS BOOK IS NOT: OTHER BOOKS

Before outlining the trajectory of this book, I think it would help to point out some excellent books that have similar scope, but are not exactly what is going on here. This will help to situate this book, but also demonstrate the structuralist insight that meaning is defined by differences—the "others" that give shape to any issue in question, much as a fence or border will define property or an area.

This book is not Charles Van Doren's book charting the *progress of knowledge*:

> "The knowledge that [...] expands and accumulates is of several kinds. We know more today about how nature works than we knew a hundred years ago, or a thousand, and we can expect to know even more a hundred years hence. It is easy to understand and accept the idea of progress in know-how, or technology, and to be optimistic about its continuing in the foreseeable future.
>
> Progress in other kinds of knowledge *may* have occurred. For example, as long as historians are free to write about the past, and readers are free to read their books (neither has always been true, as the Roman historian Tacitus reminds us), we will never forget the new ideas about just government that were advanced and fought for during the revolutions of the eighteenth century in England, America, and France. This does not mean that better governance is inevitable; the time may come when we look back with a sigh to those happy days when democracy flourished throughout much of the globe. But even then we will *know* more about governance then we once did.
>
> Similarly, the glowing examples of Socrates, Jesus, St. Francis of Assisi, and Dr. Martin Luther King, Jr., to name only a few, will not be lost while we can read or otherwise recall the stories of their lives and realize how they challenge us to live like them. This does not mean we will necessarily be better human beings, but we will know more about what human excellence is and can be" (Charles Van Doren, *A History of Knowledge: The Pivotal Events, People, and Achievements of World History*, Ballantine Books: New York (1991), p. xvi).

This book is not Richard Tarnas' Western *intellectual history*:

> "The 'man' of the Western tradition has been a questing masculine hero, a Promethean biological and metaphysical rebel who has constantly sought freedom and progress for himself, and who has thus constantly striven to differentiate himself from and control the matrix out of which he emerged. This masculine predisposition in the evolution of the Western mind, though largely unconscious, has been not only characteristic of that evolution, but essential to it" (Richard Tarnas, *The Passion of the Western Mind: Understanding the Ideas That Have Shaped Our World View*, Ballantine Books: New York (1991), p. 441).

Indeed, much of the western male history of thought has worked itself out, as if it were working on some sort of "generic" image of humanity and the world—that it didn't matter that *men* were developing these philosophies—my own view is that the masculine differs from the feminine about as much as physical bodies do—there *are* differences... but most of us have arms, legs, and heads to think with. Who would exclude women from the insights of history's lopsidedly male philosophical lineage? Yet, I think it appropriate to appropriate this history, and transform it towards more inclusive and ethical ends.

This book is not Ian P. McGreal's edited collection, *Great Thinkers of the Eastern World*:

> "The Western intellectual tradition has been most successful in its study and use of scientific and pragmatic thinking, but for centuries it preoccupied itself excessively with theological and metaphysical speculations (unfortunately with little or no empirical or logical warrant), and it repeatedly made intuitive and logical attempts (in my opinion, bound to be unsuccessful) to discover the foundations of ethics. However, with all its faults, the Western philosophical tradition has been helpful in illuminating the relations between the

uses of language and the world in which we presume to find ourselves; it has often been demanding in its examination of pretenders to wisdom; and it has valiantly defended the systematic search for knowledge, the exercise of benevolence, and the ideal of moderation in the pursuit of happiness.

> The Eastern tradition is refreshing in that it is predominantly and forthrightly ethical without purporting to prove what is a matter of commitment and tested cultural practice, and its fundamental message—often appropriately expressed in poetic language—is that one will manage best in this life if one disciplines oneself to go with the *Dao* (*Tao*), that is, to be in harmony with the universe as it is and, accordingly, in harmony with one's fellow human beings" (Ian P. McGreal (ed.), *Great Thinkers of the Eastern World*, Harper Collins Publishers: New York (1995), p. ix).

I do not believe the insights of the western male thinkers that dominate the present book's discussion, are necessarily biased if they focus on what might be called "generic" or "universal" aspects of humanity—unless such a focus on the "generic" or "universal" perspective itself is inherently male and / or western. Gender and geographic bias critiques do not invalidate "western male" knowledge: they simply call for a broader scope and inclusiveness in what is considered important.

This book is neither of Peter Watson's *histories of ideas*:

> "There *has* been a general development, a steady progress much of the time [....] But by no means all of the time. Throughout history certain countries and civilizations have glittered for a while, then for one reason or another been eclipsed. Intellectual history is very far from being a straight line—that is part of its attraction" (Peter Watson, *Ideas: A History of Thought and Invention from Fire to Freud*, Harper Collins Publishers: New York (2005), p. 2).

> "Postmodernism and relativism are still in the ascendant, but for how much longer? While the cultures of Africa, Bali and other third world countries have been recovered, to an extent, and given a much needed boost, none has so far found the widespread resonance that the classical civilizations of the Middle East once enjoyed. No one doubts that the jewels of art, learning and science have occurred in all places and at all times, and the identification and extension of this wide range has been a major achievement of twentieth-century scholarship. In particular, the vast body of knowledge concerning the early, pre-Columbus native America, has revealed a very rich set of interlocking cultures. But have these discoveries produced any body of written material, say, which causes us to re-think the way we live? Has it revealed any body of law, or medicine, or technology which leads us to change our ways either of thinking or doing? Has it revealed a completely new literature or philosophy with a new vision? The blunt answer is no" (Peter Watson, *The Modern Mind: An Intellectual History of the 20th Century*, Harper Collins Publishers: New York (2001), pp. 769-770).

Some could disagree with Watson: Native Americans may have something to teach us about our relation to the environment and ecology that a focus on technological progress can block. And Watson barely addresses astrology as a set of ideas—if there is an "openly secret" theme of this book, it is that astrology may be seen as a repressed and excluded backdrop to almost the entire history of Western thought.

This book is not Glenn Ward's *introduction to Postmodernism*. Yet note how relevant his outline of "new historicism" is to the above quotes:

> "Under the influence of deconstruction, new historicists avoid:

• overarching hypotheses, total explanations, or sweeping historical narratives

• the notion of clearly divided historical periods

• the ideas that any historical 'period' has a single 'world view'

• speaking of historical 'facts' as [if] they were accessible independently of texts

• the idea of critical objectivity or disinterest

• distinctions between high and low culture

• distinctions between literary and non-literary works

• distinctions between the social and the cultural

• distinctions between political / historical background (context) and cultural foreground, and the related idea that texts neutrally 'reflect' history

• the idea that artworks are complete, unified objects

• the author as unique source of meaning

They look at:

• non-'canonical' or 'marginal' works, and apparently trivial events / anecdotes

• how texts and activities are enmeshed in a network of social practices

• exchanges of ideas between 'literary' and 'non-literary' texts

• the circulation of meanings across a range of practices and institutions

• art's active involvement in structures of power and economy

• the production of local knowledge

• disputes over meaning

• culture and history as dynamic networks of texts

• links and coincidences between apparently unconnected events

• the historian / critic's own motives and interests

They have been criticized for:

• a badly theorized approach to history

• 'canon-bashing'

• making history up as they go along but dressing it in scholarly language

• disregarding academic standards of proof and evidence

• mixing different disciplines in a cavalier manner" (Glenn Ward, *Teach Yourself: Postmodernism*, McGraw Hill: Chicago (2003), pp. 111-112).

This book aims to incorporate what I see as useful aspects of deconstructive new historicism, including reference to the non-philosophers William Shakespeare and James Joyce—since *Hamlet* probes thinking in itself so deeply; and Joyce was a pioneer in this regard too, exploring "stream-of-consciousness" techniques in literature. Although I believe AI and Deconstruction *do* have a history, this book is not a *narrative* of such. And, although focused on the canon proper, contemporary marginal thinkers are referenced in the epilogue as outsiders and outliers ripe for inclusion. However, I do not "deconstruct" each philosopher in my accounts—my readings do aim at elucidating basic structures and themes in various thinkers' systems or anti-systematic thinking. My writing is fairly straight-forward and clear—and many of the readings put forward are backed up by copious quotation.

This book is not John C. Malone's *history of psychology* that reaches back from ancient Greek philosophy and forward to Freud's psychoanalysis and B.F. Skinner's behaviorism—Malone also differentiates his book from others too:

> "An overview based on a consensus of popular textbook presentations would run pretty much as follows:
>
> 1. Psychology is the study of mind and behavior, which are two different things.
>
> 2. The mind is almost synonymous with the brain. The mind is composed of faculties, or powers, such as attention, memory, and reason, and these faculties are localized in specific brain centers or distributed in specific neural networks.
>
> 3. The senses, such as vision, are directly analogous to input channels—sensory information enters and is 'processed.' Seeing and hearing are somehow brought about by nerve cells in the brain.
>
> 4. The mind / brain is profitably viewed as a 'wonderful computer.'
>
> It is almost impossible to entertain the ideas that this is the best conception of psychology that has ever existed. In fact, it is not greatly different than Plato's psychology!
>
> There is a lot that is good in Plato, but it's not his psychology, and we can do better if we try. It *is* difficult, because the mind / brain / computer viewpoint is pervasive and actually remolds history, as presentist writers compose new histories by selecting material that contributes to the appearance of an unbroken ascent to the currently popular model. Thus, writers find 'anticipations' of modern views in the thought

of the ancients, and Aristotle is portrayed as an empiricist / associationist, hardly distinguishable from the simple associationists of the early twentieth century!" (John C. Malone, *Psychology: Pythagoras to Present*, The MIT Press: Cambridge, MA (2009), p. 3).

Here too, I will try to avoid pigeonholing the ideas of history's philosophies; yet I hope to show that the two main themes of this book, artificial intelligence and deconstruction, do have a history!

This book is not Margaret A. Boden's comprehensive *history of cognitive science*, although I agree with her noting:

> "Every history is a narrative told for particular purposes, from a particular background, and with a particular point of view" (Margaret A. Boden , *Mind as Machine: A History of Cognitive Science*, Oxford University Press: Oxford (2006), p. xxxliv).

Indeed, Boden's history is very personalized, and I think this makes her book much more readable. Yet, this book will be less than one half the length of Boden's 1600+ pages—not due to a pithy style and lack of all but occasional logorrhea, but because the sheer number of thinkers and issues is reduced here considerably. A good pedagogical practice is to repeat core concepts—and to illustrate these from a plethora of perspectives. Although certain motifs can be cited in this book, my practice will not be to repeat and weave a common thread over and over; I've tried to keep an open mind, without a particular axe to grind.

This book is not Nils J. Nilsson's *history of artificial intelligence*; yet some of the "Clues about what might be needed to make machines intelligent" are covered in this book too; Clues:

> "From Philosophy and Logic
>
> From Life itself:
>
> *Neurons and the Brain*

Psychology and Cognitive Science

Evolution

Development and Maturation

Bionics

From Engineering:

Automata, Sensing, and Feedback

Statistics and Probability

The Computer" (Nils J. Nilsson, *The Quest for Artificial Intelligence: A History of Ideas and Achievements*, Cambridge University Press: Cambridge (2010), pp. 10-41 – section headings from chapter 2).

Although about half of this book is dedicated to historically situating artificial intelligence, you can see chapter 24 for my mini-history of what is more specific to that field rather than the broader focus of understanding intelligence via conceptual engineering.

Finally, this book is not Jean-Pierre Dupuy's *history of cybernetics* (and it's failure); although I do sympathize with Dupuy's conclusions:

> "the weakening, indeed the deconstruction of the metaphysical (i.e., Cartesian and Leibnizian) concept of subjectivity took place at the intersection of the social sciences and cognitive science on both a macro- and a microlevel. On the macrolevel, the attributes of subjectivity are not the monopoly of individual subjects: collective entities can exhibit them as well. On the microlevel, the attributes of subjectivity are not attributes of an alleged subject: they are emergent effects produced by the functioning of subjectless processes. In both cases the deconstruction of the subject proceeds from a recognition that a complex network of in-

teractions among simple entities—formal neurons in the case of the individual quasisubject, schematic individuals in the case of the collective subject—is capable of exhibiting remarkable properties. For cognitive scientists who carry on the cybernetic tradition, it is *neither more nor less* justified to attribute a mental state, such as an intention, to a human being than to a group of human beings" (Jean-Pierre Dupuy, M.B. DeBevoise (trans.), *On The Origins of Cognitive Science: The Mechanization of Mind*, The MIT Press: Cambridge, MA (2009), p.160).

Cybernetic Revelation may also be a history of a failure: the as yet failure to fully achieve artificial intelligence. Indeed, how can one deconstruct that which has yet to be constructed?

WHAT THIS BOOK IS: "DIGITAL JOURNALISM"

This book is a clear exposition of some of history's deepest thinkers—more specifically, it is a history of "conceptual engineering." A concept is a more or less clearly defined idea—an idea that can be employed in interpreting the world. In its most general use, a concept could be just about any word that we use (like "red"); at its most particular, it could be a technical term, with precise usage, like the term "zeitgeist" (the spirit of the times). But concepts, as ideas, can be more than singular terms: Nietzsche's notion of the "will to power" is a concept that he interpreted as ubiquitous throughout reality. And indeed, conceptual engineering often references a constellation of concepts—more than one concept related in a *system*; hence the notion of a general "systems theory." Systems theory, cybernetics and structuralism—these schools of thought relate to the paradigm of the paradigm. A conceptual system is something like a loose mechanism; but this looseness could be the Achilles' heel of almost the entire history of Western philosophy and systematic thinking: these concepts may aid us in interpreting the world (and hence be pragmatically useful)—but when the rubber hits the road of instantiating a theory physically—as when attempting to construct an artificially intelligent robot, much like a computer program, nothing can be left out: *every* functional detail of an engineered formal entity must be accounted for.

But human beings do not think in exact detail all the time, if ever. This could be seen as illustrated with this book's focus on a few dozen thinkers, when history has had thousands if not millions of scholars, with each one having read books, learning from and teaching other scholars, etc. The genealogy of any particular concept is not entirely unlike the etymology of a newer word—yes, we may trace back to origins, but these origins are often complex, not simple—and often newer concepts seem to be transformations of other older concepts: fusions and fissions of other ways of thinking. Certain concepts may stand out as important: Plato cited "existence," "difference," "sameness," "rest," "motion,"—ideas of the "one" and the "many" etc.—these concepts seem to pervade much of our thinking. In an analogous way, as many concepts can be seen to fall under the generalization of a single concept (the "good," "beauty," "truth," "being," etc, falling within the realm of the "one," for Plato)—we can also see a sort of representation of *many* thinkers with a certain *few* leaders in their fields. Noam Chomsky situates his views on language in relation to the rationalist philosophical tradition, and although Jacques Derrida may have been the initial practitioner of "deconstruction," he was one of many thinkers in that school of thought. Derrida too has traced his thinking to previous thinkers (Heidegger's "destruction" of the history of philosophy, etc.)—and he recognized that his texts and the concept, "deconstruction" would be shaped by their reception and use by other thinkers in the future.

The "method" and "aim" applied here are simple: *report* on the deep conceptual structures of the major systems developed by philosophers, with a focus on those elements relevant to artificial intelligence and deconstruction. David Hume was one of the first philosophers to limit knowledge based on his model of how we know. In an impossible quest for objectivity, I have tried, and failed, to see the history of the mechanizing of thought through a machine's perspective. Not only have I reduced the millions of scholarly texts to a few dozen representative examples, I have mostly "bracketed" the social and historical context of these texts (like Kant's *Critique of Pure Reason*)—there is little or no biographical detail given for philosophers; and most of the readings are original—and not indebted to secondary sources and the historical reception of these texts. Of course, my own educational background could not be avoided—and there is a sort of "internal" reference to historical context in this book, as a reading of a later philosopher may be related to the reading of a previous one in history. Despite these readings being, for the most part, original—they should not be unique, and to my knowledge, they usually are not; if I have read these philosophers "correctly"—there should be (and often is) some resonance with other scholars' readings as well. In some cases, nuance has been sacrificed, not only for the sake of clarity and brevity, but for my intention to bring forward those conceptual relations that are *important* and *relevant* to this book's focus. Too often, philosophers will go on tangents or split hairs on extended detours from what really matters. These detours can be edifying and illuminating, but going far down an endless dead end can be a waste of time and effort.

Despite *deconstruction* being a form of criticism, this book is mostly uncritical. My purpose was neither to destroy the history of philosophy from a more contemporary perspective... nor to find hidden agendas throughout—not even in the name of deconstructive affirmative action and justice. There are many lessons from deconstructive thinkers that have been ignored here—and in some ways this was an outcome of taking the history of *artificial intelligence* "seriously" (Derrida has deconstructed notions like "the serious" in philosophy). There are points of overlap between the two areas in question here—but obvious differences as well. Deconstruction is often a negative thinking—questioning more than answering—*being intelligent*; whereas artificial intelligence is all about the positive project of building or *modeling intelligence*.

It is my hope that readers will above all find this book *useful*. It is by and large, *introductory*, aimed at a general reader: possibly a literary deconstructive scholar interested in artificial intelligence, or a cognitive scientist interested in post-structuralism. In my research I've encountered less than a dozen scholars writing specifically on the intersection of cognitive science / AI and deconstruction / post-structuralism / postmodernism (and some are referenced towards the end of this book). Ideally the reader would come away from this book, not only with a grasp of the philosophical history of artificial intelligence and deconstruction, but with intellectual *tools* that could be applied in their own thinking.

LEGITIMATION & BIAS

There are three major areas that may put the legitimacy of this book in question: 1) the lack of adequate peer review; 2) my personal background and lack of authority; and 3) the tentative status of "deconstruction" in the academic system, esp. with regard to science.

With such a small potential audience, publishing this book at major publishing house is out of the question. This is not a general self-help book, and not even moderate amounts of money will not be made from it. Similarly, with less than a dozen active scholars working in this area, adequate peer review is unlikely. Several chapters have been published on my personal website (<//jdcasten.info>) and have been read by thousands of readers over the globe—especially popular were my chapters on Theodor Adorno and Walter Benjamin, Martin Heidegger, and Noam Chomsky—the latter chapter having been linked to at Chomsky's official website (<//chomsky.info>). These chapter-papers have been referenced in some web-based scholarship, but not in any books that I know of. By and large, this book is unlikely to be "vetted" by a "peer elite."

Re: meritocracy—peer review often operates like the TV show "American Idol," where three judges (one of whom started the show: and hence is somewhat self-appointed, and who appointed the other two judges) decide which acts will be judged by a public "call-in-vote" at large. These judges are the "gate-keepers" as it were; albeit that some look up to educated or established critics as having more important opinions than the public at large. Although I would appreciate peer appraisal of this work, as a DIY (do-it-yourself) independent scholar, I'm not against self-promotion—bypassing the often self-selected cadre of expert peers, and publicizing this book on any media channels that have an audience. I admire famous celebrity icon scholars like Noam Chomsky or Jacques Derrida (they are both the subject of movies, *Manufacturing Consent* and *Derrida*; yet both had and took the time to respond letters or emails I sent to them—see the epilogue for these)—again, they are like representatives or even figureheads for larger groups of scholars and the public—yet, I would like to see more people grab the public megaphone and be heard in our mass-media system. The internet has made this more of a possibility—and hence we've seen an explosion of information availability, and even challenges to copyright law, which threaten the establishment investment in monetized information. Still, media hubs—both as channels with an audience, and celebrities with followers—offer the most reliable way to recognition; hence new scholars often work with established professionals as "opening acts" if not riding on coattails—and a "lucky break" still means getting noticed by a popular journal or the like. So much is to say, it's who you know as much as what you know.

Legitimation matters get more complicated, when considering my own personal history and authority. I often deal with a debilitating barrage of foreign yet internal voices, and have suffered the experience of being harassed by a sadistic logic professor over a brain-walkie-talkie, with the consequent phenomena of "thought broadcasting" 24/7 (even in dreams). Despite my Schizophrenia diagnosis, I have managed to find a modicum of time over the past two decades to work on this book. Before sliding off the deep end, I did, however, complete my B.S. and M.A. in philosophy at the University of Oregon (at a time, the early 1990's, when that philosophy department heavily emphasized in the later Wittgenstein; my studies focused on the history of philosophy, logic and language, and philosophy of mind). I have also been programming computers since I was in the 7th grade. Back in 1980, I ordered a 1K RAM Sinclair ZX80 computer from a Scientific American magazine ad (I was looking up a "screw-driver" solution to the Rubik's Cube). I soon upgraded to a 16K Atari 400 computer, and had several popular programs published in Antic magazine (with a moderately sized 100,000+ circulation) in the mid 1980's, while still in high school. My first cover byline was for a text adventure program, which introduced me to programming a language parser and the navigation of a micro-world—that, along with a research paper written in my senior year at high school, inaugurated my studies of artificial intelligence. At that time I wondered if something like communism would be necessary to control the power of artificial intelligence at a government level. Chapter 29 outlines my current stance on technologically relevant political economics: far from totalitarianism, I am for individual human freedom and rights balanced with global and democratic consequentialist responsibilities—a radical moderate and sustainability libertarian, focused on employee ownership of businesses, and biodiversity in our environment.

The third hurdle of legitimation can be exemplified with the following quotations:

> "As Derrida has shown, Freud has recourse to the (writing) machine in order to represent the processes of the psychic apparatus itself. The 'standing in' of machines for genitals is yet another machine: dream symbolism works like a machine, diligently turning genitals into hats, ties, and increasingly complex machines—stand-ins or substitutes that circulate freely in and through dreams" (Catherine Liu, *Copying Machines: Taking Notes for the Automaton*, University of Minnesota Press: Minneapolis (2000), p. 36).

> "A *proximity dendrogram* is a dendrogram that takes into account the level of proximity where two clusters are merged *for the first time*. When a dissimilarity (similarity) measure is in use, the proximity dendrogram is called a *dissimilarity*

(similarity) dendrogram. This tool may be used as an indicator of the natural or forced formation of clusters at any level. That is, it may provide a clue about the clustering that best fits the data" (Sergios Theodoridis and Konstantinos Koutroumbas, *Pattern Recognition*, Academic Press: San Diego (1999), p. 407).

"There are many natural scientists, and especially physicists, who continue to reject the notion that the disciplines concerned with social and cultural criticism can have anything to contribute, except perhaps peripherally, to their research. Still less are they receptive to the idea that the very foundations of their worldview must be revised or rebuilt in the light of such criticism. Rather, they cling to the dogma imposed by the long post-Enlightenment hegemony over the Western intellectual outlook, which can be summarized briefly as follows: that there exists an external world, whose properties are independent of any individual human being and indeed of humanity as a whole; that these properties are encoded in 'eternal' physical laws; and that human beings can obtain reliable, albeit imperfect and tentative, knowledge of these laws by hewing to the 'objective' procedures and epistemological strictures prescribed by the (so-called) scientific method.

But deep conceptual shifts within twentieth-century science have undermined this Cartesian-Newtonian metaphysics; revisionist studies in the history and philosophy of science have cast further doubt on its credibility; and, most recently, feminist and poststructuralist critiques have demystified the substantive content of mainstream Western scientific practice, revealing the ideology of domination concealed behind the façade of 'objectivity'. It has thus become increasingly apparent that physical 'reality', no less than social 'reality', is at bottom a social and linguistic construct; that scientific 'knowledge', far from being objective, reflects and encodes the dominant ideologies and power relations of the culture that produced it; that the truth claims of science are inherently theory-laden and self-referential; and consequently, that the discourse of the scientific community, for all its undeniable value, cannot assert a privileged epistemological status with respect to counter-hegemonic narratives emanating from dissident or marginalized communities" (Alan D. Sokal, "Transgressing the Boundaries: Towards a Transformative Hermeneutics of Quantum Gravity," Social Text #46/47, spring/summer 1996, pp. 217-252).

"If one examines neosemioticist textual theory, one is faced with a choice: either reject postcapitalist desublimation or conclude that context must come from the collective unconscious. The subject is contextualised into a predeconstructive narrative that includes sexuality as a whole. Therefore, Bataille uses the term 'postcapitalist desublimation' to denote not discourse, as predeconstructive narrative suggests, but prediscourse.

'Society is fundamentally elitist,' says Derrida; however, according to d'Erlette, it is not so much society that is fundamentally elitist, but rather the dialectic, and some would say the absurdity, of society." (John P.D. d'Erlette, "The Context of Collapse: Postcapitalist desublimation and neosemioticist textual theory," section #2, "Narratives of Meaninglessness," <//www.elsewhere.org/pomo>, 2012).

These quotations above are meant to illustrate the issue of *jargon*. The history of deconstruction and artificial intelligence is replete with technical vocabularies. The first two quotes above are actually from legitimate scholarship, referencing deconstruction and cognitive science. The third quote, though was written as a hoax... and the fourth, last quote was generated by a computer program as a joke! What are terms like "phallogocentrism" and "spreading activation" supposed to mean? If you read this book, you may get an idea—but throughout, the burden of legitimacy ultimately rests on you. There are numerous quotations throughout this book, to backup and ground my exposition of what these various thinkers have thought. But this is not the "last word" on these subjects, and *you* should maintain a critical stance throughout.

BOOK LAYOUT

There are three parts to this book. Part I includes this introduction, and two other chapters to situate "my" point of view, and to give some ideas (esp. about *networks* and *consciousness*) to keep in mind when reading the heart of this work, the major philosophers and thinkers covered in Part II. Part I thus offers a fuzzy glimpse at the whole and end of the work—a tentative view of the "outcome." Just as the last chapter proper (30), before the epilogue, covers politics, the first proper chapter (2), after this introduction, covers aesthetics. Such is to situate this book in a broader context, and frame it in reference to wider applicability—pointing out the relevance of this work to literature, art, and music, as well as ethics and justice.

Part II is the heart of the book, with a history of philosophy and thinkers dating back from antiquity up to modernity. No narrative has been imposed on this history, but if a "story" were to be noted, it might be outlined as such (this outline skipping a few of the philosophers covered in the book):

- Anaximander (first western theory of the unknowable and first map)
  [1: unknown passes to known];

- Plato (complete dialectical system leading to the final revelation of knowledge)
  [2: the journey towards knowledge];

- Aristotle (disciplined logical science)
  [3: what is known / observed];

- Augustine (turns towards the self)
  [4: subject as confessor / revealer];

• Descartes (subjective doubter's perspective on knowable object)
[5: subject / object focus];

• Locke (associational cognitive mechanism)
[6: subject's mechanism of knowing: observed connections and agreement of ideas];

• Kant (from the empirically unknowable implicit to explicit knowledge via necessity and / of cognitive structure)
[7: subjective structure(ing) of (un)knowable beyond immediate observation];

• Hegel (historically situated systematic revelation of being)
[8: subject revealed / shaped through relation to object / world];

• Nietzsche (relativisitic "non"-systematic embodied prophecy critique)
[9: subjective debunking of knowledge];

• Freud (human being as sex-driven animal)
[10: subject is unconsciously directed];

• Dewey, Wittgenstein (complex context directs person's description of it)
[11: subject intertwined in world it engages using strategies];

• Heidegger (structure and activity of technical and poetic revelation)
[12: subject rejects en-webbed world for revelation of being]);

• Adorno, Derrida (deconstruction as rejection and acceptance of the revealed in order to open up new revelations)
[13: learning how to reveal the unknowable];

• Chomsky et. al. (breaks from embodied socio-habitual norms and investigates new found innate brain structures) [14: unknowable subjectivity reveals itself in contrast to knowable brain];

Part III focuses on artificial intelligence and cognitive science, outside their philosophical background, and brings together various contemporary thinkers where AI intersects with deconstruction, and includes my thinking on a few areas of philosophical interest:

* ***The Pragmatism Debate:*** A "highlighting" theory of perception (between passive and constructive perception) and the revealed knowledge occurring when new ideas fall together and pan out (metaphors of pragmatic coherence): how does one notice something new that fits in with the old, when the old is the lens through which we perceive? Loose analogies that catch fire?

* ***Deconstructing Artificial Intelligence—Questions:*** Deconstruction as not knowable / mechanical; the notion of always already being *artificially* intelligent (are we intelligent, if we don't know *exactly* what intelligence is... well enought to model it?) Robots are not a copy of original conscious life, but are the mirroring mechanical aspect of who we already are; with box-like strategies of thinking outside the box (outside of our box / robot / mechanical habits); this is related to the quality Phenomenal (sensation) / quantity Genealogical (structure) opposition.

Finally, the epilogue speculates on the "technological singularity" (where computers may surpass human intelligence), looks towards a few "outsider" outlier thinkers, and hints at some "spiritual" implications of deconstructing artificial intelligence, as well as including some of my (limited) correspondence with major 21st century thinkers.

Fur
has
has
Cat
Bear
Neuron
Dendrites
Axon
is a
Whale
Vertebra
is a
Electrical Impulses
has
Mammal
lives in
Neurotransmitter
Molecules
is an
Water
lives in
is an
Animal
Fish
Receptor
Synapse

Chapter 2

Style, Taste & Cyber-Networks

MEDIUMS

We live in a super-human society. Our television shows, movies, automobiles, clothing and houses—even our jobs—are far beyond the capacity of human individuals to conceive of. No single person can make an automobile; such complex constructions require teams of engineers, assemblers, managers, etc, all working within traditions of production which date back before the births of their own generations. Biologically, one can see that society itself is like a large organism, with inter-penetrating and over-lapping cultural forces: fads, traditions, conflicts, and solidarities of the social body arise in global forces far outside the capabilities of its individual cells and limbs—we humans. Perhaps this has always been the case, but never so much as today.

With pop-culture we see the effects of the larger forces which proliferate through an economy driven by consumer supply and demand. Beyond selling out, much of what many consider their favorite "art" was formulated from the very start as a way of making money by satisfying popular demand. We virtually have poll takers prescribing which "art" products to produce (and such is the case with focus groups). The hand-made crafts of counter-culture folks

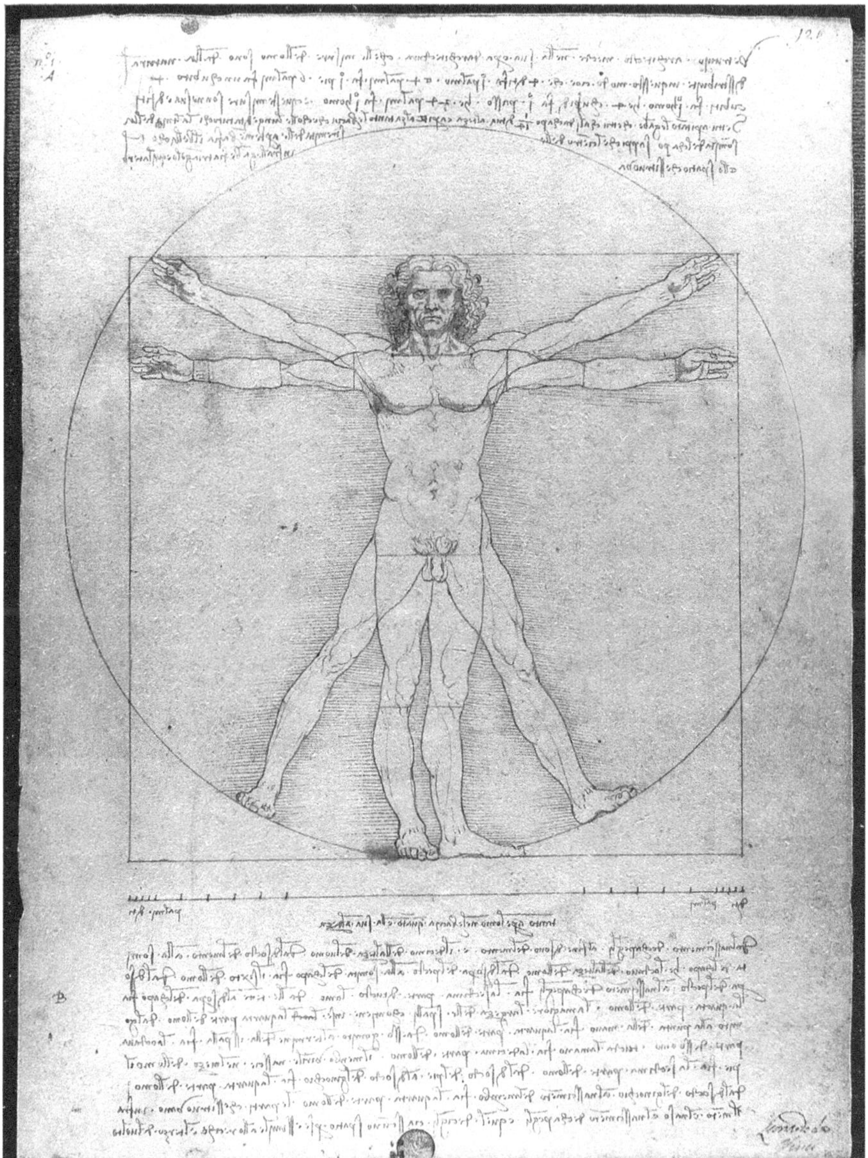

Da Vinci - Vitruvian Man (~1487)

and the folk song by the wandering guitar player are almost always eclipsed by the multi-million dollar motion picture investment. No doubt, such mega-investments of the entertainment industry have often produced quality art that make Leonardo da Vinci look like a bore. Yet, these may also contribute to our super-human reality; as a society, we have made ourselves out to be much more than we could ever be as individuals. And this has discouraged many, I believe, from turning their passive-consumerism into an active creativity. We live in an age where the medium which used to be the mere obstacle between one and one's expression has become the mass media which structures the expression of a social force so powerfully we can barely do more than gaze at its spectacle in utter fascination.

No doubt, the medium has always played a shaping role in artistic production. A piano and a guitar lend themselves to different types of musical composition, and one may wonder to what extent the history of music has been shaped by the instruments used. With painting the instrumental "interference" could be at a minimum: the simple paint brush worked as the cybernetic extension of the pointing finger. Maybe it wasn't until the cut and paste collage method was used that the limitations, or rather the coercions, of the brush or pointing implement could be fully illustrated: new techniques are needed to demonstrate the limits of older ones. And with today's technology, especially with computers, art has gone far beyond the limitations of bodily movements, allowing minds to roam through a space restrained only by mathematical possibility. But have our bodies, in the ultra-contemporary media take-over, been dissolved into mere aesthetic mediations which only serve to interface us with the social machine? Has the style of the body been cut out by the tastes of a consumer society?

AUTOMATIC STYLE

The body is one. It may couple with others, or grow out another, but in its integrated functionality it has an autonomy which guarantees the possibility of comfortable security. The body provides a sanctuary from the nightmare of global responsibility, of conscience—for the actions of our bodies are by and large unconscious. To discover this, one need do no more than listen to one's verbal soul as it bursts forth from the body. One's stream-of-conscious is not consciously intended; one cannot decide what one is going to think before one thinks—there is a perpetual movement which one neither follows nor anticipates, but which one is effectively. Such arises out of the over-determination of personality; the body is shaped and trained by cultural forces and a personal history, and the body's actions evidence this training as style.

Van Gogh - Starry Night (1889)

Hokusai - The Great Wave off Kanagawa from *36 Fuji Views* (~1829)

The style of one's body as trained is most evident in the visual arts with gesture. When one paints or draws, the trace left by bodily action evidences the body's history. When one learns to write, for example, certain muscles are developed; and if some muscle is damaged, one's handwriting may evidence this. After awhile, one's handwriting may settle upon a recognizable style—one which would be quite different than the results of initial attempts to write. And this style of gesture is not consciously producible—one may try to forge a signature, but even this would be a modification of one's own signature style, and would require practice.

Similarly, attention to gesture, and perfecting it through practice has long been a factor in painting, and has been especially prominent in Asian traditions where the stroke has been a major mode of stylistic signification. Just as the writing in Asian languages has been based on pictograms, special strokes have been developed to designate certain visual textures. For example, beyond the extended post-impressionistic strokes used by Vincent Van Gogh, Asian artists such as Wang Meng have used different types of strokes respectively for rocks, leaves, or tree bark. With practice through repetition, these strokes may gain a unique aspect akin to the style of the letters in handwriting. The published "sketch-books" of the Japanese ukiyo-e artist Katsushika Hokusai evidence practiced strokes, as with the watery splashing "fingers" on the breaking crest of his "Great Wave off Kanagawa." With Hokusai, practiced gestures go beyond mere strokes to entire shapes. More than seeing some representation of a wave, we may actually read Hokusai's artwork, which is composed of arranged signifiers which loosely resemble what they signify. Hokusai's practiced local images come close to providing a bridge between pictorial representation and pictographic language. In western art as well, bodily style can be seen with art

Wang Meng
Forest Grotto In Juqu (1370s)

Hokusai
Wave on Hato Coast (~1805)

& from *100 Fuji Views* (1834)

nouveau artists, such as William Blake and Edmund Dulac, where fluid organic lines suggest physical gesture as much as pictorial representation.

Style, as evidencing the practice and training of the body, is habitual. Repetition of certain procedures reinforces used muscles and neural pathways throughout the body and in the brain. As with martial arts, certain actions and reactions become automatic. Nietzsche, who claimed in *The Will to Power* that "all perfect acts are unconscious," (Friedrich Nietzsche, *The Will to Power*, Walter Kaufmann, (ed.), Walter Kaufmann and R.J. Hollingdale, (trans.), Random House, Inc.: New York (1967), p. 163) approximates this in his own writing style with aphorisms that suggest bursts of spontaneous thought. Automatic style can also be found in jazz music, such as with Thelonious Monk's spontaneous improvisation. Use of medium implements or tools and techniques and the internalization of cultural customs become automatic as well—as Marshall

William Blake
The Divine Image (1789)

Edmund Dulac
from *The Little Mermaid* (1911)

Hokusai - Lake Suwa from *36 Fuji Views* (~1830)

McLuhan writes in *The Medium is the Massage*, "The wheel ...is an extension of the foot"(Marshal McLuhan and Quentin Fiore, *The Medium is the Massage* (New York: Bantam Books, 1967), pp. 31-32). The leading edge of spontaneous habitual action is the stream-of-conscious; a stream-of-conscious that includes not only an internal voice, but also the full spectrum of sensations—imagined or observed images, the movement of muscles, the flows of sound, etc. Although sensation is often seen as passively observed, the brain does structure much of what is perceived.

CRITICAL TASTE

All is not, however, smooth sailing automatic style. Often the unforeseen occurs and stops style in its tracks. Hesitation sets in, consciousness elevates, and lucid choices must be made (one might recall the heightened awareness of learning how to drive a car). Here, taste refers to these moments of choice, in contrast to the automatic actions of style. Taste is in the domain of the critic, and

Hokusai - Asakusa Hongan-ji Temple from *36 Fuji Views* (~1830)

is often critical. And, as a choice, taste offers a certain amount of freedom; taste interrupts the spontaneous flow of a style that is unconsciously determined in its origin, and opens up possibilities and new potential courses of action. Taste presents the opportunity to change one's style.

For artists, taste is evident with the choice of colors, the choice of perspective or viewing angle, the arrangement of shapes, and the choice of what to illustrate. In writing, taste is evident in selecting quotations, in using the thesaurus, in editing, and in using cut and paste options with word processors. The poetry of e.e. cummings also demonstrates interruptions of style with the invention of new kinds of spacing and grammar. Critical taste can be found in music too, with the use of sampling, such as in the Beatles' song, "I am the Walrus," and in modern rap or hip hop music. The choice and treatment of subject matters may evidence a taste that can not only be aesthetically beautiful, but critical, and even sarcastic or satirical (as with works by Andy Warhol and the early Roy Lichtenstein, who both directed others in the creation of their works).

Clockwise from upper left:

Toulouse-Lautrec
Portait of Van Gogh (1887)

Gauguin
Van Gogh with Sunflowers (1888)

Van Gogh
Self-Portrait (1889)

Hokusai's selection of a specific view in his Great Wave was one of many views of Mount Fuji included in his book *Thirty-six views of Mount Fuji*. In line with Hokusai's multiple perspectives of one object, consider the various images of Van Gogh, in his own "Self-portrait with Bandaged Ear," in Paul Gauguin's "Van Gogh Painting Sunflowers," and in Henri de Toulouse-Lautrec's "Portrait of Vincent Van Gogh;" with these artworks we can see how different styles may converge on the same, or similar, selected "object" (and one may notice how much more seriously Van Gogh took himself than Gauguin or Toulouse-Lautrec did). That which may be the "object" of a painting—whether it be a view, an object, a person, a feeling, an idea, literary passage, historical event, etc.—offers itself to innumerable ways of being visualized. To a large extent the "object" of concern is, to borrow Jacques Derrida's use of the term, "undecidable;" it lends itself to possibilities, and this may force an artist to make choices. One might argue that photo-realism would offer a method of reproducing worldly objects as they "really" are; yet, Pablo Picasso's multi-perspective cubism clearly demonstrates that even methods like photo-realism already cut off certain possibilities of representing things as they "really" are. (Besides, humans do not perceive exactly as cameras do). Any type of reproduction requires choices which skew the way an object is seen.

Lichtenstein
Drowning Girl (1963)

However, the ambiguity of the "object" encountered may not be enough to force an artist to make a conscious choice and depart the status quo—one may simply treat an ambiguity in a conventional manner. For a choice to be de-automated, one's action must be doubted—style castrated by taste (not that style or taste are any more masculine or feminine). One must be stopped short and hover for awhile in an abyss of indecision before one can make a

Picasso
Weeping Woman (1937)

conscious choice. There must be a moment of having to choose because one's action has been paralyzed by uncertainty. And such a free choice is what allows one to change one's style and to affirm it with conviction as one's own. Likewise, ambiguity and techniques of de-familiarization used by artists can compel an audience to re-examine their own perspectives.

SEMANTIC NETWORKS

As the discussion has heretofore centered on a loose dialectic or juxtaposition of subjective style and taste in the arts, I will turn here to a loose, yet more scientific, analysis of objective action and its possible suspension in cognition. The second half of this chapter, from another perspective, should interlock with the first half by offering details concerning the unconscious and automatic production of some aspects of a stream of consciousness exemplified by the spreading activation theory of cognitive science.

Mentally, ideas have connections to other ideas, and physically, neurons are connected to other neurons; and in each case, the connections can be strengthened by reinforcement through use. If we were to give a visual representation of the logical space inherent in both types of connections, we would imagine a web of nodes with varying numbers of lines (arcs) connecting them; lines which would be weighted ("thicker") according to the strength of connection (and this is what neurological networks could be said to look like).

On the mental side of this interconnectivity, we can easily see that some ideas have consistently been associated with other ideas, and that there are specific types of connections. Indeed, the fact that a relationship between nodes is of a specific type requires more of our visual example than simply a line connecting two nodes—each relationship (line) could be indexed and connected to a "type of relationship node" (regular nodes could designate nouns or adjectives, and relational nodes could represent verbs). These complex webs of ideas comprise a semantic network and function as memory when developed (as introduced by Ross Quillian (Ross Quillian, "Semantic Memory" in Marvin Minsky (Ed.), *Semantic Information Processing,* The MIT Press: Cambridge, Mass. (1968); and also see Allan M. Collins, and M. Ross Quillian, "Retrieval time from semantic memory," *Journal of Verbal Learning and Verbal Behavior* 8 (1969), pp. 240-248).

We can imagine a semantic network as having first-order and fuzzy logical structures with hierarchies of concepts. However, the unconditional variability of connection types in a semantic network (there could be connections to representations of images and sounds too) and the inclusion of connection strengths in a semantic network may offer more flexibility than fuzzy or first-order logic alone; with semantic networks containing logic within their possibilities.

SPREADING ACTIVATION

Beyond a static group of nodes, a semantic network grows with use, and it can also be "animated." Each node in a semantic network can be activated into consciousness through a process called 'spreading activation,' as introduced by Allan M. Collins and Elizabeth F. Loftus (Allan M. Collins, and Elizabeth F. Loftus, "A spreading-activation theory of semantic processing," *Psychological Review* 82 (1975), pp. 407-428). The general operation for such an activity is such: the activation of one node facilitates the activation of related nodes (one thought leads to another). Here, it would be necessary that each node in a network would require a certain amount of prompting from the nodes it was connected to in order to activate—each node would have a certain threshold level which, if met by the prompting of other connected nodes, would cause a node to activate (possibly causing a term to come to mind).

Biologically, these nodes need not be represented with single neurons, but may be represented with clusters of neurons representing a single term or idea. Hence, it would not be necessary for every node that contributes to the activation of a term into consciousness to be activated into consciousness itself. If each of the facilitating nodes was called a 'cue' then the more cues for a certain term are given, and the stronger the connections are between these nodes are, the more likely that term is to come to mind.

There are many types of prompts which might be called 'cues': thought words, physically given words, or even external objects and situations. And again, it may be possible that some cues could be given that were not explicitly thought of by a person. For example, if someone came upon an animal that was furry, had four legs, barked, and was on a leash tied to a dog house, each of these cues might contribute to the conscious activation of the single term 'dog' without each cue itself being activated into consciousness (and the activation of 'dog' might also contribute to the further activation of the dog's name, 'Fido'). There are thus semi- or unconscious activations of nodes and terms hovering on the verge of consciousness that may activate other terms. (An excellent discussion of spreading activation can be found in chapter xi of Douglas R. Hofstadter, *Gödel, Escher, Bach: an Eternal Golden Braid*, Basic Books: New York (1979)).

SEQUENTIAL THOUGHT AND ACTION

Two other types of interconnection should also be discussed—contiguity, or the flow from one lexical unit to the next; and the combination of smaller units interconnected as larger wholes. On a small scale, these combined connections can be seen with the combination of letters that go to make up a whole word. In turn, several words could be combined to form idioms. This sort of sequential clustering could be seen to form even larger structures such as sentences, paragraphs, or even entire episodic memories (recalling that image and sound representations could be connected to a semantic network).

The flow from one idea to the next in a stream of consciousness requires an explanation beyond spreading activation. Although the logic or deep structure of any train of thought might be mapped out on a semantic network, the flow of thought usually follows some syntactical or grammatical form—this temporal, and unfolding aspect of grammar, as distinct from and complementing the semantic relations involved, could be called 'rolling grammatical "progression"' (I put "progression" in quotes to note that the flow of grammar may not be going anywhere). The fact that there is a grammatical process distinct from semantic processing is suggested by the results of people having an aphasia due to a lesion in the posterior part of the left hemisphere in the brain (the left hemisphere being more likely to be used in language usage). People with posterior aphasia, or Wernicke's aphasia, can produce correct grammar without substantial semantic content; grammar is distinct from semantics, although there may be an interactive activation (a kind of feedback) between the two. A simple example of rolling grammatical progression would be the sequence of "subject verb object." And, just as one can observe a child learning new associations, one can also see a learning of more complicated grammatical styles. Such a progression is illustrated by comparing the simple grammar of grade-school texts with the complex grammar of Henry James or Edith Wharton.

There are also more complex activities such as learning and employing skills, techniques, and strategies (such as playing the piano, painting, playing chess, etc). The later Ludwig Wittgenstein investigated some of these activities with his concept of "language games," which includes local practices such as reading or naming (the "performatives" of J.L. Austin, the "speech acts" of John R. Searle, and the "memes" of Richard Dawkins are also related to language games). Like terms in a semantic network, language games and skills can be associated, (as with the language games of asking questions, and those of answering them); they may have their own spreading activation; and they can be associated in hierarchies of larger wholes and parts (an account of how skills may have their own spreading activation can be found in Pattie Maes' article "How to Do the Right Thing," (Pattie Maes, "How to Do the Right Thing", *Connection Science*, 1:3 (1990)) a good discussion of which can be found in Stan Franklin's *Artificial Minds* (Stan Franklin, *Artificial Minds*, The MIT Press: Cambridge, MA (1997)). At a broader level, we have Michel Foucault's use of the term, "discourse," which designates the likes of the military, democracy, or psychiatry—practices and institutions that are made up of interrelated skills and language games. Semantic networks, skills, language games, and discourses would all also incorporate new terms and activities by connecting these with older established terms and practices. In his *Society of Mind*, Marvin Minsky claims:

> "Each new technique presumably begins by exploiting methods already learned in other, older agencies. So new ideas often have roots in older ones, adapted for new purposes" (Marvin Minsky, *Society of Mind*, Simon and Schuster: New York (1986), p. 141).

It should be noted that these language games and discourses are not tools at hand, but operate automatically; they are not used by a subject but effectively constitute it.

IMPLICATIONS AND LIMITATIONS

The outline of cognition heretofore, concerning the automatic action of the brain, has some interesting philosophical implications. Not only is there no need for abstract essences, platonic forms, signifieds, or mental prototypes, there is also nothing essential for the use of any term; there is no criterion necessary for the activation of a term into consciousness, but only sufficient prompting from any number of related nodes with specific connection strengths. A given object or situation need not be compared to a mental prototype to be recognized, as the given cues or features of objects or situations would directly activate terms in a semantic network. Similarly, skills and language games would not need essential rules of operation. As Hubert L. Dreyfus claims in his *What Computers Still Can't Do*:

> "The important thing about skills is that, although science requires that the skilled performance be described according to rules, these rules need in no way be involved in producing a performance"(Hubert L. Dreyfus, *What Computers Still Can't Do: A Critique of Artificial Reason*, The MIT Press: Cambridge Mass. (1992), p. 253).

A various number of micro-actions would be sufficient to complete a task without any one action being necessary in general; explicit symbolic rules need not be followed, as one action naturally leads to other associated actions.

Just as it was noted earlier that semantic networks provide a structure that could be broader than logic, it may be noted that skills, language games and discourses provide structures that are more broad than reason or intelligence; reason, intelligence and other goal related activities are only specific language games and discourses among many others (hence the term "artificial intelligence" already limits a scope of inquiry).

However broad cognitive structures may be though, they are limited by the structure of the brain. This means that our ability to perceive and think about our world is also limited by brain structure; our brains project their own structure and functioning on the environment. For example, the perceived singularity of objects may be a projection of a singular brain (differences that distinguish one object from another, like that between a TV and the table it is on, are relative to a judge); and semantic networks that have interactive activation with perception mechanisms help shape what is perceived (we often see what we expect to see, such as with optical illusions). Of course our world lends itself to such projections; yet, there remains the possibility of other kinds of structure, or even realms beyond what we perceive as structure, which we cannot begin to conceive of. Even our modeling of brain activity would be limited by brain structure.

In a related limitation, neural structure would not comprehend conscious qualia or sensual experience (including emotions). There is simply no place in the brain where physical structures could turn into qualia without becoming non structural, and hence leap out of the circle of structural causality (if one were to say that structure causes qualia, then why couldn't qualia effect structure?) Gottfried Wilhelm von Leibniz noted:

> "it must be avowed that perception and what depends upon it can not possibly be explained by mechanical reasons, that is, by figure and movement. Suppose that there be a machine, the structure of which produces thinking, feeling, and perceiving; imagine this machine enlarged but preserving the same proportions, so that you could enter it as if it were a mill. This being supposed, you might visit its insides; but what would you observe there? Nothing but parts which push and move each other, and never anything that could explain perception" (Gottfried Wilhelm von Leibniz, Paul Schrecker and Anne Martin (trans.), *Monadology and Other Philosophical Essays*, The Bobbs-Merrill Company, Inc.: New York (1965), p. 150).

(Here the word "perception" could designate qualitative sensation.) Such reassures us that robots would be no more sentient than a rock (they might have personality or spirit), but it raises problems concerning the claim that neurons can activate terms into consciousness. (A contrasting view point on this issue of qualia can be found in Daniel C. Dennett's *Consciousness Explained* (Daniel

C. Dennett, *Consciousness Explained*, Little, Brown and Company: New York (1991)). No doubt though, some neural activity would be consciously experienced and some would be unconscious—this might be decided, unconsciously in the brain, by an attention system; a system that would also limit the number of thoughts or perceptions entering into consciousness. Some claim that consciousness is the ability of a system to have a model of itself. If this were so, then how conscious could we be without constructing artificial intelligence, without having a complete model of ourselves? Perhaps the qualitative consciousness involved in this, however, is such that it eludes neurological comprehension—it is something radically other than structure as our brains' projections can know it—our efforts to know qualia may be like groping for air with our hands.

CONTEMPLATION AND MEDITATION

We can now bring together the two parts of this chapter. As spontaneous action, style designates the activation of nodes into consciousness and the automatic performance of skills and language games. These activities can be learned, practiced and refined, creating new neural connections and strengthening old ones. Taste most likely arises with the creation of new connections—in situations where neural habit is challenged by unexpected circumstances, choices must be made in a confrontation with uncertainty.

Taste affects the choice of which style to exercise; taste shapes style. Yet, taste can also be automatic, as simply another pre-determined action, or as a choice implied by style. In ordinary language, the use of the words "taste" and "style" are often synonymous—the difference between the two terms may collapse. But their interrelations can be made more complex, as with contemplation. When contemplation occurs there is a wavering between style and taste—inspirations arise spontaneously only to be called into question by "thinking twice" as one works thoughts over the subject matter at hand. Contemplation can be exercised if one chooses to pursue language games that employ critical strategies like being skeptical, raising questions, exploring options, finding assumptions, and identifying prejudices. One can practice language games that will question habitual norms and open up possibilities for further free choice. Moreover, use of metaphors and learning new word usages can create new semantic connections—poetry can help free the soul.

Conversely, just as there are both style and taste in contemplation, there is neither style nor taste in meditation—one may try to suspend the activity of the mind without falling into doubt. This too would require much practice. Contemplation and meditation (which are especially relevant in an approach to Zen and its koans) may provide a way to strike a balance between style and taste, a balance between action and free choice (although there is more likely to be a balance between balance and imbalance—we will always have artists, musicians, and writers that lean one way more than another).

On a greater scale, people could be considered as nodes in a social and environmental network. People have relations with other people and the environment, and these relations become stronger through reinforcement. Via communication and worldly occurrences, spreading activation crosses brain boundaries as ideas circulate and proliferate. With technology, spreading activation operates through the mass media and cyber-space, and enters the realm of super-human society where hype and mass-produced commodities explode in a dazzling array of hypnotizing spectacles. Here personal tastes and styles combine to form social trends. The individuality that is lost in super-human team projects is supplanted by new super-individuals—actors, athletes, and politicians are constructed as super-star celebrities and heroes.

All hope for individual artists to compete with super-human group collaborations is not lost though. As evidenced by Michelangelo, who worked for four years on the Sistine Chapel ceiling, and by James Joyce, who worked for over a decade on *Finnegans Wake*, it is possible for a person to create works that may even out do the super-human; yet, even these works were not created in a social vacuum; paints and books are manufactured by groups. At any rate, it still remains the obligation of individuals to (1) question or accept social styles and tastes with their own choice of conscience, a conscience that possibly bears the overwhelming weight of global responsibility, and to (2) critically direct actions accordingly with a freedom based on examining one's options rather than bulldozing one's way.

Michelangelo - Creation of Adam (~1511)

Arcimboldo - Vertumnus (~1590)

Chapter 3

Inside-Out
Consciousness

PREAMBLE

Consciousness may remain one of the most enigmatic and intriguing areas of concern for any intellectual inquiry regarding the mind. Interest in consciousness, when one considers the whole history of philosophy has more recently gained prominence due in part to the limits approached by science and logic (e.g. Heisenberg's uncertainty, Gödel's incompleteness, etc). The scope of science and logic are largely limited by reproducibility, as illustrated quite well with Wittgenstein's *Tractatus*; and today our understanding of general systems has reached new pinnacles under the rubric of cybernetics: the science of information flows, and the physical productions of complex electronic and mechanical systems. Paralleling the extent that our "maps" and machines become more complex, many aspects of human action become explainable in terms of our own biological machinery. These maps and working models are completely objective though, in that they are useful objects; which could mean that our clearer view of ourselves as objects brings us closer to the division between objectivity and subjectivity—yet it just may be that consciousness and subjectivity, and possibly objectivity too, are so intertwined as to be indistinguishable.

In this chapter I hope to present a loose discussion concerning consciousness; the discussion will revolve around such issues as stream-of-consciousness, time and change, habits, "it-ness," and the contrast between the qualitative and the traceable. I'm fairly confident that no definite conclusions will be reached here, yet hopefully the issue of consciousness will be clarified by expanding beyond the simple word, "consciousness," and employing it in connection with various other related ideas.

CONSCIOUSNESS CONTRASTED WITH UNCONSCIOUSNESS

If one were to defer to that bastion of philosophical profundity, the dictionary, one would find that "consciousness" variously involves willing, perceiving, thinking, and awareness. Intuitively, people can easily understand what consciousness is by examining their own experience, especially with regard to being unconscious. Sleep and states of stupor or delirium may provide the best areas where the border of consciousness might be found, as one can relate to differing degrees of being conscious—one may be fully awake and alert, as if just after facing a crisis situation, or when performing a skill that may be life threatening; moderately attentive, as when watching a boring television program; somewhat confused, as with drunkenness; almost in control, as in a lucid dream; or barely there as an observer, as with most dreams—or one may simply fade or snap out (of memory, at least). These levels of awareness in no way exhaust all possible modes of consciousness, yet it should provide an indicator as to how we usually understand ourselves to be more or less conscious. Such a contrast between unconsciousness and consciousness does not render any precision—it only helps us to comprehend how we usually understand our employment of the word, "consciousness." The lack of a precise definition for consciousness has led to its being used in various contexts to mean quite different things—various thinkers often have something completely different in mind when they employ the word.

SINGULARITY OF PERSPECTIVE

The confusion involved with this word, "consciousness," may arise from the fact that consciousness may be seen to have numerous aspects. From our contrast above, we can see that consciousness usually involves a sense of focused awareness, directedness, control, and lucidity. A question may arise here though, as to the separation between the object and subject of consciousness. Many academic discussions of consciousness seem to revolve around an agent who, through consciousness, "pulls together," "gathers" (in reference to Heidegger's terminology), or integrates various phenomena in a single, aimed, perspective. Often, these accounts imply a little person within a person (a homunculus)—but such accounts may lead to an infinite regress, as one is lead to ask, "Is there another little person inside the first?" and so on. The various apparatuses of the body—organs of sense, and limbs of motion—are observed or controlled from a single position (e.g. somewhere in the brain, Descartes suggested the pineal gland). Often, consciousness is separated from any type of sensation, and is designated as one's orientation towards, and awareness of sensation—something like a "vacuum cleaner spotlight" which draws together what it is aimed at. At any rate, these views of consciousness are ones which place a subjectivity in complete contrast with anything usually considered objective, including sensations.

The sense of a self, or an "I" is most obviously the result of this experience of consciousness—the ever-changing flux of sense experience is always anchored, directed, or referenced, to a single point: we experience change itself in contrast with that which does not change—our perspective. No matter how close to the surface of sensation you may feel yourself to be, if anything is to be consciously perceived at all, it is from a particular point of view; Kant called this "subjective apperception," (as opposed to the "logical objective apperception" that your computer might have when its programming causes it to refer to itself as an object). Whether this central point itself actually "moves" is probably an irresolvable metaphysical riddle: we usually grant that there are various consciousnesses, various subjective points of view centered in bodies—yet it may be that there is one (non) centered consciousness with various bodies of memory moving about it. This is much more difficult to imagine, and might be paralleled by the image of staged play where God would walk around, yet would stay in one place while the scenery would scroll by in the background. At any rate, "there is a perceiving" which, if remembered (even in dreams), is always in reference to a point in continuity with other memories. This point is not necessarily a self, and the experience may not be of the world—yet, if sensation is, it is experienced as being centered; a centrality which is not the sensation.

STREAM-OF-CONSCIOUSNESS

This continuity of experience, the flux of sense where the flux itself is relative to a point, has been well illustrated in books using the stream-of-consciousness technique, such as by Marcel Proust, James Joyce, Virginia Wolfe, William Faulkner, J.D. Salinger, Ken Kesey and others. Usually, this technique does not narrate a situation and describe it as if from a bird's-eye point of view, but tries to put you inside of the perspective of a particular individual. More than film techniques where you hear someone's thoughts, or where the camera runs around at a dog's eye view (I recall the 1958 movie "Bell, Book and Candle" showed a cat's eye view where everything was green), stream-of-consciousness writing will often follow the flow of the character's experience as their attention shifts among feelings, memories, thoughts, and immediate perceptions. In this case, we read along the words, picking one up at a time, while being linked with the flow of sentences—we are always at one place, but that place is shaped by where we were and are in the text. In actual experience, the situation may be more complex—various thoughts, feelings, memories, and perceptions among the five senses may be experienced simultaneously. Yet, there is always our present experience, a focus which may be more or less acute, which is shaped by the past, and often concerned with the future.

PARALLEL CONSCIOUSNESS

Our ability to appreciate more complex experiences can be seen in the history of music, for example, where simple tunes of the past have met with more recent complex, multi-layered orchestrations. Often, one's sense of centeredness can be swept away in a well integrated, mesmerizing "sensory overload"—one can be overwhelmed as the ear is pulled in several directions by a layered piece of music, just as the eye can swim in a sea of harmonious complexity. Meditating on a mandala, vision may scatter a bit, usually with the sense snapping back more vividly—a sort of letting relaxed lucidity spread. Or, one may choose to focus on one part, and ignore the rest—new parts might be discovered, adding a new appreciation of the whole (an old song or picture may suddenly seem new). Yet, as much as one may be swept away with aesthetic experience, (indeed, such experience can be quite sublime, or "enlightening"), as long as something is sensed, it is from some perspective. Perhaps, momentarily, one may be completely swept out the door, but the most you will ever remember will be your experience at the threshold.

TIME AND CHANGE

Contemplation of time helps illuminate the duality of subjective centrality. When one looks and concentrates on some view, at any "moment" there will be a single focal point—and one may not be able to tell if that single point is the singularity of the viewed, or the singularity of the perspective (more on this later). But, when one shifts from one view to another, from one thought, emotion, etc, to another, a single perspective remains as the view changes. Again, the flux of appearance and change itself is so only in contrast to a singular perspective; just as relativity might be itself relative to the absolute, change may be relative to the non-changing—we might not be able to judge or experience change if there was not a non-changing perceiver: such is beyond imagination, perspective, and possibly knowledge. Of course, we as perceivers do change—in position, orientation, location, etc—but even then it is a self-same perceiver that changes.

Now, the physicist in our minds will suggest that an arbitrary, non-absolute frame of reference can be set up to judge change. Although we may not know how fast the earth is zipping through an universe in motion (where everything else may be moving as well), we can say how fast it revolves around the sun—the sun being a point of reference that is not absolute, but one which can be used in contrast to the earth. The "largest known center of gravity" has been designated as a pseudo-absolute—a point from which to judge, but not the absolute center itself. We could just as easily designate the earth as non-moving, and suggest that the entire universe moves about it (even in day to day rotation). The insight here is that any judgment concerning change requires a point of reference, even if that point is not deemed absolute. These cases involve the comparison of two "objective" points (one changing relative to the other), and with our own consciousness, any objective account would need to discuss a given body and what that body was judging—the location of the body would be a non-absolute reference point. (Such would be the same with a robot, as with Kant's objective apperception—a robot would be able to make judgments about objective facts relative to its position / training—but whether or not a robot would have subjective experience is another issue.)

Subjectively, however, we only ever experience one point of view in contrast to the flux of experience. Now, if our experience was entirely consistent, we might not ever see our point of reference (our personal subjectivity) as being anything other than centered in our body, and the non-absoluteness of our perspective would be as obvious as our faith that others too have their own perspective. Our experience is not always consistent though; such is most evident with the subjective experience of time. "Time flies when you're having fun," and a painful moment may circle around for a seeming eternity. Our senses may be altered too, as with certain types of hallucination where objects may lose detail and become more abstract (like cartoons); and in both of these cases, the perceiving subject begins to lose its contact with external objects—perception becomes more subjective, and ultimately subjectivity may be completely severed from objectivity: pure creative vision.

Such experiences cannot be entirely attributed to alterations in the bodily mechanisms of perception, as it is not always a change in regularity that is involved, but often a radical loss of regularity itself: a point where consciousness is no longer of an objective world reflected through a single objective body—pure sensation. But again, a complete loss of the body as a perceptual perspective could not be remembered by that body (given that our body is entwined with our memory, in the brain, muscle memory, sense organs, etc). Subjectivity is being indirectly pointed at here, with this discussion, by bracketing the objective body, and the objective world it encounters. We are approaching the difficult notion of consciousness being "exterior" to the body (and really, now, we never do see the insides of our bodies—we only experience exterior surfaces—although these surfaces may be of our own bodies, internal organs, and even brains). And just as our epidermis is one continuous surface, we might imagine a single universal epidermis of consciousness (possibly this is something that the Deleuze and Guattari's concept "body without organs" could designate). At any rate, as we begin to remove the unity of consciousness from the particular body, we are moving towards a unity of consciousness in general—an absolute reference point (pure subjectivity), which itself might also be akin to absolute change (no objective, static, structure)—and such would never be remembered by the particular body (again, at most you will remember the threshold). A good illustration of this in the history of art can be found with Hokusai's *36 Views of Mount Fuji* (1826-1833) where the singularity of perspective is inverted with multiple perspectives pointing towards a single dormant volcano: possibly an erupting volcano can be compared to the will, or desire, and a connection could be seen with Schopenhauer's *Will and Representation* (1818-1844).

This attempt to explain the possible transcendence of a particular perspective is most difficult to convey, as it cannot be accounted for objectively. (Later, I will discuss the radical limitations of any objective explanation.) But one can clearly see the relevance to ethics—especially if the relativity of particular points of view can be contrasted to a universal absolute. However, who is to say that every prophet's or Buddhist's claim to universal god-head is not to different god-heads? (Ironically, it may be the consistency of their various inspired accounts which substantiates their claims: verification—but then maybe it is an experience of a generic non-conditioned part of the body—the human body. Yet, who wants to sacrifice their individual identity and be a generic "Buddha-Christ" all the time, even if it is one's truest human self?)

(BREAKING) CONDITIONING

Now, the flow of one's experience, being predominately tied as it is to a particular perspective, as with one's body, will be shaped in large by the very "physicality" of one's training and habits. As we grow, we find certain actions bring certain results, and our behavior is often adjusted accordingly. Defecate on your neighbor's lawn, and you may get whacked. Try eating some weird looking berries, and you might get sick. Smile and be nice, and you may get at treat, etc. Over time, our habits usually get refined to a level we feel comfortable with—no-one wants to be in a perpetual state of anxiety and discord, or stupor, although some may feel more comfortable with certain levels of distress and fogginess than others. Whether one seeks out high-attention crises ("I'm alive!"), or dreamy reverie ("where was I?") is a matter of taste (although these two are not necessarily mutually exclusive). Usually, one's habits adjust and some sort of comfort or equilibrium is maintained. And it is these habits that make one an individual, distinct from family and species genetics—people develop their own peculiar mannerisms, and it is with people's consistencies that we begin to feel we know them.

For those seeking new modes of consciousness though (as if the old ones weren't good enough), one may have to examine one's habits. Indeed, upon examination one will most likely find that one's habits are really not their own at all, but have been formed by the environment—by one's fellow world inhabitants, and the world itself. We find ourselves acting very much like our family, friends, species, etc. In a search for individuality, and / or in a rejection of the state of affairs one finds one's self in, one may decide to question all inherited habits. Certain habits will be difficult to question though, especially those concerning the "physicality," or regularity, of our perceptions—breathing is a hard habit to shake, and probably not one you would want to break. I have as yet to levitate, despite my efforts!

Social habits can "easily" be rejected by focusing on the regularity of our perception "system"—five senses, motion of body, etc. To look at ourselves in this way is also a sort of stepping out—to recognize yourself as being "in" a "mechanism" of regularity may be a way to separate from mechanism: there is vision, hearing, touching, smelling, tasting, etc, although the radical difference between these various modes of perceiving suggest that one is none of these in particular—they could be lost one by one, and who knows what's left when they're all gone—maybe nothing. (But what was it that integrated them—I see and hear you, the seeing and hearing are simultaneous in one body.)

(LOSS OF) EXTERNALLY PERCEIVED SINGULARITY

As one focuses on one or the other of one's sensory modes each sense always has, if you pay attention, its own particular centeredness. I see this here, now; tasted that there, then, etc. There is usually something, an "it" which one concentrates on (even though that "it" may be some "one"). But, using physics as one example (where it suggests at an intuitive level that real things are particles, or at least localizable energy waves, or strings, that cluster), we can see that the singularity of external objects and events is largely a product of our own thinking—it is my experience with glasses that allows me to see the glass as being a separate thing from the table it is on. Ranging from bright colored children's building blocks to sub-atomic particles, through conditioning, we become able to discern finer distinctions and singularities; although, at some point, such as with sub-atomic particles, over-refinement may become more ridiculous than helpful. Even focusing on atoms is simply trading one *thing* for another *thing* though. Might it be possible to be conscious without a thing, without an object of consciousness?

One can daydream with open eyes, in which case there is some sensation without a particular "it" or "thing," but then one may be thinking of "something." One may completely drift off though, not concentrating on anything, which might be like dreaming (although one can often see things in dreams—or even have specific thoughts). Yet, our senses, and our attention towards a thing may not necessarily be attached to each other—attention may not be necessary aspect of consciousness as one meditates on emptiness or drifts into a stupor. Meditation can also exercise the mind in a different direction: the muscles of mind, strengthened through a repetitive yet relaxed focus.

CRISIS OF THE UNFAMILIAR

A surprise crisis may snap one out of a stupor, in which case we may be overwhelmed by the opposite extreme—uncertainty, indecision, anxiety. When one's comfort is imposed on, as for example, when one experiences something unprepared for, beyond a reaction via habits, one may desperately try to name the circumstance—what is it that is going on? One has "lost the name of action," as it were, and there may be a bit of confusion as one struggles to evade the possibly dangerous circumstance. It is at these points that a serious whack may be needed to snap one out of their paralysis—as in "get a hold of yourself! You're alive! If there is an immediate threat, deal with it: move! Go with those trusty, bodily trained habits! Have a bit of style. Breathe." As one grows older, these crises of action may become fewer and less severe—you gain habits that allow you to perform in the face of disaster: the world may be falling down around you, but you've got lives to save!

So if you wake up suddenly in an inferno, such is not a time to smell the roses. On the other hand, one does not want to run into a crowded theater and yell, "Fire!" when there is none. A point being that a time of crisis is most likely a time for action, but don't make a crisis for other's when there isn't one. This is where one must take stock of our external environment, and judge the state of affairs externally, rather than internally: your particular panic may not be the world's—it has been around for quite some time now, and it's going to be awhile before the planet melts. These last remarks may have more to do with conscience than consciousness, yet it is important, I believe, to contrast one's search for a different individual consciousness, with taking on the responsibility of global or universal consciousness: we are also the eyes and ears of the planet and universe. Although possibly phenomenally irrelevant, such may have important consequences for action: to bear the power of the universe may be to feel the gravity of responsibility.

BEYOND THE GENEALOGICAL: UNSPEAKABLE PHENOMENOLOGY

Much of this discussion has revolved around the limitations of what I would call the "genealogical." This would include such categories as (physical) regularity, mediation, memory, and habits, as well as biological heritage: any structure with a history. And, as we shall see, our understanding of such may simply be a result of neurological projection—our understanding being limited by the very mechanism of understanding, the brain. Variously, the genealogical is concerned with connection, which itself is often seen as causal. One can trace the connections of any particular experience indefinitely—that is, just as one can trace one's family heritage (possibly back beyond the apes), one can trace the over-determination of any particular event experienced or evidenced. Ultimately, one might trace everything beyond electron-microscopic physical events—but that stage of detail would involve zillions of elements; and even then physicists may have as yet to reach the ultimate building blocks they search for. The genealogical fabric may be infinite—possibly extending "far" beyond the reach of our largest telescope, and being composed of ever smaller connections. If there is any limit, it may be our own size—our life forms could be the beginning and the end of the scale, or rather the center, zero point.

Although physics represents a most fundamental form of genealogical understanding, one can see progressive stages (each building on physics)—chemistry, biology, psychology, sociology, ecology (and which branch off at various points, as with anthropology, history, geology, etc). However, as we work our way up the tree, we find the standards of measure change—the smallest units change. Instead of talking about atoms, we talk about molecules, genes, cells, neurons, and information flows (cues, words, language games, discourses, etc.). Some of the unit transitions from one field to another are clearly defined, as from atoms to molecules—but some are a bit more rough, and such accounts for the more speculative nature of fields like scientific sociology. Such does not prevent us from still seeing genealogical connections—one can see certain habits spreading from a certain "source," such as when slang or phrases are picked up from media events.

This genealogical tracing of connections (regularities) does not however, require any phenomenal experience at all. At its most simple, genealogy could be described as "feeling" the threads connecting one area with others, where it is not the sensation involved which counts, but the emulation of a connection. As I move my finger from point A to point B and C, I have reproduced the connection with my motion—I have mimed a "mental picture" of the connection (although "picture" and "representation" can be problematic—there is simply a reproduced connection that is connected with the found connection). In a similar way, I can move my finger around a circle, and reproduce it—the form can be recorded, as with a memory which is composed of a network of neurological connections. The question arises here—is form and genealogical connection a product of the form of our memory? We see connections and patterns because our neurons are composed of connections and patterns of connections. But—these patterns and connections we find remain consistent outside of our brains, so there must be at least some synchronization (aside from physical connection) between us and the world, and amongst ourselves.

A blind and deaf person could make these connections, just as a person with no tactile sensation could. Again, as far as form, memory, and genealogical connection go, there is no need of phenomenal experience. A blind person can know that the sky is blue even if they can't see it—and blindness doesn't mean that they never experience the color blue; their sight mechanisms, and any memory associated with them, is probably defective (they don't see the *world's* blues). Hence we can see that objectively, our *knowledge* of the regularity of connections is deaf and blind—it has no more sense experience than a data-base on a computer. Quality (or *qualia*)—any sense experiences whatsoever are not necessarily objective: they are non-reproducible as quantitative form; and our genealogical knowledge is of form, which is not quality or qualia.

Does this entail that quality is subjective? Not in some traditional senses of subjectivity—it is not "in our heads" (e.g. as with a private language)—for it has no objective location (remember, any time you point and locate a color, as far as you *know* anything, you may as well be a blind computer. You may point to the wavelength of light associated with a color, but your *objective* knowledge is of the wavelength, not the experience). Even though we *experience* qualities directly, we simply cannot *know* anything about quality directly through knowledge (Kant knew this): someone may tell you of another color (infra-red or ultra-violet) but you will not know the subjective experience of that color, only its relation to other colors, as a blind knowledge.

Talk of an external subjectivity, the universal epidermis, does not help much. We might say that quality is experienced at the surface, and remembered, known, and talked about by bodies (and their interior brains); in which case our inability to know about quality is irrelevant, since we luckily are in sync with a genealogy that consistently coincides with our qualitative experience. To bring this back in line with this chapter though, is not quality (qualia, sensation) itself, as centrally experienced, consciousness? And hence, can we ever know anything objective about consciousness? Like an inversion of Derridian *différance* always deferring presence, our genealogy never touches present quality—in an objective search for consciousness the connecting threads of genealogy never reach it; our logic will never touch emotion. Our understanding can only mime real experience. (Wittgenstein does not entirely resolve this problem—his discourse, failing to precisely grasp the importance of this distinction between the phenomenal and the genealogical, is reducible mostly to genealogy—although there may be something like negative-theology to his approach: quality clarified for demonstrating what it is *not*).

The motivating factor for people desperately trying to connect the phenomenal and the genealogical might be reduced to the fear of their separation: fear of possible death. What good would it be to be written into the book of life if there is no sensation? And who wants sensation with no regularity? Perhaps with the genealogical and phenomenological we have the ultimate yin-yang duality infinitely intertwined, yet ultimately irreducible to the same. Or maybe we could see that any talk of the genealogical is already phenomenal: an unknowable unspeakable oneness. And so our knowledge reaches the point (and Hegel noted this) where it swallows itself up in the unknowable—a latter we pull up behind us. What was the problem?

QUESTIONING CERTAINTY

I would like to end this chapter with a few questions, questions which I see as approaching the irresolvable:

1. Is the biology of our body a habit?—evolution would suggest that it is an extremely slow changing habit.

2. Would being a rock be like having sensation with no habits (or only the physical habits of being a rock)?

3. Our brain-structure may project itself on the environment while being attuned to it, in that we understand the world in a way that it is possible for the brain via the way it is built; could the same be said of our sense-organs?

4. Given the difficulty of explaining sight to the blind (like textures for hues and temperatures for luminosity), and

sound to the deaf (like very fine vibrations), how might we ever conceive of a radically different mode of perception (like being a plant, if such has subjectivity—they do have biological integrity)?

5. Often we think of our bodies as giving us the means to perceive, but may it possibly be that our bodies limit modes of consciousness? (e.g. our biology can't handle that sixth or seventh sense).

6. Is balance a sense? Are emotions? Chagrin?

7. Finally, seeing colors, hearing sounds, and feeling vibrations or surfaces are all explained by physics as variable wavelengths of energy impinging upon our body: external stimuli are often oscillations to a variable degree. Oscillation appears to be one of the basic geometric building blocks of the mathematical genealogy of sensation. And, given that our understanding and knowledge (but not our customs!) are literally determined by our delicate brains reaching out to touch the world at a distance, Oscillation + Touch figure as the synchronization of our minds with the world: Could it be (quite in line with Georges Bataille's essay, "The Solar Anus") that physics says we are always, in a way, making love with the world?

Part II

Major Advances In The Philosophical History of Cognitive Conceptualism

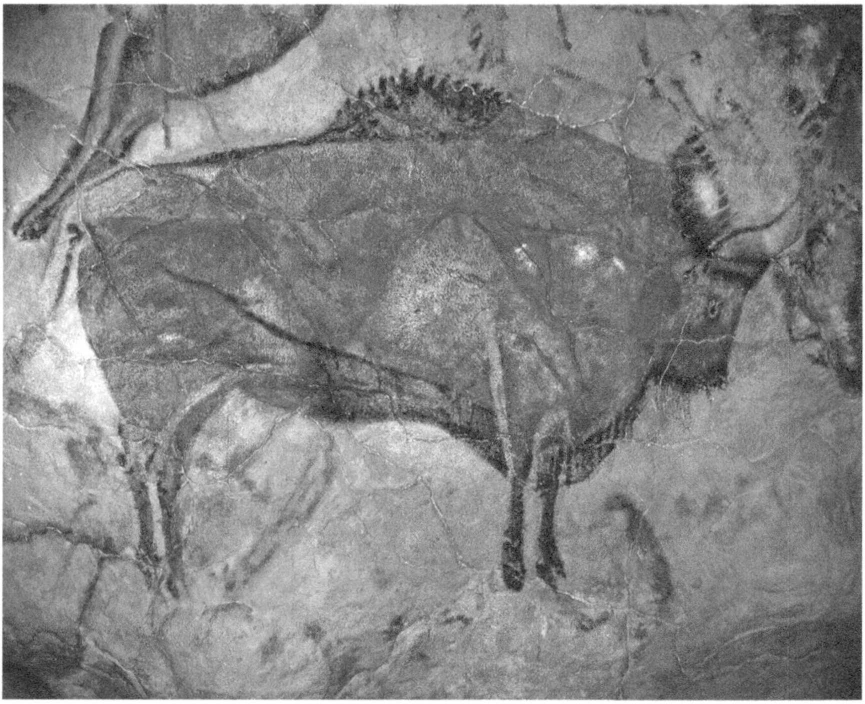

Bison at the Cave of Altamira (~14,000 BCE)

Chapter 4

The Divine

Birth of Conception

THE RECORD OF CULTURAL EVOLUTION

The Venus of Hohle Felswhich figure dates back at least 35,000 years; one of the oldest known artifacts, it demonstrates an exaggeration of female bodily features, and may indicate an early matriarchal religion. Although the paintings of animals at Lascaux date back almost 22,000 years, writing evolved slowly out of a necessity for agricultural accounting. However, many of our oldest surviving texts are religious and are to be found in tombs—the Egyptian Pyramid Texts dating back into the third millennium BCE. We see a transformation of human preoccupations on record, from sexuality and hunting to farmers and then a sort of "divine bureaucracy" that accompanied the ruler and ruled. Here are a few lines from the tomb of King Unas (~2375-2345 BCE), utterance #258:

"Osiris is Unas in a dust storm.
His horror is the earth.
Unas entered not into Geb [the earth] that he might be
    annihilated,
that he might sleep in his house upon earth,
that his bones might be broken.
His injuries are effaced.
He has purified himself with the Eye of Horus.

[...] Unas is on the way to heaven" (Alexandre Piankoff, *The Pyramid of Unas*, Princeton University Press: Princeton (1968), p. 35).

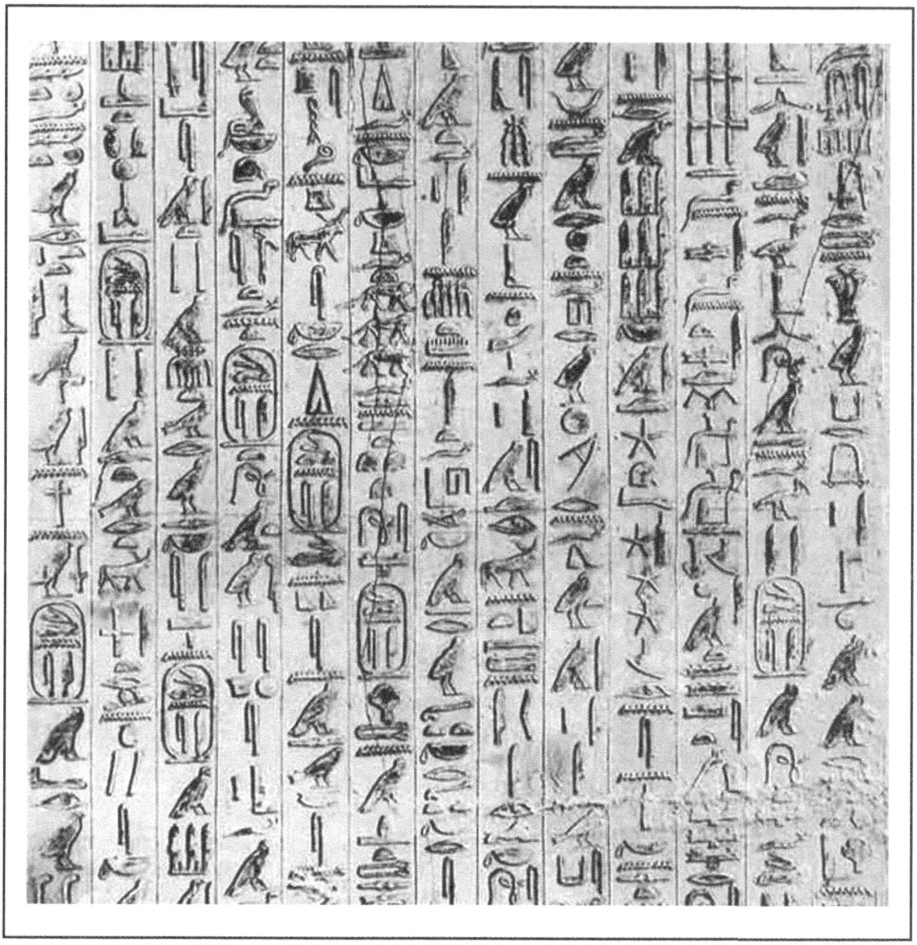

These ancient incantations, about preparing a king's way to the after-world, draw upon Egyptian gods such as Ra, Osiris, or Isis. With these diverse gods, we can see early attempts at systematic interpretation, as they represented various powers of the world and the beyond, and often had a consistent relation among each other. For instance, we have Osiris, a universal god-man and bringer of civilization, born of the sun-god Ra, and reconstructed by Isis after his death. A precursor to Christ, Osiris was the bridge kings would embody on their way to the after-world.

Conceptually, with the Egyptian gods, we have key, reoccurring archetypes—centralized concepts which have defined relations to one another. And the ancient texts often have numerous repetitions of these names followed by attributions—they served as focal subjects for predicates. The heavy repetition found in these texts, no doubt in line with the ancient rituals of sages, must have served to help ingrain these gods in habitual cultural memory—repetition being a seed of habit, and habit being a necessary aspect of evolutionary survival. The hierarchies of worshiped gods could represent transcendent powers which affected our world—which makes them precursors to Plato's transcendent forms.

With the ancient Vedas of India (mid second millennium BCE) dating prior to Moses' single personified God (~1200 BCE) till the Upanishads were started (~800 BCE), we can begin to see a shift from a concern with anthropomorphic gods to more abstract conceptions such as Brahman (an indefinable). This can be discerned in the Nasadiya sukta or Song of Creation, X.129 from the Rigveda:

नासदासीन् नो सदासीत् तदानीं नासीद् रजो नो व्योमापरो यत्
कमिावरीवः कुह कस्य शर्मन्नम्भः कमिासीद् गहनं गभीरम्

न मृत्युरासीदमृतं न तर्ह नि रात्र्या अह्न आसीत्प्रकेतः
आनीदवातं स्वधया तदेकं तस्माद्धान्यन् न परः कि चिनास

तम आसीत् तमसा गूळमग्रेऽप्रकेतं सललिं सर्वमाइदम्
तुच्येनाभ्वपहितिं यदासीत् तपसस्तन्महनिाजायतैकम्

कामस्तदग्रे समवर्तताध मिनसो रेतः प्रथमं यदासीत्
सतो बन्धुमसत निरविन्दिन् हृद प्रितीष्याकवयो मनीषा

तरिश्चीनो वतितो रश्मरिषामधः स्वदिासी दुपरस्विदिासी
रेतोधाआसन् महमिान आसन् स्वधा अवस्तात् प्रयतः परस्तात्

को अद्धा वेद क इह प्र वोचत् कुत आजाता कुत इयंवसिृष्टः
अर्वाग् देवा अस्य वसिर्जनेनाथा को वेद यतआबभूव

इयं वसिृष्टर्यित आबभूव यद वा दधे यद वा न
यो अस्याध्यक्षः परमे व्योमन् सो अङ्ग वेद यद वा नवेद

> "1. Then was not non-existent nor existent: there was no realm of air, no sky beyond it. What covered in, and where? and what gave shelter? Was water there, unfathomed depth of water?
>
> 2. Death was not then, nor was there aught immortal: no sign was there, the day's and night's divider. That one thing, breathless, breathed by its own nature: apart from it was nothing whatsoever.
>
> 3. Darkness there was: at first concealed in darkness, this All was indiscriminated chaos. All that existed was void and formless: by the great power of warmth was born that unit.
>
> 4. Thereafter rose desire in the beginning, Desire, the primal seed and germ of spirit. Sages who searched with their heart's thought discovered the existent's kinship with the non-existent.
>
> 5. Transversely was their severing line extended: what was above it then, and what below it? There were begetters, there were mighty forces, free action here and energy up yonder.
>
> 6. Who verily knows and who can here declare it, whence it was born and whence comes this creation? The gods are later than this world's production. Who knows, then, whence it first came into being?
>
> 7. He, the first origin of this creation, whether he formed it all or did not form it, Whose eye controls this world in highest heaven, he verily knows it, or perhaps he knows not" (Nicol Macnicol, Trans, Hindu Scriptures, J.M. Dent & Sons limited: London (1938), pp. 36-37).

With this hymn we can already see the use of explanatory concepts, such as "Desire," which go beyond anthropomorphic gods. Here we can also see reference to water as a possible first principle.

Such a notion is evidenced in the Egyptian Book of the Dead (a sort of general source book for incantations, mid fifteenth century BCE), which clearly predates Greece's first philosopher Thales (seventh to sixth century BCE) who, having traveled to Egypt, is known to have established western philosophy by positing water as a first principle of being—a major step in claiming that all diversity fell under one kind of element: the one of four "known" elements (air, water, fire, earth) that could take the form of liquid, gas or solid. With relevance to Plato, the Rigveda hymn is also important, for it has a cosmology which includes a single creator, gods, and abstract forces in a hierarchy much as we will see in the next chapter.

Raphael - School of Athens (~1510)

FROM MANY GODS TO A THEORY OF ONE

One can see a shift from godly to abstract concepts in Greek history as well. We have the poems of Homer (which include heroic journeys of exploring the *unknown* and encounters with the *supernatural*) and of Hesiod, which (dating back in written form to mid eighth century BCE) include a hierarchy of gods who have variously distinct characters. Hesiod's *Works and Days* (~700 BCE) includes reference to creation and subsequent ages, and is a sort of farmer's almanac promoting labor. Later, Thales who is said to have predicted an eclipse that occurred on May 28th 585 BCE; to have discovered the seasons and divided the year into 365 days; and, to have fallen in a well while star-gazing on a walk—initiated Greek philosophy. But it is not until Anaximander (~610-547 BCE), a younger contemporary of Thales, that we have a purely abstract philosophy. He was the first Greek map-maker, and the first Greek prose-writer, and perhaps these interests shaped his view:

> ἐξ ὧν δὲ ἡ γένεσίς ἐστι τοῖς οὖσι, καὶ τὴν φθορὰν εἰς ταῦτα γίνεσθαι κατὰ τὸ χρεών· διδόναι γὰρ αὐτὰ δίκην καὶ τίσιν ἀλλήλοις τῆς ἀδικίας κατὰ τὴν τοῦ χρόνου τάξιν.
>
> "it is neither water nor any other of the so-called elements, but some infinite nature, which is different from them, and from which all the heavens and the worlds within them come into being. And into that from which existing things come-to-be they also pass away according to necessity; for they suffer punishment and pay retribution to one another for their wrongdoing in accordance with the ordinance of Time" (Simplicius, quoted by Paul Seligman, *The APEIRON of Anaximander*, The Athlone Press: London (1962), pp. 19-20).

As well as inventing the sundial, Anaximander also formulated the first cosmogony that did not include gods (resembling an earth centered extension incorporating dualities). It was these new doctrines, concerned more with abstract relations than personalized gods, which paved the way for future philosophers.

MATHEMATICS & SOUL

The teachings of Pythagoras (~570-495 BCE) are buried in lore... steeped in a cult of mathematics and second hand accounts.

He was the originator of the Pythagorean Theorem (where in a right angled triangle the square of the hypotenuse c is equal to the sum of the squares of the other two sides, a and b: $a^2 + b^2 = c^2$), and was said to have discovered mathematical equations for musical notes. Diogenes Laërtius, following Alexander, covers Pythagoras in his *Lives of the Eminent Philosophers*—a few quotes from which demonstrate how Pythagoras saw mathematics intertwined with fundamental aspects of creation, and the soul as a mental concept:

> "Alexander in his *Successions of Philosophers* says that he found in the Pythagorean memoirs the following tenets as well.
>
> 25. The principle of all things is the monad or unit; arising from this monad the undefined dyad or two serves as material substratum to the monad, which is cause; from the monad and the undefined dyad spring numbers; from numbers, points; from points, lines; from lines, plane figures; from plane figures, solid figures; from solid figures, sensible bodies, the elements of which are four, fire, water, earth and air; these elements interchange and turn into one another completely, and combine to produce a universe animate, intelligent, spherical, with the earth at its centre, the earth itself too being spherical and inhabited round about. There are also antipodes, and our 'down' is their 'up.'
>
> [....]
>
> 30. The soul of man, he says, is divided into three parts, intelligence, reason, and passion. Intelligence and passion are possessed by other animals as well, but reason by man alone. The seat of the soul extends from the heart to the brain; the part of it which is in the heart is passion, while the parts located in the brain are reason and intelligence. The senses are distillations from these. Reason is immortal, all else mortal. The soul draws nourishment from the blood; the faculties

> of the soul are winds, for they as well as the soul are invisible, just as the aether is invisible.
>
> 31. The veins, arteries, and sinews are the bonds of the soul. But when it is strong and settled down into itself, reasonings and deeds become its bonds. When cast out upon the earth, it wanders in the air like the body. Hermes is the steward of souls, and for that reason is called Hermes the Escorter, Hermes the Keeper of the Gate, and Hermes of the Underworld, since it is he who brings in the souls from their bodies both by land and sea; and the pure are taken into the uppermost region, but the impure are not permitted to approach the pure or each other, but are bound by the Furies in bonds unbreakable" (Diogenes Laërtius, Robert Drew Hicks, trans, *Lives of the Eminent Philosophers,* 1925, <//wikisource.org>).

Not only do these passages suggest that Pythagoras expanded upon Anaximander's thinking about the fundamental principle of reality, making it mathematical through and through, but that he developed a notion of soul as having distinct faculties and being intermingled with, yet distinct from the body—and that there is a struggle between body and soul for control with immortal consequences: quite an advance.

THE MORE THINGS CHANGE...

Our next stop on this brief overview of major pre-Socratics is with the odd philosopher, Heraclitus (~535-475 BCE). Heraclitus is known as a philosopher concerned with change; and his cryptic aphorisms suggest paradox and pursue such ideas as the "Logos," and the "continuum," and emphasize time and individual perspectives. Consider this fragment:

> "2. We should let ourselves be guided by what is common to all. Yet, although the Logos is common to all, most men

> live as if each of them had a private intelligence of his own" (Heraclitus, William Harris (trans.), *The Complete Fragments*, <//community.middlebury.edu/~harris/Philosophy/Heraclitus.html>).

It seems somewhat clear that what is meant by "Logos," is a sort of public intelligence—or "reason" that is common to all, which we should all follow. "Logos" translates as "word," "account," "formula," or "reckoning"—and could be seen as designating thought reflecting through words. Moreover, Heraclitus claims:

> "3. Men who love wisdom should acquaint themselves with a great many particulars."

And:

> "8. I have searched myself."

Already we see a tension between a Logos "common to all" which should guide philosophers ("Philo-sophy" means "love of wisdom") and an imperative that philosophers look beyond generalities—possibly it is through learning many particulars that we learn that the "Logos" connects them all—and especially through self-examination. Perhaps the first philosopher to suggest thinking "outside the box," Heraclitus says:

> "19. Unless you expect the unexpected you will never find truth, for it is hard to discover and hard to attain."

And bringing in the notion of a temporal continuum:

> "21. You cannot step twice into the same river[....]"

This last fragment / aphorism could be metaphorical; rivers are often connected with the concept of time. But there is a linguistic paradox too, how is it that the river changes but still remains the self-same river? This may be because it embraces both change in the material world, and stasis through the Logos, or word. Hence:

> "23. It is in changing that things find repose."

What was cool is now warm, what was dry becomes wet; the properties change, but again, the Logos, or word remains the same. Moreover:

> "28. There is exchange of all things for fire and of fire for all things [....]"

Here, "fire" could be a metaphor for change or energy—and change is found to be the nature of all things found stable through the Logos. Moreover, the principle of change is found not only in particulars, but in the entire universe:

> "29. This universe, which is the same for all, has not been made by any god or man, but it always has been, is and will be an ever-living fire, kindling itself by regular measures and going out by regular measures."

Heraclitus considers the soul as well:

> "42. You could not discover the limits of soul, even if you traveled by every path in order to do so; such is the depth of its meaning."

> "43. Soul is the vaporization out of which everything else is composed; more-over it is the least corporeal of things and is in ceaseless flux, for the moving world can only be known by what is in motion."

Again, the world is in flux, as contrasted to the stasis of the Logos:

> "64. Although intimately connected with the Logos which orders the whole world, men keep setting themselves against it, and the things which they encounter every day seem quite foreign to them."

But again, worldly things can seem to be their opposites, depending on your point of view:

"108. The way up and the way down are one and the same."

Whether ladder or incline, you can go both up and down, depending on which direction *you* are travelling.

...THE MORE THEY STAY THE SAME

Parmenides, in his philosophical poem (dated ~490-475BCE), seems to have taken the Logos of Heraclitus, and found it to be the nature of being itself, not just reasoning with words. His philosophy is conveyed and revealed in a heroic journey through "the gates of the paths of Night and Day" (Parmenides, David Gallop (trans.), *Parmenides of Elea: Fragments*, University of Toronto Press: Toronto (1991), p. 51) to hear the truth from a revealing Goddess. Despite this "revelation" of truth, Parmenides' poem goes beyond dogma, to actually make *arguments* for its truths:

> "Come, I shall tell you, and do you listen and convey the
> story,
> What routes of inquiry alone there are for thinking:
> The one—that *is*, and that *cannot not be*,
> Is the path of Persuasion (for it attends upon truth);
> The other—that *is not*, and that *needs must not be*,
> That I point out to you to be a path wholly unlearnable,
> For you could not know what-is-not (for that is not
> feasible),
> Nor could you point it out" (*Parmenides*, p.55).

And moreover:

> "...because the same thing is there for thinking and being" (*Parmenides*, p. 57)
>
> "And it is all one to me" (*Parmenides*, p. 59).
>
> "A single story of a route still
> Is left: that is, on this there are signs

Very numerous: that what-is is ungenerated and
imperishable;
Whole, single-limbed, steadfast, and complete;
Nor was once, nor will be, since [it] is, now, all together
One, continuous; for what coming-to-be of it will you
seek?" (*Parmenides*, p. 65).

Parmenides seems to have discovered, or created eternity. He equates the timeless whole of all being as one and the same for both being and thinking. Hence, time and multiplicity are illusions; such is expanded upon by Plato, as we will see in the next chapter.

3rd century fragment of Plato's Republic

Chapter 5

Plato
Memory of the Gods

"Time brings everything; and dragging years alter
names and forms, nature and even destiny."

"On Time," an epigram attributed to Plato; Willis Barnstone (trans.), *Sappho and the Greek Lyric Poets*, Schocken Books: New York (1988), p. 180

Plato (~428-348 BCE) has sometimes been heralded as the founder of western philosophy, which, although partly true considering the breadth of his writings, the Academy he helped establish, and his ever growing heritage, is also ironic, bearing in mind that other older philosophers take center stage in many of his dialogues. These dialogues—a format which makes him more of a playwright than a poet—give Plato the aspect of an objective reporter, adding to the lucid sobriety of investigations into what are often obscure mysteries. This outside perspective also allows Plato to forsake some responsibility (as well as credit) for the views put forth; it leaves the possibility of Plato having a critical stance concerning his own writings. This absence from his own text, a sort of negativity, parallels the critical movement of the dialogues themselves, as the main orator, usually Socrates, frequently slips around arguments, putting most assertions in question.

Despite Plato's absence, and the plurality of orators with their seemingly wandering critiques (which dialogues, more often than not, turn out to be meticulously crafted and interlocked with related metaphorical themes), I will argue that a consistent metaphysical and epistemological structure of reality and our relation to it remains throughout his major writings. That is, arguing against some contemporary interpretations suggesting that he changed his views, I will elucidate the coherent conceptual framework which operates as a backdrop, and often as a linchpin, throughout the entirety of Plato's works—a sympathetic reconstruction of Plato's outline of reality, as it were, which, while not examining the arguments put forth in much depth, may be a critique in that it will bring weaknesses and "implied inconsistencies" within the consistent structure to the fore.

THE STRUCTURE OF PLATO'S COSMOLOGY

In various episodes throughout his dialogues—as with Socrates' account of the underworld in the *Phaedo*, in the "Myth of Er" in *The Republic*, in the story of time reversal in *The Statesman*, and in the account of the universe's creation in The *Timaeus*—we get a view of the hierarchical structure of the universe. The *Timaeus* (29e-48e; citations from Plato, Edith Hamilton and Huntington Cairns (eds.), *The Collected Dialogues of Plato*, Princeton University Press: New Jersey (1989)) spells this out in detail, as it describes how a single God created order out of chaos, gave intelligence to soul (a she), put within her a bodily "world" "animal" (a he) made from the four elements (fire, air, water, and earth), and this was a rotating globe and also a heaven—all of which was the whole encompassing all parts. Soul was created by mixing being, sameness, and difference from both the bodily and the eternal; and it was centered in the body and diffused to the exterior outside it. This notion of a bodily universal infused with a universal soul (see also *Philebus*, 30a), goes beyond the notion of an earthly Gaia, in that it included reference to the inner sphere of the moving planets, and the outer sphere of rotating stars (the earth then thought to be at rest)—the soul also had revolving inner and outer spheres corresponding to the universal bodily ones (which hints at astrology). Subsequently in creation, the single God created time (to be discussed later), and made the host of Greek gods, who were instructed to create humans with individual souls (evidently made from soul stuff) that were assigned to stars and put in bodies.

In the "Myth of Er," recounted in book x of *The Republic* (614b-621b), we find further details of the beyond, as Er, revived from the dead, told of his soul visiting various places: an earthly meadow where souls are judged and go through divine portals leading up or down; an elevated height whence they could see the inner and outer spheres (similar to those described in The *Timaeus*, yet from another perspective); and a "transcendent" land where souls draw lots for their future lives and forget previous ones before being reborn.

A different account of the journey after death is described by Socrates in the *Phaedo* (107d-115b). There he makes a remarkably accurate description of the earth as seen from space—he notes that the atmosphere ends like the surface of water, and that:

> "the real earth, viewed from above, is supposed to look like one of these balls made of twelve pieces of skin" (*Phaedo*, 110b).

He describes the underworld where the dead are judged, and claims those who:

> "have purified themselves sufficiently by philosophy live thereafter without bodies, and reach habitations even more beautiful" (*Phaedo*, 114c).

In the reversal of time story in *The Statesman* (269b-274e), Plato's Eliatic stranger tells of when after:

> "each soul had run through its appointed number of births [....] the pilot of the ship of the universe [...] let go the handle of its rudder [... and] destiny and its own inborn urge took control of the world again and reversed the revolution in it" (*The Statesman*, 272d-e).

After an initial shock, the universe went well for awhile, but began to become disordered, until the creator God noticing:

> "its troubles, and anxious for it lest it sink racked by storms and confusion, and be dissolved again in the bottomless abyss of unlikeness, [... took] control of the helm once more" (*The Statesman*, 273d).

This story is important to Plato's cosmology, in that it discusses both God's abandonment of and intervention with our universe, and the consequent disruption of our past, which is now in a process of healing.

THE IMMORTAL SOUL

In the *Phaedrus* (246a-248b), the soul is described as a charioteer struggling with two opposing desires—for and against order; while in *The Republic* (book iv, 437b-441a), there are three parts of the soul—"desire," "high spirit" (anger), and "reason." The discrepancy between these two accounts could be resolved if one sees a harmony of the three parts as following orderly desire, while a disharmony of the three parts would be following disorderly desire. At any rate, both accounts depict a soul attempting to desire the orderly in harmony with reason.

Now, at a more metaphysical level, various discussions claim the soul to be the source of all:

> "changes and transformations" (*Laws* X, 892a).

> "all actions have soul in them [....and] all things that have part in soul change, for the cause of change lies within themselves, and as they change they move in accord with the ordinance and law of destiny" (*Laws* X, 905a, c).

Also:

> "All soul is immortal, for that which is ever in motion is immortal. But that which while imparting motion is itself moved by something else can cease to be in motion, and therefore cease to live [.... The soul as] self-mover is the source and first principle of motion for all other things that are moved [.... and] a first principle can not come into being" (*Phaedrus,* 246c-d).

However, since the *Timaeus* discusses the creation of soul, one must take into account that the soul was created before the creation of time (which seems to contradict the soul's un-created immortality). Yet, having been created before the distinction of the eternal and temporal, soul never comes into being when it is created—it already exists in motion before time begins; or, possibly its "beginning" coincides with the beginning of time, the motion of soul and time itself being intimately interconnected.

THE ETERNAL, NATURAL FORMS

The distinction between the eternal and temporal marks the most fundamental structural (dialectical) division in Plato's philosophy. In the *Philebus* (53e-b), it is suggested that becoming is for being in the same way that shipbuilding is for ships. Yet, our temporal world itself was modeled after the eternal (*Timaeus,* 37c), implying that the eternal was needed for the temporal. Both accounts coincide, though, when we see that becoming strives to be that which it is modeled after—its origin is its goal. This division, between an original eternal reality, and a subsequent temporal world of shadow play, and our own striving to exit the latter to the former, forms the crux of Plato's philosophical doctrine, and will underlie the entirety of the following discussion.

The order of the eternal, for Plato, includes the original paragon for every modeled aspect of our universe. These forms are variously discussed throughout the dialogues as being ideas, essences, kinds, natures, characters, or real existences. Hence various objects are modeled after and fall short of their forms just as a painting may be modeled after and fall short of its subject. When one breaks a weaving shuttle, one does not make a new one with reference to the broken one, but to:

> "the true or ideal shuttle" (*Cratylus*, 389b).

It may seem strange that something like a weaving shuttle would have an eternal form, when one might consider such an artifact arbitrary on a universal scale. However, Plato sees such items arising as naturally as plants—that is, given our situation, and our progress towards eternal reality, certain things were bound to arise. Moreover, not only do our objects have forms, but also our actions:

> "actions are as real as well as the things [....] actions also are done according to their proper nature [....] In cutting for example, we do not cut as we please, and with any chance instrument, but we cut with the proper instrument only, and according to the natural process of cutting" (*Cratylus*, 425).

There are forms of skills too; and entire arts also serve as "superforms" where:

> "forms of skill converge" (*Sophist*, 232a).

And one must:

> "consider [... an art] in all its forms" (Statesman, 281b).

For the arts are far from artificial (see *Laws* X, 889a), and Plato maintains one should:

> "defend the claim of law itself and art to be natural, or no less real than nature" (*Laws* X, 890d).

These eternal forms can be difficult for us to discern in our temporal world—for, if we consider time as a fourth dimension, trying to understand the eternal from a temporal stand-point would be like trying to imagine a living three or four dimensional Socrates when only given a two dimensional picture. This difficulty can be illustrated further by noting that any comprehensive definition we may state for a form could not predict the possible future changes in that definition (e.g., consider how the invention of the automobile changes one's conception of what a "vehicle" can be). The eternal forms, in their simplicity or complexity, surpass any temporal understanding we could have of them.

Although each form can be isolated as a precise singularity, forms also come in groups, as there is discussion of a "family" of kinds (*Philebus*, 25d); and as the forms can have a "pedigree" (*Sophist*, 226a), and a "lineage" (*Sophist*, 268d) (Such familial relations among the forms will become more evident in the subsequent discussion on the dialectic; relations such as between the form of a knife, and the form of cutting should be readily seen though.)

The perfection of forms and their degraded temporal manifestation can easily be discerned with geometrical forms. For example, we might imagine a perfect circle, yet we will never see one in our world. We come to know of circles in our temporal experience, yet we judge the perfection of these worldly circles by a transcendent standard. Hence we can see why it is suggested that:

> "forms are as it were patterns fixed in the nature of things" (Parmenides, 132d).

The perfect forms may be as easy to see in the world as far as one can see a perfect circle in a worldly one, and these forms are transcendent to the extent that we have as yet to manifest them—a bit of the eternal may be found in every worldly art, action, and object. Indeed, it is claimed that forms actually work through us:

> "carpentering does the works of a carpenter" (*Cratylus*, 416d).

Plato is careful to note that the relation between forms and our names for them may be problematic. In the *Cratylus*, numerous etymologies are traced to show how names (of people, gods, and things like wisdom, the sun, and the soul) originally described their objects; e.g. "sun" may mean (in Greek)—that which "gathers" (cp. attention), "rolls" (through the sky), and:

> "variegates the productions of the earth" (*Cratylus*, 408e-409a).

It is suggested that language originated with hand gestures, audible syllables, and written letters which imitated the objects and actions named (as with pictographs and onomatopoeias—consider the word "crow" for example, its root, "crawe," and the sound of cawing). Hence language, with its original names, consisting of copies of originals, falls short of its object, just as our objects fall short of the forms they are modeled after. And with words changing forms and uses with cultural convention, the problem worsens. This aspect of language going on holiday can be found in the *Euthydemus* as well, where a couple of sophists use the quirks and ambiguity of language to confuse people without regard to truth—linguistic ploys may be used with language itself without regard to its connection with reality. Such a notion, that language (like all artistic imitation) is far removed from worldly objects and affairs, prompts Socrates, at the end of the *Cratylus* to suggest that we look beyond the original defining names, and investigate the full complexity of the things themselves. (Interestingly, the arguments put forth in the *Ion* and the latter parts of *The Republic*, concerning the corruptness of imitative arts, imply that language might be banned as well—Plato forgot to ban his own art.)

Our discussion heretofore has drawn a distinction between the eternal forms, and their temporal copies. Yet, in the *Timaeus* (48e-52e), a new kind of "being" is introduced on the other side of the copy, across from form: *matter*. The three types are variously described as such (note the relation between the words "paternal" and "pattern," and "maternal" and "matter"):

Pattern	Generation	Matter
Father	Child	Mother
Form	Model	Formless
Being	Intermediate	Space

This formless matter seems in accord with what is called the "infinite" or "unlimited" by Anaximander (and can be contrasted with Parmenides' claim that reality is limited by being). One can see that the temporal multiplication of forms in matter produces defective models; un-structured matter causes a loss of formal structure as real unity is dissipated in dreamy multiplicity.

It should be noted though, that the bridge between form and matter is illustrated with reference to the four elements. Since one element can become another (heated Water becomes Air), it is suggested that there is a common substratum for both forms: matter (*Timaeus*, 49c-d). Indeed, the four elements do have geometric forms or shapes (*Timaeus*, 54b-56c) and are "substantiated" (generated) in matter. This argument is actually stronger than a mere critique of the four elements would suggest—for what are our contemporary elements (hydrogen, oxygen, etc.) if not consistent geometric configurations (forms) of matter or energy? (Moreover, our sub-atomic particles and energies may be few in kind as well.) The question this leaves is, if so much can be explained with reference only to the elemental forms (as is done in the latter part of the *Ti-*

maeus) what need do we have for any other types of form? An answer to this would most likely include a reference to complex family ties between elements and "ordinary" objects and actions, and / or a purposive or intended relation between the former and latter. Reference is made to this problem, concerning the scope of causal action (and its relation to the soul as self-mover), in the *Phaedo*:

> "to say that it is because of them [the causality of the elements] that I do what I am doing, and not through choice of what is best—although my actions are controlled by mind—would be a very lax and inaccurate form of expression. Fancy being unable to distinguish between the cause of a thing and the condition without which it could not be a cause! It is the latter, as it seems to me, that most people call a cause—attaching to it a name to which it has no right. That is why one person surrounds the earth with a vortex, and so keeps things in place by means of the heavens, and another props it up on a pedestal of air, as though it were a wide platter" (*Phaedo,* 99a-b).

This problem, concerning the action of causality, is also mentioned in the *Sophist,* where causality is noted as being that which marks things as physically real, but since such actions are in becoming, they are not considered ultimately real (*Sophist,* 248c). Ironically, a thorough examination of the way causal element forms may participate with the other forms may reveal that these more complex forms are actually types of content.

MEDIATING SPIRITS

The triad structure of 1) eternal being, 2) the unlimited void of matter, and 3) the fluctuating generation in-between, occurs also in a discussion in the *Symposium*. There Socrates tells of:

> "a Mantinean woman called Diotima [....] who taught [... him] the philosophy of Love" (*Symposium*, 201d-e).

She points out that there can be a middle ground between either / or oppositions (e.g. something may be neither good nor bad), and discusses a type of agent intermediate between the divine and mundane, the spirits:

> "They are the envoys and interpreters that ply between heaven and earth, flying upward with our worship and our prayers, and descending with the heavenly answers and commandments, and since they are between the two estates they weld both sides together and merge them into one great whole. They form the medium of the prophetic arts, of the priestly rites of sacrifice, initiation, and incantation, of divination and of sorcery, for the divine will not mingle directly with the human, and it is only through the mediation of the spirit world that man can have intercourse, whether waking or sleeping, with the gods. And the man who is versed in such matters is said to have spiritual powers, as opposed to the mechanical powers of the man who is expert in the more mundane arts. There are many spirits, and many kinds of spirits too, and Love is one of them" (*Symposium*, 202e-203a).

"Love" itself is said to be the son of "Need" and "Resource" ("Resource" being a son of "Craft"). This triad, Resource, Love, and Need, parallels the triad of form, model, and matter, where Love operates between lack and supply. And in Diotima's example, the lack in question is the wisdom of the eternal, and so:

> "Love is a lover of wisdom, and, being such, he is placed between wisdom and ignorance—for which his parentage also is responsible, in that his father is full of wisdom and resource, while his mother is devoid of either" (*Symposium*, 204b).

In line with generation being placed between form and matter, Diotima also compares Love's striving for the eternal with humans' drive for procreation and desire for lasting fame: Procreation is:

> "how the body and all else that is eternal partakes of the eternal" (*Symposium*, 208b).

And:

> "men's great incentive [...] is to 'To win eternal mention in the deathless roll of fame'" (*Symposium*, 208c).

Such a conception is comparable to more modern concepts such as the *desire for reproduction*, or *repetition compulsion*; yet it differs in that Love strives for the super-temporal—not a striving for a repetition in time, but a striving to actually traverse all time.

With this concept of spirit, we can see Plato's hierarchical scheme correlates to the evolution of conceptual thinking—spirits, being anthropomorphized concepts, bridge a gap between the complex personalities of the gods, and the delineation of abstract concepts. However, the relationship between spirits and forms is not explicitly spelled out. What would the relationship be between the form of love and the spirit of love? Spirits do fall short of the eternal forms though, so one could surmise that spirits are the purest temporal manifestation of their forms. Such a hierarchy is re-enforced by Diotima's claim that:

> "we are only at the bottom of the true scale of perfection" (*Symposium*, 210a).

REVELATION AND RECOLLECTION

With Diotima's account of the spirits, we are moving away from an account of the eternal order of things towards our relation to and experience of eternity. She suggests that our striving for eternal wisdom culminates in a:

> "final revelation" (*Symposium*, 210a),

which one reaches after successive levels of "initiation;" one falls in love with one body, then with all bodies; then with the soul, the laws, institutions, and sciences; and finally with the

> "one form of knowledge" (*Symposium*, 210a-d).

Diotima describes how:

> "Whoever has been initiated so far in the mysteries of Love and has viewed all these aspects of the beautiful in due succession, is at last drawing near the final revelation. And now, Socrates, there bursts upon him that wondrous vision which is the very soul of the beauty he has toiled so long for. It is an everlasting loveliness which neither comes nor goes, which neither flowers nor fades, for such beauty is the same on every hand, the same then as now, here as there, this way as that way, the same to every worshiper as it is to every other. Nor will his vision of the beautiful take the form of a face, or of hands, or of anything that is of the flesh. It will be neither words, nor knowledge, nor a something that exists in something else, such as a living creature, or the earth, or the heavens, or anything that is—but subsisting of itself and by itself in an eternal oneness, while every lovely thing partakes of it in such sort that, however much the parts may wax and wane, it will be neither more nor less, but still the same inviolable whole" (*Symposium,* 210e-211b).

No doubt, this step by step progress towards *the one* is akin to (yet subtly different than) the progress portrayed in the "Allegory of the Cave" in book vii of *The Republic.* Ironically, that account of finding reality was Socrates' "dream" (*Republic*, 517b)). In the *Phaedrus*, Socrates claims:

"I am a seer" (*Phaedrus*, 242c).

The madness of prophecy is discussed, and again, reference is made to the "final revelation." I quote Plato at length here in respect to a poetic experience which can barely be summarized (Socrates notes that he is often on the verge of poetry in the *Phaedrus*):

> "[O]nly the soul that has beheld truth may enter into this our human form—seeing that man needs understand the language of forms, passing from a plurality of perceptions to a unity gathered together by reasoning—and such understanding is a recollection of those things which our souls beheld aforetime [....] Therefore is it meet and right that the soul of the philosopher alone should recover her wings, for she [...] is ever near in memory to those things a god's nearness whereunto makes him truly a god. Wherefore if a man makes right use of such means of remembrance, and ever approaches to the full vision of the perfect mysteries, he and he alone becomes truly perfect [.... A]nd when he that loves beauty is touched by [... divine] madness he is called a lover. Such a one, as soon as he beholds the beauty of this world, is reminded of true beauty, and his wings begin to grow [....] Now [...] every human soul has, by reason of her nature, had contemplation of true being [....] Some, when they had the vision, had it but for a moment [....] Few indeed are left that can still remember much [....] Beauty it was ours to see in all its brightness in those days when, amidst that happy company, we beheld with our eyes that blessed vision [...] whole and unblemished were we [...] blissful were the spectacles on which we gazed in the moment of final revelation; pure was the light that shone around us, and pure were we, without taint of that prison house which now we are encompassed withal, and call a body [.... And] when one who is fresh from the mystery, and saw much of the vision, beholds a godlike face or bodily form that truly expresses beauty, first there

> come upon him a shuddering and a measure of that awe which the vision inspired [.... And] with the passing of the shudder [....] by reason of the stream of beauty entering in through his eyes there comes a warmth, whereby his soul's plumage is fostered [...] the stump of the wing swells and hastens to grow from the root over the whole substance of the soul, for aforetime the whole soul was furnished with wings" (*Symposium,* 249b-251c).

This "final revelation," which is humanly experienced via "recollection," is in accord with the discussion of learning in the *Meno*. There, Socrates argues that the immortal soul has been born so many times as to have seen everything, and hence:

> "seeking and learning are in fact nothing but recollection" (*Meno*, 81c-d).

The poetical aspect of this account of revelation is also in accord with recollection, in that the mother of the muses who inspires poets is "Mnemosyne," or "Memory"—whose name implies she is a spirit; in which case it would be the spirits of Love and Memory which figure in our approach to the divine revelation of the one.

The mention, in the extended quote above, of one's being reminded of the vision of beauty by a beautiful bodily form, is also consistent with the discussion of recollection in the *Phaedo*, were it is noted that we can be:

> "reminded by similarity" (*Phaedo*, 74a).

In that discussion it is also claimed that:

> "equal objects of sense are desirous of being like it [absolute equality], but are only imperfect copies. (*Phaedo*, 75b).

That is, one worldly object is never exactly the same as another, yet each strives for the perfect single identity of the form. Although the notion of objects desiring is mysterious (is it we or the objects themselves that desire they would be perfect?) such helps to explain how our memory of the eternal can be prompted in the temporal world.

THE MAIEUTIC METHOD

Plato has Socrates note that his own method of instruction is not one of imparting knowledge, but of helping others to gain their own. In the *Theaetetus* (149a-151d), Socrates compares himself to a midwife, where the mind may be:

> "in some labor with some thought it has conceived" (*Theaetetus*, 151b).

Socrates had also noted that:

> "Diotima's own method of inquiry [was] by question and answer" (*Symposium*, 201e).

And other philosophers are noted as using the method as well, e.g. the Eleatic stranger:

> "asking questions, as Parmenides himself did" (*Sophist*, 217c).

Although never explicitly stated, such a method, where the instructor (theoretically) does not hand over knowledge, but assists the student in discovering things for themselves, is congruous with learning being a form of recollection. In this way, we can see the Socratic, or maieutic, method as being a means to agitating and prompting a recollection of the eternal.

Such does not necessitate our participation in an exterior dialogue for the gaining of knowledge, as it is noted that:

> "thinking and discourse are the same thing, except that what we call thinking is, precisely, the inward dialogue carried on by the mind with itself without spoken sound" (*Sophist*, 263e).

Thus, one might carry on a maieutic soliloquy with one's self, drawing out wisdom with one's own inner voice, as if in prayer.

THE DUAL DIALECTIC

The maieutic method has sometimes been confused with the dialectic method, the dialectic dubiously understood to be the method of question and answer. Although the two can coincide, their distinction should become clear in the subsequent discussion.

In the *Phaedrus* (265d-266c), two "procedures" are discussed; one of bringing:

> "a dispersed plurality under a single form" (*Phaedrus*, 265d).

And the other, a reverse, where one may:

> "divide into forms" (*Phaedrus*, 265e).

Socrates describes a method which was a "gift of the gods [... and] passed on," (*Philebus*, 16c-e), in which one must search for a single form, and divide it, and each subsequent division as far as possible. Socrates,

> "a lover of these divisions and collections... [calls those] able to discern an objective unity and plurality [....] dialecticians" (*Phaedrus,* 266b).

Together, this splitting and splicing, the cutting and weaving of forms, comprise the "dialectic" method:

> "the pair of arts [...] of universal scope, the art of combining and that of separating" (*Statesman,* 282b).

Socrates discusses and illustrates this dual method in the *Sophist* and the *Statesman*, reminding us that:

> "any discourse we can have owes its existence to the weaving together of forms" (*Sophist,* 260a).

And demanding that:

> "we must in every case divide into the minimum number of divisions that the structure permits" (*Statesman,* 287c).

As:

> "the philosophical method itself [...] consists in ability to divide according to real forms" (*Statesman*, 286d).

Although one or the other of these two aspects, pluralizing or unifying, is emphasized at various times, we must remember that:

> "the dialectic art never considers whether the benefit to be derived from the purge is greater or less than to be derived from the sponge" (*Sophist,* 227b-c),
>
> "binding... together... [and] separating... off" (*Sophist,* 227c)

—being equally useful.

The ultimate aim of the dialectic is, of course, a:

> "purification of the soul or intellect" (*Sophist,* 227c)

—a soul or intellect on its way to recollecting the final revelation. Yet, it has a practical aim as well: clarifying hazy distinctions and bringing the implicit out into public discourse. With a particular thing:

> "all that you and I possess in common is the name. The thing to which each of us gives [... a] name we may perhaps have privately before our minds, but it is always desirable to have

reached an agreement about the thing itself by means of explicit statements, rather than to be content to use the same word without formulating what it means" (*Sophist*, 218c).

In *The Republic*, book vi (510b-511e), a finer distinction is made concerning two types of intellection: 1) a dialectic concerned only with cutting and weaving forms in order to transcend assumptions and clarify issues by finding higher principles solely on an intellectual plane (e.g. the many are one) is contrasted with 2) an intellection which is concerned with our perceivable world, and maintains assumptions that hold true with reference to that world and from which it derives conclusions (e.g. geometry). The former is termed "Reason," the latter, "Understanding." Reason transcends the world and concerns itself with discovering the proper relations of the eternal forms, while the understanding limits itself to the formal relations found in the temporal world. This distinction would be important in considering the relation between the elements and other forms. Inevitably, any phenomenal-elemental investigation would reach assumptions surmountable only by dialectical "speculation." However, Plato would probably suggest that such would be structurally consistent and knowable, yet phenomenally un-verifiable, in the same way that we could never verify that a dog is an animal in purely physical terms.

With this discussion of dialectical splicing and splitting of forms, and the prior discussion of forms having families, and, most of all with the demonstration of the dialectical method found in the *Sophist* and the *Statesman*, we get a fairly clear picture of what would have to be a complex hierarchical network of forms. Plato does not use the metaphor of a net, though. The method of division suggests that some forms are contained within others, as with categorical logic, (or the notion of the whole containing its parts) and hence Aristotle's subsequent and questionable development. But a division of one into two does not include those two within the first: they are also divided from the first, and maintain a family tie to it;

we cannot think the whole with the parts. In a similar way, weaving two forms together leads to a new form distinct from the previous two. Ironically, the process of division creates connections, and the process of weaving creates schisms (but remember: one does not actually split and splice, but follows these implicit splits and splices).

The entire "logos" of forms may be impossible to picture; yet, with reference to a group of unities called "monads" (*Philebus*, 15b), which are associated with each other by belonging to families, we may comprehend it. Hence a form of the bulldog would be a family member of the dog form, which in turn is related to the mammal form, of which the cat form is also a member. The dialectic is often illustrated with the dialectician (mysteriously and intuitively) making a leap up a family tree, and, by division, working back down to the original form, whose form is thus made more explicit. Of course, this all becomes more complicated when we consider that actions and skills are forms, and entire arts are super-forms as well.

THE TRANSCENDENT ONE MANIFESTED

The doctrine of forms receives its greatest critique in the *Parmenides*, a dialogue that makes the philosopher Parmenides out to be more subtle than our existent texts of his indicate. It is in this dialogue that we find Parmenides noting a schism between the transcendent forms, and the names and objects in our world:

> "those forms which are what they are with reference to one another have their being in such references among themselves, not with reference to those likenesses, or whatever we are to call them in our world, which we posses and so come to be called by their several names. And, on the other hand, these things in our world which bear the same names as the forms are related among themselves, not to the forms, and all the names of that sort that they bear have reference to one another, not to the forms" (*Parmenides*, 133c-d).

In other words, there would be three different "networks" of associations for what was termed the "logos" above—the relations of forms, things, and names; and each realm would be self-referential (although names initially resemble the objects for Plato, convention changes their relationship, and they get caught up in language itself).

This problem is akin to the "third man" problem which stems from the fact that we can only understand one form in reference to another. Particulars are compared with each other by the standard of a single form, yet it would seem that we would need another form to compare the particulars with that form (*Parmenides*, 131d-132b). So beyond using the form of blueness as a standard for two blue particulars (blueness is what these two particulars have in common) another form of "super-blueness" must be used to compare the form with the particulars ("super-blueness" is what the form has in common with the particulars), and so on in an infinite regress. The issue involved here is relativity—since the forms are what they are in relation to one another (and this is illustrated by the dialectic), no form is considered absolute in itself. No form can be used as an absolute standard of reference, and any judgment needs such a standard to be real. And, moreover, since forms, things, and names are all in self-referential realms, not only can we not make an absolute judgment within these self-referential networks, but moreover, we could not connect one realm with another to judge one by the other. This difficulty, however, is resolved in the *Parmenides* and the *Sophist*, where one finds that name, thing, and form coincide in the absolute limit itself, which is the "one." One cannot go beyond this limit—it stands outside of all—and hence is the standard by which all can be judged. And, because its unity pervades all forms, things, and names, it is the single point in which all three realms converge, and are thus connected—it "grounds" the entire logos. Here I will discuss how the one is to be "understood," with mention of its relation to the dialectic.

Parmenides' account of the one may seem contradictory, although contradiction is what Plato finds to "provoke thought" and helps convert the soul (*Republic*, book vii, 523c). Such comes from the dual aspect of oneness that arises with the temporal. In describing the one as the eternal whole (*Parmenides,* 137c-142b), Parmenides variously illustrates how the one cannot be understood in reference to another—for such a comparison would have at least two, and such would not be oneness. In other words, to begin to approach the one, we must dissolve all distinctions—any distinction or relation will include a plurality which is not one. This makes the one seem transcendent, for any worldly comparison would include a multiplicity. Yet, the one even transcends the distinction between the transcendent and the mundane—beyond being a-temporal, it transcends the distinction between the temporal and eternal itself. It simply "is" everything, yet cannot be understood with reference to anything—

"it cannot have a name or be spoken of" (*Parmenides,* 142a).

One could endlessly list what it is not, yet for simplicity of discussion, this non-differentiated one might be *provisionally* understood as being a transcendent and encompassing limit (of all parts of the whole).

However, in the discussion from *Parmenides* 142b, the one which was understood as a whole as we approached the transcendent is re-manifested throughout plurality. That is, as soon as we have a division, each part is itself one singularity—the unity of oneness is distributed throughout and pervades each part. Hence, no matter how many names, things, or forms you may have, each will be only one form, one thing, or one name, in so far as it actually is such. A dog may have four legs, yet it is still only one dog.

The transition from the transcendent one to the manifest plurality of unities is complex. In the *Timaeus* (37d), when the creation of time is discussed, God is said to have:

> "resolved to have a moving image of eternity, and when he set in order the heaven, he made this image eternal but moving according to number, while eternity itself rests in unity, and this image we call time" (*Timaeus*, 37d).

In other words, the eternal one, as a whole, contains its own image; yet the image itself is a copy of eternity, and thus being separate, it moves in multiplicity. It is with this transition from the eternal to the temporal that the transcendent whole becomes manifest with a unified plurality (cp. a single prophet and multiple witnesses; or a single writer and multiple speakers). Any unity found in the world results from a manifestation of the one, yet the one always leaves a trail of plurality in its wake:

> "as coming to be one it must be combined, as coming to be many, separated" (*Parmenides*, 156b).

Hence, we can see that it is with the one that the dual nature of the dialectic, combining and separating, weaving and cutting, is resolved by and operates through the one. Such is akin to the etymology of "Apollo" found to be the:

> "single one, the ever-darting, the purifier, the mover-together" (*Cratylus*, 406a),

—or the "sun" which we saw above as gathering, rolling, and variegating. And so it is also recounted that:

> "the Eleatic set, who hark back to Xenophanes or even earlier, unfold their tale on the assumption that what we call 'all things' are only one thing [....] Later, certain muses in Ionia and Sicily perceived that safety lay rather in [...] saying that the real is both many and one [....] 'parting asunder it is always being drawn together'" (*Sophist*, 242e)

—where the one is understood to be prior to the many.

This entry and exit of the transcendent one into the manifest requires a transition from the rest of eternity to the motion of temporality, and back again, which is said to take place in the "instant." The instant:

> "is situated between the motion and the rest; it occupies no time at all [....] the one, since it both is at rest and is in motion, must pass from the one condition to the other [...] it makes the transition instantaneously; it occupies no time in making it and at that moment it cannot be either in motion or at rest" (*Parmenides*, 156d-e).

Interestingly, this instant might coincide with the "moving image of eternity" or:

> "the forms of time, which imitates eternity and revolves according to the law of number" (*Timaeus*, 38b).

This "image of eternity" mediating between the eternal and the temporal sounds akin to some instantaneous enlightenment experience.

There is also a relationship between the "one" and "being." It is said:

> "for the things other than the one it appears to be that from the combination of unity and themselves there comes to be in them something fresh, which gives them limit with reference to one another, whereas their own nature gives them, in themselves, unlimitedness" (Parmenides, 158d).

Such seems connected to the distinction made in the *Timaeus* between matter and form, where matter was said to be formless; for in both cases there is a relation between the structured and the un-structured. Yet, contrary to what one might expect, we shall see that it may not be the unlimited, or formless matter, which gives things their being.

Now, oneness as a whole, being beyond all distinctions, neither exists nor does not exist—which, practically, is as good as saying *provisionally* that it is nonexistent. Yet:

> "if it is not to exist, it must have the fact of being nonexistent to secure its nonexistence, just as the existent must have the fact of not being nonexistent, in order that it may be possible for it completely to exist [....] since the existent has not-being and the nonexistent has being, the one also, since it does not exist, must have being in order to be nonexistent" (Parmenides, 162a).

In other words, with the "first" distinction in the one, a limit is created between being and not-being, which are co-dependent; and the one is both of these. Yet (see *Parmenides* 162b-d), the one must pass from one state to another in the instant of transition: the one which transcends the distinction between being and non-being is manifested as unified parts with being in opposition to a transcendent (from the standpoint of the manifest) one with no being (as existence "in the world").

However, a question arises for me here, and that concerns the one's status as a limit—for if the transcendent one has no distinctions, then how could it be limited? Would it not be the unlimited itself? No, for it could be neither unlimited nor limited before differentiation begins, and the one is designated as that which is limited and dispenses limitation when there is differentiation. Another question might concern the one's having existence or being in the temporal, rather than the eternal, which is supposed to be real being, as opposed to the illusory shadows of our world. It may also be that the one is non-existent from a worldly point of view, where, from a transcendent perspective, the world would be non-existent, or illusory. However, in the *Timaeus*, different types of being are discussed—for forms, models, matter, and soul. Hence, a proper discussion would avoid the word "being" in discussions of the eternal in favor of something like "ultimate reality," or "real being"—it is the terminology which is inconsistent, not the theoretical structure (although, in the *Sophist*, "reality" is found to be as slippery as the undifferentiated "one").

We have seen how the one, encompassing and pervading all, ties the eternal to the temporal—as such is spun through the image of the eternal where, in the instant, transcendent oneness is fed through into multiple unities spread across time. The involvement of names in this process is so mysterious as to have possibly led the translator F.M. Cornford to have left out what may be a key sentence of this entire doctrine. This sentence, at line 244d in the *Sophist* asserts:

> "'the one' can only refer to one thing—that is to say, to a name" (*Sophist*, 244d).

The one does not have a name, the one *is* a name (which, of course, does not mean the physical manifestation of a spoken or written name, but the name itself—hence, "in the beginning was the word," etc). It is unclear whether this name is the word "one," or the "image of eternity" (unlikely, since such is already a split in the one—but then, is not being a name already a split?—could a name pervade everything?) It could be some secret, as with some Cabala traditions (Plato too would have to account for a personified creator distinct from the one), or something unknowable. At any rate, this would be where name, "thing," and form truly converge in singularity.

Again, it is this one which is absolute, and while pervading all down to the smallest particular, it is also the pinnacle of all forms. And such would mean, with reference to *The Republic*, that the one is the good; of which is said:

> "the objects of knowledge not only receive from the presence of the good their being known, but their very existence and essence is derived to them from it, though the good it-

> self is not essence but still transcends essence in dignity and surpassing power" (*Republic*, 509b).

The one too, prior to its division, would neither be essence or non-essence. Moreover, consider the following parallels in the analogy between the good and the sun (*Republic*, 508b-509b):

One	Context	Faculty	Objects
Sun	Visible World	Vision	Things
Good	Intelligible Region	Reason	Forms

It might be suggested that the good, aside from being that which creates beauty and truth, is that which actually gives quality to the things beyond their other objectively quantifiable aspects which it "nourishes"—this is merely speculation. However, this would help to explain where the generated models of our world get their vividness, seeing that forms, as ideas, are invisible ("ideas are invisible," *Republic* 507b), and that matter is itself void.

THE POSSIBILITY OF ERROR

A problem with understanding the one as the source of truth is that it pervades all existence, and therefore, it might seem that everything is true. Similarly, if what is is true, then there could be no falsity, for falsity would not exist. Obviously this is false, and an explanation of the possibility of falsity requires recourse to the notion of difference, as summarized in the *Sophist*, 259a-b:

> "(a) [...] the kinds [forms] blend with one another, (b) [...] existence and difference pervade them all, and pervade one another, (c) [... the form of] difference (or the different), by partaking of [the form of] existence, is by virtue of that participation, but on the other hand is not that existence of which it partakes, but is different, and since it is differ-

> ent from existence (or an existent), quite clearly it must be possible that it should be a thing that is not, (d) and again, existence, having a part in difference, will be different from all the rest of the kinds, and, because it is different from them all, it is not any one of them nor yet all the others put together, but is only itself, with the consequence, again indisputable, that existence is not myriads upon myriads of things, and that all the other kinds in the same way, whether taken severally or all together, in many respects are and in many respects are not" (*Sophist*, 259a-b).

Logically, it would seem problematic to call existence a form, until one realizes that any sort of logical discussion of existence would require such a form, whatever the relationship between the form for existence and existence itself might be (ditto with difference, or any other abstract concept). But, it is argued here that the forms of existence and difference, actually being what they designate, allow for the isolation of being, and its ability to be associated with some forms, and not with others. Indeed, it is implied that existence, difference, sameness, rest, and motion are something like primary forms (along with the dividing and combining of the dialectic): possibly this would be some original multiple division of the one (again, compare this to the image of time). This, coupled with the limitation that:

> "some of the kinds will combine with one another and some will not, and that some combine to a small extent, others with a large number, while some pervade all and there is nothing against their being combined with everything" (*Sophist*, 254b-c).

This suggests that just about any combination is possible (although not necessarily so). However, it means that some forms in the eternal realm could partake of, say, the form of falsity, which seems to contradict their eternal truthfulness.

The crux of the problem is this: The realm of the eternal forms is supposed to be true reality and ultimate being, while our world is illusory and becoming from and towards real being. Opinion is said to fall somewhere between this ultimate reality and its reverse—it is not knowledge of the forms, but it is better than nothing. Now, although one may form false opinions because of a failure of memory (*Theaetetus*, 190e-195b, 196d-199c), it is noted that:

> "'recollection'... [is] something different from memory" (*Philebus*, 34b).

And that recollection is the way that true opinions, "tethered" by reason, become knowledge (*Meno*, 97e-98a). So, there are false opinions, true opinions, and the tied down knowledge of the recollected forms.

Now, the problem is that, although the *Theaetetus* discusses how we might have false opinions, the argument in the *Sophist* is powerful enough to allow for false knowledge. That is, it is possible that there are forms that are inherently false, and that, via recollection, we could come to have true knowledge of something false—which would be a contradiction, if we did not know it was false knowledge. The only way out of this, that I see—a probable solution—is that the forms are somehow in harmony with our recollection of them: the only false forms would be temporal ones. This may be true, considering that the image of eternity, the form of time itself, being a copy, may be false.

Bust of Aristotle, copy after Lysippos 330BCE

Chapter 6

Aristotle
Animation Between The Frames

THE ORGANIZED ORGANON

We concluded our discussion of Plato by noting that there was a hierarchy of forms (culminating in the one / sun / good) and that Plato implied that certain forms were more fundamental or primary including the forms of "existence," "difference," "sameness," "rest," "motion," etc. Such leads us straight to Aristotle (384-322 BCE) and the opening of his "Organon," a subset of his books which detail the discovery (or invention) of formal Logic—these works have been organized (by Theophrastus) to originate with Aristotle's list of fundamental "Categories."

Like Plato's "existence," "difference," etc; Aristotle defines ten categories:

1. **Substance**: Although Aristotle defines substance in the negative—it "is that which is neither said of a subject nor in a subject" (*Categories* 5, 2a14; all references to Aristotle, Jonathan Barnes (ed.), *The Complete Works of Aristotle*, Princeton University Press: New Jersey (1995))—he gives positive examples of both *primary* specific substances ("e.g. the individual man or the individual

horse" (*Categories* 5, 2a15)) and *secondary,* more general, substances—the general genus of the specific species—e.g. "the individual man belongs in a species, man, and animal is a genus of the species" (*Categories* 5, 2a18). Clearly, the notion of "substance" is a sort of "concrete Platonic form"—for Aristotle, as we will see, rejects Plato's transcendent forms, in favor of these new sorts of instantiated forms, *substances*, which are immediately in our world. Such really sets the stage for Aristotle's thought on a whole, as we will see why the particular is more primary than the universal, and how we logically relate the two.

2. **Quantity**: This pertains to the amount or extent of a substance—how much there is of something (size, number, etc.) that can be either discretely divisible, or a continuum. It has more to do with measurement than with mathematical numbers in themselves—it is "quantity," not the "quantitative."

3. **Relation**: This category denotes that substances can be compared or related to other substances; relations can be of quantity (this substance is *larger* than that substance), or of the particulars and generals of the other Categories to be listed (Quality, Place, Time, etc). All relations are reciprocal, as one substance may be placed *below* another, that other is placed *above* the first substance. Aristotle claims relations arise when substances are something "*of* or *than* other" substances (*Categories* 7, 6a37)—hence *knowledge* is relative to its object as well: knowledge "of" something.

4. **Quality**: Aristotle claims: "By a *quality*, I mean that in virtue of which things are said to be qualified somehow" (*Categories* 8, 8b25). Quality denote *states or conditions* of a substance: "virtuous," "healthy," etc; but also abilities for *action* (e.g. a "swift runner"); and *affections* that can be subjectively experienced, like "sweetness," and "hotness." Very much like an adjective.

5. **Place**: A position in space—as being in a house, or in a city, etc.

6. **Time**: A position in the past present or future—yesterday, noon, etc.

7. **Position**: An orientation in place and time; e.g. "sitting," "standing," etc.

8. **State, or Condition**: Aristotle claims, "States are also conditions, but conditions are not necessarily states. For people in a state are, in virtue of this, also in some condition, but people in a condition area not in every case also in a state." (*Categories* 8, 9a10-12). States are said to be "more stable" than conditions (*Categories* 8, 8b29)—as the state of "being knowledgeable" is more stable than the condition of "feeling cold." Generally, States and Conditions relate to *having* something.

9. **Action**: A doing that changes something; e.g. to lance or cauterize.

10. **Affection**: The reception of an action... e.g. to be lanced or be cauterized.

These *Categories* somewhat set the "reality" or "metaphysical world" for Aristotle's foray into logic where qualified and related objects and ideas oriented in a space-time milieu can be had, acted upon, and experienced. However, as J.L. Akrill's English translation of *Categories* begins with the word, "When," *De Interpretatione* begins with the word "First,"—and seems to me to be the correct starting point of the *Organon* proper.

SENTENCES AS TRUTH KERNALS

De Interpretatione, begins by defining the elements of a *sentence* or *statement*—names and verbs; and how such sentences can affirm or negate. This is clearly a monumental step on the way towards linguistic logic. The discussion of language here is not too deep, but the orientation and further stage-setting here could not be more important to the project of logic. For here, Aristotle *isolates* the sentence—possibly taking such out of a broader context, and makes it a kernel of truth. And such is illustrated by the very process of definition that Aristotle pursues; but this is also betrayed by the fact that such definitions can take considerable time to elucidate, with many caveats, examples, etc. Aristotle clearly knows that reasoning takes extended care and thinking things through; yet he still reduces his logic to examining lone sentences. Possibly this is the smallest unit that logic could examine, and hence would be at least a good place to start.

After isolating sentences that can affirm or deny something of something, Aristotle moves on to discuss the opposition of the Universal and the Particular:

> "I call universal that which is by its nature predicated of a number of things, and particular that which is not; man, for instance, is a universal, Callias a particular" (*De Interpretatione* 7, 17a37).

Such relates to the secondary substances (universals) and primary substances (particulars)—and already we can see a hard division that some might question (noting that there may be some continuum between universals and particulars—some universals are definitely more or less specific than others—in an extreme case, if there was only one last platypus in existence, the universal "platypus" might be that only one particular "platypus.") But the logical incision here is precise—not some continuum between the more general and the more specific, but between any sort of general substance, and the singular substances.

Aristotle goes on to discuss the various combinations of Universals and Particulars in sentences... and tries to isolate those types of sentence which signify something about something such as "Socrates is a man" (a particular related to a universal); "men are mortal" (a universal related to a universal); but also introduces some odd sounding sentences like "A not-man is just," and the negative "a not-man is not-just." This odd contortion of language ties sentences in a sort of "truth-relation" to what they signify: there is a hard either / or—either Socrates is a man, or Socrates is not a man. The use of terms like "affirmation" and "denial / negation" suggest a strict truth or falsity, without shades of grey. But such ties are complicated with further discussions of the "possible," the "admissible," the "impossible," the "necessary," and "contradiction" and the interrelation of these concepts, to all of which, Aristotle concludes:

> "it is not possible for either a true belief or a true contradictory statement to be contrary to a true one. For contraries are those which enclose their opposites; and while these latter may possibly said truly by the same person, it is not possible for contraries to hold of the same thing at the same time" (*De Interpertatione* 14, 24b5-9).

Beyond reaching the law of non-contradiction (that we cannot have both A and not A)—Aristotle has claimed that no true sentence can contradict another true sentence: all true sentences must be in harmony—having two true sentences that contradict one another would be contrary to what can be logical reality.

To sum up *De Interpretation*: Aristotle has introduced the simple sentence as the logical focus—such sentences can affirm or deny one thing to be *possible* (it could be), *admissible* (it probably is), *impossible* (it can't be), or *necessary* (it must be). Again, not only can such sentences not contradict themselves, any true sentence must not contradict other true sentences.

IMPLIED LOGICAL SPACE

With the *Prior Analytics*, Aristotle shifts the terminology from sentences / statements to "propositions"—propositions are statements that can affirm or deny something of something "and this is either universal, particular, or indefinite" ("indefinite" being something not clearly designated as particular or universal—with the Aristotle's example "pleasure is not good" failing to hold the tight logically reciprocal relations that universals and particulars do) (*Prior Analytics* 1, 24a16-21). Propositions further limit the scope of language involved (to just declarative sentences that say something about something) and pave the way for what Aristotle calls "demonstrative understanding" where new knowledge is found through syllogistic deduction:

> "A deduction is a discourse in which, certain things beings stated, something other than what is stated follows of necessity from their beings so" (*Prior Analytics* 1, 24b19-20).

In discussing Aristotle's syllogistic deductions, I will make reference to Venn diagrams. Although these did not come into popular use until the 19th century—they help to clarify the logical space carved out by certain types of propositions. Consider the First Figure:

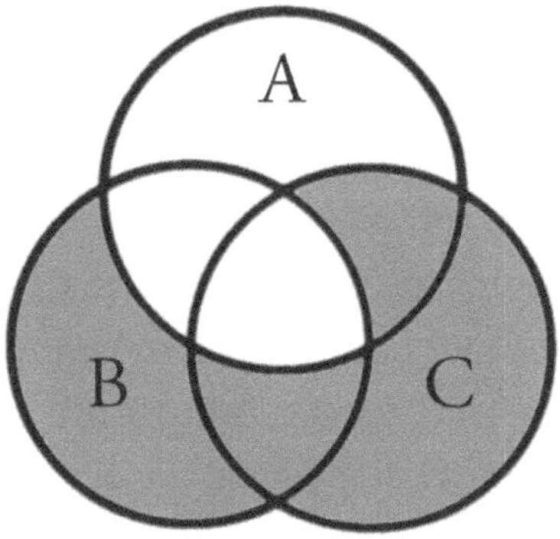

First Figure

In these diagrams, the shaded portions denote that which is eliminated or denied, the clear portions affirmed. In the first figure we illustrate: "If A is predicated of every B, and B of every C, A must be predicated of every C" (*Prior Analytics* 4, 26a1-2). Thus given two propositions, a third conclusion can be deduced: new knowledge is found to be implicit within this logical space. If all mammals are animals, and all dogs are mammals, then all dogs are animals.

In a related way, the Second Figure:

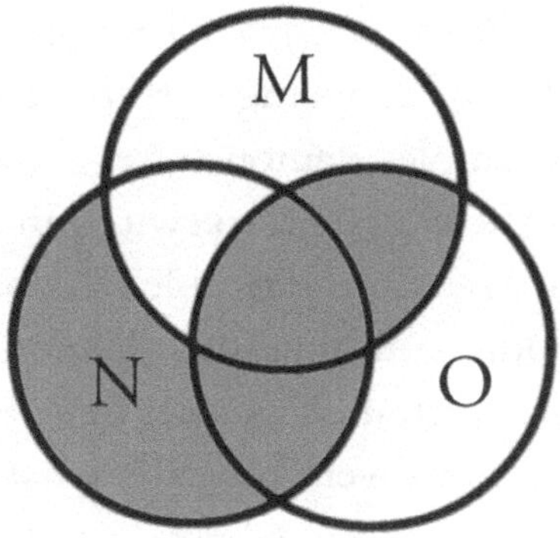

Second Figure

"If M belongs to every N, but to no O, then O will belong to no N" (*Prior Analytics* 5, 27a10-11). Again, another conclusion from deduction. If every dog is an animal, but no animal is a rock, then no dog is a rock.

And the Third Figure:

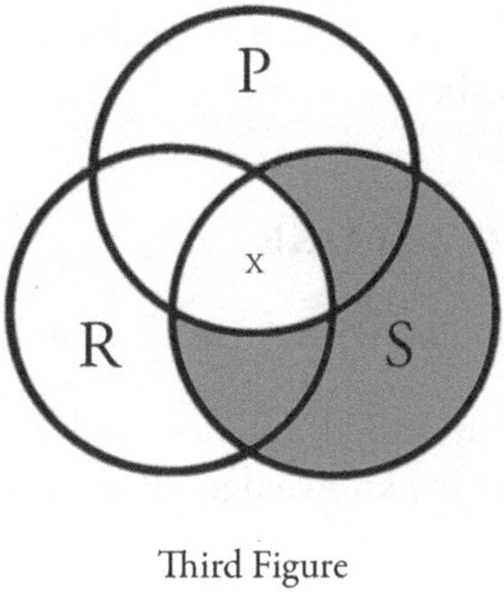

Third Figure

"If they are universal, whenever both P and R belong to every S, it follows that P will necessarily belong to some R" (*Prior Analytics* 6, 28a18-19). Once again, a deduction (although here, the existence of some S is assumed—denoted by an x). If every dog is an animal and is a mammal, then some animals will be mammals.

Throughout the *Prior Analytics*, Aristotle goes through just about every possibility of deducing a "third" conclusion from two relations as theses examples demonstrate. Things get a bit more complex though, as the we begin to switch in an out: Universals, Particulars, Necessary relations, Possible relations, affirmatives and negatives—where some combinations do not yield valid deductions. Although not diagramed as such by Aristotle, I believe the Venn diagram illustrations were basically what he had in mind: a sort of geometric logical space, where "variables" (like A and B) could overlap or be exclusive of each other—although throwing "necessity" and "possibility" into the mix can make things more complicated. The law of non-contradiction is built into these visual examples, i.e. an area could not be both shaded and not shaded (although one could imagine other colors of shading!) But, for the most part, the syllogisms stand to reason, with examples such as "Socrates is a man" "all men are mortal" therefore "Socrates is mortal." Where Plato's dialectic of weaving and cutting created a hierarchical tapestry of forms or concepts, Aristotle found, with his demonstrations through syllogistic deduction, that some relations logically implied others.

REFINING KNOWLEDGE

In the *Posterior Analytics*, Aristotle claims:

> "All teaching and all intellectual learning com about from already existing knowledge" (*Posterior Analytics* 1, 71a1-2).

Although science works best when deducing a particular from a universal, getting to this new knowledge requires working within a context. Having developed the "mechanics" of logical deduction, Aristotle turns to situate it within our worldly knowledge. We have a sort of common sense—a knowledge base from which to work from—esp. principles and "primitives" that are familiar (*Posterior Analytics* 3, 72b5). We may not have deduced these principles, but find them self-evident—and stronger than any conclusions we draw from them, since conclusions are built upon said primary principles. Moreover, Aristotle recognizes that these principles may be grounded in (often intuitive) induction where many particular examples suggest a general principle (e.g. seeing that many humans have died in the past... none living past a certain age—one would set up a sort of principle by induction (maybe not a primary principle) that "all humans are mortal.") Hence the hard 100% logical certainty of deduction is grounded, by way of non-deduced principles, in less than 100% certain induction.

The sweep of the *Organon*, at least from the *Prior Analytics* through the *Posterior Analytics*, *Topics,* and *Sophistical Refutations*, moves from a "simple" logical formalism to more and more complex use of that formalism to arrive at valid arguments, scientific conclusions, and specific definitions—the stick-figure of logical form is fleshed out by the content of worldly related facts—and the *Organon* concludes at the outer limits of logical argument, where 13 specific fallacies, erroneous arguing practices, are identified—including "Verbal fallacies" like use of ambiguity, and "Material fallacies" such as "begging the question" (simply assuming what you're trying to prove).

This sweep continues as Aristotle, following Plato's examination of the state of worldly (and cosmic) knowledge in the *Timaeus*, blazes a path enriching many sciences from *Physics* through *Meteorology*, Psychology, and the complex understanding, history and taxonomic classifications of biology (albeit often with anecdotal qualitative reasoning, and not concerned with the modern scientific method of quantitative experimental testing of hypotheses)—much of this detail is irrelevant to and quite beyond the scope of this book, and hence will not be examined here.

PHYSICS: THE CELL DIVISIONS OF REALITY

With the six books of his *Physics,* as with the *Organon*, Aristotle again organizes his investigation in a move from the general to the particular. In a reverse of modern science's looking for theories to fit particular evidence, and like some sort of biological cell division, Aristotle claims our general notions or universal principles can be analyzed regarding our diverse sense perceptions—the whole divided into its parts, just as:

> "a child begins by calling all men father, and all women mother, but later on distinguishes each of them" (*Physics*, Book I, 184a 28-29).

This refinement of knowledge of the world begins for Aristotle with "principles, causes, or elements" (*Physics*, Book I, 184a 10-11). What he has in mind is made clear with discussions of Parmenides and Melissus, who claimed that all things are "one" (infinite being)—a position that Aristotle rejects in favor of a less quantitatively abstract, and more dynamic reality of a plurality of qualitative things which are *becoming* and change through causation. He argues:

> "Now Melissus says that what exists is infinite. It is then a quantity. For the infinite is the category of quantity, whereas substance or quality or affection cannot be infinite except accidentally, that is, if at the same time they are also quantities. For to define the infinite you must use quantity in your formula, but not substance or quality. If then what exists is both substance and quantity, it is two, not one; if only substance, it is not infinite and has no magnitude; for to have

> that it will have to be a quantity." (*Physics*, Book I, 185a 33 – 185b 5).

All things are thus not "infinitely one" in principle, cause, or element—at the start of his Physics, Aristotle divides the qualitative from the quantitative; substance from magnitude or as he further claims later in Book I:

> "(The universal is knowable in the order of explanation, the particular in the order of sense; for explanation has to do with the universal, sense with the particular)" (*Physics*, Book I, 189a 6-9).

Much as with Plato's dialectic, one can picture here a sort of pyramid, yet here without a monopolizing capstone, but topped with an oligarchy of principles, or so far, at least two: a hierarchy where unchanging generalities of logical language (explanations) at the apex can be compared to a plurality of concrete sensations at the base. In many ways a precursor to the objective / subjective split as the objective is usually quantitative, and the subjective having to do with qualitative sensation.

A third principle arises when thinking of what could mediate between these two realms (also echoed in discussions of the "simple" and the "complex" (*Physics*, Book I, 190a 1-4)). Aristotle talks of "intermediates" "relations" and "what underlies" contraries... and that this:

> "underlying nature can only be known by analogy. For as the bronze is to the statue, the wood to the bed, or the matter and the formless before receiving form to any thing which has form, so is the underlying nature to substance, i.e. the 'this' or existent" (*Physics*, Book I, 191a 9-11).

Such issues of connecting the two realms of logical universals and qualitative sensations ties directly to becoming, change and time. Although senses sense the plural and that which is changing all the time, there is also another sort of change to account for: how the plural qualitative world is given a "shape" (not Aristotle's terms) by the universally reasonable. Or, how does *causality* which divides the "potential" from the "actual" come into play? Aristotle makes the suspect claim that:

> "The truth is that what desires the form is matter, as the female desires the male and the ugly the beautiful—only the ugly or the female not in itself but accidentally" (*Physics*, Book I, 192a 21-24).

Clearly harkening back to Plato's concept of matter (maternal) and form (pattern / paternal), but differing in that matter is different for different objects (*bronze* statues or *wood* beds) and distinguished from the void which Plato's matter resembled; Aristotle also finds the bridge of becoming between the two realms, here as a kind of *desire*. Desire could be understood as a temporal force for change—to actualize the potential; or in other words, causality.

MATTER, FORM, AGENT, AIM

The nature of causality is discussed in Book II of *Physics*—how things come to be; and Aristotle finds there to be four types of causality. Besides matter and form, another type of causality has to do with nature's aim, and Aristotle's observation that nature begets nature as a dog gives birth to a puppy:

> "nature in the sense of coming-to-be proceeds towards nature" (*Physics*, Book II, 193b 13).

> "nature is the end or that for the sake of which" (*Physics*, Book II, 193a 29).

> "nature is a cause, a cause that operates for a purpose" (*Physics*, Book II, 199b 32).

Here we can see that nature seems to have a purpose ("for the sake of which")—as eyes are made to see with, answering a "why" question (*Physics*, Book II, 198a 14) and providing an aim for the course of life (life's purpose seems to become itself, again and again). Long before evolution was theorized, Aristotle was looking at the "end-result" of nature, and noting that in its various incarnations, creatures, as whole and parts, and even artists making artifacts, employ a purpose in bringing something into existence.

Sometimes there is also an "agent" who helps bring something into being, as a carpenter helps build a home. This brings our number of causes to four:

> "the matter, the form, the mover, that for the sake of which" (*Physics*, Book II, 198a 24-25).

Hence a house can be made from wood (matter), in the shape of an A-frame (a form), by a carpenter (the mover), for sake of shelter (a purpose). Interestingly this fits in both with the types of questions we can pose and the parts of language: "who?" (agent), "what?" (matter), "why?" (aim), "how?" (form), "where?" & "when?" (place and time, to be discussed below)—these all fit in with parts of language as with the "nominative" (*who* as subject/agent), "accusative" (*what* as object/patient), and with the "dative's" "to" being a causal push to the "genitive's" "from" being a causal pull. Although such language categories could be said to reflect reality, one may wonder if language limits or shapes perceptions or knowledge as well.

How do the four causes fit in with our hierarchy mentioned above? Aristotle claims that although:

> "the end and the means towards it may come about by chance" (*Physics*, Book II, 199b 19),

> "[n]ecessity is in the matter, while that for the sake of which is in the definition" (*Physics*, Book II, 200a 14).

And since nature is itself the ultimate "for the sake of which," (hence we talk of the "nature" of an item, as its purpose too—"what is human nature?")—matter operates out of necessity (stronger than "desire!") to move towards an end which is a "definition" or form:

> "the necessary in nature, then, is plainly what we call the name of matter, and the changes in it. Both causes must be stated by the student of nature, but especially the end; for that is the cause of the matter not *vice versa*; and the end is that for the sake of which, and the principle starts from the definition or essence [....] Perhaps the necessary is present also in the definition" (*Physics*, Book II, 200a 31-35).

Such is to say, quoting Einstein, "God does not play dice." Necessity requires that the world be the way it is, not chance—it is in the very purposive relationship between matter and definition (or form). Nature needs matter to fulfill its purpose potential in actual formality—again, form is the aim of matter; while change is brought about by an agent (some creature or thing) upon a patient:

> "motion is in the movable. It is the fulfillment of this potentiality by the action of that which has the power of causing motion; and the actuality of that which has the power of causing motion is not other than the actuality of the moveable; for it must be the fulfillment of *both*. A thing is capable of causing a motion because it *can* do this, it is a mover because it actually *does* it." (*Physics*, Book III, 202a 13-18).

HOW SOON IS NOW?

Motion relates to efficient causes (near what we typically think of as cause)—and leads Aristotle into discussions of Zeno's paradoxes (e.g. how can an arrow move from point A to point B if the space between the two is infinitely divisible—it's as if there were an infinite number of intervals between any two points so that a moving object could never get anywhere). Aristotle sidelines such paradoxes by noting that although time is not constructed of a series of discrete "nows"—the now is rather a limit between the past and future, but not an instant "atom" of time (*Physics*, Book IV, 222a 10-12)—although objects do not make jumps on a space-time grid, and although infinity is a potential, but never actual aspect of reality—the continuous nature of space is to not be infinitely or finitely divisible (with time being divisible as "a kind of number" (Physics, Book IV, 219b 5))—moreover, time and motion are relative to each other:

> "Not only do we measure the movement by the time, but also the time by the movement, because they define each other." (*Physics*, Book IV, 220b 15-16).

Objects do not move through time as they do through space: they move through space (Aristotle would say, "places") as measured by time as number of movement: particular sensed movement is continuous, but is measured discretely and abstractly with number. But it need not be a change in place (movement)... as a color could shift, perhaps of a chameleon, and mark the passing of time while said chameleon is at rest—meaning time would more properly be measured by change.

What I think we can see with Aristotle, as resolving Zeno's paradoxes concerning motion and time, is that although things might be said to *theoretically* have to move an infinite number of intervals between point A and point B, the time between the intervals approaches zero as the interval approaches the infinitely small. That is, the "now" as limit between past and future is infinitely small (zero time length): it takes no time to cross an infinitely divided space (each infinite division being zero distance at a limit). I say *theoretically*, because in reality, Aristotle does not see the infinite as real: it is more like an extrapolation. But here, I think approaching the limit of the infinitely small flips inside out and creates an indivisible continuum. The infinitely small division is no distance or time at all, and hence not a division at all. Yet, in our perceived reality, changes in time and position can be measured, *as if* discrete jumps from point A to point B were made during "nows" that have duration. With motion, an object speeds up towards an infinite velocity as the distance to be traveled is infinitesimally smaller: and you could see this if you magnified an arrow in flight—the closer you zoomed in, the quicker the arrow would fly through your frame of reference. Extrapolate that frame to an "infinite zoom" and the frame is crossed instantaneously. Such is just to note that the sensible continuum of motion in place and time *is* divided by the frame and abstract number: the particular sensed object never "jumps" from point A to point B, but the time and place "clicks" from, say, 1 to 2.

A SOULLESS GOD?

Consider:

> "Since every motion is continuous, a motion that is one in an unqualified sense must (since every motion is divisible) be continuous, and a continuous motion must be one" (*Physics*, Book V, 228a 20-21).

And:

> "everything that changes must be divisible" (*Physics*, Book VI, 234b 10).

And:

> "Everything that is in motion must be moved by something. For if it has no the source of the motion in itself [as with natural creatures] it is evident that it is moved by something other than itself" (*Physics*, Book VI, 241b 34-35).

Hence, the continuous "movement" does not change (Aristotle does not seem to compound velocity with acceleration) but is divisible as time enumerates the movement. And any motion can be traced back to prior motions—all the way back, for Aristotle (since there is no real *infinite* regress), to a "first mover" (Physics, Book VII, 242a 54). This all ultimately leads to an "unmoved mover":

> "it is clear that the first unmoved mover cannot have any magnitude. For if it has magnitude, this must be either a finite or a infinite magnitude. Now we have already proved in our course on *Physics* that there cannot be an infinite magnitude; and we have now proved that it is impossible for a finite magnitude to have an infinite force, and also that it is impossible for a thing to be moved by a finite magnitude during an infinite time. But the first mover causes a motion that is eternal and causes it during an infinite time. It is clear, therefore, that is indivisible and is without parts and without magnitude" (*Physics*, Book VIII, 267b 18-26).

This conclusion to Aristotle's *Physics*, curiously plays up against an earlier unresolved question as to whether time is dependent on there being a soul:

> "Whether if soul did not exist time would exist or not, is a question that may fairly be asked; for if there cannot be some one to count there cannot be anything that can be counted either, so that evidently there cannot be number; for number is either what has been, or what can be counted. But if nothing but soul, or in soul reason, is qualified to count, it is impossible for there to be time unless there is soul, but only that of which time is an attribute, i.e. if *movement* can exist without soul" (*Physics*, Book IV, 223a 22-27).

So it seems there is an uncertainty, an "IF" "*movement* can exist without soul" that interlocks with the god-like first unmoved mover—*possibly* this is a deity beyond anthropomorphizing—*possibly* a vast soul frozen beyond time.

SOUL FORMS

In the three books of *De Anima* or *On the Soul*, Aristotle rejects previous notions of the soul in favor of his own idea of substances and their relation to a four-fold causality. Although he does agree that,

> "the soul is in some sense the principle of animal life" (*De Anima*, Book I, 420a 7),

and that characteristically it involves

> "movement and sensation" (*De Anima*, Book I, 403b 26),

it is not, as others had put forward, a

> "self-moving number" (*De Anima*, Book I, 404b 29);

nor is it in an astrological harmony where

> "the movements of the soul are indentified with the local movements of the heavens" (*De Anima*, Book I, 407a 1-2),

nor is it a "kind of harmony" with the body (*De Anima*, Book I, 407b 30); nor is it inhaled

> "in from the whole when breathing takes place" (*De Anima*, Book I, 410b 29).

Instead, the soul is not separate from the body and connected with it, but is rather,

> "a substance in the sense of the form of a natural body having life potentially within it" (*De Anima*, Book II, 412a 20-21).

Also,

> "it seems [...] to be the soul that holds the body together" (*De Anima*, Book I, 411b 7-8),

while the

> "incapacities of old age is due to an affection not of the soul but of its vehicle, as occurs in drunkenness or disease" (*De Anima*, Book I, 408b 22-23).

Of the four types of causality, body is the material vehicle that is shaped by the soul, as soul

> "is the source of movement, it is the end, it is the essence of the whole living body" (*De Anima*, Book II, 415b 11-12),

In analogy, Aristotle claims:

> "Suppose that the eye were an animal—sight would have been its soul" (*De Anima*, Book II, 412b 18-19).

The soul, for Aristotle is a substance that actualizes the material of the body—soul is the form of the body, but it is also its purpose and its mover. As a complex cause, the soul itself can be divided into its forms, or powers of:

> "self-nutrition, sensation, thinking, and movement" (*De Anima*, Book II, 413b 14).

Not all life forms have all these powers (plants only have self-nutrition, most animals lack thinking, or may have limited types of sensation). Self-nutrition has to do with eating, as well as reproduction: the maintaining of life. Sensation can be divided among the traditional five senses (sight, hearing, smell, taste, and touch)—and is both actual and potential as we may sense something now that only had potential to be sensed before, as Aristotle claims:

> "that the sensible objects are individual and external" (*De Anima*, Book II, 417b 26-27),

but also that:

> "nothing except what has soul in it is capable of sensation" (*De Anima*, Book II, 415b 25-26).

Of special note is the voice, which

> "is a kind of sound characteristic of what has soul in it; nothing that is without soul utters voice" (*De Anima*, Book II, 420b 5-6).

The various senses are tied together by a

> "common sensibility which enables us to perceive them [common sensibles] non-incidentally; there is therefore no special sense required for their perception [....] The senses perceive each other's special objects incidentally; not because the percipient sense is this or that special sense, but because all form a unity" (*De Anima*, Book III, 425a 27-31).

Thus the various perceptions are integrated, even though each leaves an impression isomorphic with its cause:

> "Generally, about all perception, we can say that a sense is what has the power of receiving into itself the sensible forms of things without the matter, in the way in which a piece of wax takes on the impress of a signet-ring without the iron or gold" (*De Anima*, Book II, 424a 16-18).

Thinking is like perceiving, but includes judgment (which can be wrong) and imagination (which is not sensation, since sensation is infallible; and not opinion which is tied up with a belief system). In contrast with imagination's fallibility, Aristotle says,

> "Actual knowledge is identical with its object" (*De Anima*, Book III, 430a 20);

Which is to say that thought and objects share the same formal aspects, as with above mention of *impressions*, but also since:

> "the soul never thinks without an image" (*De Anima*, Book III, 431a 17).

Hence, not even abstract thoughts can be separated from their material instantiation (much like Aristotle's substances fuse form and matter)—even mathematics requires that we have an image of what is numbered, for us to imagine it.

The last power of the soul is movement, and it arises from both appetite (desires) and thought (calculation)—which both can incorporate the imagination. Aristotle recognizes that we may act both by volition, and also by impetus. This division between spontaneous and voluntary action is also found in recollection, which unlike Plato's recollection of formal knowledge, has more to do, for Aristotle, with reminiscence:

> "Whenever, therefore, we are recollecting, we are experiencing one of the antecedent movements until finally we experience the one after which customarily comes that which we seek. This explains why we hunt up the series, having started in thought from the present to some other, and from something either similar, or contrary, to what we seek, or else from that which is contiguous with it" (*On Memory*, 451b 17-19).

And again, since thought is through images, memory operates through the imagination, and:

> "memory is a function [...] of the primary faculty of sense-perception, i.e. of that faculty whereby we perceive time" (*On Memory*, 451a 16-17).

Moreover, memory, since it is an impression of an image isomorphic with its object, is corporeal:

> "recollection is a searching for an image in a corporeal substrate" (*On Memory*, 453a 14-15).

Aristotle clears much ground in setting up his psychology—he clarifies not only what soul is in the abstract (the moving, purposeful, and formal causes of the material body), but also expands and consolidates what the various powers or faculties of the soul are: a compartmentalization of mind, which Plato had only began to develop (Plato divided understanding from reason, and also wrote on motivation and memory). Aristotle's material notion of memory clearly anticipates modern thinking on the subject—with his theory of recollecting implying an association of images in the mind. And like most thinkers to follow, his psychology ignores personality traits (a focus of Astrology), and does not get too deep into how the mind is supposed to carry out functions other than the recollection of sense impressions; sense impressions that he compares to "a picture painted on a panel" (*On Memory*, 450b 21)—hence Aristotle developed, or at least implied, an early theory of idea representation—a theory that some 20th century thinkers found fundamentally flawed (e.g. the later Wittgenstein, Richard Rorty, et. al.) A clearer portrait of subjectivity would have to wait for St. Augustine though, who will be a subject of the next chapter.

Commodilla Catacomb Christ – Late 4th Century

Chapter 7

Jesus Christ
Indirect Discovery of Subjectivity

A LIGHT THROUGH THE DARK AGES

The history of philosophy between the height of ancient philosophy (with Plato and Aristotle) and modern philosophy (often studied as beginning with renaissance thinkers such as Descartes) does not have so many super-star philosophical systems advances as much as an extended and slow progress. By and large, we see extensions of Plato (as with Plotinus' ~204-270 CE neo-Platonism) and Aristotle (as with St. Thomas Aquinas' 1225-1274 CE integration of Aristotle and Christian philosophy); but also progress towards pulling philosophy's "head out of the clouds" as it were, and orienting towards a more individualistic perspective on reason here in the historical world—advances towards an understanding of subjectivity and a more scientific, objective methodology.

Although we will not take on the philosophy of Jesus Christ directly here, we will see (1) an aspect of the philosophical climate that his teachings arose in, with Stoic and Skeptical philosophy; (2) that his philosopher contemporary Philo combined Greek philosophy with the Judaic tradition through the Logos; (3) how his student St. Augustine turned "the mind's eye" inward towards God; and (4) how his follower Friar Roger Bacon's thinking was emblematic of a shift towards scientific method and its implementation.

WORLDLY STOIC LOGIC & RELATIVE SKEPTICISM

Although connotations and denotations of the word "stoic" suggest a more or less emotionless way of being, like Aristotle the Stoic philosophers such as Zeno of Citium (~334-262 BCE—not to be confused with Zeno of Elea whose paradoxes Aristotle took up), and Chrysippus (~279-206 BCE), etc, sought *happiness* in line with a life of reason. Extending Aristotle's transition of transcendent Platonic forms down into worldly substances, the Stoics too were interested in the world we live in, and how to live ethically in accordance with the way things are here on earth. This focus on lifestyle and ethics follows the Epicurean philosophy (Epicurus 341-270 BCE) which suggested that one tend to one's own garden; but now with a renewed focus on self-control, and a duty according to nature. The stoics were determinists, and saw a harmony between individual will and causally determined nature as a goal. It is in this way that logic becomes important, as a path to side-lining troubling emotions such as desire and see the world for the way it is—our knowledge comes only from the world, and logic was seen as a way to understand how the world as we sense it directly operates:

> "Indeed, in their view this is the only world that exists; there are no incorporeal entities for us to know. Everything we know comes to us directly or indirectly through sense-perception; the cognitive powers of the human mind, itself a physical entity, are the product of a providential and purposive plan by the creative deity, so it is no wonder that our senses are adequate to the task. In the area of logic proper, the analysis of the structures of inference, proof, and discovery, the stoics (Chrysippus in particular) were second to no

> one in the ancient world, not even Aristotle. Stoic logic surpassed Aristotle's exploring and systematizing the relationships among propositions; the Stoics sought to analyze all valid inferences in terms of five basic and self-evidently valid argument forms" (Inwood, Brad & Lloyd P. Gerson, trans, *The Stoics Reader: Selected Writings and Testimonia*, Hackett Publishing Company, Inc: Indianapolis (2008), p.xiii).

With the Stoics, esp. with Chrysippus, we see an advance over Aristotle's logic, with the introduction of propositional logic. That is, instead of looking at the logical relation between terms in sentences (e.g. "Socrates is a man" "All men are mortal" hence "Socrates is mortal") the Stoics took off from Aristotle's *De Interpretatione*, and saw that sentences corresponded to facts in the world—but extended this to study how these *sentences* interrelate, rather than just the *terms* within the sentences. Again, propositions were sentences that could be affirmed or denied by states of affairs in the world, such as "it is day." Such simple propositions, that state one fact, can be made more complex by using logical connections: IF, AND, EITHER...OR, BECAUSE, and such is MORE / LESS LIKELY... THAN. These logical connectives lead to the possibility of making deductions from various sentences—here are five basic argument forms (adapted from the Wikipedia entry on Chrysippus, <//en. wikipedia/wiki/Chrysippus>):

> Modus Ponens:
> If p then q. p. Therefore q.
> (If it is day, it is light; it is day, therefore it is light)
>
> Modus Tollens:
> If p then q. Not q. Therefore not p.
> (If it is day, it is light; it is not light, therefore for it is not day)

Modus Ponendo Tollens (1):
Not both p and q. p. Therefore not q.
(It is not both day and night; it is day, therefore it is not night)

Modus Ponendo Tollens (2):
Either p or q. p. Therefore not q.
(It is either day or night; it is day, therefore it is not night)

Modus Tollendo Ponens:
Either p or q. Not p. Therefore q.
(It is either day or night; it is not day, therefore it is night)

Daybreak, twilight, full moonlit nights and solar eclipses aside, this propositional logic, clearly in line with Aristotle's view that true sentences could not be contradictory, was on the way to capturing much more of how we talk and reason about the world—but the advance was not clearly recognized later, until the 20th century, when the work of the 19th century logician Gottlob Frege, a founder of modern logic and analytic philosophy, took prominence as advocated by Bertrand Russell and others.

Another line of thought that has surfaced time and again throughout the history of philosophy is that of Skepticism. Such a view was evident in pre-Socratic sophist philosophers such as Protagoras (~490-420 BCE) with his statement "man is the measure of all things." The skeptical viewpoint, questioning claims to absolute knowledge, is closely tied with relativism, as can be seen with the tropes delineated by the later skeptics involved with the School of Pyrrho founded by Aenesidemus (1st Century BCE) and related by Sextus Empiricus (~160-210 CE); here paraphrased by Frederick Copleston:

> "Aenesidemus of Knossos (who taught at Alexandria and probably composed his work around 43 B.C.) gave ten [... tropes] or arguments for the sceptical position. They were:

1. Difference between types of living beings imply different—and so relative—'ideas' of same object.

2. Differences between individual men imply the same.

3. The different structure and presentation of our various senses (e.g. there is an eastern fruit that smells unpleasant but tastes delicious).

4. The difference between our various states, e.g. waking or sleeping, youth or age. For example, a current of air may seem a pleasant breeze to a young man, while to an old man it is a detestable draught.

5. Differences of perspective, e.g. the stick immersed in water appears bent, the square tower appears round from a distance.

6. The objects of perception are never presented in their purity, but a medium is always involved, such as air [....] For example, grass appears green at noon, golden in the evening light. A lady's dress looks different in sunlight to what it looks in electric light.

7. Differences in perception due to differences of quality, e.g. one grain of sand appears rough, while if sand is allowed to slip through the fingers it appears smooth and soft.

8. Relativity in general [....]

9. Difference in impression due to frequency or infrequency of perception, e.g. the comet, seldom seen, makes more impression than the sun.

10. Different ways of life, moral codes, laws, myths, philosophic systems, etc. (cf. Sophists).

These ten [...tropes] of Aenesidemus were reduced to five by Agrippa.

1. The variation of views concerning the same objects.

2. The infinite process involved in proving anything (i.e. the proof rests on assumptions that require to be proved, and so on indefinitely).

3. The relativity involved in the fact that objects appear differently to people according to the temperament, etc., of the percipient and according to their relation with other objects.

4. The arbitrary character of dogmatic assumptions, assumed as starting points, in order to escape the *regressus in infinitum*.

5. The vicious circle of the necessity of assuming in the proof of anything the very conclusion that has to be proved.

Other Sceptics meanwhile reduced the [...topics] to two:

1. Nothing can be rendered certain through itself. Witness the variety of opinions, between which no choice can be made with certainty.

2. Nothing can be rendered certain through anything else, since the attempt to do so involves wither the *regressus in infinitum* or the vicious circle" (Frederick Copleston, S.J., *A History of Philosophy – Volume I: Greece and Rome*, Image Books: New York (1985), pgs. 443-444).

Through these sets of tropes, we see a questioning of logical proof, as with the problem concerning Aristotle's absolute deductions being based on knowledge found from uncertain induction. We will see that such skepticism arises again in modern philosophy with thinkers like David Hume, and also with impact of relativism on postmodern thought.

PHILO'S LOGOS: THE WORD IS LOVE

As we have seen that various philosophers and philosophical eras often work in opposition to their predecessors (think Heraclitus and Parmenides, or Plato and Aristotle), developing new theories in contrast, at least to some extent, to those that came before, although often keeping lessons learned in mind; so too we see a ground breaking break with tradition in the philosophy of Jesus Christ's philosopher contemporary Philo (20 BCE-50 CE). With Philo we see the integration of Greek philosophy with Jewish biblical exegesis. He was influenced by the Stoics, who read Homer allegorically to support their philosophical perspectives. Philo was also at odds with previous biblical scholars who read the Bible literally, and his emphasis was clearly outside the stoic focus on literal logic. Beyond Philo's discussions of the "Logos," this is critical for framing the teachings of Jesus, not only for the "passion of Christ" prophet of love and divine miracles in contrast to stoic emotionless embracement of only this world, but for situating the parable as a way of teaching. There is also the rejection of *skepticism* in favor of revelation and faith. Philo may have never seen Jesus, but there was definitely a zeitgeist spirit of the times at play here in a turn from skepticism to faith and from worldly logic to divine allegory.

How does one recognize allegory? This list of 21 "rules" may help us to understand how to interpret passages, as they may have:

"1. The doubling of a phrase;

2. An apparently superfluous expression in the text;

3. The repetition of statements previously made;

4. A change of phraseology—all these phenomena point to

something special that the reader must consider.

5. An entirely different meaning may also be found by a different combination of the words, disregarding the ordinarily accepted division of the sentence in question into phrases and clauses.

6. The synonyms must be carefully studied; e.g., why λαὸς ('people') is used in one passage and γένος ('genus') in another, etc.

7. A play upon words must be utilized for finding a deeper meaning; e.g., sheep (πρόβατα) stand for progress in knowledge, since they derive their name from the fact of their progressing (προβαίνειν), etc.

8. A definite allegorical sense may be gathered from certain particles, adverbs, prepositions, etc.;

9. And in certain cases it can be gathered even from the parts of a word; e.g., from διά in διάλευκος.

10. Every word must be explained in all its meanings, in order that different interpretations may be found.

11. The skillful interpreter may make slight changes in a word, following the rabbinical rule, 'Read not this way, but that way.' Philo, therefore, changed accents, breathings, etc., in Greek words.

12. Any peculiarity in a phrase justifies the assumption that some special meaning is intended: e.g., where μία ('one') is used instead of πρώτη ('first'; Gen. i.5), etc. Details regarding the form of words are very important:

13. The number of the word, if it shows any peculiarity in the singular or the plural: the tense of the verb, etc.;

14. The gender of the noun;

15. The presence or omission of the article;

16. The artificial interpretation of a single expression;

17. The position of the verses of a passage;

18. Peculiar verse-combinations;

19. Noteworthy omissions;

20. Striking statements;

21. Numeral symbolism. Philo found much material for this symbolism in the Hebrew Bible, and he developed it more thoroughly according to the methods of the Pythagoreans and Stoics. He could follow in many points the tradition handed down by his allegorizing predecessors" (<//en.wikipedia/wiki/Philo>).

By and large, Philo's interpretation of the Bible is read though the understanding of Greek philosophers such as Plato, whom, Philo might argue, actually got their insights from biblical sources: as if Plato had clarified the original biblical revelations, which could now be understood via that clarification. For example, Philo claims in his work "On The Creation" that:

> "[Moses] when recording the creation of man, in words which follow, asserts expressly, that he was made in the image of God—and if the image be a part of the image, then manifestly so is the entire form, namely, the whole of this world perceptible by the external senses, which is greater imitation of the divine image than the human form is. It is manifest also, that the archetypal seal, which we call that world which is perceptible only to the intellect, must itself be the archetypal model the idea of ideas, the Reason of God" (Philo, C.D. Yong (trans.), *The Works of Philo:*

> *Complete and Unabridged*, Hendrickson Publishers: USA (2000), p. 5, *On The Creation*, VI - 25).

Here with "man" as an image falling short of the original God, we see a parallel with Plato's worldly things falling short of the eternal forms (and hence a similar concern, with poetry being seen as bad by Plato for its imitative nature in his dialogue *Ion*, and Moses' second commandment "Thou shalt not make unto thee any graven image, or any likeness *of any thing*..."). With Philo as well as with Plato, we find a hierarchy between an eternal intellectual realm approaching God or "the one" good, and our mundane world of bodies. Against the Stoic commitment to our world, Philo taught of another invisible world:

> "[Philo] represents Jehovah as a single uncompounded Being; unchangeable, eternal, incomprehensible, the knowledge of whom is to be looked upon as the ultimate object of all human efforts. He teaches that visible phaenomena are to lead men over to the invisible world, and that the contemplation of the world so wonderfully and beautifully made proves a wise and intelligent Cause and creator of it [....] an invisible world, appreciable only by the intellect, as the pattern of the visible world in which we live" (*The Works of Philo*, p. xx).

Here we can clearly see an integration of Plato with the Old Testament—and such is on the way to Christian philosophy. Moreover, with his concept of the "Logos" (recall in the later New Testament "In the beginning was the Word..."—"Logos" suggesting "Words" or "Reckoning" / "Reason") Philo develops a bridge between the intelligible divine realm and the visible world we live in. In fact this Logos bridge can be said to be:

> "The Utterance of God [....] The Divine Mind [.....] God's Transcendent Power [....] First-born Son of God [....] Universal Bond: in the Physical World and in the Human Soul

[....] Immanent Reason [....] Immanent Mediator of the Physical Universe [....] The Angel of the Lord, Revealer of God [....] Multi-Named Archetype [....] Soul-Nourishing Manna and Wisdom [....] Intermediary Power [.... and] 'God')" (Marian Hillar, *Philo of Alexandria*, The Internet Encyclopedia, 24 Apr 2005 <//www.iep.utm.edu/philo>).

Indeed, Logos, the word, is the very fabric of a reasonable world, and the human manifestation of a doorway to the divine. It is god-given reason, which structures the entirety of our relation to the Transcendent One: it is wisdom, God, the relation of humans to God, and the very laws of our universe and how we can understand them all rolled into one.

It is with this in mind that we can begin to see the shift of thinking, via the era of Jesus Christ, towards trusting one's own intuition as opposed to customary traditions—trust in the heart is a way of connecting with the Logos which we all share in our minds. The figure of Jesus Christ could be seen as combining the relativism of the skeptics with the absolute revelation of God: *this* man is the measure of all things. But Jesus Christ can be seen as both an historical figure and a portal that we all have inside—the Logos as connecting and binding force that keeps our minds singular, and our world coherent, but manifested in a single human being. It is as if an historical portal to another dimension dilated in our own souls with the advent of Christ as human manifestation of the Logos.

SAINT AUGUSTINE: ETERNITY'S INNER LANGUAGE

It is not by chance alone that the "discovery" of an inner self, or rather the making explicit of subjectivity, was elaborated in one of the Western world's first autobiographies. *The Confessions of St. Augustine* portrays the first three decades of St. Augustine's life (which in total spanned from 354-430 CE), relating his deep spiritual conversion and, along with his other works relating to philosophy, situates Christianity and the Bible in relation to Plato (via the Neo-Platonism of Plotinus and his student Porphyry who St. Augustine is known to have read). St. Augustine placed the Christian God at the pinnacle of Plato's hierarchy between the intelligible world (topped, by Plato, with "the good"), which is more easily understood in relation to mathematics and eternal laws, and the sensible world that Plato saw as a flawed imitation of the higher world of forms... and as we saw in Plato, the "image of time" itself is a flawed reproduction of eternity, making time illusory as well.

Saint Augustine
by Philippe Champaigne (~1650)

St. Augustine examines time in *The Confessions*, and it is fitting that this early philosopher of subjectivity (although he does not use that term), sees time as being perceived by his mind:

> "It is in you, my mind, that I measure time" (St. Augustine, Rex Warner, trans, *The Confessions of St. Augustine*, Mentor Books, New York (1963), p. 281, Book XI, Ch. 27).

> "And those who tell us about the past certainly could not tell us the truth unless they saw it in their mind's eye" (*Confessions*, p. 271, Book XI, Ch. 17).

Like Aristotle, St. Augustine finds that the present, now, has no extension:

> "No one can deny that the present time has no extension, since it passes by in a flash. But nevertheless our attention

> (our 'looking at') is something constant and enduring, and through it what is to be proceeds into what has been" (*Confessions*, p. 282, Book XI, Ch. 28).

Regarding the recitation of a psalm, St. Augustine relates:

> "So the life of this action of mine is extended in two directions—toward my memory, as regards what I have recited, and toward my expectation, as regards what I am about to recite" (*Confessions*, p. 282, Book XI, Ch. 28).

St. Augustine seems to aim, like Plato, towards explaining time as an illusion; but claims that God is outside of time, in an eternity without beginning and end, and may see history as if it were laid out in a linear continuum from past to future. However, St. Augustine does not claim certainty as to how God's omniscience works—and he here says regarding the speculation that all history could be recited like the aforementioned psalm:

> "far be it from me to say that it is in this way that you the creator of the universe, creator of souls and bodies, know all the future and the past" (*Confessions*, p.284, Book XI, Ch. 31).

Despite the knowable aspects of the future, at least from a divine perspective, in a prior section of *The Confessions* (Book VII, Ch. 6), St. Augustine derides Astrology, recalling a story of two children, born at the same instant, going on to live disparate lives as a slave owner and slave. Such a "test" of Astrology seems misplaced, as the ancient art is mostly related to personality traits, and was not necessarily or only used for fortune telling. St. Augustine also reveals his lack of insight into Astrology as a sort of psychology; for Augustine's psychology is much more about personal experience in relation to God, more about internal subjective experience than interpreting objective psychological traits that differ from person to person, and hence he ridicules the ability to forecast the future and:

> "those fools who make money out of astrology" (*Confessions*, p. 144, Book VII, Ch. 6).

Such a forecasted fate would be at odds with the younger St. Augustine's emphasis on the will—it is our personal wills that obstruct our doing what we know we ought to do, or ought to believe. It is in this way that St. Augustine anticipates later "folk psychology" which tries to explain our behavior in terms of will-like "desires" and "belief" systems. And it is with a bifurcation between beliefs and understanding that certainty becomes a central issue; and of course this all relates to Christian faith. The younger St. Augustine saw life on earth as a sort of test, where one could reject the life consumed solely by the body and world steeped in original sin after Adam's fall from grace, and choose the righteous path of intellect and enter Heaven, or choose wickedness only to be tormented in an afterlife of endless Hell. The older St. Augustine discerned a more dire circumstance, more in line with the God that is omniscient, and claimed that only a chosen few would be able to live in *The City of God* (the title of another of St. Augustine's great works, focused on a teleological history), solely at the discretion of God's given grace: predestination and providence—a sort of humble surrender to God's will over one's own (an issue we will also see addressed in Shakespeare's *Hamlet*).

Not only does St. Augustine's struggle between will and fate with Heaven and Hell in the balance anticipate Hamlet's "to be, or not to be"—but also his investigations into subjectivity anticipated Descarte's "cogito, ergo sum" (I think, therefore I am) with St. Augustine's own "si fallor, sum" (if I am fallible, I am)—here arguing against the skeptics who might claim we have no certainty, and claiming the certainty of our soul's existence even when this soul is mistaken:

> "I am most certain that I am, and that I know and delight in this. In respect of these truths, I am not at all afraid of the arguments of the Academicians, who say, What if you

> are deceived? For if am deceived, I am. For he who is not, cannot be deceived; and if I am deceived, by this same token I am" (St. Augustine, Marcus Dods, trans, *The City of God*, Barnes & Noble, New York (2006), p. 445, Book XI, Ch. 26).

This line of thinking is no accident for St. Augustine; for although he may be using the common sense and turns of phrase of his era, he is one of the earliest thinkers of what has been termed a "homunculus" or person inside a person. St. Augustine believes that our path to God's glory is a turn inward:

> "I was unable to form an idea of the light and honor and of a beauty that is embraced for its own sake, which is invisible to the eye of the flesh and can only be seen by the inner soul" (*Confessions*, pp. 133-134, Book VI, Ch. 16).

And even closer to the concept of a "homunculus," St. Augustine references:

> "*the inner man*" (*Confessions*, p. 158, Book VII, Ch. 21).

> "It was in vein that *I delighted in Thy law according to the inner man, when another law in my members rebelled against the law of my mind, and led me captive under the law of sin which was in my members*" (*Confessions*, p.169, Book VIII, Ch. 5).

This "inner man," "inner soul" or "inner self" is beyond this world—but not beyond the subjective senses, evidently—for beyond the experience of this world through our bodily "members," the inner self, referencing the five senses, is able to connect to the experience of God:

> "[God] is the light, the melody, the fragrance, the food, the embracement of my inner self—there where is a brilliance that space cannot contain, a sound that time cannot carry

> away, a perfume that no breeze disperses, a taste undiminished by eating, a clinging together that no satiety can sunder" (*Confessions*, p. 215, Book X, Ch. 6).

> "Already, Lord, in my inner ear I have heard your voice loud and strong telling me that you are eternal, *Who only hast immortality*, since you suffer no change in form or by motion, and your will is not altered by the course of time" (*Confessions,* p. 291, Book XII, Ch. 11).

The earlier cited reference to the "mind's eye," and here the "inner ear" as well as the inner self / man / soul clearly indicate St. Augustine's emphasis on subjectivity... our perspective from "within" our body rather than a "God's eye" or "bird's eye" objective view on the world, time and eternity, etc. Such is also evident when St. Augustine talks of our "awareness" and "attention"—concepts that are still relevant to contemporary cognitive science. The inward turning towards God does not mean that we are in direct contact with God though... such is beyond us, and our knowledge of truth comes indirectly from God—unlike Philo's Logos, which seems to indicate some direct connection with the Godhead via knowledge. St. Augustine follows Plato's metaphor of the sun / good illuminating our minds, as we come to move beyond the sensing of objects and the like in our world, to the forms of such held in eternity.

St. Augustine's concern with our inner self, and the separation of our access to knowledge from the divine source of knowledge, is further illustrated, when he discusses our very thinking, as in a third great work, *The Trinity*:

> "However, just because we say that thoughts are utterances of the heart, it does not mean that they are not also seeings, arising when they are true from the seeing of awareness. When these things happen outwardly through the body, speech is one thing, sight another; but when we think inwardly they are both one and the same; just as hearing and

> seeing are two things that differ from each other among the senses of the body, while in consciousness it is not one thing to see and another to hear" (St. Augustine, Edmund Hill, O.P. trans, *The Trinity*, New City Press: New York (1991), pp. 408-409, Book XV, Ch. 18).

This is to say that our "ideas" are singular despite our senses (and the aspects of ideas) being plural. Such ties into language, where physical words (spoken or written) can be seen as separate from their ideal meanings:

> "If anyone then can understand how a word can be, not only before it is spoken aloud but even before the images of its sounds are turned over in thought—this is the word that belongs to no language, that is to none of what are called the languages of the nations, of which ours is Latin; if anyone, I say, can understand this, he can already see through this mirror and in this enigma some likeness of that Word of which it is said, *In the beginning was the Word, and the Word was with God, and the Word was God* (Jn 1:1). For when we utter something true, that is when we utter what we know, a word is necessarily born from the knowledge which we hold in memory, a word which is absolutely the same kind of thing as the knowledge it is born from. It is the thought formed from the thing we know that is the word which we utter in the heart" (*The Trinity*, p. 409, Book XV, Ch. 19).

Hence:

> "Thus the word which makes a sound outside is the sign of the word which lights up inside, and it is this latter that primarily deserves the name 'word.' For the one that is uttered by the mouth of flesh is really the sound of a 'word,' and is called 'word' too because of the one which assumes it in order to be manifested outwardly. Thus in a certain fashion our word becomes a bodily sound by assuming that

> in which it is manifested to the senses of men, just as the Word of God became flesh by assuming that in which it too could be manifested to the senses of men. And just as our word becomes sound without being changed into sound, so the Word of God became flesh, but it is unthinkable that it should have been changed into flesh. It is by assuming it, not by being consumed into it, that both our word becomes sound and that Word became flesh" (*The Trinity*, pp. 409-410, Book XV, Ch. 20).

Such is to keep a strict line between the eternally intelligible, and the worldly sensible. True words are their meanings, which have a connection to the divine—and hence St. Augustine makes a sharp division between physical signifiers and something like Philo's Logos or Word—the signified meaning of words. This keeps the intelligible / sensible hierarchy intact, as the intelligible does not become sensible, but is rather expressed through the sensible—and our worldly knowledge too, tied up as it is with our worldly perceptions and memories, is informed by the intelligible realm. What we know comes by illumination from the light of God, but is not something our bodies and senses can touch directly. God's word informs the world, while not actually being of the world.

Here we can see Plato's forms in contract with the subjective soul. However, it is not clear exactly how our thoughts, tied to our inner hearts and God's word, *connect* with our physical speech and writing. For the connection is not physical, but rather something like the act of an expression. Strangely, an expression is temporal only in our world, while the intelligible is unchanging and eternal. While there might be a one-to-one correspondence between sensible signifiers and their intelligible meanings, St. Augustine seems to suggest that the connection is not a tangibly static one, but one that occurs through the process of "becoming." Through the process of communicating, a connection between the two realms of the intelligible and sensible is made: heartfelt thought and articulation themselves welding the two. The Logos of forms becomes manifest, Word becomes flesh, and human understanding becomes possible. Such a welding of the intelligible to the sensible through subjective articulation remains an issue even for 21st century philosophical thought. Such a perspective is challenged though, by those (some cognitive scientists, connectionist philosophers, etc.) who might see intelligence and the intelligible as epiphenomena arising from purely physical processes. Perhaps even some deconstructionists would see the signified as a chimera... as simply an illusory appearance suggested by the play of signifiers. Such debate will be covered in later chapters, which bear the impact of scientific method, etc.

FRIAR ROGER BACON: ANTI-AUTHORITARIAN SCIENCE

Christian philosophers after St. Augustine often turned to Aristotle for inspiration. And although one might trace many of Franciscan Friar Roger Bacon's (~1214-1294) ideas to his predecessors and contemporaries (such as his teacher, Robert Grosseteste), this irascible figure, who may in fact have been the author of the Voynich Manuscript (a mysterious 200+ page book discovered in 1912, written entirely in code, and copiously illustrated with astrological diagrams and botany images, etc.) was one of the first clear voices advocating something like the modern scientific method.

His *Opus Maius* or "Major Work," was divided into seven parts, beginning with the lines:

> "A thorough consideration of knowledge consists of two things, perception of what is necessary to obtain it, and the method of applying it to all matters that they may be directed by its means in the proper way" (quoted in Lawrence & Nancy Goldstone, *The Friar and the Cipher*, Broadway Books, New York (2005), p. 117).

In the first part of *Opus Maius*, Bacon demonstrates his anti-authoritarianism, and identifies four roots of human ignorance:

> "subjection to unworthy authority, the influence of habit, popular prejudice, and making a show of apparent wisdom to cover one's own ignorance" (Frederick Copleston, S.J., *A History of Philosophy – Volume II – Augustine to Scotus*, Image Books, New York (1985), p. 444).

The second part, entitled, "Philosophy" argues that theology is at the root of science, since "God is the active intellect" (Copleston, p. 445). It is in this section that Bacon teaches that we should also carry on tradition while slowly reforming it:

> "we should use pagan philosophy in an intelligent manner, without ignorant rejection and condemnation on the one hand or slavish adherence to any particular thinker on the other. It is our business to carry on and perfect the work of our predecessors" (Copleston, p. 445).

In the third part, "Study of Tongues," Bacon argues that we should better know foreign languages (Greek, Hebrew and Arabic) as many of his contemporaries, and even he himself, did not know how to fully translate the work he hoped to improve upon.

The fourth part, "Mathematics," discusses not only this field of knowledge, but how it could be applied, e.g. to astronomy and the calendar. It is in this part that Bacon argues in favor of Astrology (again, hinting at his possible authorship of the Voynich Manuscript). Copleston notes:

> "Astrology is regarded [by others] with suspicion as it is thought to involve determinism; but this suspicion is unjust [according to Bacon.] The influence and movements of the heavenly bodies affect terrestrial and human events and produce even natural dispositions in human beings, but they do

> not destroy free will: it is only prudent to gain all the knowledge we can and use it for a good end" (Copleston, p. 446).

"Optics," relating to the eyes, light and lenses, takes up the fifth part of *Opus Maius*, and builds upon the work of Bacon's predecessor, Grosseteste, himself influenced by al-Hazen and al-Kindi. It is in this part of his major work that Bacon claims light travels at a certain speed rather than instantaneously. (The topic of optics interested later philosophers as well, including Descartes, Berkeley, and Spinoza). Bacon even discuses eye-glasses, before they were invented.

In his sixth part, "Experimental Science":

> "Bacon considers experimental science. Reasoning may guide the mind to a right conclusion, but it is only confirmation by experience which removes doubt" (Copleston, p. 446).

Bacon writes:

> "Without experience, it is impossible to know anything completely [....] For there are two modes of acquiring knowledge, namely, by reasoning and experience. Reasoning draws a conclusion, but does not make the conclusion certain, nor does it remove doubt so that the mind may rest on the intuition of truth, unless the mind discovers it by the path of experience [....] For if a man who has never seen fire should prove by adequate reasoning that fire burns and injures things and destroys them, his mind would not be satisfied thereby, nor would he avoid fire, until he placed his hand or some combustible substance in the fire, so that he might prove by experience that which reasoning taught. But when he has had actual experience of combustion his mind is made certain and rests in the full light of truth. Therefore reasoning does not suffice, but experience does" (quoted in Goldstones, pp. 147-148).

In *The Friar and the Cipher*, the authors claim:

> "This was the final break from classical scholasticism, and particularly the brand of pseudoscience that Bacon considered to have been so recently perpetrated by Aquinas. It was not enough to reason one's way to truth, no matter how sophisticated the argument. Reason followed experiment, not the other way around. 'Hence in the first place there should be readiness to believe, until the second place, experiment follows, so that in the third reasoning may function,' he wrote. This is the first clear statement in Christian Europe of what the modern world recognizes as hypothesis-experiment-conclusion." (Goldstones, p. 148).

Moreover, Bacon writes:

> "Reasoning does not attest these matters, but experiments on a large scale made with instruments and by various necessary means are required. Therefore no discussion can give an adequate explanation in these matters, for the whole subject is dependent on experiment. For this reason I do not think that in this matter I have grasped the whole truth, because I have not yet made all the experiments that are necessary, and because in this work I am proceeding by the method of persuasion and of demonstration of what is required of the study of science, and not by the method of compiling what has been written on the subject. Therefore it does not devolve on me to give at this time an attestation possible for me, but to treat the subject in the form of a plea for the study of science" (quoted in Goldstones, pp. 151-152).

Yet, despite this emphasis on scientific method and experience (clearly anticipating the categorical split in modern philosophy between the Rationalists and the Empiricists), Bacon holds ethics as the highest science, in his seventh and last part of *Opus Maius*.

In the present chapter, covering the span of years between Ancient and Modern philosophy, we have seen no major breakthroughs in original systematic thinking; but rather, extremely important insights on the way to understanding logic, relativity, allegory, subjectivity, and the scientific method—clearly we are on the way to modern thinking. The next chapter will take a detour from traditional philosophers, in order to examine the complex of motifs, techniques, and insights employed by Shakespeare in *Hamlet*; for it is here that a stage is truly set on the way to what has been called the "Cartesian Theater" with the play's profound take on the modern mind, where certainty, if not reality itself as many have known it, is called into question, and consciousness is illuminated.

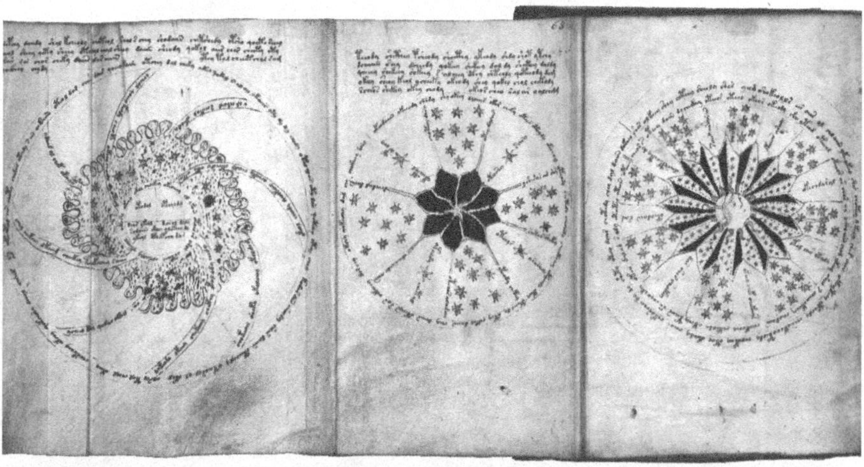

Gatefold from the Voynich Manuscript (~1421)

Sarah Bernhardt as Hamlet (~1880)

Chapter 8

Shakespeare
Hamlet's Homunculus

"What coil is that you keep?"—Thomas Kyd, *The Spanish Tragedy* (III.xiii.45) ~1584-1589

"For in that sleep of death what dreams may come
When we have shuffled off this mortal coil"
—William Shakespeare, *Hamlet* (III.i.66-67) ~1599-1601

"for in my troubled eyes
Now may'st thou read that life in passion dies"
—Thomas Kyd, *The Spanish Tragedy* (II.iv.46-47)

"A mote it is to trouble the mind's eye"
—William Shakespeare, *Hamlet* (I.i.112)

"...I regard myself as intuiting with the eyes of the mind in the most evident manner."
—Descartes, *Meditation III*, ~1641

SEDUCTION INTO GOD'S EYE VIEW

One measure of good creative writing and art may be to gauge the power it has to seduce you into its reality—this seduction may elicit a "suspension of disbelief," where, when reading fiction, or watching plays, one is immersed in a fabricated world: one's own perspective is swept away as one becomes a sort of "universal," "God's eye," third person, or audience observer / spectator in a world that is not real (or is no more than semi-real). In a Buddhist like way, one is a passive eye in the storm of an unreality on par with dreams: one's own ego ceases to exist, merging with the virtual non-existent. This sort or experience—one of belief—contrasts with the more detached and *self-conscious act* (if there is such a thing) of criticizing a text: analyzing the parts, and studying the details of plot, character, motifs, etc. With this withdrawal from the text, one confronts an artistic object, or artifact: there is a more conscious separation between subject and object. Subjectivity itself, since St. Augustine and the Renaissance philosopher Renee Descartes, has often been thought to be somewhat like a theater of the mind, where there seems to be a "homunculus," or a little person inside the person, a self-certain ("I think, therefore, I am") ego, who is the integrated observer (or single audience member) of the play of the senses. Hence, Descartes' Renaissance thinking emphasized a sort of "doubling" of consciousness, or a separation of the observer and observed senses, where the observed itself could be doubted as a dream (and which could be guaranteed as reality, only by its inclusion in God's perfect being).

OBSERVERS OBSERVING OBSERVERS OBSERVING

Unlike Bertolt Brecht's later foregrounding of stage apparatuses and musicians in the play "The Threepenny Opera," in order to break the willing suspension of disbelief; and hence attempt to cause the audience to be more distanced from, and possibly more critical of his play, I believe Renaissance playwrights such as Thomas Kyd (1558-1594) and William Shakespeare (1564-1616) used techniques to draw viewers deeper into their plays. One such technique, used in various forms, was to establish the audience or reader as part of a play by creating a situation where the play parallels the situation of the reader or the audience member. Such occurs when there is a "play within a play" (we too are part of the inner play's own audience), or when a character reads within a play, and we as readers read the text with the character. More than simply hearing what other characters hear, or what one character *thinks out loud*, with this technique of mirroring, paralleling, or representing the act of observing an object or action, the character(s) and spectator(s) become one in the act of observing. The play represents what the observer is doing (reading or watching a play); but instead of making the observers more self-conscious about their status as observers (self-consciousness has been defined as being able to explicitly model one's actions), I find this technique actually immerses the viewer or reader even further into the play.

SHAKESPEARE'S DIALOGUE WITH KYD

This chapter cites examples of this paralleling technique, in *The Spanish Tragedy*, and *Hamlet*, in part, to highlight how Shakespeare may have picked it up and used it after reading Kyd. Yet, unlike T.S. Eliot, who in his "Hamlet and His Problems" (*Selected Essays of T.S. Eliot*, Harcourt Brace & World Inc.: San Diego (1960), pp. 46-50) found that *Hamlet* was lacking an authentic emotion encoded in the text with an "objective correlative," due to its derivative relation to *The Spanish Tragedy*, I find that with *Hamlet* Shakespeare takes up an active dialogue with Kyd. *The Spanish Tragedy* may have touched Shakespeare's heart, for in that play a father loses a son. The name of the murdered son in that play, "Horatio," was used in *Hamlet*, where in a reversal the father has been murdered, and "ironically" this Horatio is one of the few principals to survive the tragic finally. Shakespeare too lost a son, most non-coincidentally named, "Hamnet;" much akin to the "revenging" son named, "Hamlet," and here we see the seeds of a deep connection between the two plays through Shakespeare's eyes (the survival of Horatio perhaps being a wish for the survival of his own "kid"—and included in *Hamlet*, not as the lead, but as a friend of the lead: *The Spanish Tragedy's* author, Kyd, could be seen somewhat as a written friend of Shakespeare). Possibly Shakespeare was demonstrating how meaning arises from relationships, not just between words (ala Ferdinand de Saussure), or characters in a play, but also with the emotional ties between writers and readers at a distance: inter-subjective meaning, rather than meaning embodied in an observed object, as Eliot might have it. The quotes that head this chapter demonstrate that the dialogue between Shakespeare and Kyd was not only plot / structural, a lifted use of a technical motif (as this chapter will illustrate), but even in the interplay of specific lines.

The motif technique of having observers observe (with) observers observing, relates to this Shakespeare, Kyd (& Eliot) dialogue, in that it highlights both the subject / object relation (with subject observing and object observed, where meaning might be embodied in the observed object); yet, subverts that distinction between observer and observed for the actual observer, and puts them in direct relation to the characters and objects within the play—suspending disbelief, and making the play all the more significant in the experiential moment. This homunculus paralleling technique puts the observer in the play, possibly letting them feel even more inter-subjective-like relations with the characters. Although Eliot might cast Shakespeare as the emotional manipulator, I believe Shakespeare writes from the heart, with an emphasis on relationships, that again, includes his own, as well as his characters'. Yes, the use of a novel technique may seem crafty, but when huge egos are involved, and Hamlet appears to have an ego, or at least some degree of honor and pride, this technique that emphasizes the subjective and the inter-subjective is germane to the issue of egos—and relations between egos; and with soliloquies, and the ego's relation to itself with the whole issue of (hesitating) self-reflecting and self-consciousness. In this dialogue with Kyd's *The Spanish Tragedy*,

Shakespeare deftly uses the technique of self-reflective modeling in *Hamlet* (a play within a play, or texts within the text of the play) to parallel Hamlet's own self-conscious self-reflecting.

PARALLEL SELF-REFLEXIVE LOVE & DEATH REVELATION

And both *Hamlet* and *The Spanish Tragedy* use the play within a play (with each lead character being a playwright—at least in part, Hamlet adds to a play, as Shakespeare transformed Kyd's), as a tool for exposing the murderers—the sought revenge upon whom drives the tragedy of the plays. This self-reflexive technique operates as a pivot for revelations—in *The Spanish Tragedy*, the "public" finally discovers the truth, when the play tragically ends with a revenge murder suicide bloodbath as the revenge in the play within the play plays out for real; while in *Hamlet* Shakespeare uses the play within the play to a more subtle end of simply confirming suspicions for Hamlet—the virtual circumstances evoke a real emotional response from the guilty spectators and expose them as the murderers (just as the virtual play may evoke (sometimes cathartic) emotions from the audience).

The core elements of love and death (central to comedy and tragedy) are often found in written letters within Renaissance plays—and also the revelation of some key fact: perhaps it is not a coincidence that the most basic elements of these plays are often represented with moments using the self-reflexive technique. Possibly there is some socio-psychological element to this condensation of the essential issues in these moments; at any rate, there remains this reduction of key elements of many plays with the use of the self-reflexive technique where an audience may become more consciously aware of said elements.

EXAMPLES OF READERS READING IN *THE SPANISH TRAGEDY*

Now, in *The Spanish Tragedy*, there are two important letters, the first from Bel(le)-imperia:

> "A letter written to Hieronimo! Red Ink.
> 'For want of ink, receive this bloody writ.
> Me hath my hapless brother hid from thee;
> Revenge thyself on Balthazar and him:
> For these were they that murdered thy son.
> Hieronimo, revenge Horatio's death,
> And better fare then Bel-imperia doth'" (III.ii.25-31; C.F. Brooke and Nathaniel Paradise (eds.), *English Drama 1580-1642*, D.C. Heath and Company: New York (1933).

Again, the central pivot in the play, involving death and the murderer's guilt, is presented using the self-reflexive technique. Even though letters are used in other plays too; there is much conspicuous discussion of letters and writing in *The Spanish Tragedy*, and *Hamlet*, making such a motif in itself; e.g. later in *The Spanish Tragedy*, Hieronimo apologizes to Bel-imperia:

> "I found a letter, written in your name,
> And in that letter, how Horatio died.
> Pardon, O pardon, Bel-imperia,
> My fear and care in not believing it" (IV.i.35-38).

Connected to the issue of the willing suspension of disbelief when reading fiction, Kyd relates the revelation of facts in writing and disbelief; and a sort of repentance for a prior lack of faith. Like Hamlet requiring his mouse trap play to further confirm his suspicions of his murderer Uncle, Hieronimo had to have the material evidence verified by a second letter, this time from Pedrigano to the King of Spain, which fell into Hieronimo's hands, reading:

> "'My lord, I write, as mine extremes requir'd,
> That you would labour my delivery.

If you neglect, my life is desperate,
And in my death shall reveal the troth.
You know, my lord, I slew him for your sake,
And was confederate with the prince and you;
Won by rewards and hopeful promises,
I holp to murder Don Horatio too'" (III.vii.33-41).

To further illustrate this self-reflexive technique motif, note Kyd bringing in yet another parallel with yet another letter:

"*Hier.* Say, father, tell me what's thy suit?
Senex. No, Sir, could my woes
Give way unto my most distressful words,
Then should I not in paper, as you see,
With ink bewray what blood begin in me.
Hier. What's here? 'The humble supplication
Of Don Bazulto for his murd'red son.'
Senex. Ay, sir.
Hier. No, sir, it was my murd'red son:
O my son, my son, O my son Horatio!
But mine, or thine, Bazulto, be content.
Here take my handkercher and wipe thine eyes,
Whiles wretched I in thy mishaps may see
The lively portrait of my dying self" (III.xiii.73-84).

Like the readers in a play mirroring the readers of the play, Hieronimo sees the old man Bazulto, with his letter of complaint, as a mirror or portrait of himself.

"TO BE OR NOT TO BE" & "I THINK THEREFORE I AM"

Hieronimo, in the prior scene (where he begins to carry a book), had commissioned a piece of art (from a painter whose son was also murdered):

> "paint me a youth run through and
> through with villains swords" (*The Spanish Tragedy*, III.
> xiia.131-132)).

which has:

> "no end; the end is death
> and madness! As I am never better than when I
> am mad; then methinks I am a brave fellow, then
> I do wonders; but reason abuseth me, and
> there's the torment, there's the hell" (III.xiia.167-171).

This seems to inform Shakespeare's own "To be, or not to be" soliloquy, with its:

> "To die, to sleep—
> To sleep, perchance to dream—ay, there's the rub"
> (*Hamlet*, III.i.63-64; *The Riverside Shakespeare*, Houghton
> Mifflin Company: Boston (1974)),

> "But the dread of something after death" (III.i.77),

> "Thus conscience does make cowards [of us all]" (III.i.82),

> "And lose the name of action" (III.i.87).

Here, we have something other than "robbed" themes—instead there is a philosophical "dialogue": Kyd compares bravery to madness, and reasoning to a self-inflicted abuse which delivers him to a hell (of knowing that his bravery is possibly madness); Shakespeare picks up this theme of reasoning between action and inaction, relating it possibly to thoughts of suicide (and Hieronimo does kill himself)—but conscience, like reason, thwarts "bravery" for fear of a possible "hell" after death. Shakespeare seems to be saying to Kyd, yes, rational contemplation (of actions often being done irrationally) may be hell, a self-abusing paralysis... but how irrational may be the final act of risking ending all acts, when the hell to come may be worse? Shakespeare, through Hamlet, seems to agree with Kyd's Hieronimo who is caught between mad bravery, and reasoned cowardice, but ups the stakes, as it were, brilliantly putting uncertain death in the mix, making action or inaction all the more vital—with a consequence of a living hell or a possibly worse death.

Shakespeare compares a possible *death* brought by action to an unknowable *dream*; while Kyd compares bravery to madness: in both instances risky action may metaphorically separate one from reality (or waking sanity) since to act is to no longer think things through. Where Kyd seems to emphasize the hell of reasoning; Shakespeare emphasizes the possible hell of suicidal action. Yet both seem to imply that reasoning brings one closer to reality ("godlike reason" (*Hamlet*, IV.iv.38)), since action sends one off into something possibly other than reality: an ego temporal entrance into the world with a roll of the dice (outside of the interior mind), which ironically results with ultimate real consequences. Both Kyd and Shakespeare may be bound to the sort of Renaissance thinking philosophically inspired by St. Augustine and exemplified by Descartes, believing that the interior mind is intertwined with divine reason's reality and that the exterior world is a possible illusion; directly connecting to Descartes' "cogito ergo sum" ego which by God given reason finally finds the doubted world to be real. Yet, Kyd and Shakespeare develop Descartes notion (before Descartes!), associating madness and suicidal risk with action in the external world, artistically and tragically embodying Descartes' abstract exercise in doubt: God may make the external world real; but fear of death and madness may have a way of preventing one from acting in the world as if one were in a dream, even without God's reality.

Hamlet is replete with a cluster of motifs that revolve around the notion of the unreal: madness, dreams, the ghost, books and the play (fictions, and reproductions), and thoughts (which are ephemeral). This too parallels Descartes' philosophical discussion in his Meditations (again the exterior world possibly being illusory like a dream, or the subjective observer being tricked by some devil, and actually being insane)—and the connection to Descartes is furthered by Shakespeare's use of the phrase "mind's eye" used by both Hamlet (I.ii.186) and Horatio (I.i.112) which brings to mind the notion of a homunculus—the homunculus being a philosophical outgrowth of St. Augustine and Cartesian philosophy. Philosophy itself is mentioned in *Hamlet* too:

> "There are more things in heaven and earth, Horatio,
> Than are dreamt of in your philosophy" (I.v.166-167);

and,

> "there is something in this more than natural, if philosophy could find it out" (II.ii.367-368);

Which two citations fix philosophy between the explainable and unexplainable. Hamlet also speaks of the "book and volume of my brain" (I.v.103) while the Queen speaks of "the very coinage of your brain" (III.iv.137; when Hamlet sees the ghost); bringing the possible materiality of mind to fore; (which however contrasts with the Queen's claiming that Hamlet "turns't my eyes into my very soul" (III.iv.89)—suggesting that the soul is mentally seated behind the material eyes, looking out through them). Hamlet often delves into the issue of what is real and what may not be, what may be merely mental, and what may be deadly fact. It is in struggling between these two realms, that Hamlet becomes concerned with honesty, and is suspicious of those around him; speaking to Polonius:

> "I would you were so honest a man" (II.ii.176).

Accosting Ophelia:

> "are you honest?" (III.ii.102).

Chastising Guildenstern:

"do you think I am easier to be play'd on than a pipe?" (III. ii.369-370).

And deriding Osric as a phony: (V.ii.187-194).

But suspicious thoughts and over-contemplation give way to action, fate (and death) with the conclusion of *Hamlet*. At the close of scene IV, Hamlet exclaims:

> "I do not know
> Why yet I live to say, 'This thing's to do,'" (IV.iv.44-45).

And the foreshadowing words of the Player King in the play within the play:

> "Our thoughts are ours, their ends none of our own" (III. ii.213),

lead to Hamlet noting:

> "And prais'd be rashness for it—let us know
> Our indiscretion sometimes serves us well
> When our deep plots do pall, and that should learn us
> There's a divinity that shapes our ends,
> Rough-hew them how we will" (V.ii.6-10).

Again, with Hamlet's initial suspicions and eventual "divinely" fated action, we can see a parallel with Descartes' methodical doubt and certain conclusion that God is and guarantees reality; Descartes is emotionally embodied in *Hamlet*, yet *Hamlet* predates Descartes' *Meditations* by some 40 years. Shakespeare builds a play that is fraught with doubt and talk of madness and dreams which end up in certain death, quite awhile before Descartes "systematically" doubts waking reality and ends up with the certainty of God. It is almost as if Descartes were responding to the tragedy of *Hamlet* with his proof of God's existence. Just as Shakespeare's "To be or not to be" indecision and doubt is resolved in an half-planed action that accords with God given fate, Descartes' "I think therefore I am" self-consciously relieves the doubting ego by showing it that it is already on a path of existence included in God's perfection. Both Shakespeare's Hamlet, and Descartes' ego face doubt, are transformed by a realization that we are already on a path not completely created from within ourselves, and conclude with a faith in divinity—even dying Hamlet uses the word "Heaven" twice in his closing lines (V.ii.332 & 343).

More important technically, that seminal turning point of modern philosophy, Descartes' *Meditations*, where the subjective ego, made explicit by St. Augustine, was isolated and opposed to a doubted theatre of objective sensation (the "Cartesian theater of the mind" opening the door to the likes of solipsism and German idealism), that turning point may have erupted in part from the play *Hamlet*, where beyond a stream-of-conscious soliloquies, the "homunculus paralleling self-reflexive technique" was combined with motifs of madness and dreams; where the act of observation was modeled, and thus made more self-conscious—with *Hamlet* subjectivity itself was made more self-conscious—and was placed in a world of dreamy madness and doubt.

In a way, the egoism of the character Hamlet, in combination with Plato's Transcendent Oneness, could be seen as opening the door to German Idealism's transcendental ego! And hence Hamlet's supposed madness, following St. Augustine's *Confessions*, could be a precursor to the philosophical schism between subjective and objective reality: a sort of philosophical schizophrenia can be traced back at least through Renaissance theatre and Shakespeare:

> "So They loved as love in twain
> Had the essence but in one
> Two Distinct, division none:
> Number there in love was slain. [...]
> Reason, in itself confounded,
> Saw division grow together,

> To themselves yet either neither,
> Simple were so well compounded:
> That it cried, 'How true a twain
> Seemeth this concordant one!
> Love hath reason, Reason none,
> If what parts, can so remain.'"

(Lines from the poem, *The Phoenix and Turtle,* which Shakespeare had printed in 1601, around the time he was working on *Hamlet*, and displays his own interest in dialectical matters).

Hamlet says:

> "Give me that man
> That is not passion's slave, and I will wear him
> In my heart's core" (III.ii.71-73):

Hamlet is very passionate, and continually struggles to channel this passion wisely, yet finds there is no steering clear of fated death: the possibility of death being that ultimate reality that may separate dreams from waking life. *Hamlet* is not

> "like the painting of a sorrow,
> a face without a heart" (IV.vii.107-108);

Even though it be a play, it grips it's audience with "real world" examples of the anxiety accompanying the possibilities of losing reality with the mind on the one hand, and finding the reality of the world through death on the other: Hamlet truly is caught between a rock and hard place, between the possible phantasms of the mind, and the deadly consequences of action.

THE SIMULACRUM FLOWS THROUGH US

The split between the interior mind of contemplation, and the exterior world of action, brings us back to the subjective and the objective, or subject / object split that informs the self-reflexive paralleling technique of using plays within plays and texts within texts. These techniques invite us to enter a fictional world, suspending our own reasoning and withdrawing critique. Yet, we don't act out in these worlds, we simply observe the actor / characters in them: we become part of a simulacrum of unreal actions that are not our own. Possibly, and this is conjecture, our own "real" world too, with its memes, and information flows, discourses, and language games, that pass from one generation to the next, and spread like diseases, or wildfires; possibly this world acts through us: we don't use language, "language speaks us;" a language and culture informed and shaped by writers of fiction—hence making our reality semi-fictional. Shakespeare and Kyd, as playwrights who model our world, make it explicit, and bring its details to consciousness may also be seducing us into their worlds in part to raise our own consciousnesses—we may suspend both our reasoning, and our own actions, only to learn more about what it means to self-consciously reason with our linguistic minds and act in our consequential world. And since self-consciousness is so bound up with language, it is no surprise that the writers, Kyd and Shakespeare, are obsessed with language. *The Spanish Tragedy* concludes with Hieronimo cutting out his tongue with only the ability to write; and tongues are referenced throughout *Hamlet*:

> "murder, thought it have no tongue will speak" (*Hamlet*, II.ii.560).

EXAMPLES OF READERS READING IN *HAMLET*

Besides indirect references to letters (such as the letter sent with Rosencrantz and Guildenstern ordering Hamlet's death in England), there are several references to writing and reading in *Hamlet*. For example, Polonius reads a letter from Hamlet to Ophelia, the text of which he momentarily breaks from:

"'To the celestial, and my soul's idol, the most beautified Ophelia'—
That's an ill phrase, a vile phrase, 'beautified' is a vile phrase. But you shall hear. Thus:
'In her excellent white bosom, these, etc.'
Queen. Came this from Hamlet to her?
Pol. Good madam, stay awhile. I will be faithful.
[*Reads the*] *letter*.
'Doubt thou the stars are fire,
Doubt that the sun doth move,
Doubt truth to be a liar,
But never doubt I love.
O dear Ophelia, I am ill at these numbers. I have not art to reckon my groans, but that I love thee best, O most best, believe it. Adieu.
Thine evermore, most dear lady,
whilst this machine is to him, Hamlet.'" (II.ii.109-124).

Here again, we have issues of *love* and faith; but also there is the interruption made by Polonius to comment on, and critique the letter. This hint at interpreting a text is further echoed later in the same scene, when Polonius addresses Hamlet, with many asides (not all quoted here) to the audience: Hamlet "[*reading on a book*]":

Pol. [*Aside.*] [...] still harping on
my daughter [...]
—What do you read, my lord?
Ham. Words, words, words.
Pol. What is the matter, my lord?
Ham. Between who?
Pol. I mean the matter you read, my lord.
Ham. Slanders, sir; for the satirical rogue says here that old men have grey beards" (II.ii.191-197).

With these asides and interchange of words, we hear Polonius comment on Hamlet to the audience, and Hamlet indirectly comment about Polonius to Polonius, by way of innuendo: both Hamlet and Polonius are engaged in a bit of interpretation, where Hamlet includes the audience in on his little joking jabs at the aged Polonius with his synopsis of the book; and Polonius draws in the audience with his direct asides (yet another technique for drawing the audience into the play). Since Hamlet's words are oblique, the audience too must, like Polonius, interpret them (another paralleling); and like other texts within plays that reference belief and truth—this text that we do not get to read directly, supposedly includes "Slanders." (Although Hamlet himself may have been sarcastic: Polonius notes that Hamlet's words are "Pregnant" (II.ii.209), suggesting that the real meaning is yet to be delivered: i.e. interpreted).

CHARACTER REVEALED THROUGH WRITINGS

We find Hamlet as both reader and writer, as when he writes the last words of his dead father's ghost, "'Adieu, adieu! remember me'"(I.v.111)—words that say something about writing itself: the author is gone, possibly dead, but the written words live on in material memory. With these written words, in a moment of self-reference, Shakespeare appears to make the play speak for him and itself, drawing the reader into an act of interpretation (and indeed, when composing letters for characters, the playwright takes on a different relation to their characters than when writing dialogue, which would be spoken spontaneously in the real world: they are put in the paralleling mode, genuinely writing as someone else). Curiously, Hamlet himself is the only author of the four instances of actual quoted text in the play: the words of the ghost, the love letter to Ophelia, a letter to Horatio recounting a sea adventure with pirates taking him prisoner (IV.vi.13-31), and a letter to King Claudius warning of his return, with the postscript, "alone" (IV.vii.43-53). These four writings refer to Hamlet's four most important relationships: his dead father, his love, his best friend, and his enemy. Possibly these four writings can help clarify the character of Hamlet a little: First, he inherits his father, the King's, legacy and all that entails. Second, his love letter to Ophelia demonstrates his sincere romantic side: he tries some sweet poetry, but confesses that his passions surpass his ability to articulate them. Third, his letter to Horatio accounts some slightly exaggerated adventurous and dubious heroism. Fourth, the letter to Claudius is mockingly polite. The four points of this constellation illuminate a complex character, a character who feels compelled to boldly and slyly defend his inflated honor, uses wit and eloquence to mask his aggressions, and yet still surpasses the limit of his capacity to express his emotions. No doubt, with this complexity comes some contradiction (as with verbal exaggerations falling short of earnest feelings); these complexities and contradictions make for a much more subtle character than Kyd's Hieronimo, and beyond a suspicious literary embodiment of and precursor to Descartes' doubting abstract ego—and perhaps it is this factor—Hamlet being a complex, human, character, that seduces one into a willing suspension of disbelief, and draws an audience into the play more than any self-reflexive, paralleling, homunculus technique.

And character, personality and personality traits evidenced by our expressions, which in turn issue form our streams-of-consciousness, much like Hamlet's soliloquies—the expression of character may be a greater challenge for computer simulations than artificial intelligence itself; and again, such may be necessary for the full equivalent of human intelligence.

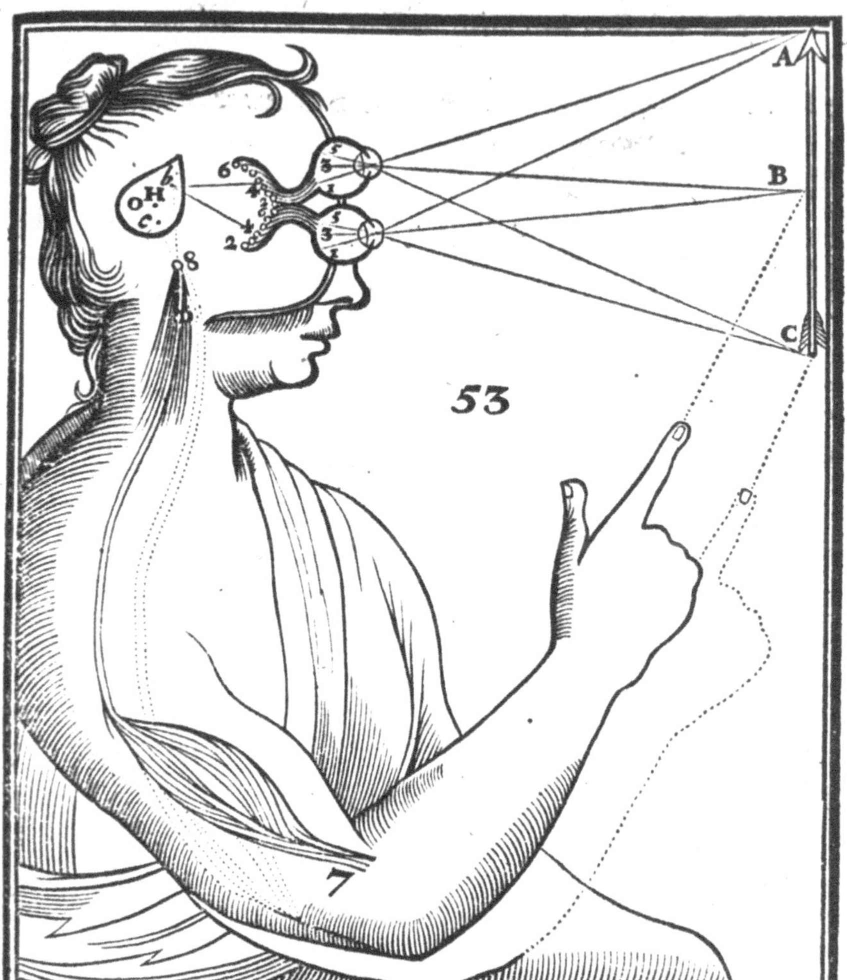

Diagram by René Descartes (~1630)

Chapter 9

Descartes
God Beyond Rote Memory

RELATIVE POINTS OF VIEW

René Descartes' (1596-1650) pivotal philosophy exemplifies modern philosophical emphasis's on subjectivity—bringing a more scientific bent rooted in mathematics, but also turning away from established canons in favor of thinking from the ground up.

Academically, Descartes came from a scientific and mathematical background (think of the Cartesian coordinate system, which integrated algebraic equations with geometric visualization). Philosophically he sought to build more complex ideas upon clear and distinct simple ideas, much as a geometrical proof builds upon given axioms that themselves cannot be proven, but must be taken as self-evident via innate intuition.

A second angle of departure for Descartes, as illustrated in his *Discourse on Method*, was developed from a relativism found in the history of philosophy: so many thinkers had so many different stances on issues, making a stable truth difficult to discern. Hence we have phrases such as "the diversity of our opinions" (René Descartes, Norman Kemp Smith (trans.), *Descartes Philosophical Writings*, The Modern Library: New York (1958), p. 93), "It may be that in this I am deluding myself" (p. 94), and "each of my readers may be able to judge for himself" (p. 94). Right from the start of his *Discourse*, Descartes claims we are all fallible; that we come from different personal histories, which form different opinions:

> "that our thoughts proceed along different paths, and that we are, therefore, not attending to the same things" (*Descartes*, p. 93).

He clarifies this "perspective" notion by noting that he has traveled to various countries, and read thinkers from various ages:

> "To hold converse with those of other ages is almost, as it were, to travel abroad; and travel, by making us acquainted with the customs of other nations, enables us to judge more justly of our own, and not to regard as ridiculous and irrational whatever is at variance with them, as those ordinarily do who have never seen anything different" (*Descartes*, pp. 96-97).

We find Descartes an "open minded" thinker—willing to drop dogma; but not in order to better compare previous thinkers and weigh their insights, nor to embrace relativity, but rather to build his own perspective. Unlike Plato reconstructing dialogues of previous philosophers, or Aristotle being concerned with the history of philosophy, but more like St. Augustine in his *Confessions*, Descartes speaks from a first-person "I" perspective, recalling his own experiences that have lead to his philosophical breakthroughs. Since everything in previous philosophy was subject to interminable argument, Descartes took it as doubtable:

> "there is not a single thing of which it treats which is not still in dispute, and nothing therefore, which is free from doubt" (*Descartes*, p. 98).

> "I judged that nothing solid can have been built on foundations so unstable" (*Descartes*, p. 98).

Moreover, Descartes is not interested in "the predictions of an astrologer" (*Descartes*, p. 99) and the like, but rather sought:

> "no other science than that which can be found in myself and the great book of the world." (*Descartes*, p. 99).

Not satisfied with "example or custom" (*Descartes*, p. 100), Descartes sought a more self-reliant philosophy, one uniquely constructed by his own singular hand:

> "there is less perfection in the works composed of several parts and the product of several different hands, than in those due to a single master-work-man" (*Descartes*, p. 101).

> "My design has all along been limited to the reform of my own thoughts, and to basing of them on a foundation entirely my own" (*Descartes*, 104).

RADICAL DOUBT, SELF-CERTAINTY

Descartes prefers the orderly and elegant to the messy and complex, inventing a new method that does not build on old foundations (Descartes repeatedly uses the analogy of a "building"—as if philosophy and theory building were analogous to architecture and carpentry). His method is one of doubting everything that is not self-evident, clear and distinct; analyzing problems into their parts; starting from the simple and building to the complex; and generalizing and analyzing so completely that nothing is left out. Here, the exception does not prove a rule, but rather must be accounted for: again, a simple perfection is sought, as with a comprehensive and elegant scientific law that applies in each case—as evidenced by his claim (dubious to those who know more than one way to "skin a cat"):

> "on each particular issue there is but one true solution, and that whoever finds it knows all that can be done regarding it" (*Descartes*, p. 109).

Descartes' radical skepticism is tempered morally by a few principles (abide by laws; be resolute; change yourself, not the world; and try to learn); and with this in mind, he claims:

> "I did nothing but roam about the world, seeking to be a spectator rather than an actor in all life's dramas" (*Descartes*, p. 115).

Yet regarding science, he talks of:

> "difficulties that I was able to make almost mathematical" (*Descartes*, p. 116)

The mention of "life's dramas" brings Shakespeare's *Hamlet* to mind—and Hamlet's deliberations on whether or not to act on his suspicions. Descartes' philosophy clearly favors observation over action—yet his mentioning that he can "make" non-mathematical disciplines "almost mathematical"—"making" them that way, rather than discovering them, suggests not only action, but a bit of the "vanity" that he often derides; albeit that such may be in part true. And despite many caveats, Descartes is egotistical (while discussion fame, he mentions not wanting it, but does not really question whether or not he deserves it).

At any rate, using his method(s), Descartes doubts his senses, his prior reasoning, and even the distinction between waking and dreaming. What he cannot doubt even if in a dream, however, is that he is doubting, which leads to the "implication":

> "*I think, therefore I am*" (*Descartes*, p. 119).

Analogous to Shakespeare's mousetrap play within a play, Descartes thinks about thinking—drops back a step and *observes* his thinking with the birth of the Cartesian Theater. Although not an account of a "stream-of-consciousness" (which Hamlet is on the verge of with his soliloquies), Descartes has brought an ego's mental activity to the fore: an individual thinker's thinking, rather than thought and reason in general:

> "I had only to cease to think for an instant of time and should then (even although all the other things I had imagined remained true) have no ground for believing that I can have existed in that instant. From this I knew that I was a substance whose whole essence or nature consists entirely in thinking, and which for its existence, has no need of place, and is not dependent on any material thing; so that this I, that is to say, the soul by which I am what I am, is entirely distinct from the body" (*Descartes*, p. 119).

What we have here is something approaching an invention or discovery of philosophical subjectivity confronted with a world radically doubted—a subjectivity "prior" to the world, and found only implicit in observing the thinking of the ego. Clearly on his way from St. Augustin's "si fallor, sum" to Kant's concept of "apperception," Descartes had not yet divided this ego-unity into a sensing observer, and a logical reasoning entity: which raises a question: is thinking for Descartes both sensing and reasoning? Why didn't he further doubt reasoning as simply a thinking that is only observed? That is, although he later doubts mathematics as well, why not consider the very ideas that are observed in observing thought as fallible too? This notion of "Reason" as epiphenomenon would have to wait for later philosophers.

GOD'S PERFECTION: A DREAM COME TRUE

Not satisfied only with self-certainty, Descartes finds that he is not perfect, but has a conception of perfection as found with mathematics. In the world we only find imperfect circles, but a perfect circle does exist, at least mentally. For Descartes, less perfect things can only come from more perfect things:

> "I resolved to inquire whence I had learned to think of something more perfect than I myself was; and I saw clearly that it must proceed from some nature that was indeed more perfect" (*Descartes*, p. 120).

And, the highest "containing" perfection, or the highest link on what was later termed "The Great Chain of Being" (Arthur Lovejoy, 1936), for Descartes, is God. This God must exist, for how perfect would a God be if God did not exist? And a perfect God would be good, not a deceiver, and hence would not let Descartes, or any other radical skeptic, be completely wrong about the world. Solipsism escaped via a Deus Ex Machina.

However, despite the power of mind, soul, and reason to command philosophical attention, more practical investigations, like cosmology, physiology or mathematics also attract Descartes' extended attention. And although he notes that "the body is regarded as a machine" (René Descartes, Elizabeth S. Haldane & G.R.T. Ross (trans.), *Philosophical Works of Descartes Vol. I*, Dover Publications: New York (1955), p.116), that humans are in part mechanical, he also claims that animals and automata do not have thinking souls:

> "if there were machines which bore a resemblance to our body and imitated our actions as far as it was morally possible to do so, we should always have two very certain tests by which to recognize that, for all that they were not real men. The first is, that they could never use speech or other signs as we do when placing our thoughts on record for the benefit of others [....] it never happens that it arranges its speech in various ways, in order to reply appropriately to everything that may be said in its presence, as even the lowest type of man can do." (*Descartes* (1955), p. 116).

Such presages the Turning test for artificial intelligence which will be discussed later in a later chapter. The Second test:

> "And the second difference is, that although machines can perform certain things as well as or perhaps better than any of us can do, they infallibly fall short in others, by which means we may discover that they did not act from knowledge, but only from the disposition of their organs. For while reason is a universal instrument which can serve for all contingencies, these organs have need of some special adaptation for every particular action. From this it follows that it is morally impossible that there should be sufficient diversity in any machine to allow it to act in all the events of life in the same way our reason causes us to act" (*Descartes* (1955), p. 116).

Machines of Descartes' time were definitely solution specific; whether or not contemporary artificial intelligence research may produce a flexible thinking machine has yet to be seen. So far, most AI experiments, even such successes as IBM's Watson winning the game show *Jeopardy!* are quite task specific. But then again, our brains may be more task-specific than has been thought too.

MADNESS, DREAMS, CHIMERAS

In his *Meditations on First Philosophy – In which the Existence of God and the Distinction in Man of Soul and Body are Demonstrated*, Descartes fleshes out many of the arguments of the *Discourse*, heretofore discussed. Using his doubting method, which is similar to the "bracketing" method of later phenomenological philosophers such as Husserl, Descartes eliminates categories of experience and prior knowledge in order to get at the core ego as observing thinker.

Of the senses, he says:

> "senses I have sometimes found to be deceptive; and it is only prudent never to place complete confidence in that by which we have even once been deceived" (*Descartes* (1958), p. 177).

Such illustrates that Descartes was not interested in the probable, but the absolute. For a robust history of experience does indeed give us the ability to differentiate waking from dreaming, *at least most of the time.* For Descartes, however, exceptions throw out the rule; possible deception means 100% untrustworthiness. He does limit this however, noting of the insane who think:

> "their head is made of clay and their body of glass, or that they are pumpkins. They are mad; and I should be no less insane were I to follow examples so extravagant" (*Descartes*, p. 177).

Nonetheless, Descartes hypothesizes that if God wanted to deceive him on matters as diverse as perceptions (which can also be imagined when reconfiguring past experiences—a Griffin being made of various real animals, etc), deceptions concerning extension, and even mathematics—could be made—and also possibly by a "malignant genius exceedingly powerful and cunning" (*Descartes*, p. 181). A contemporary example of this epistemelogical situation is the notion of a "brain in a vat" that is stimulated into thinking it was actually in a person in a world; somewhat as portrayed in the Matrix movies. How do we know that we're not in a virtual reality—that we are not dreaming, or deceived by an evil genius?

WHAT IS AN "I?"

Again, as fulcrum to his later re-establishment of knowledge of the outer world, Descartes turns to the "Ego sum, ego existo" (*Descartes*, p. 183)—"I am, I exist" (this time avoiding the dubious "implication" that I exist because I think). But what is this "I?":

> "Sensing? There can be no sensing in the absence of body; and besides I have seemed during sleep to apprehend things which, as I afterwards noted, had not been sensed. Thinking? Here I find what does belong to me; it alone cannot be separated from me. *I am. I exist.* This is certain. How often? As often as I think" (*Descartes*, p. 185).

> "What then is it that I am? A thinking thing. What is a thinking thing? It is a thing that doubts, understands, affirms, denies, wills, abstains from willing, that also can be aware of images and sensations" (*Descartes*, p. 186).

With these two quotes we see the crucial mind / body spilt taking place at "sensing." The bodily organs are needed to perceive the world... however "awareness" of images and sensations is of the mind—a sort of vanishing point of experience, but not the experience itself; the I that "can apprehend images" (*Descartes*, p. 187):

> "that I see, that I hear, and that I am warmed. This is what in me is rightly called sensing, and used in this precise manner is nowise other than thinking" (*Descartes*, p. 187).

UNDERSTANDING THINGS, THOUGHTS, IDEAS

And "things?" Descartes considers our various exposures to a piece of wax. It can melt, which causes all its properties to change. With this "subject / object" opposition between ego observer and wax thing, Descartes claims that we cannot know wax by its properties, even that of extension, because these can all change, yet it is still wax—we may know it as such only through the mind:

> "I cannot by way of images comprehend what this wax is, and that it is by the mind alone that I apprehend it" (*Descartes*, p. 189).

> "bodies are not cognized by the senses or by the imagination, but by the understanding alone" (*Descartes*, p. 191).

Throughout these discussions on wax, Descartes begins to name several "faculties" of the mind: "the faculty of imagination" (*Descartes*, p. 189); "the faculty of judgment" (*Descartes*, p. 190); "memory" (*Descartes*, p. 216); "faculties of willing, sensing, understanding, etc." (*Descartes*, p. 244); the sort of notions that were later developed by Kant and were initiated in part by St. Thomas Aquinas who divided Plato's three parts of the soul (desire, high spirit, and reason) even further. For Aquinas:

> "In the powers or faculties there is a certain hierarchy. The vegetative faculty, comprising the powers of nutrition, growth and reproduction, has as its object simply the body united to the soul or living by means of the soul. The sensitive faculty (comprising the exterior senses, of sight, hearing, smell, taste, touch, and the interior senses of *sensus communis*, *phantasia* or imagination, *vis aestimativa* and *vis memorativa* or memory) has as its object, not simply the body of the sentient subject bur rather every sensible body. The rational faculty (comprising the active and passive intellects) has as its object, not only sensible bodies but being in general" (Frederick Copleston, S.J., *A History of Philosophy – Volume II: Augustine to Scotus*, Image Books, New York (1985), p. 377).

Also, anticipating Kant's "Copernican revolution," and developing what was later termed the "Cartesian Theater," Descartes discusses "images" or "representations"—here distinguishing types of thoughts:

> "Some of my thoughts are, as it were, images of things; and to them alone strictly belongs the title 'idea,' e.g., when I represent to myself a man, or a chimera, or the sky, or an angel, or even God" (*Descartes*, p. 196).

"Ideas" or images as one type of thought are contrasted with "judgments." "Ideas" in turn are divided into "innate" (from within), "adventitious" (from without) and "invented" ideas—that seem to correlate with the understanding, the senses, and the imagination. But even the invented ideas may be adventitious, in that the imagination works with material previously perceived before. Careful consideration of representing ideas reveals that, as images, they can portray their subjects more or less accurately:

> "Those which represent substances are without doubt something more, and contain in themselves, so to speak, more objective reality (that is to say participate by representation in a higher degree of being or of perfection) than those which represent only modes or accidents; and again, the idea by which I apprehend a supreme God, eternal, infinite, immutable, omniscient, omnipotent, and the creator of all things which are in addition to Himself, has certainly in it more objective reality than those ideas by which finite substances are represented" (*Descartes*, p. 199).

Moreover:

> "what is more perfect, i.e., contains more reality, cannot proceed from what is less perfect" (*Descartes*, p. 200).

For Descartes, representational ideas can be caused by something more perfect than they are: somewhat like Plato's perfect forms, ideas come not from reflecting their objects but from something higher:

> "if an idea is to contain one objective reality rather than some other, it must undoubtedly derive it from some cause in which there is to found as much formal reality as in the idea there is objective reality" (*Descartes*, pp. 200-201).

> "Just as the objective mode of existence belongs to ideas by their very nature, so the formal mode of existence apper-

> tains to the causes of these ideas, at least to the first and chief of their causes, by the very nature of those causes. For although, it may be, one idea gives birth to another, the series of ideas cannot be carried back *in infinitum*; we must in the end reach a first idea, the cause of which is, as it were, the archetype in which all the reality or perfection in that is in the idea only objectively, by way of representation, is contained formally. In this way the natural light makes it evident to me that the ideas are in me in the manner of images, which may indeed fall short of the perfection of the things from which they have been derived, but can never contain anything greater or more perfect" (*Descartes*, p. 201).

From this, Descartes concludes:

> "if the objective reality of any one of my ideas be so great that I am certain it cannot be in me either formally or eminently, and that consequently I cannot myself be the cause of it, it necessarily follows that I am not alone in the world and that there is likewise existing some other thing, which is the cause of this idea" (*Descartes*, p.201).

There are hence, for Descartes, ideas which are "archetypes"—primary ideas that are more perfect than the partial notions we have of the objects and such in the world around us. Our image / representation ideas fall short of their possible perfection as exemplified by the archetypes—and for Descartes, this means that the archetypes "contain" and cause our more mundane ideas which represent the world around us.

A PERFECT CIRCLE

The highest perfection—the most perfect idea for Descartes, is God. It naturally follows that God is the container and cause of all other ideas—nothing can come from something less perfect than itself, as far as ideas are concerned: the archetype of all chairs must be more perfect than our particular representations of chairs, or else we would have no way to tie all these particular experiences into a single idea.

Descartes wonders if this highest perfection can be found in his own self as a potential, but lacking a clear theory of the unconscious, finds his conscious self quite short of being able to obtain complete perfect omniscience. God has not forgotten God's power by inhabiting the mind of and being Descartes. Moreover, God is decidedly a unity as Descartes claims this to be a "chief" perfection (*Descartes*, p. 209), and God has placed the idea of God in the mind of Descartes: "the mark of the workman imprinted on his work" (*Descartes*, p. 210).

How Descartes came to a *certain* definition of God and perfection is beyond me. He rejects custom, but his definitions of God smack of tradition, not some innate idea—do not other cultures have different concepts of God (e.g. Buddhism)? In his Dedication to the *Meditations*, Descartes suggests people could claim that theologians of his time were "reasoning in a circle" (*Descartes*, p. 162) when suggesting that God is revealed by scriptures that God inspired—the authority of the scriptures is based on what it assumes—so why should I believe the God inspired scriptures relating that God exists, if I don't already believe in God? But it seems to me that Descartes' own use of "perfection" is much like a use of "scripture" to prove God's existence: Why should I believe the idea of perfection implying God's existence if I don't already believe in perfection? Couldn't the notion of perfection be a non-existent extrapolation of varying degrees of better or worse towards a limit? Some might say that perfection never exists in our world, but only exists as an idea—and hence God as perfect being would be so only as an idea.

Then again, ideas, as extensions of the mind, or at least the archetypes they come from, are more real for Descartes than the world around him: in fact, in a strange reversal, Descartes can prove the world around is real only in reference to the ideal! Hence we can see later idealists taking inspiration from the mind of Descartes.

BRAIN MALFUNCTIONS?

But if God is perfect, and guarantees that Descartes is not deceived on the whole (since a perfect God would not be a deceiver), how is that Descartes can be fallible? To this, Descartes claims that there is clearly an idea contrary to the idea of God and being, which is nothingness and imperfection—and that he finds himself caught in-between the two. Why did God make him this way? Although the motivation might be unfathomable and beyond understanding, Descartes suggests that the part may not be perfect in itself, but as part of a whole—and he himself is only a part of the whole.

Error arises for Descartes, out of a disjunction between will and understanding—when one desires to take non-clear and distinct ideas as true, when in fact they are false.

Regarding memory, Descartes seems to have two opposing views:

> "I can yet by attentive and oft-repeated meditation so imprint it [a thought] on my memory that I shall never fail to recall it as often as I have need of it, and so can acquire the habit of not erring" (*Descartes*, p. 220);

> "on beginning to discover them [the modes of extended things, e.g. shapes, number, etc,] it does not seem to me that I am learning something new, but rather that I am recollecting what I already knew, i.e., that I am for the first time taking note of things that were already in my mind but to which I had not hitherto directed my attention" (*Descartes*, p. 222-223).

Following Plato's theory of recollection, we can see that Descartes sees his innate clear and distinct ideas as being *recalled.* And when discussing God, again and again, as existing by his very essence of being perfect—we see that "existence" is not necessarily what we would take as "worldly" existence, but, for Descartes, following Plato—God's existence is a more real formal existence. And indeed, as this world falls short of perfection in many ways, it is less a part of Being than God—and more entrenched in nothingness.

To best distinguish the mental and the material, Descartes goes to the division between them found between his mind and his body. He finds that he is not "a pilot in a ship" (*Descartes*, p. 239), but rather the mind is intermingled with the body. He makes several distinctions, between intellect and imagination (*Descartes*, p. 230-232); emotion and sensation (*Descartes*, p. 233); and importantly between passive and active sensing:

> "Now there is, indeed, a certain passive faculty of sense, i.e., of receiving and knowing the ideas of sensible things, but this would be useless to me if there did not also exist in me, or in some other being, an active faculty capable of producing or effecting these ideas. This active faculty cannot, however, be in me—not at least in so far as I am only a thinking thing—since it does not presuppose intellection, and since the ideas present themselves to me without my contributing in any way to their so doing, and often even against my will. This faculty must therefore exist in some substance different from me—a substance that, as already noted, contains, either formally or eminently, all the reality which is objectively in the ideas produced by the faculty, and this substance is either body, i.e., corporeal nature, in which there is contained formally, i.e., actually, all that is objectively, i.e., by representation, in those ideas; or it is God Himself, or some creature nobler than body, in which all of it is eminently contained" (*Descartes*, p. 238).

But Descartes finds that it must be the body and not God that has this active sense, since these senses can be deceiving and God does not deceive.

Descartes identifies himself with this thinking thing called mind which he finds, despite the differing faculties, is just one thing; while the body with its feet, hands, etc, is many. Even though a foot could be amputated, and still the mind perceives pains in a phantom limb that does not exist—mind is only the passive receiver of such pains—the pains as *sensations* are not mental, but the *experiences* of them are. How could this be so? Because the mind is affected *only* by the brain:

> "I take note that the mind is immediately affected, not by all parts of the body, but only by the brain, or rather perhaps only by a small part of it, viz., by that part in which the *sensus communis* is said to be" (*Descartes*, p. 244).

Although the brain can be a source of erroneous judgment (my amputated leg feels an itch), the willing mind can derive that the leg is indeed not there, or err, and think it really is. Although the senses can deceive, it is the mind that can err.

BIOLOGY OF MEMORY

Descartes delves deeper into the functioning of the brain in *The Treatise on Man* (René Descartes, Stephen Gaukgroger (trans.), *The World and Other Writings*, Cambridge University Press: Cambridge (2004)). It is in this work that Descartes considers the physiology of the human body as a body like that of animals—i.e. the workings of the human body that have little or nothing to do with the reasoning mind. But there is more than reflex going on in this body as a:

> "statue or machine made of earth, which God forms with the explicit intention of making it as much as possible like us" (*Descartes* (2004), p.g 99).

Although Descartes discusses the body in more mechanical terms than his predecessors (explaining the various activities of the body without relying on various "organ souls," or non-mechanical processes), when it comes to the brain, which is nourished by the heart, he claims:

> "As for those parts of the blood that penetrate as far as the brain, they serve not only to nourish and sustain its substance, but above all to produce there a certain very fine wind, or rather a very lively and very pure flame, which is called the 'animal spirits'. For it should be noted that the arteries that carry these from the heart, after having divided into countless small branches and having composed the little tissues that are stretched out like tapestries at the bottom of the cavities of the brain, come together again around a certain little gland which lies near the middle of the substance of the brain" (*Descartes* (2004), p. 105-106).

To the modern cognitive scientist, these "animal spirits" might correlate to the electrical signals of neurons... but Descartes has a more hydraulic conception of how nerves conduct impulses from one part of the body to another. He sees the brain largely as a network of nerves carrying impulses from the central pineal gland (that "certain little gland") to the sense organs and bodily limbs and back—the brain itself being like a network of nerve bundles:

> "Consider its surface AA [in figures illustrating the central pineal gland 'H' surrounded by a network of nerves radiating outward], which faces cavities EE [the network of nerves close to the center], to be a somewhat dense, compact net or mesh all of whose links are so many tiny tubes through which the animal spirits can enter and which, since they always face gland H from where these spirits originate, can easily turn this way and that toward the different points on this gland [....] Assume also that the chief qualities of these tiny fibres are the ability to be flexed readily in every

> way simply by the force of the spirits that strike them, and the ability to retain, as if made of lead or wax, the last flexure received until a contrary force is applied to them" (*Descartes* (2004), pp. 143, 145).

Boldly, Descartes sets out to consider:

> "how ideas of objects are formed in the place assigned to the imagination and to the common sense, how these ideas are retained in the memory, and how they cause the movement of all the bodily parts" (*Descartes* (2004), p. 146).

And:

> "I say 'imagine' or 'sense'. For I wish to apply the term 'idea' generally to all the impressions which the spirits are able to receive as they issue from gland H. And when these depend on the presence of objects they can all be attributed to the common sense; but they may also proceed from other causes, as I shall explain later, and they should then be attributed to the imagination" (*Descartes* (2004), pp. 149-150).

It is somewhat clear here, that given the previous notion of a *sensus communis* cited above, as where the mind interacts with the brain, that the pineal gland 'H' operates as an intermediary between mind and body.

Descartes also notes that some ideas can be implanted in the blood from one's mother—but also that memories of sense impressions can be "encoded" in the brain:

> "I could add something here about how the traces of these ideas pass through the arteries of the heart, and thus radiate throughout the blood; and about how they can sometimes even be caused by certain actions of the mother to be imprinted on the limbs of the child being formed in her womb. But I shall content myself with telling you more about how

the traces are imprinted on the internal part of the brain marked B [the outer half of the brain], which is the seat of memory" (*Descartes* (2004), p. 150).

Descartes explains memory:

> "Imagine that after issuing from gland H spirits pass through tubes 2, 4, 6 and the like, into the pores or gaps lying between the tiny fibres making up part B of the brain. And suppose that the spirits are strong enough to enlarge these gaps a little, and to bend and arrange any fibres they encounter in various ways, depending on the different ways in which the spirits are moving and the different openings of the tubes into which they pass. And they do this in such a way that they also trace figures in these gaps, corresponding to those of the objects. At first they do this less easily and perfectly here than on gland H, but they gradually improve as their action becomes stronger and lasts longer, or is repeated more often. Which is why in such cases these patterns are no longer easily erased, but are preserved in such a way that the ideas that were previously in this gland can be formed again long afterwards without requiring the presence of the objects to which they correspond. And this is what memory consists in" (*Descartes* (2004), p. 150).

Moreover, memories can arise spontaneously, as if a stream-of-conscious were stimulated by reflex:

> "it should be noted that when gland H is inclined in one direction by the force of spirits alone, without the aid of either the rational soul or the external senses, the ideas which are formed on its surface derive not only from inequalities in the tiny parts of the spirits causing corresponding differences in the humours, as mentioned earlier, but also from imprints of memory. For if the figure of one object is imprinted much more distinctly than that of another at the place

> in the brain towards which this gland is properly inclined, the spirits issuing from it cannot fail to receive an impression of it. And it is in this way that past things sometimes return to thought as if by chance and without the memory of them being stimulated by any object impinging on the senses" (*Descartes* (2004), p. 155-156).

What these passages makes clear is that, considering the bifurcation of memory cited earlier (between memorization through repetition, and recollection of clear and distinct ideas)—what is typically considered memory is wholly a bodily process distinct from the reasoning mind:

> "the effect of memory that seems to me to be most worthy of consideration here is that, without there being any soul present in this machine, it can naturally be disposed to imitate all the movements that real men—or many other similar machines—will make when it is present" (*Descartes* (2004), p. 157).

Descartes' account of the animal-mechanical aspects of body, and how such can operate in an animal that has, for Descartes, no soul, limns with his prior accounts of the mind-body split. That so much is accounted for in terms of the bodily (sensation, memory —and even the activation of thoughts through involuntary images arising, e.g. esp. in sleep)—that images and the whole apparatus of representation is bodily—suggests not much is left for the mind except passive experiencing, and a reasoning connected to archetypical forms or essences as clear and distinct ideas. But it is this process of elimination that helps us see how Descartes defines mind—much like Kant will later find limits of knowledge and certainty itself.

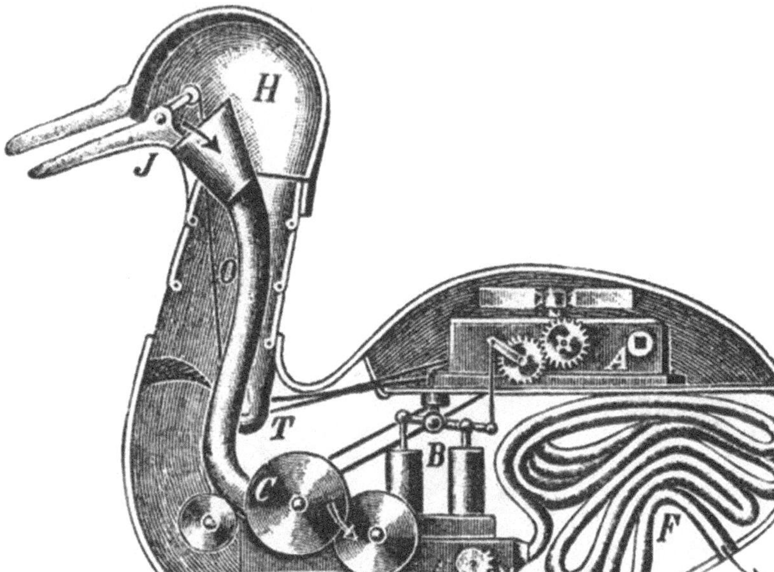

Misrepresentation of the Duck of Vaucanson (1738)

Chapter 10

Rationalists

Necessarily Contingent

SPINOZA: THE DESIRE OF BEING ONE

Benedict De Spinoza (1642-1677) was a rationalist philosopher in the tradition of Descartes, his greatest work, *Ethics*, being published upon his death. Like Descartes, Spinoza admired the mathematical method, and his own work is laid out like a geometrical proof, with definitions, axioms, and demonstrations. Unlike Descartes, who aimed at reconstructing a philosophy over the ruins left by his radical doubt, Spinoza begins with his idea of God, and demonstrates how the implications of his understanding of God can be seen to explain our human experience and knowledge. Despite a focus on metaphysics, epistemology and psychology, the *Ethics* is aptly entitled, due to its ultimate aim of clarifying how we can overcome our inclinations and live in harmony with God:

> "*...no one, to my knowledge, has determined the nature and powers of the affects, nor what, on the other hand, the mind can do to moderate them. I know, of course, that the celebrated Descartes, although he too believed that the mind has absolute power over its own actions, nevertheless sought to explain human affects through their first causes, and at the same time to show the way by which the mind can have absolute dominion*

over its affects. But in my opinion, he showed nothing but the cleverness of his understanding..." (Benedict Spinoza, Edwin Curley (ed. & trans.), *Ethics*, Penguin Books: New York (1996), p.69; Heretofore cited by volume and page number of the standard edition: *Ethics*, II/137-138).

"Affects" are those feelings we have of love, hate and other emotions and motivations, and the overall aim of Spinoza's *Ethics* is to show how we can free ourselves from our inclinations and come closer to God. But here we are skipping too far ahead, and should begin as Spinoza does, with God.

The first of the five parts of *Ethics* explains what Spinoza sees God to be, with a proof of God's existence. In a pantheistic way, God is seen to be nature, and to be an infinite, eternal, self-caused, single substance—possibly what other philosophers have termed "Being." Spinoza's God / nature is just this one substance—an abstract, non-anthropomorphic, yet immanent God—an (at the time heretical) idea that Spinoza may have gotten from Cabala texts:

> "The idea of God as the infinite Being which expresses itself in and yet comprises within itself the world seems to have been suggested at least to Spinoza by his reading of Jewish mystical and Cabalistic writers" (Frederick Copleston, *A History of Philosophy – Volume IV: Descartes to Leibniz*, Image Books: New York (1985), pp. 208-209).

This monism is opposed to the dualism of Descartes, as Spinoza claims that there are an infinite number of ways for this substance to be: it has an infinite number of attributes, two of which we humans partake in; namely, thought and extension. There are also the modes of these attributes, namely the *things* in the world. Being a mode of substance, human beings are one with God, albeit we are bodies that perceive this nature through the limitations of the attributes of thought and extension—our thoughts and bodies do not interact causally though, as they are simply two aspects of the same thing, operating in parallel. For Spinoza, we are, literally, a portion of God experiencing God.

Spinoza is a rationalist, due not to his metaphysics however, but to his epistemology. As with his "geometric proofs," he sees true knowledge arising out of necessity from God / nature—and we can form "universal notions" in the following ways:

> "I. from singular things which have been represented to us through the senses in a way which is mutilated, confused, and without order for the intellect; for that reason I have been accustomed to call such perceptions knowledge from random experiences.
>
> II. from signs, for example, from the fact that, having heard a read certain words, we recollect things, and form certain ideas of them, like those through which we imagine the things; these two ways of regarding things I shall henceforth call knowledge of the first kind, opinion of imagination;
>
> III. finally, from the fact that we have common notions and adequate ideas of the properties of things. This I call reason and the second kind of knowledge.
>
> IV. In addition to these two kinds of knowledge, there is (as I shall show in what follows) another, third kind, which we shall call intuitive knowledge. And this kind of knowing proceeds from an adequate idea of the formal essence of certain attributes of God to the adequate knowledge of the essence of things" (*Ethics*, II/122).

Such might roughly divide into perceptions, beliefs, common sense / science, and knowledge of essences. This hierarchy moves from the senses, through our imagination, to our social understanding, on into a sort of divine knowledge—from fallible to infallible knowledge. Although it may be our public ways of knowing, and scientific methods that lead to apprehending essences, like a trigger, ultimately these are "intuitions" that arise from within—making such knowledge innate, prior to experience (a priori), and thoroughly rationalist.

After situating the human mind as part of God / nature, Spinoza goes on to make a considerable breakthrough in understanding humans as psychological beings, as the third part of *Ethics* examines the "affects" of the human mind, esp. the "appetites" (*Ethics*, II/138). Spinoza develops a clear binary dichotomy between positive and negative emotions (e.g. love and hate)—and connects such emotions with the motivation towards self-preservation (interlocking with Utilitarianisms' later concern with desire for pleasure and avoidance of pain, and also anticipating Freud's later studies of unconscious desires):

> "*We strive to affirm, concerning ourselves and what we love, whatever we imagine to affect with joy ourselves or what we love. On the other hand, we strive to deny whatever we imagine affects with sadness ourselves or what we love*" (*Ethics*, II/159).

Spinoza sees humans as slaves to such motivations—the fourth part of *Ethics* is entitled "Of Human Bondage, or the Powers of the Affects" (*Ethics*, II/205). However, such bondage may be surpassed, at least to some extent, by knowledge (where knowledge may not alter our determined motivations, but give us the power to recognize them)—as the final part of *Ethics* is "On the Power of Intellect, or On Human Freedom" (*Ethics*, II/277). Such falls in line with a history of philosophers extolling the virtues of philosophy, and claiming that wisdom leads to enlightened happiness, or what Spinoza calls "Blessedness" (*Ethics*, II/307). "Blessedness" may be seen as a contentment with the way things are—as we become closer to God / nature, we get closer to a perfection which Spinoza sees as more active—and hence more powerfully free—a "surrender" to God and nature's way is in a sense the way to partake in that power:

"*The more perfection each thing has, the more it acts and the less it is acted on; and conversely, the more it acts, the more perfect it is*" (*Ethics*, II/306).

Also:

"*God loves himself with an infinite intellectual love*" (*Ethics*, II/302).

And:

"*The mind's intellectual love of God is the very love of God by which God loves himself, not insofar as he is infinite, but insofar as he can be explained by the human mind's essence, considered under a species of eternity; that is, the mind's intellectual love of God is part of the infinite love by which God loves himself*" (*Ethics*, II/302).

Such seems to suggest not becoming one with God / nature subjectively though, but possibly to approach the very "intellectual" laws of the world as we know it: if one would say that God expresses nature through the laws of physics, that the power of determination is a physical process, becoming one with nature is a way to be part of that determining physical process, and hence identifying with the power of the universe. But intellect for Spinoza may surpass what we understand as physics and even mathematics or geometry—as the mind is eternal:

"He who has a body capable of great many things has a mind whose greatest part is eternal" (*Ethics*, II/304).

LEIBNIZ: PRINCIPLES, MONADS & THE ALPHABET OF THOUGHT

"To see a world in a grain of sand,
And a heaven in a wild flower,

Hold infinity in the palm of your hand,
And eternity in an hour" (from William Blake, "Auguries of Innocence," 1803).

If ever there was a philosopher that William Blake's 19th century poem "Auguries of Innocence," or at least its opening sentence quoted above, is relevant to, it would be the 17th-18th century philosopher Gottfried Wilhelm Leibniz (1646-1716). For Leibniz's theory of "monads" says that the whole of existence is reflected in each of its parts, and that each part contains its entire history within itself. But Leibniz's philosophy was also scorned by Voltaire, in the 1759 satirical novella *Candide*, where Voltaire, through illustrating the world's sufferings, suggests dropping the "Leibnizian optimism" that we live in the "best of all possible worlds," in favor of a wisdom where "we must cultivate our own garden." It is probably a good thing that great writers and poets took up Leibniz's themes, as his own writing is often logically terse. He was something of an eccentric polymath, with a vast array of writings spread over numerous letters, essays, and a few books; and with his having discovered calculus independently of Newton, having engineered an early mechanical calculator, having been an early student of binary numbers, and also having developed an idiosyncratically unique rationalist philosophy drawn from his first principles.

As with Descartes, who built his philosophy on the foundation of what was implied in the ego's indubitably deduced existence "I think therefore I am" (and a conception of God as perfect); and as with Spinoza who demonstrated or proved that his metaphysics was implied by a perfect God or nature as only one substance; Leibniz too starts with a few basic principles, that he shows to imply a whole metaphysical system. Two of his fundamental principles are:

1. The Principle of Identity / (Non-) Contradiction: A proposition cannot be both true and false: not both A and not A (a principle we saw with Aristotle as well).

2. The Principle of Sufficient Reason: Nothing is without reason or cause; that is, there is an explanation for everything; even if one cause leads to another, all the way to a prime mover.

Besides these two principles (identified by Leibniz as the foundation of reason, see *The Monadology,* sections 31-32, in G.W. Leibniz, Roger Ariew and Daniel Garber (eds. & trans.), *Philosophical Essays*, Hackett Publishing Company: Indianapolis (1989), p. 217), Leibniz claims:

> "There also two kinds of *truths*, those of *reasoning* and those of *fact*" (Leibniz, p. 217; *The Monadology*, section 33).

Leibniz also finds there are "simple ideas" and "primitive principles" that are self-evident in contrast to the contingent truths of fact. These principles include:

1. The Principle of Predicate-in-Notion: What is said of a subject is a property of that subject, and is actually innate or implied by the very idea of that subject. E.g. rationality is implied by humanness, if humans are defined as rational animals.

2. The Principle of the Identity of Indiscernibles: No two things have everything in common—yet although space and time cannot differentiate two things with identical properties—if something is in a different place or time compared with another thing with otherwise identical properties, it is still a different thing: e.g. two lions are not the same, but are unique creatures.

3. The Principle of Continuity: There are no great leaps in the continuum of nature—as with the principle of sufficient reason, nothing comes out of nowhere; and changes occur over an infinitely dividable number of intervals.

4. The Principle of Best / Optimism: God, as the "ultimate reason of things" (*Monadology*, section 38) always acts for the best. We live in the best of all possible worlds.

5. The Principle of Plentitude: Everything that can happen will happen, eventually.

6. The Principle of Pre-established Harmony: Everything is independent, yet internally interconnected and synchronized for Leibniz—despite the autonomy of each thing, or monad, each monad internalizes the whole, and acts in harmony with every other monad which also internalizes the whole. God guarantees this.

As we are summarizing the philosophy of Leibniz here, I will include at length Leibniz's own summaries of the 37 sections in his 1686 *Discourse on Metaphysics*:

"1. On divine perfection, and that God does everything in the most desirable way.

2. Against those who claim that that there is no goodness in God's works, or that the rules of goodness and beauty are arbitrary.

3. Against those who believe that god might have made things better.

4. That the love of God requires our complete satisfaction and acquiescence with respect to what he has done without our being quietists as a result.

5. What the rules of the perfection of divine conduct consist in, and that the simplicity of the ways is in balance with the richness of the effects.

6. God does nothing which is not orderly and it is not even possible to imagine events that are not regular.

7. That miracles conform to the general order, even though they may be contrary to the subordinate maxims; and about what God wills or permits by a general or particular volition.

8. To distinguish the actions of God from those of creatures we explain the notion of individual substance.

9. That each singular substance expresses the whole universe in its own way, and that all its events, together with all their circumstances and the whole sequence of external things, are included in its notion.

10. That the belief in substantial forms has some basis, but that these forms do not change anything in the phenomena and must not be used to explain particular effects.

11. That the thoughts of the theologians and philosophers who are called scholastics are not entirely to be disdained.

12. That notions involved in extension contain something imaginary and cannot constitute the substance of body.

13. Since the individual notion of each person includes once and for all everything that will ever happen to him, one sees in it the a priori proofs of the truth of each event, or why one happened rather than Another. But these truths, however certain, are nevertheless contingent, being based on the free will of God or his creatures, whose choice always has its reasons, which incline without necessitating.

14. God produces various substances according to the different views he has of the universe, and through God's intervention the proper nature of each substance brings it about that what happens to one corresponds with what happens to all the others, without their acting upon one another directly.

15. The Action of one finite substance on another consists only in the increase of degree of its expression together with the diminution of the expression of the other, insofar as God requires them to accommodate themselves to one another.

16. God's extraordinary concourse is included in that which our essence expresses, for this expression extends to everything. But this concourse surpasses the powers of our nature or our distinct expression, which is finite and follows certain subordinate maxims.

17. An example of the subordinate maxim, or law of nature; in which it is shown, against the Cartesians and many others, that God always conserves the same force but not the same quantity of motion.

18. The distinction between force and quantity of motion is important, among other reasons, for judging that one must have recourse to metaphysical considerations distinct from extension in order to explain the phenomena of bodies.

19. The utility of final causes in physics.

20. A noteworthy passage by Socrates in Plato against the philosophers who are overly materialistic.

21. If mechanical rules depended only on geometry without metaphysics, the phenomena would be entirely different.

22. Reconciliation of two ways of explaining things, by final causes and by efficient causes, in order to satisfy both those who explain nature mechanically and those who have recourse to incorporeal nature.

23. To return to immaterial substances, we explain how God acts on the understanding of minds and whether we always have the idea of that about which we think.

24. What is clear or obscure, distinct or confused, adequate and intuitive or suppositive [symbolic] knowledge; nominal, real, causal, and essential definition.

25. In what case our knowledge is joined to the contemplation of the idea.

26. That we have all ideas in us; and of Plato's doctrine of reminiscence.

27. How our soul can be compared to empty tablets and how our notions come from the senses.

28. God alone is the immediate object of our perceptions, which exist outside of us, and he alone is our light.

29. Yet we think immediately through our own ideas and not through those of God.

30. How God inclines our soul without necessitating it; that we do not have the right to complain and that we must not ask why Judas sins but only why Judas the sinner is admitted to existence in preference to some other possible persons. On original imperfection before sin and the degrees of grace.

31. On the motives of election, on faith foreseen, on middle knowledge, on the absolute decree and that it all reduces to the reason why God has chosen for existence such a possible person whose notion includes just such a sequence of graces and free acts; this puts an end to all difficulties at once.

32. The utility of these principles in matters of piety and religion.

33. Explanation of the union of soul and body, a matter which has been considered as inexplicable or miraculous, and on the origin of confused perceptions.

34. On the difference between minds and other substances, souls or substantial forms, and that the immortality required includes memory.

35. The excellence of minds and that God considers them preferable to other creatures. That minds express God rather than the world, but that the other substances express the world rather than God.

36. God is the monarch of the most perfect republic, composed of all minds, and the happiness of his city of God is his principal purpose.

37. Jesus Christ has revealed to men the mystery and admirable laws of the kingdom of heaven and the greatness of the supreme happiness that God prepares for those who love him" (Leibniz, pp. 35-62; headings only, with italics and capitalizations omitted).

With the above section summaries, we can see that Leibniz developed the notion of singular substances: a notion transformed from Aristotle's notion of substances, and which Leibniz later called "monads." Leibniz argues that every thing in existence is a hermetically sealed world unto itself—the whole world is in fact represented in each monad just as each monad is a thing "in" the world. Yet, these monads are immaterial and timeless: "metaphysical points" (Copleston, p. 297); not at all like physical atoms, and more like ideas, or even souls, as each has perceptions (their representations of the world) and appetites (possibly influenced by Spinoza's ideas concerning motivations, but more like an inner tendency or force). Monads do not have parts or act on one another, but Leibniz writes of "aggregates" of monads, "composite substances" and "organic machines":

"If you take mass to be an aggregate containing many substances, you can, however, conceive in it one substance that is preeminent, if that mass makes up an organic body, animated by its primary entelechy [its inner tendency or force]. Furthermore, along with the entelechy, I don't put anything into the monad or the complete simple substance, but the primitive passive force, a force corresponding to the whole mass of the organic body. The remaining subordinate monads placed in the organs don't constitute a part of the substance, but yet they are immediately required for it, and they come together with the primary monad in a corporeal substance, that is, in an animal or plant. Therefore I distinguish: (1) the primitive entelechy or soul; (2) the matter, namely, the primary matter or primitive passive power; (3) the monad made up of these two things; (4) the mass or secondary matter, or organic machine in which innumerable subordinate monads come together; and (5) the animal, that is, the corporeal substance, which the dominating monad in the machine makes one" (Leibniz, p. 177 – "From the Letters to De Volder").

Moreover:

"Since monads have no parts, they can neither be formed nor destroyed [....] a monad, in itself and at a moment, can be distinguished from another only by its internal qualities and actions, which can be nothing but its *perceptions* (that is, the representation of the composite, or what is external, in the simple) and its *appetitions* (that is, its tendencies to go from one perception to another) which are the principles of change [....] in a *center* or point, though entirely simple, we find an infinity of angles formed by the lines that meet there [....] each distinct simple substance or monad, which makes up the center of a composite substance (an animal, for example) and is the principle of its unity, is surrounded by a *mass* composed of an infinity of other monads, which constitute the *body belonging* to this central monad, through whose

> properties [*affections*] the monad represents the things outside it, similarly to the way a center does. And this body is organic when it forms a kind of automaton or natural machine, which is not only a machine as a whole, but also in its smallest distinguishable parts [....] each monad is a living mirror or a mirror endowed with internal action, which represents the universe from its own point of view and is as ordered as the universe itself" (Leibniz, p. 207 – "Principles of Nature and Grace, Based on Reason").

Again, every monad has a nature or aim that outside of time has already been fulfilled—and since each monad mirrors the universe inside itself, the whole history of the universe is within each monad. But unlike atoms, and more like ideas, monads can belong together under a "dominating monad" or "central monad" which makes an animal or plant an organic unity. Leibniz also sees a difference between the subjective perception that all monads have, the self-awareness of animal souls, and the true reasoning of rational animal minds:

> "since I conceive that other beings can also have the right to say 'I', or that it can be said for them, it is through this that I conceive what is called *substance* in general" (Leibniz, p. 188 – "Letter to Queen Sophie Charlotte of Prussia");

> "when a monad has organs that are adjusted in such a way that, through them, there is contrast and distinction among the impressions they receive, and consequently contrast and distinction in the perceptions that represent them [...] then this may amount to *sensation*, that is to a perception accompanied by *memory*—a perception of which there remains an echo long enough to make itself heard on occasion. Such a living thing is called *animal*, as its monad is called a *soul*" (Leibniz, p. 208 – "Principles of Nature and Grace, Based on Reason").

> "Thus it is good to distinguish between *perception*, which is the internal state of the monad representing external things, and *apperception*, which is *consciousness*, or the reflective knowledge of this internal state" (Leibniz, p. 208 – "Principles of Nature and Grace, Based on Reason").

> "Cartesians [...] believe that only minds are monads, and that there are no souls in beasts" (Leibniz, p. 208 – "Principles of Nature and Grace, Based on Reason");

> "*true reasoning* depends on necessary and eternal truths, such as those of logic, numbers, and geometry, which bring about an indubitable connection of ideas and infallible consequences. Animals in which these consequences are not noticed are called *beasts*; but those who know these necessary truths are those that are properly called *rational animals*, and their souls are called *minds*" (Leibniz, p. 209 – "Principles of Nature and Grace, Based on Reason").

Here we see that the subjectivity of all monads does not include the sensation and memory that animals have, and that "beasts," although having a soul, do not have reason. Monads are like ideas having a subjective existence with the whole mirrored in each (somewhat like holographic images)—a sort of inverted context. Like ideas, they can come in organized hierarchies and have no extension; like human subjects, they have aims and represent their environment—yet they are also timeless and hence their aims have ultimately already been played out. A hierarchy of mentality is preserved, as things have perceptions and appetites, animals have sensations, memory, and souls, and humans have rational minds directly connected with God.

Such brings us back to Leibniz's principles that ground our reasoning. But there is more than conscious reasoning at work—more to our thought than we are consciously aware, as Leibniz anticipates the work of Freud and others on the unconscious:

"at every moment there is an infinity of perceptions in us, but without apperception and without reflection—that is changes in the soul itself, which we do not consciously perceive, because these impressions are either too small or too numerous or too homogeneous, in the sense that they have nothing sufficiently distinct in themselves; but combined with others, they do have their effect and make themselves felt in the assemblage, at least confusedly. It is in this way that custom makes us ignore the motion of a mill or a waterfall, after we have lived nearby for some time [....] All attention requires memory, and when we are not alerted, so to speak, to pay heed to some of our own present perceptions, we let them pass without reflection and without even noticing them" (Leibniz, p. 205 – "Preface to the New Essays").

"It can even be said that as a result of these tiny perceptions, the present is filled with the future and laden with the past, that everything conspires together (*sympnoia panta*, as Hippocrates said), and that the eyes as piercing as those of God could read the whole sequence of the universe in the smallest of substances [....] These insensible perceptions also indicate and constitute the individual, which is individuated by the traces which these perceptions preserve of previous states, connecting it up with his present state [....] it is these tiny perceptions which determine us in many situations without our thinking of them" (Leibniz, p. 296 – "Preface to the New Essays").

"In short, *insensible perceptions* have as much use in the philosophy of mind as corpuscles do in physics; and it is equally unreasonable to reject the one as the other under the pretext that they are beyond the reach of the senses. Nothing takes place all at once, and it is one of the greatest and best verified maxims that *nature never makes leaps*; this is what I call *the law of continuity*" (Leibniz, p. 297 – "Preface to the New Essays").

These "insensible perceptions" seem mostly to reference individuals' interactions with the world—what it is possible to sense, but which we do not pay attention to. But given that the two basic attributes of monads are representative perception and natural tendencies or appetites—it would not be too far of a stretch to see certain appetites as unconscious too: leading to Freudian ideas like libido, etc.

On the flip side of Leibniz clearing ground for the notion of the unconscious, we find his early logical work on a "Universal Characteristic" or alphabet of thought:

> "From the time of Pythagoras, people have been persuaded that enormous mysteries lie hidden in numbers. And it is plausible that Pythagoras brought this opinion to Greece from the Orient, as he did many other opinions. But since they lacked the true key to this secret, the more inquisitive slipped into futility and superstition. From this arose a sort of vulgar Cabbala (a Cabbala far distant from the true one) [....] Meanwhile, people have retained their inherent ability to believe that astonishing things can be discovered through numbers, characters, and through a certain new language that some people call the Adamic language, and Jacob Böhme calls the 'natural language'" (Leibniz, p. 5 – "Preface to a Universal Characteristic");
>
> "no one has put forward a language or characteristic which embodies, at the same time, both the art of discovery and the art of judgment, that is, a language whose marks or characters perform the same task as arithmetic marks do for numbers and algebraic marks do for magnitudes considered abstractly. And yet, when God bestowed these two sciences on the human race, it seems that he wanted to suggest to us that a much greater secret lies hidden in our intellect, a secret of which these two sciences are but shadows" (Leibniz, p. 6 – "Preface to a Universal Characteristic");

> "we must go beyond words. Since, due to the wonderful interconnection of things, it is extremely difficult to produce the characteristic numbers of just a few things, considered apart from the others, I have contrived a device, quite elegant, if I am not mistaken, by which I can show that it is possible to corroborate reasoning through numbers [....] By using these numbers I can immediately demonstrate through numbers, and in an amazing way, all the logical rules and show how one can know whether certain arguments are in proper form. When we have the true characteristic numbers of things, then at last, without any mental effort or danger of error, we will be able to judge whether arguments are indeed materially sound and draw the right conclusions" (Leibniz, pp. 9-10 – "Preface to a Universal Characteristic").

Clearly, here Leibniz was on the way to organizing language towards a formal logic. And given that he designed a mechanical calculator—we can see that he was moving towards the entire idea of mechanizing reason: making Leibniz one of the earliest thinkers of symbolic artificial intelligence.

Leibniz's separation of necessary truths of reason and contingent truths of fact roughly splits the emphasis of rationalists versus empiricists—rationalists find that higher knowledge comes innately, much as with Plato's theory of reminiscence. Empiricists on the other hand find knowledge comes from our experience in the world.

Mrs. Lavery Sketching - John Lavery (1910)

Chapter 11

Empiricists
Contingently Necessary

LOCKE: A BLANK SLATE IN A BLACK BOX

In his *An Essay Concerning Human Understanding*, John Locke (1632-1704) immediately differentiates between his perspective, which has come to be known as "Empiricist" and the philosophy of some of his peers called "Rationalist." The lines are drawn between the two based on the notion of "innate ideas," which Locke rejects. In the previous chapter we saw that Spinoza and Leibniz developed their philosophies from first principles that were considered to be self-evident, much like proofs based on the axioms of geometry. Descartes, as well, reconstructs the real world not simply from the self-certitude of the "I think," but also from a series of implications derived from God being perfect, and what that perfection entails. As we will see, Locke does not so much as deny self-evident principles, but claims instead that we should reject the notion that they come from within our souls... the "innate ideas" view going back at least to Plato, with his theory of recollection, whereby we "re-call" knowledge from a prior, higher existence. Instead, and much in line with the scientific revolution of his time, e.g. Isaac Newton (1642-1727), where scientists sought to explain as much as could be explained in reference to mechanics, mathematics and induction from experimental evidence (an advance we saw with Roger Bacon, but also furthered with Francis Bacon's (1561-1626) *Novum Organum Scientiarum* or "New Instrument of Science"), Locke looked towards our experience of the world for knowledge (note the connection between the words "experiment" and "experience.")

INTUITIVE, NOT INNATE IDEAS

Here, in a rough summary of his view of how we come to ideas, Locke notes that no innate ideas are necessary:

> "The Senses at first let in particular *Ideas*, and furnish the yet empty Cabinet: And the Mind by degrees growing familiar with some of them, they are lodged in the Memory, and Names got to them. Afterwards the Mind proceeding farther, abstracts them, and by Degrees learns the use of general Names. In this manner the Mind comes to be furnish'd with *Ideas* and Language, the Materials about which to exercise its discursive Faculty. And the use of Reason becomes daily more visible, as these Materials, that give it Employment, increase. But though the having of general *Ideas*, and the use of general Words and Reason usually grow together: yet, I see not, how this any way proves them innate" (John Locke, Peter H. Nidditch (ed.), *An Essay Concerning Human Understanding*, Oxford University Press: Oxford (1979), Book I, Chap. II, Section 15).

The term "idea" used by Locke includes both of what are often currently termed *perceptions* and *concepts*. Locke argues that even ideas like that of the law of non-contradiction (not both A and not-A) or of elementary mathematics are not universally acknowledged, and hence not innate, as they do not arise spontaneously in children or "idiots":

> "For although a Child quickly assent to this Proposition, *That an Apple is not Fire*; when by familiar Acquaintance,

> he has got the *Ideas* of the two different things distinctly imprinted on his Mind, and has learnt that the Names *Apple* and *Fire* stand for them: yet it will be some years after, perhaps, before the same Child will assent to this Proposition, *That it is impossible for the same thing to be, and not to be*" (Locke, Book I, Chap. II, Section 23).

However, note that Locke does not deny the self-evidence of intuitive knowledge:

> "For if we will reflect on our own ways of Thinking, we shall find, that sometimes the Mind perceives the Agreement or Disagreement of two *Ideas* immediately by themselves, without the intervention of any other: And this, I think, we may call *intuitive Knowledge*. For in this, the Mind is at no pains of proving or examining, but perceives the Truth, as the Eye doth light, only by being directed toward it. Thus the Mind perceives, that *White* is not *Black*, That a *Circle* is not a *Triangle*, That *Three* are more than *Two*, and equal to *One* and *Two*. Such kind of Truths, the Mind perceives at the first sight of the *Ideas* together, by bare *Intuition*, without the intervention of other *Ideas*; and this kind of knowledge is the clearest, and the most certain, that humane Frailty is capable of [....] 'Tis on this *Intuition*, that depends all the Certainty and Evidence of all our Knowledge" (Locke, Book IV, Chap. II, Section 1).

The distinction between innate ideas and intuitions is that innate ideas are supposed to exist prior to all conceivable experience and come from within our minds, whereas Locke finds intuitions to be a seeing the world and mind as they are: a recognition rather than a recollection. For although Locke argues that there are no innate ideas, he does argue that we humans are born with various mental faculties—much like those discussed by Descartes.

THE DARK ROOM OF IDEAS

Although not divided in such a way, Book II of *An Essay Concerning Human Understanding* focuses on three major areas: 1) the perceiving and thinking self, 2) the physical world, and 3) the principles of a thinkable reality. Bridging all three realms is the idea of ideas. Ideas for Lock start with perceptions; they originate with experience, or more specifically, with sensation; but we can also reflect on these ideas, via the "*Perception of the Operations of our own Minds* within us" (Locke, Book II, Chap. I, Section 4). Locke finds, like Descartes, that we must be self-reliant when it comes to real knowledge:

> "Consciousness is the perception of what passes in a Man's own mind. Can another Man perceive, that I am conscious of any thing, when I perceive it not my self? No Man's knowledge here, can go beyond his Experience" (Locke, Book II, Chap. I, Section 19).

The concept of "consciousness" broached here is situated between the passive observing of Descartes thinking "I" and the later, more technical concept of "apperception" developed by Leibniz and Kant. For Locke, one is conscious of both an external world through the senses and also of the "operations" of the mind with a sort of introspection whereby our ability to reflect upon our experiences is experienced in itself. On the ideas coming from only one of the five senses:

> "There are *some* Ideas, *which have admittance only through one Sense*, which is peculiarly adapted to receive them. Thus Light and Colours, as white, red, yellow, blue; with their several Degrees or Shades, and Mixtures, as Green, Scarlet, Purple, Sea-green, and the rest, come in only by the Eyes: All kinds of Noises, Sounds, and Tones only by the Ears: The several Tastes and Smells, by the Nose and Palate. And if these Organs, or the Nerves which are the Conduits, to

convey them from without to their Audience in the Brain, the mind's Presence-room (as I may so call it) are any of them so disordered, as not to perform their Functions, they have no Postern to be admitted by; no other way to bring themselves into view, and be perceived by the Understanding" (Locke, Book II, Chap. III, Section 1).

Evidently through introspection Locke himself was able to discern various powers or faculties of the mind. And in the mind (and most probably also from the history of philosophy) Locke finds ideas; ideas here numbered by the chapters found in Book II of Locke's *Essay*:

1. Ideas (perceptions, thoughts, notions)

2. Simple Ideas (via sensations and reflections)

3. Ideas from one sense (from each of the five senses, colors, sounds, etc.)

4. Solidity (from the sense of touch)

5. Simple ideas of diverse senses (space, figure, rest, motion)

6. Simple ideas of reflection (understanding, will and other faculties: memory, reason, judging, etc.)

7. Simple ideas of both sensation and reflection (pleasure, pain, power, existence, succession, unity).

8. Primary and Secondary Qualities (primary extension vs. secondary color, sounds, etc.)

9. Perception (first faculty of the mind: impressions from sensation)

10. Retention (second faculty of the mind: memory)

11. Discerning (a faculty of comparing, judging, naming, abstracting, etc.)

12. Complex ideas (from simple ideas: Modes, Substances, and Relations)

13. Modes of space (distance, place, extended vacuum)

14. Modes of duration (succession from stream of consciousness, time)

15. Modes of space with time (belong to all beings, inseparable)

16. Modes of number (simplest idea, used to measure)

17. Modes of quantity: Infinity (only applies to number, space and time)

18. Modes of the senses (colors are blue, red, etc; motion is to roll, tumble, slide, etc.)

19. Modes of thinking (perceiving, remembering, contemplating, reverie, attention, etc.)

20. Modes of pleasure and pain (love, hate, desire, joy, etc.)

21. Modes of power (active and passive, liberty, freedom)

22. Mixed Modes (invented modes: hypocrisy, sacrilege, murder, etc.)

23. Substances (pure substance in general and particulars: gold, horse, bread, etc.)

24. Collective Substances (army, universe, constellation, etc.)

25. Relations that are objective (bigger, smaller, older, younger, etc.)

26. Relation of cause and effect (originating or created)

27. Relation of identity and diversity (self-same beings or a plurality)

28. Relations that are practical (morality and laws, good and evil)

29. Clarity of ideas (vs. confused ideas)

30. Real vs. Fantastical Ideas (found in nature / conform with being or not?)

31. Adequate vs. Inadequate ideas (archetypical, or only partly so?)

32. True and false ideas (judged by patterns of nature and real existence)

33. The association of ideas (possiblity of accidental connections and making up ideas)

Again, Locke claims:

> "That external and internal Sensation, are the only passages that I can find, of Knowledge, to the Understanding. These alone, as far as I can discover, are the Windows by which light is let into this *dark room*. For, methinks, the *Understanding* is not much unlike a Closet wholly shut from light, with only some little openings left, to let in external visible Resemblances, or *Ideas* of things without; would the Pictures coming into such dark Room but stay there, and lie so orderly as to be found upon occasion, it would very much resemble the Understanding of Man, in reference to all Objects of sight, and the *Ideas* of them" (Locke, Book II, Chap. XI, Section 17).

"OPERATIONS OF THE MIND"

Locke goes further than Descartes, when talking about the activity of the faculties, esp. regarding the Understanding. In Book II, Chapter XI, "Of Discerning, and other Operations of the Mind," Locke discuses the necessity of being able to distinguish among ideas (*discerning*), of being able to make sound *judgments* (in contrast to mere wit), of being able to *compare* ideas, to be able to put ideas together in *composition*, to be able to extrapolate or *enlarge*, and the ability to *abstract*:

> "the Mind makes the particular *Ideas*, received from particular Objects, to become general; which is done by considering them as they are in the Mind such Appearances, separate from all other Existences, and the circumstances of real Existences, as Time, Place, or any other concomitant *Ideas*. This is called ABSTRACTION, whereby *Ideas* taken from particular Beings, become general Representatives of all the same kind; and their Names general Names, applicable to whatever exists conformable to such abstract *Ideas* [....] Thus the same Colour being observed to day in Chalk or Snow, which the Mind yesterday received from Milk, it considers that Appearance alone, makes it a representative of all of that kind; and having given the name *Whiteness*, it by that sound signifies the same quality wheresoever to be imagin'd or met with; and thus Universals, whether *Ideas* or Terms, are made" (Locke, Book II, Chap. XI, Section 9).

Since Locke has no Platonic recollection to depend on to inspire the various ideas of the mind, he spells out how the mind comes up with ideas in more detail than his Rationalist peers—yet not in as much detail as later thinkers like Kant—in his explanation of abstraction, Locke is approaching Kant's theory of synthetic a priori judgments, where a general idea can be deduced from the complex of experience. A problem for understanding Locke is that one may get the impression of the idea of a "dog" being generated simply by a single instance of seeing a dog in the world. Not so, as Locke's explanation of abstraction shows—one must come across many instances of complex perception before abstracting a general idea (an idea as simple as "whiteness" for example). Yet there may be an unbreakable circularity concerning the difference between general ideas and simple ideas, as that very idea of "whiteness" that Locke deduced from abstraction, is also an idea he used to exemplify simple ideas:

> "Though the Qualities that affect our Senses, are, in the things themselves, so united and blended, that there is no separation, no distance between them; yet 'tis plain, the *Ideas* they produce in the Mind, enter by the Senses simple and unmixed. For though the Sight and Touch often take in from the same Object, at the same time, different *Ideas*; as a Man sees at once Motion and Colour; the Hand feels Softness and Warmth in the same piece of Wax: Yet the simple *Ideas* thus united in the same Subject, are as perfectly distinct, as those that come in by different senses. The coldness and hardness, which a Man feels in a piece of *Ice*, being as distinct *Ideas* in the Mind, as the Smell and Whiteness of a Lily; or as the taste of Sugar, and the smell of a Rose: And there is nothing can be plainer to a Man, than the clear and distinct Perception he has of those simple *Ideas*; which being each in it self uncompounded, contains in it nothing but *one uniform Appearance*, or Conception in the mind, and is not distinguishable into different *Ideas*" (Locke, Book II, Chap. II, Section 1).

How is it that "whiteness" is both a simple and a general idea? It may suffice to say that whiteness is both a percept and a concept—the percept is an immediately simple idea from a single instance and the concept is a general idea deduced from several instances. Since ideas are both percepts and concepts though, we may need to split hairs to maintain consistency in Locke's philosophy (even though

Locke, as just quoted, calls an "Appearance" a "Conception!") An additional angle towards seeing Locke as consistent would be to see that each and every appearance can be a separate idea. A key to understanding Locke may be seen in the quote above about abstraction where Locke talks of "*Ideas* or Terms." In Book III of his *Essay*, Locke discusses language with chapters on both general terms and names of simple ideas:

> "That then which general Words signify, is a sort of Things; and each of them does that, by being a sign of an abstract *Idea* in the mind, to which *Idea*, as Things existing are found to agree, so they come to be ranked under that name; or, which is all one, be of that sort. Whereby it is evident, that the *Essences of* the *sorts, or* (if the Latin word pleases better) *Species* of Things, are nothing else but these abstract *Ideas*" (Locke, Book III, Chap. III, Section 12).

And:

> "The *Names of simple* Ideas *and Substances*, with the abstract *Ideas* in the Mind, which they immediately signify, *intimate* also *some real Existence*, from which has derived their original pattern" (Locke, Book III, Chap. IV, Section 2).

"White" is both a general word (that is a sign of an abstract idea) and a name of a simple idea (which names an abstract idea *plus* the intimation of some experience). But how can we have an idea of "whiteness" if the general idea comes from diverse experiences of the simple idea, and the simple idea has not yet been distinguished as a general idea? In other words, how can we come to the idea of whiteness, if we don't already have an idea of whiteness? Why would we need a general idea of whiteness if we already had a simple idea of it? Innate theorists could simply state that simple ideas are innate, and that an experience of whiteness would trigger the recollection of the idea of whiteness. But for Locke, the experiences themselves are also ideas, albeit percepts and not concepts—so how can we begin to distinguish whiteness from blackness if we don't already have some notion that there is *some idea* to distinguish? What Locke fails to convey, following Descartes' self-reliance, is that we are taught these lessons, from other human beings—a child is shown various objects, with one property in common (e.g. a white ball, a white cube, and a white liquid) and TOLD that what they have in common is "white." It is in this way that we can see that, without innate ideas, we might not come up with even simple ideas *on our own*. So yes... we may be born with a "blank slate" replete of any ideas (this is still debated today); but even Locke would admit that we need certain in-born faculties allowing us to even think or learn—and moreover, we may need a culture of teachers and parents to start to teach us how to discriminate certain aspects of our experience from others. To his credit, Locke notes:

> "those who have Children, or the charge of their Education, would think it worth their while diligently to watch, and carefully to prevent the undue Connexion of *Ideas* in the Minds of young people" (Locke, Book II, Chap. XXXIII, Section 8).

Possibly there is some middle ground between getting ideas only from experience, or also from innate recollection: something like the perceptual ability to adjudicate colors is built into our sense-organs, brain, and world—this would be an innate-like organization of our senses—where simple ideas like "whiteness" are naturally perceived directly without necessarily being understood as an abstraction. But it seems to me that our first experiences would be complex, not simple—that Locke reverses the way we really come to ideas: the vast variety of concrete experience slowly gives way to more manageable generalizations. We don't begin with simple ideas; we arrive at them after much experience, learning, and discrimination.

Although Kant explains how we can make synthetic a priori judgments (i.e. come up with simple ideas from a complex of experiences—somewhat like Locke's abstraction), we will have to wait for Hegel to trace the chronology of discrimination in his dialectic: where one starts by discriminating between perceiver and perceived, then moves on to discriminate a perceived object from a perceived environment, then the properties of an object from the object itself, etc. Locke however, has made a monumental advance over his Rationalist peers, in that he *tries* to explain important aspects of cognition without recourse to the "magic" of Platonic recollection. The contemporary "nature-nurture" debate is somewhat different in that innate ideas are said to be hard-wired in the brain via DNA, not accessed through some God-given mental realm of ideas—with thinkers like Steven Pinker suggesting that education is not as powerful as some social-constructivists claim (to be discussed in a later chapter).

TAXONOMY OF COMPLEX IDEAS

Locke finds that unlike the passive experiencing of simple ideas, we also have the active ability to construct complex ideas. Locke makes a further taxonomy of ideas with three types of complex ideas: *modes, substances,* and *relations*. *Modes* are ways of being—for sensed ideas like colors this would include specifics like blue, red, yellow etc; modes of reflection include perceiving, remembering, etc; there are also emotional modes, modes of power, and finally, mixed modes—those human conventions like hypocrisy and sacrilege. Space, duration and number also have their modes. All these modes are the detailed or finer offshoots of the major categories of ideas. The complex idea of *substance* could be a general kind of stuff, a pure "matter" (that we could not know directly from experience!) or it could be a particular sort, like man, horse, water, etc, or collections of particulars, like an army or a constellation. *Relations,* the third type of complex idea, include comparisons of ideas that are: more or less, newer or older, cause or effect, the relation of a thing to itself (identity), or moral relations to law, good and evil, etc.

For Locke, complex ideas can be more or less clear, distinct, real, adequate, or true. We have the ability to make proper or improper associations between ideas; and such leads to the possibility of knowledge or the lack thereof. Since Locke has no recourse to a Platonically recollected knowledge, and all we have to work with are our ideas, the *agreement* of ideas leads to knowledge. What exactly are ideas to agree with? Each other, of course... but also with real existence, and our perception of the world. As noted earlier in this chapter, Locke rejects innate ideas, but retains the notion of *intuition*. This intuition includes recognition of necessary connections between ideas (e.g. that white cannot be black)—connections that we do not demonstrate, but immediately see—it is indubitable, unlike some notions that require a *demonstration* before doubt is removed by a proof, as with mathematics. A third kind of knowledge, besides *intuitive* and *demonstrative*, is *sensitive* knowledge, whereby we may know external objects:

> "There can be nothing more certain, than that the *Idea* we receive from an external Object is in our Minds; this is intuitive Knowledge. But whether there be any thing more than barely that *Idea* in our Minds, whether we can thence certainly inferr the existence of any thing without us, which corresponds to that *Idea*, is that whereof some Men think there may be a question made, because Men may have such *Ideas* in their Minds, when no such Thing exists, no such Object affects their Senses. But yet here, I think we are provided with an Evidence, that puts us past doubting: For I ask any one, Whether he be not invincibly conscious to himself of a different Perception, when he looks on the Sun by day, and thinks on it at night; when he actually tastes Wormwood, or smells a Rose, or only thinks on that Savour, or Odour? We as plainly find the difference there is between any *Idea* revived by our Minds by our own Memory, and actually coming into our Minds by our Senses, as we do between any two distinct *Ideas* [....] I think, we may add

to the two former sorts of *Knowledge*, this also, of the existence of particular external Objects, by that perception and Consciousness we have of the actual entrance of *Ideas* from them, and allow these *three degrees of Knowledge*, viz. *Intuitive, Demonstrative, and Sensitive*" (Locke, Book IV, Chap. II, Section 14).

THE MARK OF QUALITY

When discussing simple ideas of perception, Locke introduces the idea of "Quality"—qualities are in objects:

> "Whatsoever the Mind perceives in it self, or is the immediate object of Perception, Thought, or Understanding, that I call *Idea*; and the Power to produce any *Idea* in our mind, I call *Quality* of the Subject wherein that power is. Thus a Snow-ball having the power to produce in us those *Ideas* of *White*, *Cold*, and *Round*, the Powers to produce those *Ideas* in us, as they are in the Snow-ball, I call *Qualities*; and as they are Sensations, or Perceptions, in our Understandings, I call them *Ideas*: which *Ideas*, if I speak of sometimes, as in the things themselves, I would understood to mean those Qualities in the Objects which produce them in us" (Locke, Book II, Chap. VIII, Section 8).

These qualities can be either primary or secondary for Locke:

> "Such *Qualities*, which in truth are nothing in the Objects themselves, but Powers to produce various Sensations in us by their *primary Qualities*, *i.e.* by the Bulk, Figure, Texture, and Motion of their insensible parts, as Colours, Sounds Tasts, *etc.* These I call *secondary Qualities*" (Locke, Book II, Chap. VIII, Section 10).

Thus we have a division between qualities denoting shape and extension (primarily discerned from touch, but also from sight, etc.)—and qualities that produce sensations in general (colors, sounds, etc.)—but it is the primary quality... extension, or matter that is thought by Locke to cause the secondary qualities; making the secondary qualities like color, "epiphenomenal," to use a modern term. Even though an empiricist focused on experience, Locke finds that something beyond experience, the "insensible parts," produce our sensations. If one were to take empiricism farther, they might question this hierarchy of qualities, and this is exactly what one of Locke's foremost critics, Bishop George Berkeley, does.

BERKELEY: CONTEXTUAL SURFACES

Although he does develop some positive ideas, the philosophy of George Berkeley (1685-1753), is largely a critical project, aimed at Locke, Descartes, and a few others. Berkeley is especially critical of the notion of matter—and introduces the idea that "esse is percipi" (George Berkeley, David M. Armstrong (ed.), *Berkeley's Philosophical Writings*, Macmillan Publishing Co.: New York (1965), p. 62, "Principles of Human Knowledge," section 4)—or, "to be is to be perceived." Berkeley arrives at this conclusion through a thinking influenced by "Ockham's Razor" (which suggests that the simplest solution to a problem is most likely the correct one)—for if we have spirits that perceive, the perceptions themselves, why add matter as a cause, when we never perceive it directly? Although some more modern thinkers might question the notion of "spirits" in favor or "matter" as the anchor of being, one would be hard pressed to deny that we do *experience qualities*. Berkeley, relying solely on experience, claims we can do without the notion of matter—that it is an abstract invention that we have no evidence of. This notion, that there is no matter, is fleshed out in his works, with arguments against the existence of abstract ideas, and his theory of vision.

In his work, "Towards a New Theory of Vision," Berkeley claims:

> "a man born blind, being made to see, would at first have no ideas of distance by sight: the sun and stars, the remotest objects as well as the nearer, would all seem to be in his eye, or rather in his mind. The objects intromitted by sight would seem to him (as in truth they are) no other than a new set of thoughts or sensations, each whereof is as near to him as the perceptions of pain or pleasure, or the most inward passions of his soul. For, our judging objects perceived by sight to be at any distance, or without the mind is [...] entirely the effect of experience" (Berkeley, p.300, "Towards a New Theory of Vision," section 41).

What Berkeley notes here, is that our sensation, esp. vision, is always of the surface—we do not see beyond a surface; in fact, depth itself is something we must surmise. That we cannot see matter beyond our experience might be acceptable to Locke... but Berkeley goes further, and claims that even the sensation of touch is always only a surface experience—we never sense anything beyond a surface that could be matter—we simply have experiences of the surface itself, in the mind.

Moreover, Berkeley claims that abstract ideas do not exist either:

> "I own myself able to abstract in one sense, as when I consider some particular parts or qualities separated from others, with which, though they are united in some object, yet it is possible they may really exist without them. But I deny that I can abstract from one another, or conceive separately, those qualities which it is impossible should exist so separated; or that I can frame a general notion, by abstracting from particulars" (Berkeley, p. 49, "Principles of Human Knowledge," section 10).

And also:

> "Colour, figure, motion, extension, and the like, considered only as so many *sensations* in the mind, are perfectly known; there being nothing in them which is not perceived. But, if they are looked on as notes or images, referred to *things* or *archetypes existing without the mind*, then are we involved all in skepticism" (Berkeley, pp. 96-97, "The Principles of Human Knowledge," section 87).

It is the idea of things in general (like an archetypical bird, or bird prototype) that Berkeley questions, not ideas like "whiteness." Yet, although Berkeley claims one can't even imagine an abstract archetypical idea, we do have words that are general. The confusion arises, when the generality of words is mistaken for a generality in ideas themselves, which ideas must always be anchored in sensation.

But what is it that holds the world together in a consistent manner, if not things or matter? Berkeley claims we know of this consistency:

> "not by discovering any *necessary connexion* between our ideas, but only by the observation of the *settled laws* of nature; without which we should be all in uncertainty and confusion [....] this consistent uniform working [...] so evidently displays the Goodness and Wisdom of that Governing Spirit whose Will constitutes the laws of nature" (Berkeley, p. 73, "Principles of Human Knowledge," sections 31-32).

In other words, God, as a spirit, makes sure that, if a tree falls in the woods, and there is no one there to hear it, it does make a sound, at least as perceived by God. God becomes the perceiving anchor of ideas and sensations, rather than things and matter being anchors—a prototypical example of what the deconstructionist philosopher Jacques Derrida will later call a "transcendental signified"—here a stopping point that anchors reality with something unknowable—in this case, the unknowable matter displaced by God as spirit. But what would anchor our sensation, if not matter or God? Perhaps

Berkeley's philosophy without a God (or matter) to anchor reality would resemble a Buddhist philosophy, where our sensations are said to be "empty." And indeed, this meditative observing—throwing away philosophically dogmatic ideas in favor of simply *looking* at our perceived world, as it is, rather than as we have historically come to know it, is taken even further by David Hume, who even questions causality—extending Berkeley's critique of "necessary connexion."

HUME: KNOWLEDGE LIMITED BY THE MODEL OF KNOWING

David Hume (1711-1776) carries on and refines the empiricist tradition, giving an account of the organization of experience much in the way that Locke did, while being skeptical like Berkeley as to what we could actually come to know from experience—but Hume refines these trajectories, e.g. by splitting Locke's idea of ideas into *impressions* and *ideas* (where the ideas as copies of impressions held in memory or the imagination are less lucid and present than the immediate impressions themselves), and by taking Berkeley's critique of abstraction further (and noting that induction and the notion of cause and effect are not absolutely known either, but are rather useful habits of thought). Although Aristotle, Aquinas, and Descartes all forayed into examining the various "faculties" of mind (like memory, imagination, judgment, etc.)—it was not until the empiricists rejected Platonic recollection and innate knowledge of fundamental principles that philosophers began to relate the limitations of a knowledge based not only on experience, but on how intelligence and reason are understood. That is, without innate ideas, not only do we need to explain how we know what we know, but we begin to limit what is knowable based on what we can explain by way of our models. It is debatable, whether a lack in what is considered knowable is due to deficiencies in such models or due to our inability to really know some things.

At any rate, Hume again limits our ideas to being *copies* of our immediate impressions. These copies are held in (fallible) memory, and can be combined and rearranged by the imagination to make new ideas (like unicorns). Besides making connections with the imagination, we have the ability to observe associations in our ideas, limited by Hume to "RESEMBLANCE, CONTIGUITY in time or place, and CAUSE and EFFECT" (David Hume, V.C. Chappell (ed.), *The Philosophy of David Hume*, The Modern Library: New York (1963), p. 34, *A Treatise of Human Nature*, Book I, Part I, Section IV); moreover, on cause and effect, Hume claims:

> "There is no relation which produces a stronger connection in the fancy and makes one idea more readily recall another than the relation of cause and effect between their objects" (Hume, p. 34, *A Treatise of Human Nature*, Book I, Part I, Section IV).

Hume calls these observed associations

> "the principles of union or cohesion among our simple ideas [... that] in the imagination supply the place of that inseparable connection by which they are united in our memory. Here is a kind of ATTRACTION, which in the mental world will be found to have as extraordinary effects as in the natural, and to show itself in as many and as various forms. Its effects are everywhere conspicuous; but as to its causes, they are mostly unknown, and must be resolved into *original* qualities of human nature, which I pretend not to explain" (Hume, pp. 35-36, *A Treatise of Human Nature*, Book I, Part I, Section IV).

Like Locke, Hume claims that it is through the association of simple ideas that we make complex ideas—of which there are three types: relations, modes and substances. Relations are subdivided into seven types: resemblance, identity, space / time, quantity, quality, contrariety, and cause / effect. An eighth relation could have been "difference," but Hume claims this is rather a negation of relations—yet could be a difference in number or kind (Hume, p. 38, *A Treatise of Human Nature*, Book I, Part I, Section V). Of modes and substance:

> "The idea of a substance as well as that of a mode is nothing but a collection of simple ideas that are united by the imagination and have a particular name assigned them, by which we are able to recall, either to ourselves or others, that collection" (Hume, p. 39, *A Treatise of Human Nature*, Book I, Part I, Section VI).

A *substance* is seen as an "unknown *something*" (Hume, p. 39, *A Treatise of Human Nature, Book I,*, Part I, Section VI), that holds various qualities (as gold is yellow, heavy, etc); while *modes* (like dancing or beauty) are *ways* that ideas can be. Note that it is the imagination that Hume believes unites particular qualities into a whole thing. And following Berkeley, Hume believes that abstraction is possible only for words and not ideas, as words may be connected to various particular instances of experience, but ideas are always connected to singular experience. Such is not to throw out the "distinction" ability though; as with Locke's "abstraction," Hume believes we are able to identify particular qualities via various instances:

> "when a globe of white marble is presented, we receive only the impression of a white color disposed in a certain form, nor are we able to separate and distinguish the color from the form. But observing afterwards a globe of black marble and a cube of white, and comparing them with our former object, we find two separate resemblances in what formerly seemed, and really is, perfectly inseparable. After a little more practice of this kind, we begin to distinguish the figure from the color by a *distinction of reason*; that is, we consider the figure and color together, since they are in effect the same and undistinguishable, but still view them in dif-

> ferent aspects according to the resemblances of which they are susceptible" (Hume, p. 47, *A Treatise of Human Nature*, Book I, Part I, Section VII).

But again, we can talk of such aspects with words, but never actually perceive them in an isolated manner—much as we could never imagine a bird in the abstract—but only a typical bird like a Robin, which would not be an abstract idea, but a specific impression that is somewhat average among impressions. What we see here in Hume (and in Berkeley and Locke as well) is a distrust of the way we use language in favor of our direct perceptions. But as noted earlier, in my questioning of the ability to generalize simple ideas or impressions, such a perspective has trouble getting off the ground, without recourse to social instruction, as with parents and teachers (or, on the other hand, innate ideas—"social instruction" presents a sort of chicken and the egg problem: which came first—instruction or the ability to generalize simple ideas from a complex of experience; how did instruction begin if we didn't have some *social* knowledge to begin with—if language is taught, were did the "first word" come from?)

But what real knowledge (again, concerning our experience, not our inherited language, etc) can an individual come up with, given the powers of perceiving impressions, remembering copies, imagining unified combinations, and observing relations? Of the seven relations, Hume claims,

> "there remain only four which, depending solely upon ideas, can be the objects of knowledge and certainty. These four are *resemblance*, *contrariety*, *degrees in quality*, and *proportions in quantity or number*. Three of these relations [resemblance, contrariety, and degree of quality] are discoverable at first sight, and fall more properly under the province of intuition than demonstration" (Hume, p. 62, *A Treatise of Human Nature*, Book I, Part III, Section IV).

This leaves knowledge of quantity mostly unexplained (Hume finds geometry problematic, due to its connection to perceived ideas—a triangle is rarely geometrically perfect when perceived in the world, in contrast to the precision of algebra and arithmetic). In what has come to be known as "Hume's Fork," matters of observable fact (resemblance, etc.) are split from self-referential relations that hold only among ideas (algebra, etc, not relating to worldly facts, but relating numbers to numbers and operations, etc). But also, space / time, identity, and causal relations are a matter of contingency, rather than immediately perceivable as knowledge. That is, objects can change location in space / time (from here to there), change identity (a caterpillar becomes a butterfly), and even cause and effect may not be consistent: hence such knowledge is not absolute. Why is it that Hume claims that what we do not recognize in an instant, besides math, cannot be knowledge? It is not due to the fallibility of memory, but rather, due to the possibility of change in the future. It is with this insight—that probability from induction cannot be 100% certain, since an entire set of relevant circumstances involves a future that is as yet to be determined, and since we are making a prediction of the future based on the past rather than an observation of the future, we cannot be absolutely sure that the future will turn out in the same way as the past. Since predictive knowledge is based on past experience, and since we cannot experience the future, we cannot predict what will happen with absolute certainty—even that the sun will rise tomorrow (someday it will explode), or that the laws of physics, such as with causality, are un-revisable (what if our universe collided with another parallel universe with different physical laws?) No... what we have instead is a belief system built by past experience: our habits that pragmatically help us navigate our world with a modicum of success—but also beliefs and habits based on social customs and information relayed culturally. A question arises though: why is it that our simple ideas from impressions are unchangeable? Couldn't blue change to green someday? Remarkably, Hume does suppose that we could have an idea that is not from experience:

> "Suppose therefore a person to have enjoyed his sight for thirty years, and to have become perfectly well acquainted with colors of all kinds, excepting one particular shade of blue, for instance, which it never has been his fortune to meet with. Let all the different shades of that color except that single one be placed before him, descending gradually from the deepest to the lightest; it is plain that he will perceive a blank where that shade is wanting, and will be sensible that there is a greater distance in that place between the contiguous colors than in any other. Now I ask whether it is possible for him, from his own imagination, to supply this deficiency, and raise up to himself the idea of that particular shade though it had never been conveyed to him by his senses? I believe there are a few but will be of opinion that he can, and this may serve as a proof that the simple ideas are not always derived from correspondent impressions" (Hume, pp. 29-30, *A Treatise of Human Nature*, Book I, Part I, Section I).

What this exception demonstrates is that there are other ways to gain knowledge than direct experience... other operations of the reasoning mind than copying impressions in memory, comparing their relations of resemblance, etc, or even rational distinctions, faulty verbal abstractions, and dubious inductions. Hume's model of thinking has not accounted for all the ways the mind might operate, and hence with his dismissal of necessary connections (that any *connection* could be altered in an open future, albeit pragmatically we act day to day as if such consistency where reliable), misses the possibility of knowledge that is not derived directly from experience, but is implied in it—Hume does not read between the lines, as it were, and ask the question, which we will see Kant ask in the next chapter: not "what are necessary connections?" but "what is necessary for us to even make the connections in the first place?" Possibly a post-Kantian Hume would have questioned the consistency of immediate experience too. Hume does question the notion of a "self" (noting that all we observe is a bundle of perceptions, and never perceive the perceiver):

> "we [have no] idea of a *self* [....] For from what impression could this idea be derived?" (Hume, p. 173, *A Treatise of Human Nature*, Book I, Part III, Section VI).

Possibly something like a self is *implied* by our perceptions... or maybe Hume has made a Buddhist-like insight: beyond post-Berkeley empty perceptions—no ego as well. Yet Hume is no enemy of desire. Like the rationalist Spinoza, Hume discuses emotions and their relation to motivation; but unlike Spinoza, Hume finds our passions in line with reason—much as it is reasonable to fear pain and seek pleasure:

> "It is obvious that when we have the prospect of pain or pleasure from any object, we feel a consequent emotion of aversion or propensity, and are carried to avoid or embrace what will give us this uneasiness or satisfaction. It is also obvious that this emotion rests not here, but making us cast our view on every side, comprehends whatever objects are connected with its original one by the relation of cause and effect. Here then reasoning takes place to discover this relation, and according as our reasoning varies our actions receive a subsequent variation. But it is evident in this case that the impulse arises not from reason, but is only directed by it. It is from the prospect of pain or pleasure that the aversion or propensity arises towards any object, and these emotions extend themselves to the causes and effects of that object as they are pointed out to us by reason and experience" (Hume, p. 226, *A Treatise of Human Nature*, Book II, Part III, Section III).

It is in this way that Hume sees that the *passions direct reason* via emotions associated with experience, not the other way around as claimed by previous philosophers from the ancient Greeks to Spinoza and beyond:

> "Reason is and out only to be the slave of the passions" (Hume, p. 227, *A Treatise of Human Nature*, Book II, Part III, Section III).

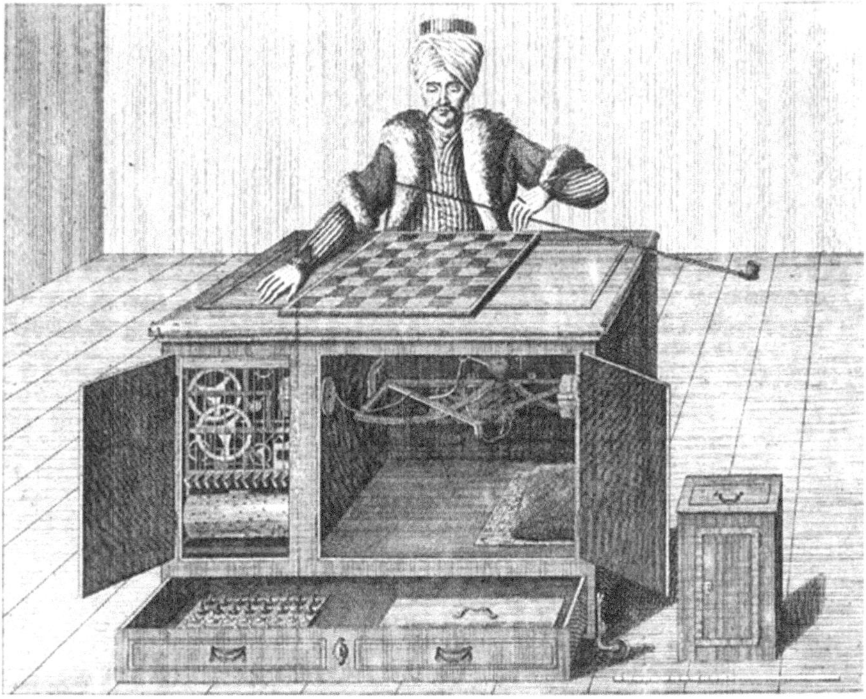

Illustration of the Kempelen Turk Chess Automaton (1789)

Chapter 12

Kant

Imagining Apperception

FACULTIES NECESSARY FOR COGNITIVE PERCEPTION

It would not be too far off the mark to suggest that in his *Critique of Pure Reason*, Immanuel Kant (1724-1804) implies a loose mechanism of perception and cognition. But even to use terms such as "mechanism" or "model" would do Kant's project an injustice, as he neither claims to even attempt this, nor that such would necessarily be possible. Kant does, however, attempt to separate what he considers to be various parts, or aspects of perception and cognition, and these various parts are said to interrelate, much as different parts of a mechanism might relate. But the various subdivisions of perception and cognition that Kant employs could in no way be said to generate perception and cognition, but should rather be viewed, as Kant would have it, as the absolutely *necessary* logical requirements for our ability to even think through our perception and cognition.

Thus we find Kant discussing Sensibility, Understanding, and other aspects (and subdivisions) of the mind that in total comprise a sort of technical description of the faculties we use in ordinary perceiving and thinking. A good bulk of his technical vocabulary is employed with special attention to the necessary relations between the specific faculties in the "Transcendental Deduction of the Pure Concepts of Understanding." It is in this section of the Critique that Kant claims to expose:

> "the a priori grounds of the possibility of experience" (A95 – citations from Immanuel Kant, Norman Kemp Smith (trans.), *Immanuel Kant's Critique of Pure Reason: Unabridged Edition*, St. Martin's Press: New York (1965)).

Which indeed is the very essence of the Transcendental project (in that the pure concepts deemed as necessary in this discussion indeed transcend all experience, yet are required for this experience to be intelligible).

It is my hope in this chapter to outline the major elements of the deduction and their necessary interrelations, in order to get at least a fuzzy grasp on what Kant considers to be the necessary grounds of experience. This discussion will include reference to Apperception, The Understanding, Imagination, Inner Sense, and The Intuition, which I hope to show constitute a sort of hierarchal relation between the perceiver (the "I") and the world of appearances. After such a broad (and, I must admit, overly ambitious) discussion, I hope to focus on what seems to be a problematic area for Kant—the Imagination, and what he holds as the highest faculty, Apperception. Indeed, it is the Imagination which mediates between the intelligible and the sensible, and thus it suggests a sort of bridge between two radically different aspects of perception; and it is Apperception which brings everything together.

OUR BASIC PHENOMENAL EXPERIENCE OF THE WORLD

Before diving into Kant's technical descriptions, however, it might be helpful to discuss some of the problems that naturally arise when we think about our perceiving. This preliminary discussion will not be too philosophically profound, yet it should help to set the stage for the sort of problems that Kant is treating.

Now, it would seem, first and foremost, that each of us finds ourselves as a perceiver in a world of perceptions. From one's earliest memories, one can note that there is a continual progression of sensations as one moves about the world, or as the world moves about one. For example, if you walk on a trail through the woods, it would seem as if sense perceptions were continually changing in a continuous manner such that your own location seems to be constant, or at least in a smooth movement from one location to another in time. In an analogous way, if one is recording such a walk with a film or video camera, the resulting footage would suggest that only one camera was used, and that it was focused on certain things and then directed towards others. The footage would not, as with certain television commercials, cut rapidly from one scene to another—the transitions would be smooth and coherent. Thus, by simply looking at the footage shot by a camera, one can hypothesize that only one camera was used in a continuous filming without a cut; and in a similar way, one could suggest by the continuity of one's own sense perceptions, that there is indeed only one perceiver, or "I," who is having those specific perceptions. So much, it seems, would be obvious—all of a person's perceptions are unified through a single perspective for a single perceiver.

Now we, as "single perceivers" find ourselves in an environment full of physical "things"—objects. These objects could be other people, animals, trees, or even parts of other objects, like hands and eyes. In each case one usually has little trouble in distinguishing one object from another; sometimes you can pick something up, a bottle for instance, and it seems to be a unified whole. There can be problematic cases though, such as when one thing turns out to be two things (what looked like one big tree turns out to be two entwined smaller ones). And there would also be cases where it would be difficult to distinguish a thing from the environment one found it in; such as a chameleon which changes colors to hide itself—seeing such a thing might require an acute focusing of one's attention.

The view just accounted thus describes a sort of perception where a perceiver finds themselves as the sole subject of a multiplicity of perceptions, perceptions often of objects (which might be related). Two types of singularity or unity have been mentioned—the singular unity of one's perspective as continuous through time, and the singular unity of objects which are distinguishable from all that are not those specific objects: a single subject perceiving single objects.

Returning to Kant's *Critique of Pure Reason*, it should be noted that Kant questions the absolute reality of objects (or, rather, our ability to know objects in-themselves), and the ultimate validity of our *peculiar* perspective (in space and time). Yet, such a "naïve" empirical situation certainly describes to some extent what we perceive given our position within the natural world. Thus, we have a rough sketch of experience in general.

STAGES OF PERCEPTUAL KNOWLEDGE: THE FACULTIES

I will now attempt to account for Kant's inquiry into the conditions he considers necessary for such experience to be possible. To do so we must first outline the various aspects of perception that Kant relies on to describe the mind; and foremost among these aspects we must distinguish the fundamental distinction between the Understanding and the Sensibility. Roughly, this split distinguishes the activity of knowing from being receptive to sensations. From our above discussion, we might suggest that objects are sensibly given to us through the Sensibility and that we as perceivers think about them via the Understanding. For Kant, the picture is not so clear though, as there are multiple stages of necessity between the "I" of the understanding perceiver and the given sensations. And, to even further complicate matters, we will be discussing two separate arguments for "the deduction of the pure concepts of understanding," from both the first and second editions of the *Critique*. The basic framework of experience described in both editions is similar though, and only a few distinctions between the second and first will be pointed out.

Throughout this discussion we must keep in mind that a major theme of the Transcendental Deduction concerns the possibility of having knowledge of empirical objects. What is knowledge for Kant? Most specifically Kant says,

> "knowledge is [essentially] a whole in which representations stand compared and connected" (A97).

And, concerning the specific knowledge that is of concern in the Transcendental Deduction, Kant says:

> "among the manifold concepts which form the highly complicated web of human knowledge, there are some which are marked out for pure a priori employment, in complete independence of all experience; and their right to be so employed always demands a deduction" (A85, B117).

Knowledge, for Kant, is what in modern terminology could be called a semantic network. Basically, it is number of different types of concepts with various types of connections: a "web" of connected concepts (for example the concept of "cinnabar" is connected with the concept "red"). Thus, an increase in knowledge may simply require the addition of a concept, or the connecting of concepts in a new way. And hence, we have the discussion of "combination" in the second deduction, which is an activity of the Understanding: the synthesis of new knowledge.

IMAGINATION CONSTRUCTS EXPERIENCE BY RULES OF THE UNDERSTANDING

Given the above remarks on the nature of knowledge, we must now begin to show how we can add to knowledge through perceiving the natural world. In the first deduction, Kant suggests that there are three phases in which our receptive sensibility can be actively incorporated into our web of knowledge, including:

> "the apprehension of representations as modifications of the mind in intuition, their reproduction in imagination, and their recognition in a concept" (A97).

Now, our mode of Intuition, which is our subjective Sensibility, offers, as it were, an appearance of an object which we sense, but do not know ("raw sense data.") And necessarily, the appearance given in the Intuition:

> "must... be run through and held together" (A99),

by the synthesis of apprehension—the given sensation (the manifold) must be a unified singularity.

Given this unified and singular manifold of sensation in the intuition, the Imagination must be able to reproduce the appearance according to a rule or rules:

> "[The] law of reproduction presupposes that appearances are themselves actually subject to such a rule, and that in the manifold of these representations a co-existence or sequence takes place in conformity with certain rules" (A100).

The rules in question here concern our knowledge-web of concepts. As opposed to the purely passive nature of our sensible intuition, imagination is an activity which (re)produces everything we can possibly know about what we perceive. The difference between the intuited manifold and the imagined manifold might provisionally be illustrated by the difference between a picture of a house and a drawing of it—indeed, the picture might contain microscopic details that one might never have noticed (and thus known) but the drawing would reflect only what we knew about the object (or its relation to other objects):

> "the affinity of appearances, and with it their association, and through this, in turn, their reproduction according to laws, and so [as involving various factors] experience itself, should only be possible by means of the transcendental function of imagination" (A123).

Of course, we do experience much before we seem to know anything (especially as children). There are rules for reproduction that the Imagination uses which are in the Understanding (implicitly), but not yet in our (explicit) knowledge—this will be discussed in more depth later.

PRELIMINARY DISCUSSION OF UNIFYING APPERCEPTION

Now, given the sensible unified manifold in intuition, which is reproducible via rules by the imagination, we must now discuss the Synthesis of Recognition in a Concept. This stage is tricky, because we must recognize the appearance of the manifold as an object, yet we do not actually perceive an object but rather a manifold of separate sensations—the object will thus be a hypothetical object = x:

> "since we have to deal only with the manifold of our representations and since that x (the object) which corresponds to them is nothing to us—being, as it is, something that has to be distinct from all our representations—the unity which the object makes necessary can be nothing else than the formal unity of consciousness in the synthesis of the manifold of representations" (A105).

This unity of consciousness is the singular and constant perspective which remains throughout all our perceptions—the "I" which is always the same subject; Kant declares:

> "This pure original unchangeable consciousness I shall name transcendental apperception" (A107).

Moreover:

> "This transcendental unity of apperception forms out of all possible appearances, which can stand alongside one another in one experience, a connection of all these representations according to laws" (A108).

Thus, this transcendental unity of apperception, the "I" that remains constant for all perceptions, is the single factor which unifies all experience, and makes such understandable (that is, inter-connectable within the web of knowledge) under concepts:

> "The unity of apperception in relation to the synthesis of imagination is the understanding" (A119).

In a strange way, this unity of consciousness somehow allows for the recognition of a manifold as singular object—Kant simply asserts:

> "this unity of possible consciousness also constitutes the form of all knowledge of objects: through it the manifold is thought as belonging to a single object" (A129).

The singular unity of an object is thus dependent on the unity of the "I" of our apperception.

The basic structure of experience explicated in the second (B) Transcendental Deduction is quite similar to the first (A)—it avoids some troublesome discussions of "objects" and gives a fuller explanation of apperception. Most importantly, there is a full discussion which explains that it is the "I" of apperception which allows for the comparison of multiple representations:

> "The thought that the representations given in intuition one and all belong to me, is therefore equivalent to the thought that I unite them in one self-consciousness, or can at least so unite them; and although this thought is not itself the consciousness of the synthesis of the representations, it presupposes that possibility of that synthesis" (B134).

Such a synthesis would also incorporate the transcendental synthesis of imagination. Also, rather than discussing the mysterious object = x as in the first deduction, Kant says in the second deduction:

> "knowledge consists in the determinate relations of the given representations to an object; and an object is that in the concept of which the manifold of a given intuition is united.... [and] it is the unity of consciousness that alone constitutes the relation of representations to an object" (B137).

Since we have seen that knowledge is simply a web of concepts, it might be suggested that an object is the presence in intuition of the criteria for a concept (its relation to other concepts) as recognized by the "I" of apperception via the structuring of the imagination (by rules of the understanding).

There are, of course, many subtleties in both the first and second deductions, and differences between the two which I have ignored for the sake of brevity. However, the overall general structure of experience and *recognition* remains similar in the two, and might be summarized as follows: *A raw appearance given in the intuition is structured by the imagination as governed by the rules of understanding and thought through the unity of apperception as an object.* I have developed this rather coarse over-generalization not as an end in itself, but rather as a way of situating a discussion of the imagination and apperception. As noted earlier, the imagination is critically placed in-between the active understanding and the receptive sensibility; and moreover, we have seen that it has the ability to structure our experience through a part of our understanding that we ourselves do not know! How the imagination functions within this crucial position will occupy the next section.

SPLIT BETWEEN CONSTRUCTION OF EXPERIENCE AND EXPLICIT KNOWLEDGE

In the above discussion we have surmised that the imagination is able to construct our experience of an appearance by certain laws that would be found in the understanding. In this way, far from developing our notion of an object from the accumulation of knowledge gained from outer appearances (as with empiricists) our experience of an appearance is predetermined by the structuring activity of our own understanding. In the second edition of the deduction, Kant claims:

> "The understanding, that is to say, in respect of the manifold which may be given to it in accordance with the form of sensible intuition, is able to determine sensibility inwardly. Thus the understanding, under the title of a transcendental synthesis of imagination, performs this act upon the passive subject, whose faculty it is, and we are therefore justified in saying that the inner sense is affected thereby" (B153-B154).

In our prior discussion, we noted that the imagination was the ability to structure an experience by way of rules of the understanding. The comparison of a drawing with a photograph illustrated the distinction between what could appear to us, and what we could construct with our imagination, again, via the rules of our understanding. Such conformed to the view that knowledge consisted of our accumulation of associated concepts—these associations that we learned could serve as rules for constructing an image (e.g., an "orange" could be associated with a "sphere" of a certain size and color, and could be drawn as such). However, it could also be noted that we definitely see things in complex ways that we might not be able to construct, nor be too articulate about—e.g. at a young age we see humans as being more complex than the stick figures we might be able to draw. Thus, it would seem that the imagination would draw upon some part of the understanding that was not part of our explicit and conscious knowledge; Kant confirms such in the first deduction:

> "Thus the understanding is something more than a power of formulating rules through comparison of appearances; it is itself the lawgiver of nature [....] All appearances, as possible experiences, thus lie a priori in the understanding, and receive from it their formal possibility" (A126-A127).

The understanding supplies the imagination with all the rules that it needs to construct our experience in all its full complexity!

There is therefore a split in the understanding, between that part which is:

> "the faculty of knowledge" (B137),

and that which via the categories:

> "grounds the possibility of all experience in general" (B167).

The imagination has the ability to draw upon both aspects of the understanding, which is distinguished by Kant in the second deduction:

> "In so far as imagination is spontaneity, I sometimes also entitle the productive imagination, to distinguish it from the reproductive imagination, whose synthesis is entirely subject to empirical laws, the laws, namely, of association" (B152).

The productive imagination therefore structures our experience in general, while the reproductive imagination can only be used to imagine images as limited by our explicit knowledge. So, with the photograph and drawing example, our experience of the photograph would be structured by productive imagination (down to the microscopic details) while our ability to construct a drawing would be limited by the powers of our reproductive imagination. With the afore mentioned split in understanding, the ability of productive imagination would therefore be to construct the entirety of all implicit human knowledge in an experience; which in turn would be the approachable limit of an individual's ability to use their explicit reproductive imagination: the limit of our obtainable knowledge is already contained in our understanding.

As the imagination structures our experience of the *sensible* by laws of the *understanding*, we can thus see how it mediates between these two realms. However, the constructions of the imagination always remain an activity of understanding, and therefore the imagination never really "touches" the sensible. In regard to this distinction, Kant suggests in the preface to the second edition:

> "In order to determine to which given intuitions objects outside me actually correspond, and which therefore belong to outer sense (to which, and not to the faculty of imagination, they are not to be ascribed), we must in each single case appeal to the rules according to which experience in general, even inner experience, is distinguished from imagination—the proposition that there is such a thing as outer experience being always presupposed" (Bxli).

However, since imagination represents the absolute limit of the understanding in its extension towards the sensible, we must conclude that our knowledge of outer sense can only be defined in the negative—as lying just beyond the grasp of imagination.

QUESTIONS CONCERNING THE EXTENT OF FOREKNOWLEDGE

Of course, there are some serious questions that arise for me concerning Kant's account of the grounds of experience. First and foremost, I find myself dissatisfied with the notion that all types of structuring of experience arise out of our own human understanding. It is one thing to say that our understanding shapes how we perceive things, but quite another to say the "arrangement" of things is determined by the understanding (which Kant seems to imply). In other words, what is the relationship between the understanding and how objects are related to each other in the world? If there are things in the world that have relationships to each other (e.g. Europe is north of Africa), if these sorts of relationships do occur outside the understanding—then how could they break into it? Would Kant suggest that even the whole factual arrangement of objects (past, present, and future) is already predetermined by the understanding?

Another serious problem concerns the ability of imagination to pick out one object from a manifold of the intuition. How is the imagination able to distinguish a chameleon from its surroundings? Through the unity of apperception? This sort of detail is inadequately dealt with in Kant's account of experience. Such criticisms do not deal a lethal blow to Kant's type of project though, as he was not attempting to build a workable model of the mind. I might suggest that he could have simply added a few more transcendental necessities.

APPERCEPTION: SOME HISTORY OF THE CONCEPT

> "...consciousness, as far as ever it can be extended, should it be to ages past, unites existences and actions very remote in time into the same person, as well as it does the exis-

> tence and actions of the immediately preceding moment, so that whatever has the consciousness of present and past actions is the same person to whom they both belong" —John Lock, Peter H. Nidditch (ed.), *Essay on Human Understanding*, Oxford University Press: Oxford (1979), p. 340, Book II, Chapter XXVII, Section 16).

There is a definite shift of emphasis in the Transcendental Deduction between the first and second edition of Kant's *Critique of Pure Reason*; and again, although this shift could not be called a complete shift from a concern with objects to a concern with the subject, one can easily notice that Kant abandons a certain lengthy discussion pertaining to the transcendental object = x (some unknown object on the other side of the "theater screen" of sensation) in favor of a more in depth discussion of apperception. Such a shift evidences Kant's high regard for this later concept, which at one point he calls:

> "the highest principle in the whole sphere of human knowledge" (B135).

It will be the aim of the second half of this chapter to elucidate this highest principle, and also to show some of the influences that previous rationalist philosophers had on Kant.

The first major philosopher to use the term "apperception," as far as I can trace, is Leibniz in his *New Essays on Human Understanding*, a work obviously influenced by Lock's own *Essay on Human Understanding* (note the above quote). The French word, "apercevoir," used by Leibniz, roughly means, "to catch sight of, to notice, to perceive, to understand, and to comprehend;" and it was used by him most specifically to distinguish the multiplicity and massive detail of our given perceptions from our limited ability to focus on particular parts of these perceptions. From my interpretive viewpoint, I would suggest that what Leibniz was pointing out with the concept of "apperception" was our ability to perform a pre-reflective recognition. Indeed, before we can reason about our perceptions, we need to be aware that we perceive something. Some critics, however, have suggested that apperception also includes a sort of self-consciousness, as one critic claims:

> "apperception is made to apply not only to awareness of perceptions but to awareness also of the I: 'The immediate apperception of our existence and our thoughts furnishes us with the first truths a posteriori, or of fact.' (NE, 4, 9, 2)" (quote from Robert McRae, *Leibniz: Perception, Apperception, & Thought*, University of Toronto Press: Toronto (1976), p.33).

In this way, one might discern hints also, of the Cartesian cogito, or thinking ego.

No doubt, Descartes' radical distinction between the perceiver and perceived in some way influenced Leibniz's, and subsequently Kant's use of the term "apperception." And in this way, we can see a correlation between Descartes' indubitable "I think" as the ground of all certainty, and Kant's use of apperception and its generated representation, the "I think," as the pivot of all understanding. Now, Descartes' use of the "I" to distinguish mind from sensible matter brings along with it not only subjectivity, but reason as well—mind, for Descartes includes both the subject and reason. However, although Kant claims that the:

> "faculty of apperception is the understanding itself" (B134n).

It should be noted that Kant often makes a distinction between apperception and the understanding. Indeed there is a sort of one dimensional scheme of faculties that might be listed as such:

Apperception	↓	(experience & judgment unity)
Understanding	↓	(web of connected concepts)
Imagination	↓	(construction of perceptions)
Inner Sense	◆	(theater screen of experience)
Intuition	↑	(indistinct experience notions)
Outer Sense	↑	(the world of "objects" we don't directly know)

A general working out of this scheme might, again, be as follows: an intuition is given in inner sense, which in turn is constructed as experience by the imagination; the imagination in turn, draws its rules of construction through the understanding; and the understanding presupposes, and uses the unity of apperception.

UNITY REMAINS THE SAME IN THE DIVERSE

Of what sort, then, is the unity of apperception? A clue to answering this question might be found in Kant's discussion of how we can become aware of this unity by considering its relationship to the diversity of representations that we are conscious of. Of this, Kant states:

> "Only in so far... as I can unite a manifold of given representations in one consciousness, is it possible for me to represent to myself the identity of the consciousness [i.e. throughout] these representations. In other words, the analytic unity of apperception is possible only under the presupposition of a certain synthetic unity" (B133).

Kant discusses this distinction between analytic unity and synthetic unity further in a footnote:

> "only by means of a presupposed possible synthetic unity can I represent to myself the analytic unity. A representation which is to be thought as common to different representations is regarded as belonging to such as have, in addition to it, also something different" (B133-B134n).

The point being made (in a footnote!) seems rather obscure, but nonetheless represents one of the most critical claims of the transcendental deduction. Following Locke's "abstraction" and Hume's "distinction," Kant is suggesting is that we receive no pure or distinct concepts in our experience of the world—our experiences are always a mixture of many ingredients, and this mixture would always precede our knowledge of any specific part. For example (following Kant's), we do not have a singular experience of the color red, from which we get the analytic conceptualization—"red" is the color red; but rather, we find red as one aspect among many in the variety of representations we encounter. In other words we get our ability to form the concept of red from our ability to distinguish this singular quality from different qualities in different representations (two objects share little other than the color red, and hence we can identify the color red as a distinct aspect knowable in itself: redness). In order to form this distinct concept of red, we must find that it is something which is the same in different representations, a connecting thread; and in each of these representations, red must be something distinct from the rest of the representation. Thus, there is a sort of "putting together" of certain aspects of distinct representations—a synthesis ("red" is something disparate objects have in common); and a synthesis which can "produce" something a priori, something true prior to experience (the concept "red" denotes redness, and redness exists as a possibility prior to experience). Hence we have something approximating the Kantian compromise between a priori innate rationalists, and synthetic experiential empiricists with synthetic a priori knowledge.

APPERCEPTION'S A PRIORI UNITY KNOWN VIA SYNTHESIZED EXPERIENCES

Now, Kant claims:

> "The first pure knowledge of understanding... upon which all the rest of its employment is based, and which also at the same time is completely independent of all conditions of sensible intuition, is the principle of the original synthetic unity of apperception" (B137).

For Kant, the most essential concept that must be discerned from experience (as being at least implied or presupposed in it) is the unity of apperception—however, it must also be noted that this apperception is required a priori for there to even be experience. Despite the special status of this concept, it is still synthesized in a way similar to the example using "red" above. In this case, however, what is seen to remain constant in a given diversity of representations is the singularity of the "I think" which can accompany them all. (B132) This point might be better illustrated by considering a given segment of film footage (as with the example of general experience given above). Now, if we view a film made by someone carrying a camera on a walk, the resulting footage would most likely evidence that only one camera was used. If there were no cuts from one perspective to another, the continuous flow from one view to another would make it obvious that indeed only one camera was used—even though we never actually see the camera itself in the film. In the same way, we can see that all our personal representations are not fragmented presentations for diverse perceivers, but that there is only one "I" which accompanies all of one's representations. In this way, we can see how Kant shows our knowledge of original apperception to be a synthesis—we are not simply given the knowledge of apperception, but must discern it from the diversity of our representations; and necessarily prior to this knowledge of the unity of apperception is the unity of synthesis. It is the unity of apperception which is necessary for the synthesis of representations to be possible, yet it is this very synthesis which gives evidence for there actually being an apperception: thus the knowledge of apperception is (loosely) synthetic a priori (this knowledge of a unity necessary prior to any knowledge at all is pulled together from a plurality of experiences).

With the above development, a definite problem can be discerned, as evidenced by Henry Allison's discussion of a point made by Dieter Henrich:

> "Henrich notes that, in his various formulations of the apperception principle, Kant sometimes refers to the unity of consciousness and sometimes to its numerical identity. He also points out that Kant himself distinguishes between these two conceptions, and he maintains that only the latter conception, construed as 'moderate' rather than 'strict' identity, ultimately provides the basis for the successful proof of the categories" (Henry Allison, *Kant's Transcendental Idealism an Interpretation and Defense*, Yale University Press: New Haven, CT (1983), p. 139).

The problem here seems to be that Kant often discusses apperception as if it referred to the subject (the "I think") of representations. However, any sort of logical "judgment" would not require an *empirical* subject (cp. the calculations of computers), but only the unity (numerical unity) of the concepts involved, and a common reference point—the *logical* "I;" each of which is guaranteed by apperception. It is this distinction, between a logical-transcendental necessity and the empirical unity of consciousness, which might be used to distinguish between the transcendental unity of apperception and empirical apperception. Of this distinction, Kant writes:

> "The transcendental unity of apperception is the unity through which all the manifold given in an intuition is united in a concept of the object. It is therefore entitled objec-

> tive, and must be distinguished from the subjective unity of consciousness.... the empirical unity of apperception... has only subjective validity" (B139-B140).

A question that arises for me here would be: how has Kant deduced a universal principle of transcendental apperception when the only singular unity that he can deduce is that of his own "I think." Has not Kant developed his entire argument concerning logical objectivity with reference only to his own subjectivity?—how is Kant certain that his logic is not only solipsistically his own? How did Kant know that the coherent logic implicit in his subjective experience was universal? In a footnote he writes:

> "that all the variety of empirical consciousness must be combined in one single self-consciousness, is the absolutely first and synthetic principle of our thought in general" (A117a).

Evading this rather thorny question, and turning to a new issue, it should be noted that the transcendental unity of apperception is necessary to discern objects in various representations. Kant claims:

> "it is the unity of consciousness that alone constitutes the relation of representations to an object." (B137)

This suggests a dilemma picked up on by Allison:

> "we can infer from the apperception principle that there can be no representation of objects apart from the unity of consciousness, because without such a unity there can be no representation of anything at all. It would seem, however, that we cannot similarly infer that whenever there is a unity of consciousness there is a representation of an object. Yet this is precisely what Kant appears to be claiming" (Allison, 146).

An important point to note is that it is apperception's activity of synthesis which unifies the various aspects of an object ("found" / "constructed" via schemas of the imagination) into a singularity. For example, given aspect cues such as fur, four legs, barking, etc, the unity of apperception can synthesize all these diverse aspects into the single concept, "dog" (just as it could draw particular occurrences of the color red into the singular concept of red). Thus, an object can only be thought through the unity of apperception. (This sort of notion, that the unity of the object requires the unity of a subject might have parallels with Descartes noting that the unity of an object beyond its appearances, e.g. wax, can only be discerned by the mind. See Descartes' *Meditation II*)

UNIFYING APPERCEPTION AND CONSTRUCTING IMAGINATION

In conclusion, I cite a general problem that I have with Kant's notion of synthesis, and a problem which thus concerns Kant's most fundamental theme of apperception—it is that the issue of memory is completely ignored. Kant often writes as if all the various components of a judgment are all present to the "I think" in a single instance. Such is not necessarily the case though, and I wonder how Kant would account for the combination of present and past representations—the use of apperception seems to be more than simply an act of unification, but also an act of remembering! However, although memory is not discussed explicitly, it may be implied by the activity of the imagination. It is these two concepts, unifying apperception and constructing imagination that form two major foci of Kant's Copernican Revolution, where the apperceptive subject becomes the new solar center of knowledge, rather than the earthly world of objects; due to the imagination's constructing (via the understanding), rather than reflecting our knowable experience. By connecting the double activity of the imagination (constructing our complex worldly experience through possibly unknown implicit rules of the understanding, and working with our explicitly known knowledge to think about worldly possibilities)—connecting this with the unifying action of apperception (where with the understanding, unified connections of diverse experiences are discerned) we may imagine that the implicit knowledge of the understanding that shapes our inner sense becomes explicit and hence workable by the imagination through constructing and connecting the similarities found in the plurality of experiences. Kant sees this new knowledge not to be entirely empirical and not only synthetic, in that what would be considered the knowable empirical world is already present in the understanding a priori. Kant falls somewhere between the parallel between Plato's theory of recollection and Chomsky's early notions of innate universal grammar in the brain—his Understanding has access to a priori knowledge that may be out of this world, like (but not necessarily identical to) Plato's forms, yet his faculties of logical perception and judgment resemble the compartmentalization of thought found in cognitive science. Do connected experiences trigger an "aha!" moment of learning an identifiable aspect of our environment that was always implicit in our understanding's construction of our perception of that environment prior to that moment? Maybe we could compromise with Kant, agreeing that some aspects of our engagement with the world are shaped by the structure of the understanding—a structuring of perception without which we could not perceive as we do—but holding out the possibility that the multitude of *facts* of an ever changing world plug into that fore-knowing structure. The core structure of the understanding might be flexible enough to negotiate a complex world of facts, events, ways of being, etc, all of which one would be hard pressed to say completely existed in an individual's ability to comprehend everything entirely. Unless, of course, knowledge and the real world are actually one and the same, and our particular perspectives fall short of that absolute effacement of the difference between all the known and all the unknown.

Saint Mark's Clock (1496)

Chapter 13

Hegel
Zodiac Robot

TWO "GENEALOGICAL" STRANDS OF *THE PHENOMENOLOGY OF MIND*

G.W.F. Hegel's (1770-1831) *Phenomenology of Mind* presents a rather imposing, yet profoundly enticing way of thinking; and beginning to read it may seem like diving into the deep end of a pool, with Hegel offering no lessons in treading water. I must admit that I have as yet to completely fathom its depths, but I have done a bit of dog-paddling here and there. Indeed, I have found some of the ideas presented in this work to have resonances with other thinkers, ways of thinking, and personal experiences; and to that extent, even though I do not feel completely at home in the work, I have not found myself completely alienated.

Two angles, from which I have approached this work in a struggling effort to "make it my own" as it were, are angles which I believe may have influenced Hegel himself in the outlaying of this book. Most critically would be Hegel's important predecessor Kant. Indeed, many of Hegel's notions seem to be reactions against, extensions of, and outgrowths of Kant's work. Such is especially evident in Hegel's almost overly technical vocabulary—a vocabulary that often echoes Kant's. One can also see Kant's influence, in that

Hegel's writing often seem to take for granted his reader's familiarity with the work of the former—it seems that Hegel saw Kant as breaking much ground which did not need to be re-worked in its entirety, but rather needed to be refined and re-situated within a larger framework.

The second angle, from which I hope to view The *Phenomenology of Mind*, regards astrology. This may sound a bit weird at first, as astrology has gained a reputation among many as being a charlatan endeavor. And indeed, Hegel mentions it in the *Phenomenology* only in connection with palm reading (p. 342, citations from G.W.F. Hegel, J.B. Baillie (trans.), *The Phenomenology of Mind*, Humanities Press Inc: New York (1977)), and his questioning of physiognomy and phrenology. He finds the claims made by the "science" of astrology troubling, for he sees the motions of the planets, etc, as having a completely exterior relation to the interiority of the psyche—and there seems to be no middle term to connect exterior astronomical events to the interior psychological attitudes. My purpose of including astrology here, though, is not to propose a defense of its validity—it should be noted however, that this "science" or "art" is indeed much more complex and intricate than most people assume, and has a long history. However, I do think that the structure of the Zodiac has strong reverberations with the cyclical progressive structure of the *Phenomenology*; the parallels found seem so deep that anyone having a familiarity with the two would be bound to recognize them. And it is my hope that, by elucidating some of these parallels, some of Hegel's insights and possible inspirations will become clearer.

I should note here that I will obviously not be covering the entire *Phenomenology* in depth, or even attempting to summarize it. In reference to Kant's work, I hope to focus mostly upon the first three of the *Phenomenology's* eight chapters—those concerned with Consciousness. Yet, even limiting one's self to these first moments, one can see the embryonic aspects of the work's progression;

and hopefully I will be able to explain how the later parts of the text expand some of these initial aspects (the whole being partially present in its initial parts), and also show how the works progression parallels the movement through the cycle of the Zodiac.

INSIDE KANT'S "ROBOT"

Kant's arguments and deductions in his *Critique of Pure Reason* undoubtedly constitute one of the most monumental advances in the logical analysis of the human mind. Indeed, it would seem that he thought he had truly found the foundational requirements for logical judgments to be made about the world. And in this inquiry into our ability to make sense of the world around us, he established a thorough explication of a (or *the*) logical structure of thinking. In his quest for discovering the possibility of synthetic a priori judgments, and through his Copernican revolution of de-centering one's mode of perceiving from perception itself, Kant unified the two distinct traditions of empiricism and rationalism in such a way as to clear the ground for new ways of thinking and questioning.

Kant did not answer each and every possible question concerning the mind though. And this is especially evident in one of the most slippery and instigating moves in philosophy—his choice to investigate what is necessary, rather than what would be sufficient for thought. (This latter objective, far from being achieved, has occupied cognitive scientists, especially those interested in artificial intelligence.) Through his discussion of various necessary faculties, such as the imagination, the understanding, and apperception, Kant did formulate a "workable" sketch (as opposed to, say, a full sculpture) of the logical operations of an individual mind. Kant's project was thus, in a way, an exposition of the form of an individual's mind.

Now, given Kant's thorough discussion of the mind's formal aspects, one might see much of Hegel's philosophy as being concerned with its content. Kant seems to have developed an empty shell, able to work with the world, but he rarely discusses what passes through that framework, besides classical philosophical problems. And indeed, Kant's discussion of what drives, and directs the mind, Reason, is quite limited, and focuses mostly on the needs of a logical progression (e.g. syllogisms).

If we direct our attention to the content of our minds, we can see that much of what is "in" it comes from what we learn and experience from "without." Our lives progress from infancy towards old age, and along that journey we mature and develop—the "content" of our minds increase, and our perspectives become broader. And, in a similar way, humanity itself advances through history: history itself seems to be the content of humanity's mind. However compelling a study of this alone might be—the content of the mind, and the "logic" of its progressive expansion—Hegel goes beyond even this and re-situates Kant's separation of the workings of the mind from the world it perceives within this "logic" of progression.

It is the first three chapters of Hegel's *Phenomenology* which most closely correlate with the thought of Kant—it is here that Hegel discusses sensation, perception, and the understanding—key elements of a description of the individual mind's form—and aspects touched on by Kant. In the following I hope to show how Hegel's perspective on these issues seems to be an outgrowth of seeds sown by Kant, and how Hegel's differences from Kant demonstrate a *dialectical* logic which recurs throughout the *Phenomenology*.

HEGEL'S OPENING: PURE EGO

In Kant's discussion of the various faculties of the mind, he devoted much attention to apperception, which he praised as the most important of all mental faculties. Roughly, this faculty of apperception designated the unity of a subject perceiving—the "I", or ego which was the logical and empirical subject of all perception. Kant also found this apperception to be that which provided for the unity of an object or thing—an object is understood as one thing as it is confronted by the unity of a person's mind.

Now, for Hegel, the opening moment of experience—the primal given from which one departs—precedes this singularity of sensation, as found with Kant's unitary mind confronting a unitary object. One begins with immediacy, that which simply is: consciousness / sense-experience. This experience is not yet understood, it simply appears; one is this appearance:

> "Consciousness, on its part, in the case of this form of certainty, takes the shape merely of pure Ego" (149).

However, we find that:

> "there is a good deal more implied in that bare pure being" (150).

And here we have the beginning of a dialectic driven by contradiction. A central contradiction which is played out through the entire *Phenomenology* is the play between the one and the many, the whole and the parts, the same and the different. The contradiction here being that for Hegel, the one is many, the whole is its parts, the same is different; or rather, each is striving to become its opposite. And, with our given singularity of the experience of pure ego, pure being in itself, we find already that:

> "pure being at once breaks up into two 'thises'" (150).

The "this I", and the "this object"—pure immediacy turns out to be a mediated relation between the subject and the object, each of which is dependent on the other. For this "I" is distinct from pure being only with its distinction from "what" is experienced, the object: the mere surface of pure sense-experience splits an interior in opposition to an exterior, where one without the other might be a mere surface without an individual perspective, or recognized object. With the breaking apart of the subject and the object, a host of other aspects of pure being fall into contradiction—the "Now", "Here", and "This" which all point towards the immediately experienced become their opposites: Now becomes not Now, Here becomes not Here, This becomes not This.

And it is in this way that Hegel distinguishes one's universal modality from the entirety of sense certainty. The Now, it would seem, is always Now—I am always in the present—indeed, I existed before, but that is not Now; and my experience always occurs within the Now. I am always where I am, Here, when I am here—Now; whatever time it is, I experience it only Now; where ever I go, when I am experiencing, I am Here. And whatever I direct my attention to, it is This. The I is always accompanied by a Now, a Here, and a This. Yet, what is Now, is also not Now as it passes; what was Now then is not Now now; what was Here before me then may not be before me now; and in that way, the Now becomes not Now, but then; the Here becomes not Here, but there. So there is what seems an endless string of Nows, and Heres, which are not the Here and Now which I presently experience. And it is the distinction between all the Nows and Heres, and the fact that my experience is always Here and Now which distinguishes my *modality* (my term, not Hegel's) from the entirety of sense-certainty (being):

> "This pure immediacy, then, has nothing more to do with the fact of otherness, with Here in the form of a tree passing into a Here that is not a tree, with Now in the sense of day-time changing into a Now that is night-time, or with there being an other I to which something else is object. Its truth stands fast as a self-identical relation making no distinction which is essential and non-essential, between I and object, and into which, therefore, in general, no distinction can find its way" (155).

In other words, there simply is the complete presence of everything to itself, all the Nows, in one Now, all the Heres in one Here; and our simple modality of I-Now-Here-This is always limited to a singularity letting the Now pass into the not Now, which distinguishes the modality, or universal medium of all egos from the complete presence of immediacy to itself.

RECOGNITION OF THE OBJECT

With Hegel's discussion of perception, what was already implicit in an experience of the universal medium becomes explicit in the complex object, as the dialectic struggling between the Now and not Now, the Here and not Here, the This and not This is superseded into a higher sphere of the perception of the complex object. For an object can be both This object Here before me Now, and also the object that was before me then and there. I see This tree, here before me Now—but then later, that same tree is not This before me Now and Here—the tree can be both This Here and Now, and not This Here, not This Now:

> "The This, then, is established as not This, or as superseded, and yet not nothing (simpiciter), but as determinate nothing, a nothing with a certain content, vis. the This. The sense-element is in this way itself still present, but not in the form of some particular that is 'meant'—as had to be the case in immediate certainty—but as a universal, as that which will have the character of a property. Canceling, superseding, brings out and lays bare its true twofold meaning which we found contained in the negative: to supersede (aufheben) is at once to negate and preserve" (163-164).

One withdraws from the contradiction between the present moving into the non-present with the object which remains through both.

However, no sooner is this contradiction superseded (aufheben), than another contradiction appears within the perceived object, this being that it has many properties, yet is one thing. For with perception, we actually only experience individual properties—a sugar cube is white, cubical, sweet; but we simply see a white color, feel a cubical shape, taste a sweetness (cp. the wax observed by Descartes). Our perception itself is not of a thing, but of a multiplicity of properties which we may experience side by side, and not singularly. Sweetness is distinct from whiteness, each of which is distinct from a cubical shape—each of these properties is distinctly different from the others. However, we do find that, despite the plurality of properties that we encounter, we can distinguish one object from another by a special property:

> "It is through the determinate characteristic that the thing excludes other things. Things themselves are thus determinate and in and for themselves; they have properties by which they distinguish themselves from one another" (170).

However, our ability to distinguish one thing from another no longer belongs to the realm of perception, which is always directed to a specific cluster of properties. And here, we find that it is the understanding which provides the unity of a cluster of properties by distinguishing one object from others.

Again, the disparity between the This Here Now and the not This Here Now was superseded by the This (thing) which can be both Here and Now, and not Here and Now. And the resulting conflict of recognizing a This as something rather than a plurality of properties is superseded by the understanding's ability to distinguish one object from another. Perception of a specific This provides a bridge between the Now-Here and the not Now-Here; distinguishing a This thing from other things through the understanding offers a bridge between the plurality of properties and a single thing. Pure-being is broken into a distinction between the Now and not Now, which is superseded by the This which is Here

Now and not Here Now; the This is broken into a distinction between its properties and its unity, which is superseded (aufheben) by its being distinguished in the understanding as an one object with properties among many objects. In each case, there is a breaking apart of unity into a plurality of differences which are overcome (yet still remaining) within a new, more subtle perspective. The contradictions of sense-certainty are superseded by perception; and in turn the new contradictions of perception are superseded in the understanding. Each realm is framed by a new one—sensation remains within perception, perception remains within the understanding: each new level maintains the prior one, yet re-situates it in a larger more comprehensive context. Yet, these new circumscriptions are found to already be implicit within what they enclose, just as a square and a circle (may) have always been geometric shapes before, being grouped as such by geometry.

THE UNDERSTANDING'S SELF-UNDERSTANDING

Now, what occurs at the next stage within Hegel's discussion of the understanding becomes quite complex, and my brief summary will most surely not do it justice. The new principle which came with the understanding was the differentiating between objects. However, this differentiating is itself the dialectical process which has heretofore been described. But with the understanding, the dialectical process of breaking apart into contradiction itself becomes understood, and this occurs through yet more breaking asunder into contradiction. The understanding of the understanding's differentiating will be understood through its sundered distinction between force and law:

> "the elements set up as independent pass directly over to their unity, and their unity directly into its explicit diversity, and the latter back once again into the reduction to unity. This process is what is called Force. One of its moments, where force takes the form of dispersion of the independent

> elements each with a being of its own, is the Expression of Force; when, however, force takes the form of that wherein they disappear and vanish, it is Force proper, force withdrawn from expressing itself and driven back into itself" (183).

And the primary law that accompanies force is difference:

> "What is found in this flux of thoroughgoing change is merely difference as universal difference, or difference into which the various opposites have been resolved. This difference as universal, consequently, is what constitutes the ultimate simple element in the play of forces, and is the resultant truth of that process. It is the Law of Force" (195).

Now, once this distinction between force and law is established, (and indeed, one can—and Hegel does—discuss the variety of manifestations of force, and of different types of laws) it is then superseded (aufheben):

> "They are no doubt distinct, but their distinction is at the same time explicitly stated to be not a distinction of the fact itself [that 'law is bare and simple force'], and consequently is itself again straightway canceled and transcended. This process is called Explanation. A law is expressed; from this its inherently universal element or ground is distinguished as force; but regarding this distinction, it is asserted that it is no distinction" (200).

Moreover:

> "The world of appearance, or the play of forces, already shows its operation; but it is in the first instance as Explanation that it comes openly forward. And since it is at length an object for consciousness, and consciousness is aware of it as what it is, consciousness is in this way Self-consciousness.

> Understanding's function of explaining furnishes in the first instance merely the description of what self-consciousness is" (210).

The operations of force and law have thus always existed implicitly. Yet, it is with this superseding (aufheben) of these two into explanation that they become explicit—implicitly and explicitly distinct, they lose this distinction in passing from the implicit to the explicit. However, this dialectical force, differentiating into contradiction, and superseding (aufheben) into higher unity is itself understood to be merely a mediated play of the understanding:

> "understanding merely experiences itself" (212).

The understanding therefore raises consciousness beyond sensible perception only to erase itself:

> "the vision of the undistinguished selfsame reality, which repels itself from itself, affirms itself as divided [....] what we have here is Self-consciousness" (212).

We thus have consciousness raised to the level beyond mere subjective experiencing on into intellection, where it is ready to engage in a dialectical relation with other consciousnesses (as is discussed in chapter four of the *Phenomenology*—e.g. the master slave relation) and is well on its way to further self-knowledge and a more comprehensive understanding of itself. Again, I will not be summarizing this grand journey, but will rather examine its character.

Before doing such, the differences between Kant's and Hegel's projects should be clearly noted. Most clearly evident is that Kant was quite cautious when speculating as to what was beyond the understanding. No doubt, Kant's and Hegel's attitude towards the understanding differ to a great extent. Kant embraced its certainty, while Hegel sees it as a ladder to be nearly discarded once climbed.

Moreover, Kant takes more seriously the particular workings of the understanding, where Hegel seems to be narrating its progress. Most importantly though, is the fact that Hegel sees the understanding as a crucial factor in forcefully mediating a differentiation into the undifferentiated being of pure ego. It is in this way that Hegel (along with others) inflated Kant's transcendental necessity for apperception (the "I") into an immanently transcendent ego.

Whereas Kant forged the path of analytic logic towards the unity of consciousness, Hegel claims to begin with this consciousness being interrupted by the logic of dialectics. And in this way, Kant may have provided the ladder which Hegel claims to have reconstructed from the other side.

THE UNFOLDING OF THE ZODIAC

What has been discussed so far has been limited to the very first stages of the unfolding of consciousness in its progressive journey of maturation. Yet we have already seen some of the ways in which Hegel sees a process in this progression: his dialectic of unity breaking asunder into contradiction with diversity, and the contradiction striving for a higher unity where the implicit is externalized into an explicit framework which re-situates the prior contradictions. What we have here is not a simple resolution of opposites, as with Kant who in his dialectics tried to resolve antinomies into unknowable, yet probable unities. Hegel's dialectic seems more like a fountain, or a flower whose inner petals burst forth, pushing the outer petals aside in a new comprehension. Inversely, this unfolding provides a progressively wider and broader scope with which to access the prior stages, as each new stage re-frames the prior within a new context. Ultimately this context extends beyond the individual that Kant was so often preoccupied with, showing the individual to be part of whole historical social realm, and a part of the whole of absolute Being as well (at which point the part again becomes whole).

Now, this sort of progression, which composes a narrative journey of consciousness, can be seen in numerous myths which reflect the maturation of a character. But one of the most ancient formulations of this journey—one of the most concise and coherent, expositions—came from a time before science and myth had been severed: astrology's Zodiac. The Zodiac presents one of the most detailed and coherent systems of myth and mythological progression. It constitutes a system of interpretation that applies to the body, character, society, and more, and integrates them all. And, against disregarding it as being outdated and refuted by science, it might behoove us to realize that the genealogy of our sciences indeed trace back through / to astrology: astrology lies at the roots of the very sciences which have sought to deny this ancestor. The whole has been broken apart, and one piece has claimed authority—religious divisions aside, society at large has to a great extent granted science mastery over myth.

Now, the structure of the Zodiac should be somewhat familiar to most. As we have twelve months (twelve tribes, twelve apostles, etc.) we have twelve signs of the Zodiac. There has been speculation that at one time there were six signs, and that these had been sundered into twelve. This seems interesting in that many of the signs could be paired together; the signs oscillate between active and passive (classically—and problematically—between masculine and feminine) and such a prior Zodiac might designate one without this active/passive scission. Nevertheless, the current Zodiac's twelve signs form a progression which loops back into itself, much as the seasons perpetually recur. To facilitate discussion it may be helpful to list the signs of the Zodiac in order, with brief key aspects of each sign given, (and also with its symbol, associated planet(s), body part, element, mode, and polarity listed in parenthesis):

1. ♈ ARIES – I AM: ego, will, being, initiative, aggression, (the ram, Mars, the head, fire, cardinal, active).

2. ♉ TAURUS – I HAVE: property, substance, stability, determination, attachment, (the bull, Venus, the neck, earth, fixed, passive).

3. ♊ GEMINI – I THINK: communication, intellect, the mercurial, flexibility, connectivity, nervousness, (the twins, Mercury, the hands, air, mutable, active).

4. ♋ CANCER – I FEEL: nurturing, growth, domesticity, sensitivity, binding, (the crab, the Moon, the breast, water, cardinal, passive).

5. ♌ LEO – I ACT: power, pride, love, leadership, expression, (the lion, the Sun, the heart, fire, fixed, active).

6. ♍ VIRGO – I JUDGE: analysis, improvement, hygiene, study, precision, modesty, (the virgin, Mercury, the stomach, earth, mutable, passive).

7. ♎ LIBRA – WE BALANCE: partnership, idealism, harmony, dependence, the social, high art, (the scales, Venus, the pelvis, air, cardinal, active).

8. ♏ SCORPIO – I CREATE: passion, sex, death, secrecy, suspicion, transcendence, (the snake, scorpion and eagle, Mars and Pluto, the genitals, water, fixed, passive).

9. ♐ SAGITTARIUS – I ABSTRACT: exploration, honesty, athletics, philosophy, enthusiasm, broad perspective, (the centaur, Jupiter, the thighs, fire, mutable, active).

10. ♑ CAPRICORN – I USE: responsibility, efficiency, practicality, status, strength, realism, (the sea-goat, Saturn, the knees, earth, cardinal, passive).

11. ♒ AQUARIUS – I KNOW: social expression, humanity, science, observation, insecurity, eccentricity, (the water bearer, Uranus and Saturn, the calves and ankles, air, fixed, active).

12. ♓ PISCES – I BELIEVE: unconsciousness, self-sacrifice, compassion, imagination, worry, indecision, (tethered fish, Neptune & Jupiter, the feet, water, mutable, passive).

As to the planets, one may get some sense of their significance if one thinks of the character of the roman gods they represent (jovial Jupiter, etc). With the elements, fire is dynamic, earth is "down to earth," air is ethereal, and water is emotional. And, with the modes, the cardinal is initiatory, the fixed carries through, and the mutable harvests. Here, one should easily be able to discern a sort of progression through the signs. This progression is additionally evident in that the first six signs are more self-oriented; while the last six are more social (a child may become more socially conscious with age). Moreover, the signs become more "refined," yet less focused throughout the Zodiac progression. Such is not to suggest, however that any sign is better than another—each sign has its positive and negative traits, and is simply different than the others. (Moreover, one's personal astrological chart always relates to many signs, not just one!)

A correlation with the progress of Hegel's *Phenomenology* may now also be evident. Each sign in the Zodiac builds upon the previous one; conflicts are resolved, and new problems begin. For example, the pride which is inherent in Leo is resolved with the modesty of Virgo; yet this too can become a problem, as with over fastidiousness. Also, the even split between the individual and the social signs in the Zodiac is evidently paralleled with the distinctly individual and social part of the *Phenomenology*.

Most relevant to this chapter might be a parallel between the progress of the first three signs of the Zodiac and of the first three chapters of the *Phenomenology*. One might recall from our discussion of those chapters, that the unity of pure ego was split up into an opposition with things which were then grasped by the understanding. This movement is also evident in the progress of the Zodiac in the shift from the emphasis on the ego in Aries, to the identification with things (property) in Taurus, and its stratification through the intellect in (thinking) Gemini. Again, in both cases there is a movement from a pure ego, to things, and to thinking about them. And in this progression we can see, in both cases, a splitting up of the primal into more subtle relations. (By the way, the next signs are Cancer and Leo, which may suggest a Freudian emphasis on mothers and fathers as the primary inter-subjective relationship, rather than the Hegelian master and slave interpretation).

Trying to draw a parallel between the Zodiac and the entirety of the *Phenomenology* would no doubt be problematic, although a rough comparison probably does remain throughout. A parallel is especially evident with the closing parts of the *Phenomenology* though, where the individual gives way to the social spirit and dissolves through religion into the absolute. In the Zodiac, as well, with the closing signs of Aquarius and Pisces, one respectively loses one's self in the social and the religiously unknowable. Thus, just as the Zodiac culminates in the reconciling self-abandonment of Pisces, the *Phenomenology* ends where:

> "Spirit is engulfed in the night of its own self-consciousness; its vanished existence is, however, conserved therein; and this superseded existence—the previous state, but born anew from the womb of knowledge—is the new stage of existence, a new world, and a new embodiment or mode of Spirit" (807).

With these last claims, we have a parallel to the Zodiac's re-feeding into itself, and beginning its cycle anew.

It may be noted that astrology does not simply use the Zodiac for personal character evaluation. The same structure of signs is also used to make social predictions, and at a very large scale, it assesses the character of an age. No doubt, you have heard that we have passed into the "dawning of the age of Aquarius." An astrological age lasts for *about* 2000 years, and the ages go through the Zodiac in reverse. The last transition was that from Aries to Pisces—approximately at the time of Christ, which has determined the base of our calendar year. Symbolically, this is quite profound, as the transition from Aries to Pisces would be one of the grand Ego's self-sacrifice—the absolute will of the Aries ego gives way to the unknowable, redemptive, self-sacrifice of Pisces. To further add to this symbolic age interpretation, Moses appeared at the transition from Taurus to Aries—from the iconic this-ness of the golden calf (Taurus is the bull) to the absolute will of the Aries ego ("I am that I am"). No doubt, the transition from Aries to Pisces may be accounted as a more crucial cusp, for it is the actual re-feeding of the Zodiac back into itself. To this extent, I would predict the "Second Coming" as occurring at around 24000 CE. Of course, this brings a question as to what our recent transition from Pisces to Aquarius might be understood as bringing—a self-sacrifice at the social level, or a resolution of sacrifice through the social? A science of the unknowable? I am not sure, but I find the astrological schema for interpretation quite intriguing. At any rate, with both astrology and Hegel's dialectic, we can see the application of a similar circular evolutionary interpretational scheme to both the individual and the social.

Such brings me to my criticism of Hegel. It would seem that he has drawn upon many sources for his work—maybe astrology, most likely also Taoism's principle of the yin-yang inter-dependence, and obviously Kant, Plato, etc. If he did indeed draw upon these sources, why did he not say so? As an artist, such might not be necessary; but as a philosopher who's works have gone on to help shape the destiny of millions (e.g. with Marx and communism) this sort of twisting around and deviousness is quite questionable. Did Hegel take the original elegance of the Zodiac and twist it around to his own purpose: did he appropriate what may have been some divine revelation only to reshape and sell it in a disfigured form? Most likely, Hegel's intentions where more sincere—his insights stray far from previous doctrines, and he brings them together in fascinating ways (not many things could seem farther apart than Kant and astrology!) It is the comprehensiveness, and audacious aim of the *Phenomenology* which makes it such an amazing piece of writing—maybe a more appropriate title, used for a book by Douglas Adams, would be *Life, the Universe, and Everything.* However, it also seems quite different than an answer to everything: it inspires and opens up a fountain of possibilities—which is quite evident in Hegel's subsequent influence.

Friedrich Nietzsche (1882)

Chapter 14

Nietzsche
Friend or Fiend?

INTELLECTUAL MARTIAL ARTS

I first encountered Friedrich Nietzsche (1844-1900) as a freshman sitting at the University of Oregon's Erb Memorial Union Skylight Refectory—there was a bit of graffiti on the back of a seat with the phrase, "'God is dead'—Nietzsche", followed bellow by a counter-phrase, "'Nietzsche is dead'—God." I remember being a bit shocked at the audacity of the claims, and also a bit humored. Since then, after having read through much of Nietzsche, I find that this kind of reversal seems somewhat in line with the sort of intellectual martial arts performed by this rather strange figure. A concern with reversal can be seen, for instance, with his revaluation of all values, and with the characterization made of him by other critics as being a reverse Platonist.

Indeed, Nietzsche often practices reversal as a strategy, or intellectual move in his writings; and it is his repeated exploitation of such techniques that helps distinguish Nietzsche from previous philosophers who so often seem to build static systems. For Nietzsche, these practices are used in an engagement with the specific issues found in his aimless intellectual roaming—a solitary and nihilistic combat which leaves a trail of ruins with little or no systematic organization. Such is not to claim that we could make a comprehensive list of his moves—it is important here simply to note that for Nietzsche "philosophizing" is an activity, and often a seemingly violent one pushed forward through an aphoristic bursting which repeatedly reverses, overturns, transforms, and transfigures.

One can see the influence of previous thinkers on Nietzsche's writings, however, and particularly a few German philosophers (Spinoza seems to be a favorite, too). The influence of Hegel and Schopenhauer, for example, are clearly evident in the strategic use of dialectical oppositions found in his first important work, *The Birth of Tragedy*—a few of the discussed oppositions are as follows:

Apollo	Dionysus
Plastic Arts	Music
Dreams	Intoxication
Homer	Aeschylus
Illusion / Naiveté	Truth / Terror / Pain
The Individual	The Whole
Antigone	Cassandra
The Epic	The Lyrical
History / The Future	The Now
Word	Tone
Hero-Vision	Mass-Spectator
Phenomena	The Will
The Soul	The Body
Socrates	Nietzsche

With this list, one may be able to discern a repetition and expansion of Schopenhauer's opposition between the Will and Representation (respectively, the Dionysian and the Apollonian for Nietzsche) which in The Birth of Tragedy are ultimately sublimated with a Hegelian Auheben into tragedy; Nietzsche recognizes such in his own appraisal of the work in his *Ecce Homo*:

> "it smells offensively Hegelian, and the cadaverous perfume of Schopenhauer sticks only to a few formulas [....]

the antithesis of the Dionysian and the Apollonian [...] in tragedy this antithesis is sublimated into a unity." (Friedrich Nietzsche, Walter Kaufmann (trans.), *Basic Writings of Nietzsche*, The Modern Library: New York (1992), p. 727 (*Ecce Homo*))

Dialectical opposition remains an important strategy throughout Nietzsche's works (the very titles of his books often suggest opposition, as with *The Birth of Tragedy* or *The Joyful Wisdom*); and keeping this in mind can be helpful in understanding a possible impetus for writing the likes of *The Anti-Christ*, or for Nietzsche calling himself an immoralist (rather than, say, an a-moralist).

One can conceivably see a parallel with Marx here, in that contrary to a Hegelian passive observation of history's progressive unfolding out of time into being, Nietzsche seems to be attempting to create new oppositions to force stagnation and decadence into a revitalized progression into the future. However, contrary to a Marxist focus on economic domination and theory inspired practice, Nietzsche concerns himself with moral domination—and he also evidences that he believes his perspective to be more descriptive of reality than prescriptive of how it ought to be (this attempt to describe and affirm "reality" will be discussed more in depth later in this chapter).

Nietzsche wears his influence from other philosophers in a new way; rather than subsuming or destroying their explanatory systems with a new one, he draws upon them for a repertoire of critical strategies. And in this way, he conducts a genealogical performance—more than simply tracing the roots of ideas, he enacts philosophical moves. Indeed, much of what could be said about Nietzsche's philosophy could also be said about Hegel's or Schopenhauer's—discerning what is original in Nietzsche can be difficult. With another reversal, though, Nietzsche does not concede an indebtedness, but writes of feeding on:

> "those who were unknowingly my workers" (Friedrich Nietzsche, Walter Kaufmann and R.J. Hollingdale (trans.), *The Will to Power*, Vintage Books: New York (1968), #464, p. 255).

This is not to say that he is uncritical of previous philosophers. With regard to Hegel, for instance, Nietzsche would deny that any type of absolute description could be made concerning the progress of a history centered on notions such as the zeitgeist:

> "the overall character of existence may not be interpreted by means of the concept of 'aim,' the concept of 'unity,' or the concept of 'truth.' Existence has no goal or end; any comprehensive unity in the plurality of events is lacking: the character of existence is not 'true,' is *false*" (*The Will to Power*, #11, p. 13).

It is important to note the word "interpretation" here, as when any absolute criterion for truth is lost, there can be nothing left but various competing perspectives, or projected interpretations.

Ironically, Nietzsche's rejection of absolute truth seems to have been inspired somewhat by Kant and science, as evidenced in his "On Truth and Lie in the Extra-Moral Sense." Nietzsche takes very seriously Kant's claims in the *Critique of Pure Reason* that human reason has no access to the in-itself, or what might be objective truth; as Nietzsche claims any idea

> "is anthropomorphic through and through, and does not contain one single point which is 'true-in-itself,' real and universally valid, apart from man" (Friedrich Nietzsche, Oscar Levy, ed, *The Complete Works – Vol. Two*, The MacMillan Company: New York (1911), p. 183).

Moreover, he claims:

> "Everything which makes man stand out in bold relief against the animal depends on this faculty of volatilizing the concrete metaphors into a schema, and therefore resolving a perception into an idea" (*Complete Works*, p. 181).

Such claims follow his description of metaphor making:

> "A nerve-stimulus, first transformed into a percept! First metaphor! The percept again copied into a sound! Second metaphor! And each time he [the creator of language] leaps completely out of one sphere right into the midst of an entirely different one [....] Every idea originates through equating the unequal" (*Complete Works*, p. 178).

These claims, concerning the metamorphosis of metaphorizing, use a "schema" terminology that is clearly Kantian; and the crux of Nietzsche's metaphorical transfiguration resorts to neuro-biology, leading to his bold claims:

> "What therefore is truth? A mobile army of metaphors, metonymies, anthropomorphisms: in short a sum of human relations which became poetically and rhetorically intensified, metamorphosed, adorned, and after long usage seem to a nation fixed, canonic and binding; truths are illusions of which one has forgotten that they are illusions; worn-out metaphors which have become powerless to affect the senses; coins which have their obverse effaced and now are no longer of account as coins but merely as metal" (*Complete Works*, p. 180).

In a radical scientific Kantianism, truth is lost the moment perception is hypostatized—ideas never directly correspond to the intricate detail of the changing world which metaphor making and schematizing severs them from, but are always caught up within reified human customs and connections; and in this way, Nietzsche puts Kant's Understanding on as dubitable grounds as Kant put the faculty of Reason. The ironic aspect of this hyper-critical (hypo-critical?) claim that all truth has the same status as lies, is that it is based on the hyper-extension of the discourses—scientific and Kantian—most concerned with and "grounded in" truth (i.e, how can Nietzsche make this scientifically valid, and true claim?)

It is with this destruction of any claim to absolute truth (and any real unity, final aim, etc.) that one may begin to wonder about the integrity of Nietzsche's writing. Indeed, a bold confidence resonates through his writings, suggesting that all should blindly follow him, as if he were directly asking, "how could you trust yourself more than you trust me?" Nietzsche explores some of life's most important aspects, often in new ways which appear to be experimental wanderings on un-trodden ground—we must ask, in the face of a claimed lost *truth*, what would make Nietzsche a *trustworthy* guide? Such a hesitance is not at all abated when Nietzsche claims on the one hand:

> "the truth speaks out of me.—But my truth is terrible; for so far one has called lies truth" (*Basic Writings of Nietzsche*, 782 (*Ecce Homo*)).

But remarks on the other:

> "Have I thereby harmed virtue?—As little as the anarchists harm princes: only since they have been shot at do they sit securely on their thrones again—For thus has it ever been and always will be: one cannot serve a cause better than by persecuting it and hunting it down—This—is what I have done" (*The Will to Power*, #329, pp. 179-180; modified also in *The Portable Nietzsche*, pp. 71-72 (*The Wanderer and his Shadow*), see also *The Will to Power*, #361, p. 197).

As suggested by the title *Twilight of the Idols*, much of Nietzsche's writing is iconoclastic—he targets any sort of given ideal, or ideology; it is this anti-ideological activity of Nietzsche which makes him impossible to pin down, as he will not rest at any given point or ideal. Yet, if he believes that resistance strengthens the opposition, what then are we to believe—that Nietzsche is attempting to strengthen ideals?

Possibly, if Nietzsche's stance were consistently evil (he says of *Twilight of the Idols*, no book has been "more evil" (*Basic Writings of Nietzsche*, 770 (*Ecce Homo*)) an appraisal of his work would be easier. His comments advocating slavery, his misogyny, and his racial stereotyping are obviously abhorrent; yet, would we want to throw out his affirmation of life on earth as well? (A Buddhist might reply, "yes!") Moreover, Nietzsche does on occasion make statements which are blatantly contradictory. This obstacle to interpretation is made even more problematic in that when reading Nietzsche one risks coming to "admire" him for his clear character and profound soul; this tends often to blind one to his "darker" side—one wishes "good" intentions upon him, despite his re-envisioning and shifting of the categories of good and evil (a shifting which through subtle seductions may also re-situate the reader's own perspective on morality). Indeed, it is all too easy to project one's wishes onto Nietzsche, a danger he points out:

> "Whoever thought he had understood something of me, had made up something out of me after his own image—not uncommonly an antithesis to me; for example an 'idealist'" (*Basic Writings of Nietzsche*, 717 (*Ecce Homo*)).

Here, I feel I have to be vigilant of this myself—one seems challenged by Nietzsche not to be reactive towards him, yet he is so polemical! Maybe it is easy to be duped into Nietzsche's game—yet he stacks all the cards in his favor, as with the paradoxical order: Don't Follow Me!:

> "I bid you lose me and find yourselves; and only when you have all denied me will I return to you" (*Basic Writings of Nietzsche*, 676 (*Ecce Homo*)).

If we are to play this game by his rules, then he has already won. And one must ponder his aim—to dominate? To invent a game where he always wins? What might have motivated him to create such a game? Does he have ulterior motives? It is often difficult to even discuss Nietzsche without sounding like an admirer or an idiot! His remarks elude summary dismissal; yet whether they deserve extended discussion is another matter.

Nietzsche does return again and again to various themes though, such as the Will to Power, the Over-man, the Eternal Return, and revenge; and while these themes may not constitute any system or ideal, they seem to provide focal points for a rather consistent, even if loose, schema of interpretation. Before indirectly discussing these themes that might constitute a "positive" aspect of Nietzsche's writing, it seems important to note that despite their extremely simple (some might say, "elegant") theoretical structure, it is Nietzsche's use of these themes which has such tremendous rhetorical force and critical insight. Moreover, his opinions and style often imply a megalomania that demonstrate how "seriously" he takes his writings:

> "who I am. Really, one should know it, for I have not left myself 'without testimony'" (*Basic Writings of Nietzsche*, 673 (*Ecce Homo*)).

THE GREAT MOTHER (OVERMAN?)

The rhetorical force and beautiful art of Nietzsche's writings might often make one forget his academic connections. His writings are, after all, often concerned with famous academic writers, and with traditionally important academic themes. One finds a telling break from what might be called Nietzsche's "performance" in a note at the end of the first section of his *Genealogy of Morals*. It is here that he modestly proposes a small contest, with the prompt:

> "What light does linguistics, and especially the study of etymology, throw on the history of the evolution of moral concepts?" (*Basic Writings of Nietzsche*, 491 (*The Genealogy of Morals*)).

One can hardly ignore the fact that Nietzsche began his academic career as a Greek philologist. And, with his question's limitation of using etymology, one can see that possibly only a few words might be applicable—it seems as if Nietzsche were trying to predetermine any answer with his question. For example, the word "morality" comes from the Latin word for "custom" or "habit"—hence Nietzsche's concern with tracing the genealogy of moral customs. And also—and this is even more telling—the word "virtue" has its origins in the Latin word for "man." Of course, I'm tracing English words to their Latin origins rather than treating German words and Greek origins—but I am sure that other cases would show parallels with Nietzsche's themes.

Now, I think the connection between virtue and manliness may be what Nietzsche really intended to load his question with; as he has cited Christianity as castrating the strong in favor of making truth feminine:

> "it becomes female, it becomes Christian" (Friedrich Nietzsche, Walter Kaufmann (trans.), *The Portable Nietzsche*, Viking Penguin Inc.: New York (1987), p. 485 (*Twilight of the Idols*)).

His nostalgia for pre-Christian times, circa the pinnacle of Greek history is obvious; Nietzsche longs for the times when "men were men" and manliness was goodness. It is from this perspective that he sees Christianity as a terrible turn of events: it was the destruction of Greek morality, the culture that was so obviously the obsession of the young Nietzsche. This Christian turn is further seen by Nietzsche (after Hegel, et. al.) as a type of revenge on the part of the mass of slaves; prior to Christianity, so it goes, there was a master morality exercised by the strong. The mass of slaves, weak as individuals but strong as a group, striped the strong individuals of their powerful claim to moral supremacy. It is because of this shift to calling the weak "good" that Nietzsche castigates Christianity.

In fact, Nietzsche is so critical of this Christian turn, that one might say he viewed it as something equivalent to the second fall of humankind (Christ has been called the second Adam) which instead of forfeiting paradise, loses for Nietzsche the "true" distinction between good and evil. Hence, there is a sense of a lost origin, a true morality which Nietzsche so ardently advocates a return to—a return from his era which he sees as so thoroughly decadent. This sort of logic of moral decay suggests that history evidences the sustained devolution of morality which should rightfully be reversed.

However, a point that comes to my mind is, "why not go further back than the Greeks?" Indeed, if we look towards Nietzsche's opinion of a morality prior to the Greeks, at least prior to any Greek history that would be subject to any genealogy or etymology, we find that there could be no global logic of "moral decay." In Nietzsche's essay "Homer's Contest" for example, we find he views such ancient times as a

> "pre-Homeric abyss of a terrifying savagery of hatred and the lust to annihilate" (*The Portable Nietzsche*, 38 (*Homer's Contest*)).

—Times which were:

> "evil...cruel...vengeful...godless" (*The Portable Nietzsche*, 39 (*Homer's Contest*)).

However, contemporary research into pre-Homeric times has demonstrated that there where many matriarchal based societies—the likes that might be found with an *archeology* of morals. Nietzsche's lack of scholarly insight into this area is seen in these comments:

> "Christianity only takes up the fight that had already begun against the classical ideal and the noble religion. In fact, this entire transformation is an adaptation to the needs and the level of understanding of the religious masses of that time: those masses which believed in Isis, Mithras, Dionysus, the 'Great Mother'" (*The Will to Power*, #196, p. 115).

What Nietzsche here fails to recognize is that interest in the likes of Isis or the Great Mother were not reactions against a Hellenic noble (master) morality—these cults have pre-Homeric origins! And hence, one might see that the so-called "noble" master morality of Greek Gods—the manliness of high-Greek culture—may have indeed been a *reaction* against the earlier Goddess religions:

> "Today's scholars habitually call all female and male deities of the ancient world 'gods,' as they also call humanity 'man.' Yet the supreme deity of the world was usually a Goddess, the creatress or Mother of the gods; and the very [English] word 'man' used to mean 'woman,' an incarnation of the same lunar Mother, in its original language" (Barbara G. Walker, *The Woman's Encyclopedia of Myths and Secrets*, Harper and Row: New York (1983), p. ix).

Even more puzzling is Nietzsche's lumping of Dionysus with Isis and the Great Mother. Could it be that Nietzsche saw the Dionysian as being feminine? (Even though the god Dionysus was male, his cult was female—and as a mythical predecessor to Christ, the Dionysian might have been viewed by Nietzsche as being as feminine as the Christian.) And, to speculate further, with Nietzsche's association of the Dionysian with the whole, as opposed to the Apollonian with the individual, might there also be a parallel with the Buddhist being / nothingness beyond the ego? Could it be that the term "Over-man" designated some feminine state beyond the ego? (Such is not to make the ridiculous assertion that women have no egos, but to explore the possible implications of Nietzsche's claims!) At any rate, such might account for Nietzsche's praise of losing one's self in Dionysian intoxication. It is more likely that Nietzsche used the term "Over-man" to designate some future man much more sophisticated and life-affirming than those of his time.

NATURE (THE WILL TO POWER)

In section #462 of *The Will to Power*, we find Nietzsche's own appraisal of his "firsts":

> "Fundamental innovations:
>
> In place of 'moral values,' purely naturalistic values. Naturalization of morality.
>
> In place of 'sociology,' a theory of the forms of domination.
>
> In place of 'society,' the culture complex, as my chief interest (as a whole or in its parts).
>
> In place of 'epistemology,' a perspective theory of affects (to which belongs a hierarchy of the affects; the affects transfigured; their superior order, their "spirituality").
>
> In place of 'metaphysics' and religion, the theory of Eternal Recurrence (this as a means of breeding and selection)" (*The Will to Power*, #462, p. 255).

Now, it would be quite ridiculous to suppose that Nietzsche advocated a master-morality simply because it was evidenced in one of his favorite periods of history; it seems more likely that he preferred the pre-Christian Hellenic world because it evidenced the morality he advocated. Where, then, might Nietzsche have gained the courage to claim that the values of good vs. evil should be replaced by the values of strong vs. weak? Indeed, as this transvaluation of values stands at the heart of much of Nietzsche's thought, his basis for making this revaluation would give at least some basis to evaluate his trust-worthiness. As evidenced by the quote above, it would seem that he has chosen to follow nature rather than culture—nature chronologically precedes culture, and thus may be genealogically "superior" (primordial)—more so than even the Greeks.

It is with this insight that we can see how Nietzsche saw his views as inevitable and necessary. For, with humanity's turning away from religion to science with respect to what was generally cited as reality and truth, morality too would have to be "scientific", or rather, exposed as fraudulent:

> "Wherever one has not yet been capable of causal thinking, one has thought morally" (*The Will to Power*, #327, p. 179).

Here "science" itself is understood to be the study of nature; the laws that nature obeys are seen as real. Much of science is concerned with what I would call the theoretical reproduction of nature, where with the likes of physics and chemistry nature is understood to be something like a machine—a machine we understand to the extent that we could reconstruct it according to the blue-prints of scientific theories. Now, when biology (and we have already seen evidence of Nietzsche's interest in neuro-biology) is combined with history (note Nietzsche the philologist and genealogist) we can derive theories of evolution (cp. "volition" as *will*). Such theories, including references to natural selection and the survival of the fittest seem to designate areas of the science of the historical or temporal progress of biological organisms.

The "story" of evolution through time, so much more scientific than any Hegelian "description" of history, has one clear message: the "strong" survive. Indeed, despite what Nietzsche may say about Darwin or science, it seems that he could have hardly developed his theory of noble values from any other kind of source: the morality of nature may be seen as strength through domination! The more one dominates, be it through brute strength, or subtle evasions, the more likely one is to survive and procreate. Couple this with a Buddhist or Schopenhauerian suggestion that all is Will (or desire), and one can find no other conclusion than that nature goes beyond the desire to survive and is one with the Will to Power.

Connections of *evolution* to Nietzsche's concerns with 1) a striving for the Over-man, 2) breeding through the Eternal Recurrence, and 3) a morality praising strength and hence the Will to Power seems too obvious and strong to be defeated by any of his claims to the contrary. If these connections are to be questioned at all, it could only be to see Nietzsche's schemas as being more efficiently elegant, and thus circumscribing and determining or situating other theories of evolution. To reject theories of evolution, or even science itself, in favor of the Will to Power would not be to throw them away, but to see how the Will to Power is a further development of these theories and a superseding of them. Indeed, the Will to Power can not only describe evolutionary forces, but psychological and physiological motivations as well:

> "Physiologists should think before putting down the instinct of self-preservation as the cardinal instinct of an organic being. A living thing seeks above all to discharge its strength—life itself is will to power; self-preservation is only one of the indirect and most frequent results" (*Basic Writings of Nietzsche*, 211 (*Beyond Good and Evil*)).

Moreover, for Nietzsche, the Will to Power may not only be found to be a better theory of life motivation, but may also be the ulterior motive behind all other theory making!

However, one can easily question Nietzsche's noble interpretation of nature. Indeed, although one can see that the "strong" have the best chance for survival, strength also comes in various modes and through collectivity. Nietzsche's preference for the lone predator over "the herd" or the vegetable is completely inexplicable—one may postulate the projection of a loner attempting to justify himself. To glorify the centralization of the Will to Power in the single individual—is such not a crude praise of the likes of Christ, one of the most individualized loci of power in European history? Besides, who strives for power—who is more filled with the Will to Power—those who have it, or those who don't? Indeed, the Will to Power might only describe the aspirations of those who have not achieved absolute domination. And it is in this way that the Will to Power is a "theory" infused with "ressentiment": the Will to Power describes the attempted revenge of those without absolute power. Perhaps Nietzsche, denying all absolutes, would suggest that all beings are victim to a perpetual struggle of revenge for the sake of the Will to Power.

This Will to Power is possibly beyond natural selection (the subjective experience of which, may sound a little like aiming for "eternal comfort" if not being "love" connected with "chance"). "Natural selection" seems a concept aiming towards "stasis" in an ever changing environment; whereas the concept of a "Will to Power" designates a temporal *force* dynamic, where "Power" may not necessarily be absolutely static (or possibly potential), but rather, kinetic. Note that although the temporal measure of power, the "Watt" was named after James Watt in 1889, a little after the time that Nietzsche was writing most of his works, the idea of power as technical term had already been developed somewhat as that which designates the intensity or amount of energy per unit of time. Possibly the Will to Power concept could be seen as a bridge between the mechanical understanding of energy, the natural selection of evolution, and the subjective experience of that bio-mechanical evolutionary process ("Will" designating both an "objective component" often attributed to an organism—like Freudian "desires" or Kantian "purposiveness"—*and* subjective experience, as we feel desires in an emotional way). The quote above about discharging strength suggests that Nietzsche is not placing "Will" as a desire for what is lacking, but more like the Kantian "purposiveness," "Will" could be seen as a positive orientation. One might wish that Nietzsche had rather coined the phrase, "Will to Empowerment"—

and really, what is Nietzsche's broader aim as an author trying to inform the public, if not an attempt at empowering others? Possibly Nietzsche had this concept in mind as a replacement for "God" as omnipotent will—possibly Nietzsche saw some Will to be all-Powerful as a desire to be God, where the human mind might play as intermediary between God's omnipotence and the traditionally "Satanic" desire to be God. With more than a desire for the messianic, that "human" mind might desire itself all the way to becoming that God, which in turn expresses that subjective desire as objective power (the hunger becomes discharge—wish becomes action; imagined freedom turns to some place between perfectly reasoned fate and wild random chance).

Much of this has taken the Will to Power purely as a description of nature. This sort of description goes beyond the reproductive aspects of concrete science; the attempts to grasp the "mechanistic" laws repeated and reverberated throughout all nature, just as the "purposiveness" of Kant's Reason goes beyond the concrete judgments of the Understanding. Nietzsche has to admit that the "truth" of the Will to Power, or any other schematics of purposive force, such as the Desire for Reproduction, are metaphorically severed from the concrete. And such speculations must explain how any centralized force, be it termed "evolution", "desire", "purpose", or "will to power" can have a relationship with concrete chains of causality. However, despite the fact that the Will to Power does not offer a mechanistically sufficient explanation of life, Nietzsche's texts offer a thorough demonstration of its descriptive potency; the Will to Power may be a more effective description of nature than even causality.

Nietzsche might view such a focus on the descriptive aspect of the Will to Power as too Apollonian, and consequently, an adequate discussion of it would have to turn towards it's more dynamic Dionysian forces; a shift from *describing* (plastic art) the phenomenal to a concern with the experience of (musical) *expression*. However, just as much science strives to purely describe nature from a perspective devoid of individual and cultural bias, here we must consider the possibility of a pure expression of nature free from the culturally directed individual will.

REVELATION (THE ETERNAL RECURRENCE)

There seems to be little doubt in Nietzsche's mind that he accomplished his highest poetic achievement with his *Thus Spoke Zarathustra.* With regard to this work, I here quote at length some comments from *Ecce Homo* concerning inspiration:

> "Has anyone at the end of the nineteenth century a clear idea of what poets of strong ages have called inspiration? If not, I will describe it.—If one had the slightest residue of superstition left in one's system, one could hardly reject altogether the idea that one is merely incarnation, merely mouthpiece, merely a medium of overpowering forces. The concept of revelation—in the sense that suddenly, with indescribable certainty and subtlety, something becomes visible, audible, something that shakes one to the last depths and throws one down—that merely describes the facts. One hears, one does not seek; one accepts, one does not ask who gives; like lightening, a thought flashes up, with necessity, without hesitation regarding its form—I never had any choice.
>
> A rapture whose tremendous tension occasionally discharges itself in a flood of tears—now the pace quickens involuntarily, now it becomes slow; one is altogether beside oneself, with the distinct consciousness of subtle shudders and of one's skin creeping down to one's toes; a depth of happiness in which even what is most painful and gloomy does not seem something opposite but rather conditioned, provoked, a necessary color in such a superabundance of light; an instinct for rhythmic relationships that arches over wide

> spaces of forms—length, the need for a rhythm with wide arches, is almost the measure of the force of inspiration, a kind of compensation for its pressure and tension.
>
> Everything happens involuntarily in the highest degree but as in a gale of feeling of freedom, of absoluteness, of power, of divinity.—The involuntariness of image and metaphor is strangest of all; one no longer has any notion of what is an image or a metaphor: everything offers itself as the nearest, most obvious, simplest expression. It actually seems, to allude to something Zarathustra says, as if the things themselves approached and offered themselves as metaphors ('Here all things come caressingly to your discourse and flatter you; for they want to ride on you back. On every metaphor you ride to every truth.... Here the words and word-shrines of all being open up before you; here all being wishes to become word, all becoming wishes to learn from you how to speak')" (*Basic Writings of Nietzsche*, 756-757 (*Ecce Homo*)).

Similarly, in *The Will to Power*:

> "the extreme sharpness of certain senses, so they understand quite a different sign language—and create one—the condition that seems to be a part of many nervous disorders—; extreme mobility that turns into an extreme urge to communicate; the desire to speak on the part of everything that knows how to make signs—; a need to get rid of oneself, as it were, through signs and gestures; ability to speak of oneself through a hundred speech media—an explosive condition. One must first think of this condition as a compulsion and urge to get rid of the exuberance of inner tension through muscular activity and movements of all kinds; then as an involuntary co-ordination between this movement and the processes within (images, thoughts, desires)—as a kind of automatism of the whole muscular system impelled

> by strong stimuli within—; inability to prevent reaction; the system of inhibitions suspended, as it were. Every inner movement (feeling, thought, affect) is accompanied by vascular changes and consequently by changes of color, temperature, and secretion. The suggestive power of music, its 'Suggestion mentale';—"(*The Will to Power*, # 811, pp. 428-429)

Indeed, during a climactic episode in the third part of *Thus Spoke Zarathustra* entitled "On the Vision and the Riddle", we find an exposition of the work's:

> "fundamental conception" (*Basic Writings of Nietzsche*, 751 (*Ecce Homo*)).

—The Eternal Recurrence:

> "Behold... this moment! From this gateway, Moment, a long eternal lane leads backward: behind us lies an eternity. Must not whatever can walk have walked this lane before? Must not whatever can happen have happened, have been done, have passed before?" (*The Portable Nietzsche*, 270 (*Thus Spoke Zarathustra*)).

What Nietzsche understood by what could be called this "revelation of the Eternal Recurrence" may also be evidenced by some remarks noted in *The Will to Power*:

> "in a reprehensible world reprehending would also be reprehensible—And the consequences of a way of thinking that reprehended everything would be a way of living that affirmed everything—If becoming is a great ring, then everything is equally valuable, eternal, necessary.—In all correlations of Yes and No, of preference and rejection, love and hate, all that is expressed is a perspective, an interest of certain types of life: in itself, everything that is says Yes" (*The Will to Power*, #293, p. 165).

With the above quotes, we can see an attempt to describe the experience of un-willed expression; an expression that, although occurring through an individual, "scientifically" suspends the individual's perspective, and lets nature force articulation. Such a suspension might be further understood as unconscious:

> "All perfect acts are unconscious and no longer subject to will; consciousness is the expression of an imperfect and often morbid state in a person" (*The Will to Power*, #289, p. 163).

> "To the extent that it is willed, to the extent that it is conscious, there is no perfection in action of any kind" (*The Will to Power*, #434, p. 238).

Nietzsche also comments:

> "Becoming-conscious is a sign that real morality, i.e., instinctive certainty in actions, is going to the devil" (*The Will to Power*, #423, p. 228).

And:

> "Intensity of consciousness stands in inverse ratio to ease and speed of cerebral transmission [....] We must in fact seek perfect life where it has become least conscious [....] Genius resides in instincts; goodness likewise. One acts perfectly only when one acts instinctively" (*The Will to Power*, #439, 440, pp. 242-243).

Such a concern with involuntary and instinctive bodily action seems quite in line with the immediacy of the moment of expression, the "now."

Despite this immediacy of revelatory expression, "something" is expressed, as with Nietzsche's disclosure of the Eternal Recurrence: the moment of expression leaves its trace which in turn is interpreted—Nietzsche's moment abysmally stretches out beyond any beginning or end, looping into itself in a perpetual repetition. Such a "ring" of becoming, beyond being evidenced in the cycle of seasons, can also be seen with the Buddhist's wheel of existence, or with the Zodiac. And it is with this thought of the Eternal Recurrence that any hope of an afterlife is lost—we are already within the eternal, our actions reverberating with infinite consequences. This possibly dreadful thought forces an affirmation of life here and now—and as Nietzsche sees life as the Will to Power, one must affirm it as well, even though it is thoroughly infused with unsatisfied ressentiment: but at its limit, the Will to Power reprehends its reprehending and is affirmed in Eternal Recurrence. Nietzsche thus prescribes the affirmation of nature, both at large and as bodily instinct (compare William Blake and Walt Whitman): but again, is Nietzsche's description of nature always valid?

Despite Nietzsche's rather zealous proclamation of the eternal return, he criticizes all attempts at religious idealization. And curiously, he has traced the origin of religions to inspirations similar to those cited above:

> "Among intelligent, strong, and vigorous races it is mainly the epileptic who inspires the conviction that a strange power is here at work; but every related condition of subjection, e.g., that of the inspired man, of the poet, of the great criminal, of passions such as love and revenge, also leads to the invention of extra-human powers. A condition is made concrete in a person, and when it overtakes us is thought to be effected by that person. In other words: In the psychological concept of God, a condition, in order to appear as effect, is personified as cause.
>
> The psychological logic is this: When a man is suddenly and overwhelmingly suffused with the feeling of power—and this is what happens with all great affects—it raises him a doubt about his own person: he does not dare to think him-

self the cause of this astonishing feeling—and so he posits a stronger person, a divinity, to account for it.

> In summa: the origin of religion lies in extreme feelings of power which, because they are strange, take men by surprise: and like a sick man who, feeling one of his limbs uncommonly heavy, comes to the conclusion another man is lying on top of him, the naïve homo religiosus divides himself into several persons. Religion is a case of 'alteration de la personalite.' A sort of feeling of fear and terror at oneself—But also a feeling of extraordinary happiness and exaltation—Among the sick the feeling of health is sufficient to inspire belief in God, in the nearness of God" (*The Will to Power*, #135, pp. 85-86).

Here we can see that Nietzsche's critique of religion applies to his Zarathustra as well:

> "the origin of the holy lie is *The Will to Power*" (*The Will to Power*, #142, p. 92).

How now can we trust Nietzsche, as he seems to have appraised his own "truths" as lies?

It would seem, in conclusion, that Nietzsche does not want us to trust him—or anyone else for that matter. Rather, he teaches that one must take the burden of truth upon oneself—one must reveal one's own truth, and creatively interpret nature from one's own perspective, possibly in a counter-cultural affirmation of one's own body. Yet, these following words of Nietzsche may be telling:

> "One should not conceal and corrupt the facts of how our thoughts have come to us" (*The Will to Power*, #424, p. 229).

Maybe Nietzsche would claim he has been honestly dishonest, where others before have simply been dishonest.

Rider-Waite tarot card (1909)

Chapter 15

Freud & Jung
Classified Unconscious

CHAOS, PILOT, LAW

Through his examination of a mountain of anecdotal evidence, Sigmund Freud's (1856-1939) work falls somewhere between philosophy and science. Lacking much statistical evidence, or experimental verifiability beyond interpretation of various psychical phenomena or case histories, and with conclusions that can strike one immediately as misogynistic, Freud nonetheless developed a loose conceptual theory of mind that has influenced many philosophers following him, including its appropriation by feminist thinkers. His vocabulary, including the "id," "ego," and "superego" has entered the cultural vocabulary at large; and his organized investigation of the unconscious continues to provoke thought.

The conceptual triad which Freud developed to categorize the human psyche, the id, ego, and superego, roughly translate into unconscious desires, preconscious mediation between the internal and outer worlds, and social ideals. What we have is a bit more complicated than an ego trying to judge how to express id-desires in a super-ego approved manner; and it will help to go into each realm with a little more depth.

Of the id, Freud claims:

> "It contains everything that is inherited, that is present at birth, that is laid down in the constitution—above all, therefore, the instincts, which originate from the somatic organization and which find a first psychical expression here in forms unknown to us" (Sigmund Freud, James Strachey (ed.), *An Outline of Psycho-Analysis*, W.W. Norton & Company: New York (1949), p. 14).

The id has blind desires (only the ego can see)—drives that can be divided into two basic instincts: Eros and Thanatos (the later, Freud called the "death instinct," and these two terms might be parallel to "desire" and "fear"). Eros aims to unify on ever greater levels, and preserve such unity. The death instinct aims to sever connections and destroy. Eros is directly connected with "libido" with all its sexual associations. Here we see a connection with Plato's dialectic of splicing and slicing—but in this case, directly intertwined with drives. At any rate, the id is involved with tensions; tensions that must be maintained at some level for life to persist, but which must be released or discharged, as with catharsis (or possibly, analogously, with orgasm) for the achievement of satisfaction and pleasure. Since the id is deaf, and blind as it were, it can't tell whether that which satisfies it is indeed real or not—even though the later structuralist psychologist Jacques Lacan equated the id with "the Real" (which Lacan saw as an unknowable flux). The id may indeed designate some deeper reality (cp. Jung), but it is not directly connected with the reality of the outer world—being a drive primarily of the body and brain—but again, its desires can be satisfied by memories and dreams as well as current perceptions.

The ego on the other hand, is an alteration of the id that operates on a pre-conscious level, and mediates between desires and consequences. Of the ego, Freud explains:

> "In consequence of the pre-established connection between sense perception and muscular action, the ego has voluntary movement at its command. It has the task of self-preservation. As regards *external* events, it performs that task by becoming aware of stimuli, by storing up experiences about them (in the memory), by avoiding excessively strong stimuli (through adaptation) and finally by learning to bring about expedient changes in the external world to its own advantage (through activity). As regards *internal* events, in relation to the id, it performs that task by gaining control over the demands of the instincts, by deciding whether they are to be allowed satisfaction, by postponing that satisfaction to times and circumstances favorable in the external world or by suppressing their excitations entirely" (*Frued* (1949), p. 15).

To Lacan, such translates to the "Imaginary" (or subjective flux)—the ego is on a journey of lowering tension to increase pleasure, and avoiding tension and pain—foreseeable pain bringing anxiety. Operating through the pre-conscious (the preconscious designates all that is available to conscious recall at will—all our memories, known "folk-psychology" desires, knowledge, etc.), ego limns the edge of our conscious experience—yet the pre-conscious ego (and even unconscious id) erupt into consciousness with speech:

> "Conscious processes on the periphery of the ego and everything else in the ego unconscious—such would be the simplest state of affairs that we might picture. And such may in fact be the state that prevails in animals. But in men there is an added complication through which internal processes in the ego may also acquire the quality of consciousness. That is the work and function of speech, which brings material in the ego into firm connection with mnemic residues of visual, but more particularly auditory, perceptions" (*Freud* (1949), p. 35).

Consciousness, for Freud, is mostly limited to the phenomenal experience of our sense perceptions—mostly of the external, but also of the internal: and our internally generated thoughts must be reality tested against the external world—lest we project our wishes, even to the point of hallucination.

Again recalling Lacan's triad, Freud's super-ego corresponds to the "Symbolic" (language). The super-ego arises out of the ego as a residue of the childhood experience of having parents. The super-ego seems to act as a sort of barrier that, like a mirror, reflects a persons' desires at themselves:

> "When the super-ego is established, considerable amounts of the aggressive instinct are fixated in the interior of the ego and operate there self-destructively. This is one of the dangers to health by which human beings are faced on their path to cultural development" (*Freud* (1949), p. 35).

Moreover:

> "It is hard to say anything of the behavior of the libido in the id and in the super-ego. All that we know about it relates to the ego, in which at first the whole available quota of libido is stored up. We call this state absolute, primary *narcissism*. It lasts until the ego begins to cathect the ideas of objects with libido, to transform narcissistic libido into object-libido" (*Freud* (1949), p. 35).

Libido, is looking for a true love "object" as it were; and such cathexis energy can be shunted from the self to one object and on to another—displacement. This mobility can be contrasted with the fixation of libido on certain objects as with fetishes.

FAMILY DRAMA & FORGOTTEN TRAUMA

The connection between the idea of libido as sex drive and super-ego as parent naturally connects with the Freudian family drama. Freud claims to have found that there are three phases of sexual development—the first beginning soon after birth until about age five whence a second "non-sexual" latent stage occurs until the third stage and second sexual wave with puberty. We are said to forget the earlier sexual wave with an "infantile amnesia." We are born, or so it seems, with no delimited erogenous zones—the entire body, not just the genitals, are potentially erotic, but are limited by stages of sexual development, when the oral phase (connected with breast-feeding) which, by way of teeth transitions to a sadistic-anal stage (where the death instinct connects with libido), and then develops into the phallic stage (where the clitoris is seen as a phallus, albeit later seen, Freud claims, as an insufficient one). The phallic stage hence sees the differentiation between male and female sexuality:

> "The boy enters the Oedipus phase; he begins to manipulate his penis and simultaneously has phantasies of carrying out some sort of activity with it in relation to his mother, till owing to the combined effect of a threat of castration and sight of the absence of a penis in females, he experiences the greatest trauma of his life and introduces the period of latency with all its consequences" (*Freud* (1949), pp. 25-26).

Whereas:

> "The girl, after vainly attempting to do the same as the boy, comes to recognize her lack of a penis or rather the inferiority of her clitoris, with permanent effects on the development of her character; as a result of this first disappointment in rivalry, she often begins by turning away altogether from sexual life" (*Freud* (1949), p. 26).

Freud's completely questionable attribution of "penis envy" to young girls makes one suspect more of his theorizing, and more than one psychoanalyst has questioned his objectivity by way of psychoanalyzing Freud, and psychoanalyzing psychoanalysis itself.

At any rate, for Freud, sexual development reaches a fourth and final stage with puberty: the genital phase:

> "A state of things then is established in which (1) some earlier libidinal cathexes are retained, (2) others are taken in to the sexual function as preparatory, auxiliary acts, the satisfaction of which produces what is known as fore-pleasure, and (3) other urges are excluded from the organization, and are either suppressed altogether (repressed) or are employed in the ego in another way, forming character-traits of undergoing sublimation with a displacement of their aims" (*Freud* (1949), p. 26).

Again and again, Freud draws a picture of energy being pent-up, connecting with an object, being directed at the self, or being displaced to other objects. There is a sort of plumbing involved, where energy flows through unconscious pipes, twisting this way and that in a network of both unconscious and perceived ideas; finding release in pleasure, tension in pain. One gets a picture of neurons transmitting signals across networks, getting stuck in loops around problem areas, and finding release through opening up into the actual (or imaginary) world. These problem loops are usually unconscious, in what are called the primary processes, as opposed to secondary processes of the ego (as Lacan discusses Freud's maxim "Wo Es war, soll Ich werden" or "where the id was, there ego shall be" in Jacques Lacan, Alan Sheridan (trans.), *Écrits: A Selection*, W.W. Norton & Company: New York (1977), p. 128). These unresolved areas are like scars retained from the first wave of sexual development in the family sexual drama.

Caught in the middle of this unconscious flow of energies, the ego develops various "ego defense mechanisms": repression, projection, reaction formations (where one instinct, e.g. love, is hidden by the opposite, e.g. hate), and regression (a flight from realistic thinking). Such show, again, that the ego is not fully conscious—and for such, psychoanalysis has tried to bring hidden knots of the mind to the fore through the "confessional" talking cure. And indeed this talking cure might have a history in Catholic confessionals, where one tries to relieve their guilt through anonymous verbalization of their "sins" (where the Church itself seems to be like some new super-ego).

The unconscious flows of energies are also found in dreams, which Freud claims are always aimed at wish-fulfillment. Dreams are good avenues to approach the unconscious with since the ego is not on guard to defend against the unearthing of painful formative memories (although we may have anxiety dreams, where the wishes of the id are in conflict with the ego):

> "We find our way to the understanding ('interpretation') of a dream by assuming that what we recollect as the dream after we have woken up is not the true dream-process but only a *facade* behind which that process lies concealed. Here we have our distinction between the *manifest* content of a dream and the *latent* dream-thoughts. The process which produces the former out of the latter is described as the *dream-work*. The study of the dream work teaches us by an excellent example the way in which unconscious material from the id (originally unconscious and repressed unconscious alike) forces its way into the ego, becomes preconscious and, as a result of the ego's opposition, undergoes the changes we know as *dream-distortion*" (*Freud* (1949), p. 39).

Despite some archaic heritage and ancestral memories (which relate to Jung's archetypes and will be discussed in a moment),

Freud sees the dream-world as highly idiosyncratic and representative of each person's unique past, yet viewed through his stable and universal schemas (Oedipus complex, etc). Freud finds there is an incredible amount of "condensation" where many latent elements are presented in one manifest element (the "manifest" being what is recalled in the dream, and the "latent" being an interpretation of what that manifest element means for a specific personal history)—and alongside "condensation" comes "displacement"—where one element stands in for another, as if by way of allusion of metaphor. Years of dream-interpretation seemed to have confirmed Freud's theoretical apparatus... at least for many psychoanalysts—as many dreams seemed to play out the early family sexual drama, and repressed traumas.

Again, with Freud we have encountered a triplet of triads:

Id	Ego	Super-Ego
Unconscious	Pre-conscious	Consciousness
Oral	Anal	Phallic

And a few dyads:

Eros-libido	Thanatos-Death-destruction-drive
Pleasure	Pain
Release	Tension
Latent	Manifest

These triads and dyads not to be seen as clean-cut, but can overlap. Yet the real insight—that psychic energy is constantly on the move, displaced from one object to another—or from latent to manifest elements; and that certain elements are repressed from conscious recall or comprehension connects directly with later developments in both deconstruction and cognitive science: with

Derrida's introduction of the endless deferring of ultimate meaning in "*differance*," and with the theories of semantic networks and spreading activation (to be discussed at greater length in a later chapter). Indeed, Lacan claimed that the unconscious is structured like a language... bringing psychoanalysis closer to cognitive science. Even Freudian slips can be seen as a result of semantic networks, where the association of words and ideas can result in our speech betraying our thoughts despite ourselves—when we say more than we want to despite our censoring egos. It is in this way that psychoanalysis and the interpretation of unconscious intent has lead to such practices as analysis of literary texts, to find meaning that the authors did not intend—and these unintended elements of novels, poetry, and such can expose not only the author's unconscious, but the unconscious aspects of culture and humanity.

THE FORGOTTEN WHOLE

In his *The Archetypes and the Collective Unconscious*, Carl Jung (1875-1961) defines the collective unconscious as such:

> "The collective unconscious is part of the psyche which can be negatively distinguished from a personal unconscious by the fact that it does not, like the later, owe its existence to personal experience and consequently is not a personal acquisition. While the personal unconscious is made up essentially of contents which have at one time been conscious but which have disappeared from consciousness through having been forgotten or repressed, the contents of the collective unconscious have never been in consciousness, and therefore have never been individually acquired, but owe their existence exclusively to heredity. Whereas the personal unconsciousness consists for the most part of *complexes*, the content of the collective unconscious is made up essentially of *archetypes*" (Carl Jung, R.F.C. Hull, trans, Joseph Campbell, ed., *The Portable Jung*, The Viking Press: New York (1971), pp. 59-60).

These archetypes are mythological motifs, like Greek Gods, Tarot card figures, or astrological signs, hero-cycles, etc. For Jung the unconscious is symbolic—and archetypes are the most basic kinds of symbols—religious, but also moral and cultural, philosophical, scientific—they are the base of concepts, or ideas applied to reality. A "complex" in contrast has more to do with a cluster of associations around an idiosyncratically problematic area for an individual—and when one practices "word association" emotions were found to slow one down around these complex constellations around painful ideas.

For Jung, a primary dichotomy is between a healthy unity and unhealthy dissociation or splitting (as with schizophrenia, or the differentiation between races, e.g. with racism). Following Hegel, Jung sees our individual consciousnesses on a journey of self-discovery in relation to an absolute whole—the eternal archetypes are part of that absolute and beyond our physical space / time reality as individuals. For Jung, libido is not just sexual, but a psychic energy in general; and dreams are unconscious communications, not just wishes. It is through the unconscious that we might have access to the eternal whole—which slips through to our reality with phenomena like "synchronicity" where meaningful coincidences suggest the bringing together of disparate events connected at a higher level that is in touch with the whole: The whole is the soul which appears thorough unconscious meanings and "coincidental" connections.

Neurosis for Jung is not formed in the past, as with Freud, but in the present—the past is reactivated, but not a root cause. Rather, here libido is redirected from the world to the self; a self which tries to compensate for imbalances—and imbalance problems are brought to attention just as a pain tells of a wound. Therapy for Jung aims to help individuals find a balance between the whole and the individual—to find meaning in life, esp. through the use of active imagination found in fantasy and art. Such a healthy individuation allows for successful relationships with others.

Along the lines of archetypical motifs, Jung finds various psychological types: the introvert and the extrovert, as well as those who emphasize thinking, feeling, sensation, or intuition. Such resonates with astrology, with each sign oscillating between introversion and extroversion, and having various psychological traits based on their element (fire, earth, air, water), etc.

Jung appears to have followed, or anticipated, the notion of "art therapy" with his recently published *Red Book*, of which he claimed in 1957:

> "The years of which I have spoken to you, when I pursued the inner images, were the most important time of my life. Everything else is to be derived from this. It began at that time, and the later details hardly matter anymore. My entire life consisted of elaborating what had burst forth from the unconscious and flooded me like an enigmatic stream and threatened to break me. That was the stuff and material for more than only one life. Everything later was merely the outer classification, the scientific elaboration, and the integration into life. But the numinous beginning which contained everything, was then" (quoted in Carl Jung, Mark Kyburz, John Peck and Sonu Shamdasani (trans.), *The Red Book*, W.W. Norton & Company: New York (2009), p. vii).

This mammoth endeavor, completely written in his own calligraphic hand—and marvelously illustrated by Jung—charts the journey of his soul through many mystic encounters on the way to profound self-discovery—the exploration of universal archetypes as expressed by his own individual soul—his own personal Odyssey—here speaks the Magician, Philemon (always spelled with Greek letters by Jung, and suggesting a "lover" of "one"):

> "Happy am I who can recognize the multiplicity and diversity of the Gods. But woe unto you, who replace this incompatible multiplicity with a single God. In so doing you

> produce the torment of incomprehension, and mutilate the creation whose nature and aim is differentiation. How can you be true to your own nature when you try to turn the many into one? What you do unto the Gods is done likewise unto you. You all become equal and thus your nature is maimed" (*Jung* (2009), p.351).

And our protagonist later asks:

> "Illustrious one, you teach that man is a gateway? A gateway through which the procession of Gods passes? Through which the stream of life flows? Through which the entire future streams into the endlessness of the past?" (*Jung* (2009), p. 354).

Again, we see a concern with the dialectic of the diversity and individuality of the many and the whole one—as symbolically embodied by archetypes—a dialectic which resonates throughout this book, and which we will see enacted again, in the next chapter on James Joyce's *Finnegans Wake*.

Salvador Dali - Dali Atomicus (1948)

Chapter 16

Joyce
Engineering Mythological Dialectics

A SIMPLE MACHINE

As our earlier diversion from the history of philosophy towards Shakespeare's "Hamlet" demonstrated that literature can inform and anticipate philosophical breakthroughs, this chapter will bring forth some important developments in the evolution of human thought on the way towards machine intelligence, or a mechanical understanding of intelligence, found in James Joyce's (1882-1941) *Finnegans Wake.* Such an avenue has been explored by a few others, especially in Jorn Barger's online "The Robot Wisdom pages" (<//www.robotwisdom.com>). Barger argues that social justice requires a computer simulation of humanity that is more along the lines of literature. And in fact, Joyce also compared *Finnegans Wake* to a machine:

> "... I really am one of the greatest engineers, if not the greatest, in the world besides being a musicmaker, philosophist and heaps of things. All the engines I know are wrong. Simplicity. I am making an engine with one wheel. No spokes of course. The wheel is a perfect square" (Postcard to Har-

riet Shaw Weaver, 16 April 1927; from Richard Ellmann (ed.), *Selected Joyce Letters,* The Viking Press: New York (1975), p. 321).

However, Joyce critic William York Tindall notes:

> "The movement of washing is up and down. This, aside from cycling, is the general movement of *Finnegans Wake*, which goes round and round, up and down, ceaselessly, like some great engine—though this comparison, while faithful to all the clicking and meshing and reciprocating, is otherwise alien to so humane an intricacy" (William York Tindall, *A Reader's Guide to James Joyce*, The Noonday Press, Inc: New York (1959), pp. 250-251).

PHILOSOPHY BETWEEN THE LINES

In the first quote by Joyce above we see him acknowledge his philosophical tendencies; as he also says in the *Wake*:

> "This is me Belchum sneaking his phillippy out of his most Awful Grimmest Sunshat Cromwelly" (James Joyce, *Finnegans Wake*, The Viking Press: New York (1968), pp. 8-9).

Here we find, as one does upon opening any page of *Finnegans Wake*, something like Lewis Carroll's *jabberwocky* neologisms and portmanteaus abound; "Phillippy" only *sounding* like "philosophy" —and indeed, if *Finnegans Wake* is a philosophical work, it is done in a sneaky way—one has to look beyond the literal "narrative" and prose and interpret the work, as Joyce speaks through his hat, before finding the profound philosophical insights within. Some additional quotes from the wake concerning philosophy include:

> "He ought to blush for himself, the old hayheaded philosopher, / For to go and shove himself that way on top of her" (*Finnegans Wake*, p. 47).

"Hayheaded" suggests a fallacious straw man argument, possibly the scarecrow from the Wizard of Oz in search of a brain he already possesses; while the blush of the *philosopher* brings to mind the shame arising from the tree of knowledge of good and evil—an evil here of the possible sexual transgression of the *Wake's* main character Humphrey Chimpden Earwicker (or HCE)—and one is reminded of the Platonic practice of placing "pattern" and father above "matter" and mother.

> "we lost and plundered of it even to the hidmost coignings of the earth and all it has gone through and by all means, after a good ground to kiss Terracussa and for wars luck our lefftoff's flung over our home homoplate, cling to it as with drowning hands, hoping against hope all the while that, by the light of philophosy, (and may she never folsage us!) things will begin to clear up a bit one way or another within the next quarrel of an hour and be hanged to them as ten to one they will too, please the pigs, as they ought to categorically, as, stricly between ourselves, there is a limit to all things so this will never do" (*Finnegans Wake*, pp. 118-119).

Again, here a transgression with regard to *Philosophy* is suggested: plundering the hidden—with a "light" to "clear up" by way of a "quarrel" (philosophical argument). As Anthony Burgess comments on the *Wake*, which can be described as a "universal dream":

> "Let us not be too much tempted to drag the big dream up to the light: shadowiness, confusion, the melting of one personage into another, of youth into age, friend into enemy—these are the essence of the dream" (Anthony Burgess, *Re-Joyce*, Ballantine Books: New York (1966), p. 259).

But here philosophy, tied to the Old Testament fall of humanity, is also tied to possible New Testament salvation and resurrection ("never folsage us!" having the word "sage" in it, and also suggesting the words of Christ on the cross).

> "I am not hereby giving my final endorsement to the learned ignorants of the Cusanus philosophism in which old Nicholas pegs it down that the smarter the spin of the top the sounder the span of the buttom, (what the worthy old auberginiste ought to have meant was: the more stolidly immobile *in space* appears to me the bottom which is presented to use in time by the top primo-mobilisk &c.)" (*Finnegans Wake*, p. 163).

In this last passage, there seems to be a rejection of a hierarchical dualism between space and time, with time on top and space solidly on the bottom. This space / time dualism is also personified in the wake between two characters, the rival sons of HCE, Shem (time, the ear), and Shaun (space, the eye).

> "And the phyllisophies of Bussup Bulkeley" (*Finnegans Wake*, p. 435).

This last clear reference, to the philosopher Bishop Berkeley, ties in also with a contrast of color games with geometry homework found in the first two chapters of Book II in *Finnegans Wake*; as Berkeley was concerned with so-called "secondary qualities" (colors, sounds, etc.) actually being more primary than the primary quality of geometrically structured extension.

In sum, these references to "Philosophy" in *Finnegans Wake* suggest that it has to do with the head (speaking through a hat, hayheaded), that it is tied to the fall and resurrection (knowledge), that it is concerned with clarifying issues through argument; and that it has plundered the hidden, and possibly feminine, only to establish erroneous hierarchies—presaging some of the themes "deconstructive" philosophy.

NARRATIVE: CHRONOLOGY & STRUCTURE

Now, in his previous two novels, *A Portrait of the Artist as a Young Man*, and *Ulysses*, the narrative is much clearer: in *A Portrait* we have an account of a young artist growing up—the narrative is driven by the progress of life from youth to young adulthood. In *Ulysses*, the narrative follows the course of events in a single day, and also parallels the heroic journey of Homer's *Odyssey*. Again: very clear temporal narratives that are chronologically driven from a starting point to a finishing point, even though minds wander in these books that depict streams-of-consciousness. *Finnegans Wake*, on the other hand, takes the stream-of-conscious approach right into dreamland, and like the Zodiac or 18th century philosopher Giambattista Vico's conception of history, is cyclical: the ending sentence loops back into the first sentence, making the whole book circular (the "wheel" that Joyce mentioned in the above cited letter). On Vico and *Finnegans Wake*, Samuel Beckett claims:

> "It is first necessary to condense the thesis of Vico, the scientific historian: In the beginning was thunder: the thunder set free Religion, in its most objective and unphilosophical form—idolatrous animism: Religion produced Society, and the first social men were cave-dwellers, taking refuge from a passionate Nature: this primitive family life receives its first impulse towards development from the arrival of terrified vagabonds: admitted, they are the slaves: growing stronger, they exact agrarian concessions, and a despotism has evolved into a primitive feudalism: the cave becomes a city, and the feudal system a democracy: then an anarchy: this is corrected by a return to monarchy: the last stage is a tendency towards interdestruction: the nations are dispersed, and the Phoenix of Society arises out of their ashes" (Samuel Beckett, *Our Exagmination Round His Factification For Incamination of Work in Progress*, New Directions Books: New York (1972), p.5).

And indeed the *Wake*, taking place largely in Phoenix park in Dublin, Ireland, cites a rainbow, and sounds thunder early on with one of ten occurrences of 100 letter words relating to thunder:

> "arclight and rory end to the regginbrow was to be seen ringsome on the aquaface. / The fall (bababadalgharaghtakamminarronnkonnbronntonnerronntuonnthunntrovarrhounawnskawntoohoohoordenenthurnuk!)" (*Finnegans Wake*, p. 3).

The first 100 letter thunder word draws upon various languages: the Japanese "kaminari," Hindu "karak," German "brontao," French "tonnerre" Italian, "tuono," Swedish "aska," Irish "tornach," Portuguese "travao," old Romanian "tun," and Danish "tordenen" are all words for the English "thunder" (Roland McHugh, *Annotations to Finnegans Wake*, The John Hopkins University Press: Baltimore, Maryland (1991), p. 3). Although readable as English, the *Wake* is a melting pot of world languages—the very first word, "riverrun" sounding like the French word for "reverie" or "dream":

> "'RIVERRUN': We begin the book in French: *Riverain*, river-dweller; *Reverons*, let's dream, *Reverrons,* we will see, again" (John Gordon, *Finnegans Wake: a Plot Summary*, Syracuse University Press: New York (1986), p. 106).

"Riverrun" is also where the *Wake* is "riven"—the circle of the end looping to the beginning is torn or split apart. It is also the river Liffey, or river of life and time, and also associated with a mother figure, Anna Livia Plurabelle (ALP), and a rivalry between the brothers Shem and Shaun. The river and the rainbow with thunder signal another cycle—that of rivers flowing down to the ocean, only to evaporate, turn to clouds, and rain back into rivers.

The thunder is also the sound of a fall of a wall and Finnegan (who is also HCE), and the Old Testament fall of humankind—James Joyce's brother Stanislaus Joyce notes:

> "The strange doctrine of actual and sanctifying grace and its relation to original sin, which last was to be the subject of *Finnegans Wake*, had puzzled and fascinated my brother, as he found it in the teaching of the Church, and in his reading of St. Augustine, and even viewed from outside the Church it held his interest" (Stanislaus Joyce, *My Brother's Keeper*, The Viking Press: New York (1969), p. 227).

Moreover, the early major work on *Finnegans Wake*, Joseph Campbell and Henry Morton Robinson's *A Skeleton Key to Finnegans Wake*, introduces us to the *Wake* with these words:

> "Running riddle and fluid answer, *Finnegans Wake* is a mighty allegory of the fall and resurrection of mankind. It is a strange book, a compound of fable, symphony, and nightmare—a monstrous enigma beckoning imperiously from the shadowy pits of sleep. Its mechanics resemble those of a dream, a dream which has freed the author from the necessities of common logic and has enabled him to compress all periods of history, all phases of individual and radical development, into a circular design, of which every part is beginning, middle and end" (Joseph Campbell & Henry Morton Robinson, *A Skeleton Key to Finnegans Wake*, Viking Press: New York (1968), p. 3).

The falls and rises of HCE: a perpetual motion machine moving up and down, but also round and round—and again, following Vico, the four parts or books of *Finnegans Wake*, correspond to the circularity of the four ages:

> "Part 1 is a mass of past shadow, corresponding therefore to Vico's first human institution, Religion, or to his Theocratic age, or simply to an abstraction—Birth. Part 2 is the lovegame of the children, corresponding to the second institution, Marriage, or to the Heroic age, or to an abstraction—Maturity. Part 3 is passed in sleep, corresponding to

the third institution, Burial, or to a Human age, or to an abstraction—Corruption. Part 4 is the day beginning again, and corresponds to Vico's Providence, or to the transition from the Human to the Theocratic, or to an abstraction—Generation" (Beckett, pp. 7-8).

The four parts / books are divided into chapters, the unnamed titles of which can be somewhat reconstructed from Joyce's letters to Harriet Shaw Weaver after 1923, regarding his "Work in Progress"—here with short chapter descriptions:

BOOK I

1. "**Old Earwiker's Funeral**": A character, "Finnegan" dies in a fall while constructing a wall but is resurrected at his wake—and morphs into HCE (a father archetype).

2-4. "**Earwicker / HCE 1-3**": HCE ascends in society, only to descend as he betrays his guilt and a rumor of his mysterious transgression spreads throughout Dublin. Dreaming that he is buried, and then on trial, he is to be exonerated by a letter written by his wife ALP.

5. "**Untitled**" (begins with a list of titles): ALP dictated this letter / manifesto to her son Shem, a writer, while her son Shaun, a postman is to deliver it. However, the letter may never reach its destination.

6. "**A Picture History from the Family Album**": The major and minor characters of the book are portrayed in a series of questions.

7. "**Shem**": Shem is described as a sibling rivalry victim of his brother Shaun's accusations, yet is defended by his mother ALP.

8. "**Anna Livia Plurabelle**": Joyce says of the ALP chapter which flows like river to a widening delta:

"It is a chattering dialogue across the river by two washerwomen who as night falls become a tree and a stone" (Letter to Harriet Shaw Weaver, 7 March 1924; *Selected Joyce Letters*, p. 301).

BOOK II

1. "**Twilight Children's Games**": The focus shifts to the children—with a color guessing game that Shem cannot compete against his brother Shaun at due to poor eyesight. Shaun clearly "shines" in public, whereas Shem has his private "shame."

2. "**Night Studies – One Squared**": The three children, Shem, Shaun (as two sides of marginalia), and sister Issy (as footnotes), are engaged in a geometry lesson, Shem tutoring Shaun—and then there is a writing task. This schooling takes place above a pub where HCE works.

3. "**A Scene in the 'Public'**": In the pub, HCE hosts, while a TV and a radio broadcast stories relating to HCE's marriage and Oedipal demise at the hands of his two sons, which draws the ire of the 12 pub customers, inducing a confession of crimes from HCE. HCE then passes out after a policeman clears the pub.

4. "**Mamalujo – 4 Old Men**": Passed out, HCE dreams of four men (Matthew, Mark, Luke and John: mamalujo), making repetitive interpretations of a Tristan and Isolde like story.

BOOK III

1. "**Watches of Shaun**": The focus shifting to Shaun, who awakens floating down the Liffey river in a barrel, giving elliptical answers to questions asked about ALP's letter that Shem had written.

2. "**Doctor (2nd Watch)**": Shaun then sermonizes to his sister Issy and her schoolmates, while aging in reverse.

3. "**Melodious (3rd Watch)**": Reverse-aged all the way back to being his father, HCE, he defends his own life story.

4. "**Last Watch of Shaun**": After attempting sex while their children sleep, one child (Shem) rises from a nightmare, and is comforted by ALP, who then returns to her husband HCE's bed, and they climax in conjunction with a rooster crowing.

BOOK IV

0. "**K's Orisons**": Daybreak brings vignettes on Shem and Shaun (as Kevin and Patrick), and the close of the book reveals ALP's letter, or at least a version of it, with her engaging in a monologue and trying to wake her husband up as she drifts like a river into the ocean—and *Finnegans Wake* circles back to its beginning.

FREUD FOR THOUGHT: DUALITY & MATRIX

Beyond a simple narrative, with *Finnegans Wake* we find a mythological structuralism—the names change, but the primary structure of a father, mother, two sons, and a daughter persists. These are "everyman" and "everywoman": archetypes of a primary family. Shades of Freud and Jung are apparent here: Freud saw latent meaning in the manifest content of dreams, and how the association between words, and a repetition compulsion, could be significant for deep psychological reasons; while Jung was deeply invested in his theory of archetypes (as found in the Zodiac or Tarot cards, etc.) Clearly, even by Joyce's own assessment, *Finnegans Wake* takes place mostly through a dream or dreams (and dreams within dreams) and he was familiar with Freud and Jung's work on the unconscious.

Such as not to apply Freud to Joyce, as William Carlos William warns against in his essay, "A Point for American Criticism" (William Carlos Williams, et. al, *Our Exagmination Round his Factification for Incamination of Work in Progress*, New Directions Books, New York (1972), p 179), but to see how Joyce applied Freud. Beyond Freud and Jung, and more like Nietzsche exploring a dialectic between Dionysus and Apollo—Joyce explores his own dual mythological dialectics, especially evident with the opposition between Shem the experimental writer, and Shaun, the showy postal carrier. Here we find an introvert / extrovert opposition, but also an opposition between Shem representing a stylistic wielding of a pen; whereas Shawn is a carrier who gets along well with girls. Shem makes, Shaun takes. Shem writes, Shaun speaks. Shem is afraid, Shaun is brave. Shem is modest, Shaun is a braggart. And the two are one in HCE, who is full of guilt, and can't keep his mouth shut about it, as if sleep-talking on Freud's psychiatrist couch. But like Orion in the night sky, HCE is of colossal proportions: a universal man.

On the other side, we have the rambling "stream"-of-conscious ALP, and sister Issy, who multiply into many more than just two women. Where Shem and Shawn might be seen as a yin and yang opposition; the women of *Finnegans Wake* form a hierarchical matrix—or at least a *delta* from the one to the many.

Joyce works "dialytically" (*Finnegans Wake*, p. 614), following the dialectical process of synthesizing opposites found in Hegel. Tindal notes:

> "This, briefly, is the family process: the father begets twin sons, who quarrel. Uniting at last against father, they replace him as he falls. The rising son (a union of the quarreling twins) becomes father in his turn and begets two sons, who quarrel, unite, and, after his fall, become father. Tempting him and attracting the divided sons, the daughter is a cause of contention and fall; for Earwicker's family, like Freud's, is more or less incestuous. But the daughter becomes the mother, who gathers the pieces after war and fall in order to put them together again. The father rises and falls, divides and unites; but the mother, appeasing and renewing, is constant. This process, which is that of any

family, suggests problems that have fascinated philosophers and theologians: the one and the many, the creator and his descent into creation, the happy or creative fall, relativity, recurrence, permanence and change, fall and renewal" (Tindal, p. 243).

And Joseph Campbell claims in his *The Hero with a Thousand Faces* (also with a footnote quote from the *Wake*):

> "The hero, whether god or goddess, man or woman, the figure in myth or the dreamer of a dream, discovers and assimilates his opposite (his own unsuspected self) either by swallowing or by being swallowed. One by one the resistances are broken. He must put aside his pride, his virtue, beauty, and life, and bow or submit to the absolutely intolerable. Then he finds that he and his opposite are not of differing species, but one flesh." (Joseph Campbell, *The Hero with a Thousand Faces*, Princeton University Press: Princeton, New Jersey (1973), p. 108).

> "equals of opposites, evolved by one-same power of nature or of spirit, as the sole condition and means of its himundher manifestation and polarised for reunion by the symphysis of their antipathies" (*Finnegans Wake*, p. 92).

DISPLACEMENT & DEFERRAL

As *Finnegans Wake* reflects deep thinkers of the mind prior to its writing, it also presages complex thinkers to come, such as Gilles Deleuze and Jacques Derrida—and Derrida has suggested as much in his essay, "Two Words for Joyce" found in *James Joyce – A Collection of Critical Essays,* Mary T. Reynolds (ed.), Prentice Hall: New Jersey (1994). Although also owing to thinkers prior to Joyce, it is highly unlikely that we would see books quite like Deleuze's *Difference and Repetition*, and Derrida's concept of "differánce" with its endless deferral of meaning without there having been *Finnegans Wake*. To see this, we need to read *Finnegans Wake* not only as a repetition of mythological structures, but more as a matrix than as a narrative.

It should be clear by now, that *Finnegans Wake* is replete with a multitude of motifs. Among the motifs noted by Tindall are "Numbers, the Cad, the girls and soldiers [....] fog, tree, and clothing [....] hat, stick, race, hunt, and tea" (Tindall, pp. 253-254). These motifs repeat heavily throughout the *Wake*. And indeed, like the characters and their travails, words are very often repeated throughout the *Wake* with a slight difference: names get slightly, and not so slightly, altered; words like "Tunc" transform into "Punc," "Hush" becomes "Hiss," etc. Like our experiences of dreams, nothing is stable as people morph into objects and other people and plot points repeat with a difference—there is little surface consistency, yet a stable deeper structure seems to appear again and again in different guises (much like Freudians reading the Oedipus complex into so many scenarios). Attempts to pin down one name or word (HCE itself is an acronym with many manifestations; many ECHoes) will not get you to a point of understanding—a veil is not lifted to reveal the truth, but rather one veil gives way to another, in a process that breeds familiarity with the primal family.

And this chain of veils can be seen if one follows the thread of specific words that repeat (although slightly altered) throughout the *Wake*. For instance, on page 122 of the *Wake* we see a mention of the "*Tunc* page of the Book of Kells." Indeed there is a "Tunc" page in the *Book of Kells*, but, self-referentially (and the *Wake* is often about the *Wake*) there is also a "Tunc" page in *Finnegans Wake*: Page 611 has a page and half long paragraph that begins: "Tunc." Turn the page to 612, and the next paragraph begins with "Punc." Similarly on page 563, a paragraph begins with "Hush!" And Following through, page 587 has a paragraph starting with "Hiss!" And note how these two "pages" ("Hush!" and "Hiss!") have "constellations" of word associations; as if the threads of words were actually webs of word clusters:

"Hiss!"	–	"Hush!"
"sinnerettes"	–	"Cigalette!"
"snake"	–	"stake"
"twitch"	–	"wretch!"
"sugar"	–	"sweets"
"pet"	–	"cat"
"mouf"	–	"mouse"
"tomorrow"	–	"tosorrow!"
"waitresses"	–	"brackfest"
"pepped"	–	"pep"
"between"	–	"between"
"first"	–	"first"
"life's"	–	"life's"
"out"	–	"out"
"see"	–	"see"
"told"	–	"told"
"two"	–	"two"

This list just begins to chart the connections between these two passages, just scratching the surface, with the connections not just being between repeated words, but between concepts, and opposites too. And these connections multiply when we look for them on the self-same page, rather than on connected pages. One can also follow words throughout the *Wake* and find other words occurring near them in high frequency: the words "red" and "green" are found on at least 67 pages together. "Green" and "gold" are found on over 59 pages together. What this demonstrates is a network of clustered concepts and words. One can follow these trails of concept constellations and discover miraculous insights along with Joyce time and again.

Like an Aladdin's lamp—"alladim lamps" (*Finnegans Wake*, p. 560) or a "lamp of Jeeshees" (*Finnegans Wake*, p. 612)—a genie spirit of Joyce might be summoned by "rubbing" the *Wake* in a proper way. With a concordance—which can be found online, or the one prepared by Clive Hart (Clive Hart, *A Concordance to Finnegans Wake*, University of Minnesota Press: Minneapolis (1963))—one can look up words, people, concepts, events, things, etc. and by reading about a page before and after the occurrence of the word, find Joyce-like insights into these notions. For example, looking at the surrounding context of the word "philosophy" as cited several times at the beginning of this chapter (pgs. 9, 47, 119, 163, and 435)—one can find these words (besides common words like "who," "what," "where," "when" and "how") repeated several times (and these here are just the *exact* matches):

big, call, cat, full, good, green, home, local, Mr., night, point, poor, quite, show, time, top

Searching for the word "Freud" one finds it loosely cited four times, and related to the words "fear" or "dread" ("freudened" (*Finnegans Wake*, p. 115) and "freudful mistake"(*Finnegans Wake*, p. 411) or "friend." Also these words were repeated near "Freud" on pages 115, 337, 411, and 579:

dear, friend, gay, hand, hear, letter, love, matter, old, past, point, pray, sea, shaun, soft, walk, woman

Although all these connections may not be immediately clear, why Joyce associates "Freud" with "fear" (and "love") should be obvious to anyone familiar with Freud's concern with drives and motivations like libido, etc.

PARANOID POLYSEMY

In fact, while researching *Finnegans Wake* with this "cue-cluster" strategy, I found that certain artworks, poems, etc. not only get referenced, but *compared* in insightful ways—and this even led me to discover two "in the world" images, drawn by a pen and pencil, which, it seems to me, are highly relevant to the *Wake*. Note the images in Figures A, B and C at right:

Like the multiple meanings of Joyce's portmanteaus and neologisms, these figures have several ways of being interpreted. Figure A (a pencil mark found in a book of Edgar Allan Poe's writings left to me by my maternal grandmother) looks like a cross, or a pointy devil's tail, an underlined, "J" or a "K," a hair glancing an ear or, upside-down, a nose to the grindstone... or best yet, a shut eye facing up to the left, as if in sleep.

Figure B (a blue pen mark found at the University of Oregon library, in Robert Stanley Forsythe, *Shirley's Plays and the Elizabethan Drama*, Columbia University Press: New York (1914), p. 63), which is a reversal of the originally discovered Figure C, holds a symmetry with Figure A, and obviously looks like a teary shut-eyed gentleman with a mustache and beard exhaling smoke. The original reverse image could be an actor taking a bow while losing his head in front of a curtain or fart; or in connection with Jesus, an Xr (Chi Rho beginning the Greek word for "Christ") signature (as found in the Book of Kells), whipping the backside of some sort of horned beast turning back. Or it could be a Sea-goat, Capricorn being the sign anyone born on Christmas would have. Perhaps in figure C we can also see the letters, "J" "O" "Y" "C" & "E?"

I may be out on a limb here, as paranoid interpretation can go far afield.

At any rate, far beyond the formal aspects of *Finnegans Wake* outlined here, the depth of content is astounding. Joyce took 17 years to write his book, and he has weaved a tale, tail or fin if you will, that will leave a wake in the sea of history of literature for ages. And with his matrix of motifs, Joyce anticipated the concept of "spreading activation" which we will discuss later, in relation to neural networks and artificial intelligence.

Figure A

Figure B

Figure C

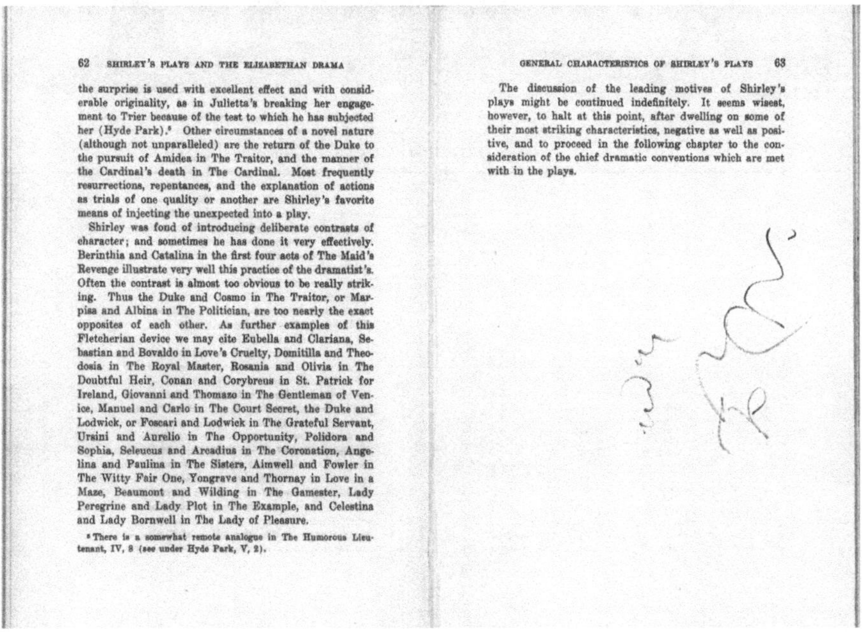

62 SHIRLEY'S PLAYS AND THE ELIZABETHAN DRAMA

the surprise is used with excellent effect and with considerable originality, as in Julietta's breaking her engagement to Trier because of the test to which he has subjected her (Hyde Park).[5] Other circumstances of a novel nature (although not unparalleled) are the return of the Duke to the pursuit of Amidea in The Traitor, and the manner of the Cardinal's death in The Cardinal. Most frequently resurrections, repentances, and the explanation of actions as trials of one quality or another are Shirley's favorite means of injecting the unexpected into a play.

Shirley was fond of introducing deliberate contrasts of character; and sometimes he has done it very effectively. Berinthia and Catalina in the first four acts of The Maid's Revenge illustrate very well this practice of the dramatist's. Often the contrast is almost too obvious to be really striking. Thus the Duke and Cosmo in The Traitor, or Marpisa and Albina in The Politician, are too nearly the exact opposites of each other. As further examples of this Fletcherian device we may cite Eubella and Clariana, Sebastian and Bovaldo in Love's Cruelty, Domitilla and Theodosia in The Royal Master, Rosania and Olivia in The Doubtful Heir, Conan and Corybreus in St. Patrick for Ireland, Giovanni and Thomazo in The Gentleman of Venice, Manuel and Carlo in The Court Secret, the Duke and Lodwick, or Foscari and Lodwick in The Grateful Servant, Ursini and Aurelio in The Opportunity, Polidora and Sophia, Seleucus and Arcadius in The Coronation, Angelina and Paulina in The Sisters, Aimwell and Fowler in The Witty Fair One, Yongrave and Thornay in Love in a Maze, Beaumont and Wilding in The Gamester, Lady Peregrine and Lady Plot in The Example, and Celestina and Lady Bornwell in The Lady of Pleasure.

[5] There is a somewhat remote analogue in The Humorous Lieutenant, IV, 8 (see under Hyde Park, V, 2).

GENERAL CHARACTERISTICS OF SHIRLEY'S PLAYS 63

The discussion of the leading motives of Shirley's plays might be continued indefinitely. It seems wisest, however, to halt at this point, after dwelling on some of their most striking characteristics, negative as well as positive, and to proceed in the following chapter to the consideration of the chief dramatic conventions which are met with in the plays.

Thomas Edison's paper-filament lamps of 1870

Chapter 17

Dewey
Methods Beyond Recogntion

SITUATING INSTRUMENTALISM

Given that John Dewey (1859-1952) belonged to the philosophical school of thought called "Pragmatism"—and that the pragmatism perspective takes into account the social milieu in contextualizing knowledgeable practices, it seems appropriate to fill in some background for his thinking. Pragmatism focuses on practice in relation to theory, and stems from a more scientific understanding of what it means to know—it looks towards the effects of action, and how effective action can be refined through experiment. In this way it is a consequentialist philosophy, drawing not only upon science, but upon the Utilitarianism that was developed in the 19th century by such thinkers (and economists) as Jeremy Bentham (1748-1832) and John Stuart Mill (1806-1873). Utilitarianism claims, following philosophers like Spinoza who suggest that we are primarily motivated towards loving pleasure and detesting pain, that we should pursue the greatest good for the greatest number of people—and that goodness is seen as aiming towards pleasure and the absence of suffering. Such a notion was implied by the economist Adam Smith's (1723-1790) notion of an "invisible hand" where *The Wealth of Nations* (the title of his most famous work) increased on the whole via the uncoordinated, if not laissez faire, pursuit of profit by individuals acting for the most part in self-interest: the good of the whole was still seen as a highest value. There are problems with reducing ethics to just the maxim of pursing the greatest good, such as with the possible infringement on the *rights* of minorities; e.g. what about enslaving or exploiting a few for the whole? Of course pleasures can be quite nuanced, not necessarily in line with your typical notion of hedonism, and the pursuit of such seems in line with a nature and evolution more clearly understood since Charles Darwin (1809-1882) conceived of "natural selection." Natural selection references a changing environment including other species, which would eliminate species that could not survive such change, and where species that had a natural fit in their niche could thrive. Adam Smith and Charles Darwin both had theories of "emergent" global phenomena from uncoordinated behavior of individuals—the macro-economy with Smith, and the ecological system with Darwin; and along with Hegel's breakthrough of historicizing philosophy, these thinkers emphasized the relation between individuals and the whole.

The term "Pragmatism" was originally used by the scientist and philosopher Charles Sanders Peirce (1839-1914) and was popularized by the psychologist and philosopher William James (1842-1910). In his essay "The Fixation of Belief" (found in Charles Sanders Peirce, Morris R. Cohen, ed, *Chance, Love and Logic: Philosophical Essays*, University of Nebraska Press: Lincoln, Nebraska (1998), pp. 7-31), Peirce claims that the best way to fix a belief is through the scientific method. Such may not seem like such a bold claim, yet the history of philosophy as we have seen, has often been an "armchair event" where thinking reflects on itself without experimental verification. Peirce was a wide-ranging thinker, also being a founder of semiotics (the systematic study of signs), and having worked on logic to the point of seeing (in 1886!) that electrical switching circuits could perform logical operations. James took up Peirce's focus on the scientific method as a way of determining not just fixed beliefs, but truth, which James saw as being determined by its "cash value" or effectiveness—pragmatism thus grew out of the scientific method, but sought to appeal to the wider frame of usefulness in general. James combined the correspondence theory of truth (that we have ideas that correspond to the world) and a coherence theory of truth (that ideas agree with other ideas), but saw that the ultimate test of the truth of an idea was its effect in the world. Hence, god or religion might not be true based on revelation or theoretical proof, but due to the positive effect that having faith might have upon the lives of believers. Such is to claim that there might not be absolute truths, that truth is more a contingent approximation based on utility.

DECONSTRUCTING DUALISTIC DOGMA

John Dewey takes the insights of Peirce and James and goes farther, fairly eliminating the notion of a foundation for knowledge in anything like a correspondence theory of truth. He examined the history of philosophy, and found there to be a chronic prejudice against change, the material, bodily, methods, practice, etc, in favor of the stable, eternal, spiritual, theoretical, etc. Dewey emphasizes the reality of material change, and claims that the notion of a truth prior to experience that we must discover intellectually is a chimera; that the reality of the human situation is better understood as a biological existence, where an organism within an environment is challenged by changes in that environment, and must adapt its actions to achieve its ends. These ends could be utilitarian, or simply our highest ideals—but even these ideals should be tested, as to their effect. Dewey sees previous philosophers as being psychologically motivated—quite naturally—by a desire for control over an often chaotic environment: security seems to be the natural aim of humans. But instead of focusing on effective methods of dealing with this change, philosophers have historically sought superficial refuge in a realm of unchanging thought. Hence we have rationalists developing their notions of inherent principles believed to exist prior to thought (as with mathematics), and even "recalled" truths from a prior existence in the case of Plato; but such does not wash up with the later developed scientific method. In fact many problems of philosophy are seen by Dewey to arise with the conflict between the scientific method and dogma inherited from a time before its use. Even the empiricists held on to the search for static knowledge—including Kant's transcendental deductions—seeking to account for the ability to recognize objects, as if recognition were enough in itself to define what is true. Dewey rejects such stable "spectator" theories of knowledge, again in favor of methods for that organism situated in an environment which it might adapt to its own ends.

TEMPORAL METHODS

With his advocacy of the scientific method of experimental verification of the effectiveness of our ideas, Dewey does not start with the individual's perspective, as with a blank slate, but rather situates the knower as already within a history that has been shaped by pragmatic adjustment to circumstances. Here, Dewey enumerates four features of experimental philosophy:

> "1. In experimental knowing, the antecedent is always the subject-matter of some experience which has its origin in natural causes, but which, not having been controlled in its occurrence, is uncertain and problematic. Original objects of experience are produced by the natural interactions of organism and environment, and in themselves are neither sensible, conceptual nor a mixture of the two. They are precisely the qualitative material of all our ordinary untested experiences. 2. The distinction between sense-data and interpretive ideas is deliberately instituted by the process of inquiry, for sake of carrying it forward to an adequately tested conclusion, one with a title to acceptance. 3. Hence each term of the distinction is not absolute and fixed, but is contingent and tentative. Each is subject to revision as we find

> observational data which supply better evidence, and as the growth of science provides better directive hypotheses to draw upon. 4. Hence the material selected to serve as data and as regulative principles constantly check one another; any advance in one brings about a corresponding improvement in the other. The two are constantly working together to effect a rearrangement of the original experienced material in the construction of a new object having the properties that make it understood or *known*" (John Dewey, *The Quest for Certainty*, Perigee Books: New York (1980), pp. 172-173).

There is thus a refinement of knowing which comes through testing a hypothesis; we start with common sense, but should reconstruct our knowledge based on verifiable experiment with an eye towards utility. In fact, Dewey claims that action precedes knowledge: we find ourselves in a world of practice—and true knowing comes through examining the results of our actions—a knowledge not existing in the abstract for us to discover, but developed in orientation to future use.

We can anticipate some connections here with Martin Heidegger's thought, esp. in reference to technological forms of knowing which are useful. This could raise the question about other forms of knowing (as with poetry); but for Dewey, truth is not revealed by turning inward and discovering one's authentic being possibly in contrast with society; truth is rather found in the efficacy of ideas and methods that have real world results—ideas which are reformations of prior social practices. Hence Dewey also focuses on education, democracy, and institutional reform, as he sees these social aspects of knowledge as shaping our lives in a real tangible way. But Dewey also sees such phenomena as qualities (like the color "red") as being epiphenomenal—his emphasis on science as effective knowing sidelines consciousness itself as not being that important.

IMPACT ON THE FUTURE

Dewey's thought, and pragmatism in general, forms an important advance in the history of thinking—and we have had more contemporary practitioners of pragmatism as well. Without such thinking we might not have Thomas Kuhn's 1962 book *The Structure of Scientific Revolutions*, an analysis of the history of science, which emphasized the history of and future possibilities of "paradigm shifts"—as with the shift from Newtonian physics to that of Einstein. However, Kuhn seems to see an ever closer approximation towards scientific truth, whereas the 20th century thinker Richard Rorty saw people as coping with reality rather than ever really knowing it. Like Dewey, Rorty was against a correspondence theory of truth in favor of discussing pragmatic actions of habit. Ideas are tools—not prior essences to be discovered, and even language does not mirror the world, but connects to it: beliefs are about the world, not reflective of it. There is no transcendent (God-like) point of view outside of history—no absolute truth that we as historically and socially situated persons can rely on: meanings change within the interconnected web of the world we are part of. What we find in the world are connections, relations and events... and if we're skillful: probabilities if not mechanistic understanding; but never an end to innovation or a stopping point at absolute knowledge.

Chapter 18

Wittgenstein
Sliding Down the Sky Rope

OVERVIEW

Ludwig Wittgenstein's (1889-1951) *Tractatus Logico-Philosophicus*, as it is named, is a logical philosophical tract—a course to follow, to logically identify some fundamental aspects of logic including its representational and linguistic character, ontological substantiation, self-referential nature, and limit at "my" world of experience. Wittgenstein had training as an engineer, and the *Tractatus* looks somewhat prophetically (it first appeared in 1921) like a computer program of numbered concise logical statements. Yet, unlike a computer program, which is composed largely of commands, the *Tractatus* is descriptive. And it is not necessarily argumentative—no doubt one may take issue with the *Tractatus*, as I will, but Wittgenstein's intent seems to be expository—he does not engage much in conflicting points of view (e.g. Frege and Russell), as Plato's does in his dialogues—what Wittgenstein is concerned with here is the logical point of view, as he saw it.

It is in this way that we can see Wittgenstein's project here as related to Kant's *Critique of Pure Reason*, where Kant tried to establish the faculties necessary for logical judgments (Apperception, Understanding, Imagination, etc.): Kant's Robot, if you will, seems to be a shot at the apparatus of subjective experience and judgment.

Now, while Hegel tries to frame the being of consciousness within that robot (pure ego framed by a world of things, other people, social history, etc), Wittgenstein tries to refine the logical apparatus of that "robot perspective." (I call it a "robot" because, by discussing such a perspective and representing it, we create a model that is not quite the flesh and blood of a real subjective experience: it is a replica of functional essentials.)

There are seven major divisions in the numbering of the *Tractatus*, with each major division (save the last) having subdivided numbering (and, the first division having a total of seven divisions itself). The number seven is symbolically significant—the sun having been considered the seventh orb in the heavens as well as the source of en-light-ened reason found outside of Plato's cave. Roughly the seven major divisions concern the following:

1. **The world as logical facts**
(being is composed of the consistently related)

2. **A state of affairs of things & our ability to picture them**
(division of the objective world, and our "subjective" ability to represent it)

3. **A shift from pictures to propositions**
(linguistic ability to think of—not about—the objective world)

4. **Truth possibilities of propositions**
(sense of language given by relation to objective world)

5. **Truth functions, operations, and their limits**
(related propositions in a self-referential and descriptive "system")

6. **Generation of numbers, mathematics, laws & limits**
(pressing logic up to the world of "geometrical" experience)

7. **The absolute limit**
(can't speak of that beyond the experiential surface)

PERCEIVING REALITY THROUGH PICTORIAL FORM

In his *Tractatus*, Wittgenstein uses pictures to represent a distinction between perceptual thinking, and "reality": (Kantian) "representation" identified as actual representation. Throughout the *Tractatus*, Wittgenstein shifts the meaning of "picturing reality," from a mirroring of reproduced concrete sensation to a denoting of abstracted logical form essentials—picturing becomes both a starting point, and a metaphor for logical representation: a representation of representation. We will briefly now plot out the course of that metamorphic journey in order to elucidate the running thread of a problem.

Just as the word "reflection" can mean "thinking," and given that early modern philosophers like John Locke, wrote of "reflection" to describe how "ideas," from the outside world are internalized, Wittgenstein is concerned with "pictures" that represent "the state of affairs." One should note that for Wittgenstein, "the world," or "the case," is made up of "facts," all being composed of "the states of affairs," which in turn are made of up combined things, the entirety of which comprise "reality." This terminology and hair-splitting used by Wittgenstein to re-describe what we take for granted as our world with consistently related objects sets the stage of a more precise logical space.

Within that logical space, as a part of it, we can have "pictures" that represent their own world with consistently related objects; and the pictures may or may not map to reality via "pictorial form." In 2.15 Wittgenstein claims:

> "The fact that the elements of a picture are related to one another in a determinate way represents that things are related to one another in the same way.
>
> Let us call this connexion of its elements the structure of the picture, and let us call the possibility of this structure the

pictorial form of the picture" (Ludwig Wittgenstein, D.F. Pears & B.F. McGuinness (trans.), *Tractatus Logico-Philosophicus*, Routledge: London (1988), #2.15).

It is not clear to me if there is some odd reversal here, were pictorial form precedes the ability to distinguish objects in a state of affairs—but at any rate, there is a proposed correspondence between pictures and objects.

FROM VIRTUAL REALITIES TO PROPOSITIONS

To take Wittgenstein's approach a step further, using a more modern technological analogy, imagine that you have been hooked up to sophisticated virtual reality gear, possibly with video goggles, a tactile suit, headphones—and that this virtual reality simulation is giving you a live sensory feed from a sensory robot in a white cubical room, with a table in the center, and an apple on it. In this case, there is a reality: the actual room, table, and apple; and there is a representation of it being displayed directly to your sense organs, so realistic that you might not be able to tell that it was not real (you might have also been placed in a sensory deprivation tank, with computers hooked up to your very nervous system with a similar sensory feed). Here we would no doubt say that the reproduction was a "picture" that shared a very minutely detailed pictorial form with reality—again, so much so that you might not be able to distinguish the two—the pictorial form would be nearly identical to the form of reality which is the logical form (see #2.18).

Whereas pictorial form is purely representational, logical form need not be. Logical form concerns the identity of what is pictured. This can be illustrated in a range of representations, from the virtual reality; to a holographic representation of the apple on the table; a photograph of it; a hyper-realistic painting; an impressionistic painting; a more abstract painting (a red circle with a line / stem on a crude stick figure table); a hieroglyph depicting a red circle on a square; and in turn simply an "O" representing the apple, an "N" representing the relation of "this on top of that," and a "T" representing "table," we might have the highly abstract, and linguistic picture that looked like this: ONT (as in ontology). Following Wittgenstein, I would suggest that while more perceptual pictorial form fades into the background with our abstract progression, logical form involved with thinking comes to the fore: perceptual mirroring gives way to logical identity (although Wittgenstein rarely uses the words "mirror" and "identity.") Wittgenstein does, however, claim that there is continuity from the pictorial to the propositional (ONT, like aRb, may be an elementary proposition), along the lines of hieroglyphics and sign language—where logical form is maintained well into phonetic notation and visual picturing is absent.

Along with propositions come representational truth values. Wittgenstein claims that the sense of propositions come from their logical correspondence with reality. If the proposition says "the apple is on the table," and the apple is on the table, the sentence is true (and even if the apple was not on the table, the very possibility of an apple on a table would give the proposition "the apple is on the table" sense).

PARALLEL HISTORIES OF PICTORIAL AND LOGICAL FORM

However, there is definitely a problem for Wittgenstein trying to maintain that propositions are representational due to their speculated "evolution" (my term). Was there not a point in history, or in our example, where convention enters the picture, when hieroglyphs metaphorically shift due to a context where, had we not been present at the entire evolution of transformations, we never would have known that the abstract proposition did indeed connect with reality via its logical form? In other words, although we might be able to see a connection between the Latin usage of the word "video" to mean "I see," and the latter use of "video" to mean "I understand," what happens to the logical identity of the word, "video," where there is a significant shift of denotation? Wouldn't Wittgenstein have to claim that representations can be ambiguous and that their *disambiguation*, necessary for a sense given his use of the possibility of the representational truth or falsity of a proposition, would require a context that is part and parcel of the entire fabric of the history of a specific word's ("video") evolution? For example, the proposition, "I see the central point of your picture," differs in contexts such as (a) understanding the main meaning of a work of art, or (b) actually visually noticing a dot that is camouflaged among many off-center dots. How does that example compare to a picture of an apple on a table being used in a multitude of situations—where the same logical form applies to different apples on tables? In each case the same representation or proposition can be used in different contexts; but in the latter, the logical form remains consistent, while in the former, the "pictorial form" of the proposition remains consistent, while the logical form differs.

Although it would seem that logical form as thought is born of pictorial form, and hence that we might have a contradiction here, due to a deep schism between the two types of form, it might be resolved by noting that nowhere is there a hard and fast rule that pictorial form must always have only a one to one correspondence to a particular logical form—they need only have a connection. While the pictorial form is material, through and through, the logical form was always mental thought, or at least has its own material evolutionary history parallel to the history of the more apparently material pictorial form (e.g. something like a "meme" history—a cultural use that is communicated from brain to brain in parallel to the material signifier, or pictorial form—more on this later).

SEVERANCE OF PERCEIVED AND LOGICAL REALITY

Wittgenstein catches this parallel when switching his terminology to that of "signs" and "symbols" in division 3. Propositions are composed of physical signs, yet multiple propositional signs can refer to singular symbols (or expressed thoughts); and single propositions also have the ability to express multiple symbols. Yet, he does not go into much depth concerning propositions having a physical-representative connection to "physically perceived" reality, in contrast to these logical symbolic conveyances of the propositions that "represent" an abstract logical reality. There is a sort of "bait and switch" here. The intuitively sensible idea of a proposition picturing a sensible reality has a relatively arbitrary relation to the idea of logical symbols of thought picturing logical objects of reality. Logical symbols could not really be said to "represent" logical objects in a pictorial way, as they have only correspondence of points of "logical thought" to points of "logical space" in reality. The connection between pictorial form and logical form in "logico-pictoral-form" (2.2) is severed; and there is the question of whether it really ever was there in the first place.

CONSTRUCTING LOGICAL FUNCTIONS: GOD'S DICEY LOGIC

Once Wittgenstein has his logical symbol / objects correspondence however, much of the logical core of the *Tractatus* falls into place: Elementary propositions, of the form "a is related to b by relation R – aRb" (much like our "ONT"), can range from tautologies at one extreme to contradictions at the other, with sensible true or false propositions in-between (and more complex propositions can be built out of these elementary propositions). To unfold three examples from the chart of division 5.101:

1234

(TTTT) (p, q) Tautology (if p then p and if q then q)

(TTFT) (p, q) in words: if p then q

(FFFF) (p, q) Contradiction (p and not p, & q and not q)

Each column contains an outcome based on a relationship between p and q:

in column 1: p:T, q:T

in column 2: p:F, q:T

in column 3: p:T, q:F

in column 4: p:F, q:F

There are four sets of binaries:

(a) Two proposition variables (p, q)

(b) Two possible values for the first (p: T or p: F)

(c) Two possible values for the second (q: T or q: F)

(d) Two possible function output values (T or F)

Now, Truth Functions present another monumental leap from one realm to another, with seemingly little connection: as with the leap from pictorial form to logical form, there is the leap from related elementary propositions (e.g. p, q) to the consequent complex proposition (e.g. if p then q): where logic holds by virtue of God! (5.123—or "a god" rather). Interestingly, following Hume, Wittgenstein proximally makes the claim that:

> "Superstition is nothing but belief in the causal nexus" (*Tractatus,* 5.1361).

Temporal causality is "superstition," while "a god" insures implication. The truth grounds (5.101) of the implication if p then q that make that proposition true are simply the truth possibilities of elementary propositions p and q that make if p then q true. This is a completely circular way of saying that what defines if p then q as true is a certain combination of truth values for p and q—in other words if p then q will be true by definition of (TTFT) (p, q). It is similar for every other truth function (i.e. "p v q", "p . q", "~p", "~q", etc): the 16 total possible variations from (a) two variables (p, q) with (b) two possibilities for p (T, F) (c) two possibilities for q (T, F) (d) and two possible outcomes (T, F) [or (2 ^ 4)] defines all logical relations (prior to their being identified with logical operators). Although the binary division is the most elegant, one wonders: Why not a logic based on threes?: True, False, or Maybe for each of p, q, r—would result in a table with [3 ^ 5] 243 possible variations: would the "modal logic (with a maybe) and 'mediated comparison/connection' (three variables)" given by those combinations be less valid?

Has Wittgenstein read too much into a set of binary possibilities, or is there something fundamental about the relationship between the binary and logic? How do we bridge the gap between a table of binary possibilities, and propositional logic? The real world application of each of these four binaries needs to be examined: (a) two propositions (p, q)—because at least two are needed for any comparison or connection; (b) & (c) two factual possibilities for two propositions (T, F)—because they independently may or may not represent a situation (more on this below) and (d) two possible results as the outcome for the relations of the two propositions (T, F)—because the comparison or connection of two true or false propositions will result in a true or false outcome. But there is definitely a problem with conflating the first (b & c inputs) and second (d output) types of truth and falsity. One type, for Wittgenstein, is either about representational (synthetic) truth or truth by definition (analytic)—while the other type is about the necessity of logical relationships: a sort of hyper-analytic (or a priori) hypothesis. I might call these first level truths (those of mirroring the world), and second level truths (those concerning relationships between mirror / propositions). No wonder a god is required here for Wittgenstein, for it must be some power exterior too his own observation (the first level truths) that would make the world's consistency deductively (by implication) absolute (the second level truths) rather than inductively "probable." In other words, although induction may be a matter of probable certainty, the "hardness" of deduction is an unknowable faith. Maybe "God does not play dice with the universe," to quote Einstein, but God is a matter of faith, not certitude. A little more on this is discussed below, concerning truth functions vs. operations. A serious question remains: what is the relationship between the analytic / synthetic truth of propositions, and an a priori structured truth of logical connections, and are the two incommensurable? Is implication ever more than a highly certain induction? An induction that in circular fashion, is deeply intertwined and based on implication itself (If something happens regularly, then it may happen again)?

No doubt, the binary logic here tied to a minimum of comparing and connecting two independent statements, and their truth or falsity, ties into representation: propositions may or may not represent reality, and hence are true or false. The whole notion of truth and falsity is "illustrated" by the notion of pictorial form corresponding to a perceived reality of logical form—but do we perceive a reality of logical form? When the complex of a picture that completely maps a reality is simplified towards hieroglyphics and propositions on the one hand; and objects are singled out in a visual field on the other—what maintains a connection between the signs of the picture and the logical object in reality? A logical symbol? Where did that come from? If propositions do not indeed represent reality, absolutely, as I am suggesting, then basing logic on truth values of propositions as representations is suspect. A logical symbol may connect to reality—but does it represent it? And if not, what is the nature of truth and falsity, other than a coherence of logical symbols associated with other logical symbols that have a consistent relation to logical objects associated with other logical objects. Wouldn't there be a sort of "indeterminate translation" between the realm of logical symbols and logical objects? In other words, might there be idiosyncratic logical symbols for each person (remember, logical symbols are not tied to logical objects via pictorial representation) in such a way that each person's understanding of a proposition would be slightly skewed from every other person: making logic a matter of "statistical convergence" among populations rather than absolute identity? Each person has their own experiences that are shared with others only in part: Language is indeed public, yet every person's use of it is semi-private, given their personal idiosyncratic histories. The "logical identity" of public "objects" (which may only exist in the minds of individuals—more on this later) might simply be as sort of holographic chimera projected by individual (human) cells. (A hologram not perceived by a who, but an ideal that has no real existence).

THE DIFFERENTIATING SHEFFER STROKE

Strategically, Wittgenstein discusses implication, both absolute (if p then q) and probable (prior evidence of p with q may mean p if q is given) before moving on to truth operations. A truth function relates only to the (VVVV) (p, q) parts of the chart (where "V" = "T" or "F"); but operations, such as "or v" and "not ~" are added and discussed later. The distinction is important, for Wittgenstein implies that truth operations (e.g. p v ~q) are "semi-arbitrary" (my words) in relation to truth functions. We could have an indefinite number of truth operations with a given truth function, given (p, q) of various combinations, and (VVVV). The truth function—which gives the truth value of a representational propositional relation—is primary, while the truth operation—a sort of propositional logical equation—is somewhat derivative and can take a number of forms:

much as single symbolic logical expressions can find form in a multitude of material sign propositions, single truth functions can find expression in a number of truth operation notations. However, it is through truth operations that more complex truth functions are constructed. Wittgenstein's notation for generating propositions is an expansion of his elementary truth functions, where beyond the simple (FTTT) (p, q) which indicates "not both p and q" (or "p|q" from which all other truth functions may be derived by use of recursive truth functions) we have (-----T) (ξ,...) (hereafter I will alter the notation from "ξ" to "E") where "E....," or "Ē" is a set of propositions, and hence (-----T)(Ē) or N(Ē) is the negation of all values of the indicated propositions—and the possibility of deriving, or generating, generalized propositions. N(Ē) is like a super Sheffer stroked ("|") series of propositions, which stroke itself visually illustrates the boundary between elements—differentiates them—and can act like a law of non-contradiction where we cannot have both p and ~p—or any contradictory logical propositions at all.

IDENTITY OF SIGN AND SYMBOL?

Now in (5.526) Wittgenstein claims:

> "We can describe the world completely by means of fully generalized propositions, i.e. without first correlating any name with a particular object.
>
> Then, in order to arrive at the customary mode of expression, we simply need to add, after an expression like, 'There is one and only one x such that...', the words, 'and that x is a'" (*Tractatus,* 5.526).

Interestingly, he soon adds:

> "Identity of object I express by the identity of sign, and not by using a sign for identity. Difference of objects I express by difference of signs" (*Tractatus,* 5.53).

Wittgenstein clearly identifies his two logical spheres in these quotes; and although one realm (symbolic propositions) might map the other (logical objects), in its entirety, or at any single point, once he says "x is a" or rather "name x is object a," or "'Fido' is Fido," he is speaking of the logical identity of name and object. Maybe he could have said, "'Fido' represents Fido," and maybe that's what "x is a" means, but then we're back to the problem of just how a logical symbolic proposition can represent or picture a logical object. Following Richard Rorty, we might claim that a symbolic proposition is "about" the world it is connected to. Either that, or "represents" means something radically other than "picturing."

All this logical "representation" takes place in a phenomenological world—where "my" body is in "my" world that is completely intertwined with logic and language; but which is observed from solipsistic subject perspective that is the philosophical self at the limit of said world.

CLIMBING THE LADDER

What sorts of aspects of the world does the logical self find? And what can it do with logic? Using $N(\bar{E})$, they can generate propositions. As we discovered when examining the truth grounds of implication:

> "The propositions of logic are tautologies" (*Tractatus* 6.1).

Which are to be distinguished from being merely general—and create a purely self-referential system of analytic propositions. Similarly, with mathematics, which Wittgenstein calls "a method of logic," we have equations which resemble tautologies, and we can prove its propositions without recourse to facts of the world—following intuition with calculation and substituting one expression for another. Moving now, from the general function of proposition generation, through the tautologies of logic, towards the more "worldly" realms of mathematics (at least with geometry) we pass on to the realm of "laws." Laws, by their very nature are more "specific" than logic itself (which is beyond being general, and is transcendental); laws sync up with the structures of the world, and with particular mechanical generalizing systems, describe the world completely (with some systems being more elegant than others). Some laws refer to systems of laws, and not the world, and mechanical laws will never point to a specific object. In fact, many candidates for laws don't wash up completely, like causality, and induction. And, aspects like the will, ethics, death, and god are literally out of this world. Answers to questions about these later issues are lacking (as they are not scientific questions of logical description). Even the *Tractatus* itself, leads the way to "transcendental logic," where the reader:

> "must transcend these propositions [...and...] throw away the ladder after he has climbed up it" (*Tractatus*, 6.54).

SLIDING DOWN THE SKY ROPE

But slipping back through the cracks of the sky, as it were, we might return to earth again with a clamorous electrical storm of the nervous system—as we gently slide down a rope from the sky of logic and see if we can touch ground.

For, as we saw earlier, Wittgenstein has problems grounding his logic in the representations of the world. He climbs the ladder of pictorial representation, only to throw it away in favor of logical symbols. His notion of truth or falsity rests on this representational non-ground (truth or falsity is defined in relation to depicting a possible world); and hence his truth functions rest on an illusion as well. The entire house of cards may fall. His logical symbols may mirror his logical world of objects, but both, as described, may be conjured out of thin air—the former as simply a mirror of smoke invented to suit the needs of "representing" the latter—the latter being a possible "continuum" not sliced up into logical objects in the first place.

But we can use logical propositions, such as "the cat is on the mat," so this must be explained. What follows is an abbreviated alternative to Wittgenstein's phenomenological-representational foundations.

UNCONSCIOUS BIO-LOGICAL DICTIONARY

Reversing through the *Tractatus*, we 21st century philosophers can hardly ignore cognitive science—taking Wittgenstein near to his word, we might try to address the scientific question of logic. First, it might be noted that our ability to think is not constrained so much by logic, as by the structure of the brain itself. The brain is composed of a chemical, electrical, and structural system—a structure of interconnected nodes. These nodes activate in parallel sequence in partly genetically grown, partly conditioned, neural pathways; pathways that are often "self-referential," but also that open out to the world via the senses and motor actions. This is a mechanical system, operated by laws, which itself is flexible enough to understand logic—and imagine what is beyond it.

This is the embodied brain as conscious agent with a robust ability to reason intelligently, where intelligence might be defined as the ability to model the world, and make connections among its various parts that work to accomplish diverse purposes. What Wittgenstein contrasts with symbolic logical thought, propositional signs, and the objects of the world, may indeed exist: but all materially, and differently than envisioned by him. One can only speculate on the first type of full conceptual language used by humans, if there was one—I would hazard the sign language used by some apes as a candidate; but even insects have signaling powers—and hence the genealogical ancestry of language may lead back to the origins of life itself. Yet, we do have denotative powers: as much as language may change, much seems to remain constant, not only for dictionaries to remain somewhat consistent during a lifetime, but also for the etymological ability to trace "meanings" of words back to the beginnings of writing, and speculatively, before (e.g. with the reconstruction of a proto-language). Although the signs have changed, some symbols, no doubt altered in some ways by changing context, may retain some of their original character. Along with the evolution of language, has come an evolution of cultural practices—which might survive based on their aptness in the human milieu and environment: memes.

It is a matter for research to determine which aspects of language and cultural practices are genetic (cp. Chomsky), which culturally indoctrinated (and possibly very old), and which freely learned.

HIGHLIGHTING PERCEPTION FOCI

Now, in contrast to the genealogy of language, we also have a phenomenology of linguistic perception: when the human body confronts an environment of objects, most likely there is in interaction between the brain and the stimuli of the environment, which might not be differentiated into specific objects yet. The activity of the brain, would neither passively receive "object" images (the visual field, for example, might only experience patches of hue, luminosity, and texture passively), nor would it completely hallucinate an object (there does seem to be a consistent world "out there" that we all share). This brain activity might rather highlight "objects" in the environment. This highlighting might occur when environmental stimuli activate parts of the brain that have prior experience with similar stimuli. The brain may focus on the familiar aspects of a possible "object" by automatic reflex, or consciously, and "identify" the object by a processing of a "cluster of stimuli cues" that activate a cluster of neurons associated with that stimuli cluster; which cluster of neurons in turn feedback to allow concentration on the relevant stimuli and differentiate an object from its milieu (a pattern may have various ways of being interpreted, but the brain may identify certain sub-patterns as relevant, and hence distinguish them as identifiable objects). Expectation from prior experience and novel stimuli would feedback with each other, as the object would become clear. Moreover, the clusters of stimulated neurons could be associated with words and propositions (among many other associations), and hence an instance of a cat on a mat, could be associated with the term "a cat on a mat." This use of words would constitute the intersection of the genealogical history of language itself, and the phenomenological epistemology of the perceiving brain and subject.

IDIOSYNCRASY AND FOCAL LOGIC

Every person has their own idiosyncratic history of experiences, experiences that may color the various uses of words, and make language semi-private and semi-public. These experiences would color the symbolic propositions, and not the material propositional signs (not as much at least). The symbolic propositions would, as sentences and words (and a variety of other uses), be instantiated in the brain, as cross-referenced by the various witnessed usages of certain signs (hence a central concept, or symbol, might have several ways of being named or described; and a single sign could bring to mind several different concepts or symbols). Indeterminate translation is not a problem, as each person may have a consistent internal language that syncs up with a parallel consistent public language—enough for the principle of charity, where everyone can communicate effectively, but nobody means exactly the same thing by what they say. For many symbols, there would be a one-to-one correspondence to a focal public sign (word or proposition), but each would be situated in the entire fabric of an individual's private vernacular. Most likely, due to the robust over-determination and over-lapping cross-references of public discourse, private languages would remain only semi-private.

"Family resemblances" among concepts need not multiply indefinitely until there is no longer an identifiable family (as some philosophers have critiqued the later Wittgenstein), but would have strong focal clusters, that have associations among themselves that are stronger than their associations with other families (although certain concepts and practices would belong to several families, they would not belong to every family—"six degrees of separation connecting every node" does not mean that we cannot distinguish network hubs). It follows from this that logic may have fuzzy boundaries: some propositional and cultural practice relations would hold strong and fast, possibly being genetic in origin, (Quine might say they are towards the stable center of a network of associated knowledge—Donald Davidson might suggest such are the core elements involved with the principle of charity); while other propositional relations would be quite contingent (on the periphery of knowledge). Yet, for me, "logical knowledge" is not simply a matter of statistical coherence of propositions among themselves—and coherence with the world's connections to which they stand in relation; for there also appears to be a logical structure to representing the world: representing can actually model (partly publicly, partly idiosyncratically) the world, by deconstructing it (meant here to designate the projection of a structure onto another structure that provides the material for the projection: turning over-all structure on itself), singling out a few of many possible interpretations: hence Derrida's claim that the copy precedes the original (we see through the copy or representation that structures the original). Of course, logic seems to limn the edge of neuro-biology, so thinking outside of logic—a logic which is flexible enough to imagine the false, and the fuzzy—and the breakdown of identity and consistency—a logic of probability and statistical foci rather than a strict Sheffer stroked identity—such a logic is framed by the general usage of propositions which can point beyond themselves to the (im)possible.

Bach - Cello Suite #5 - handwritten (~1720)

Chapter 19

Wittgenstein
Singing the Blues

A FEW OPENING WORDS ON A FEW OPENING WORDS

The "later" work of Ludwig Wittgenstein contrasts with his "earlier" work in the *Tractatus Logico-Philosophicus* in many directions simultaneously; moving from a more static logical representational systemization of language to a more dynamic, contextualizing, and edifying approach to the way we use it. Although his "methodology" changes, his subject matter remains similar: the concern with language, its structures, and their relation to that which we have a difficult time approaching within the limits of such language, e.g. subjective experience. Here, we will be examining his early work in the later style, *The Blue Book*, which he edited from notes for his student lectures, and felt comfortable enough to present to his friend Bertrand Russell.

If one is familiar with his earlier work, with its numbered affirmations, one will notice the less ordered nature of the later work, yet one still finds well crafted ideas coming in bursts—a sort of organization of aphorisms persists, although the later work may not aim for the sky as directly as the prior work did. No rungs on a ladder here, but rather a more meandering trail of crumbs on the earth. But, being aphoristic, both works lend themselves to explication and investigation into particular remarks: they lend themselves to close reading. I begin my delving into *The Blue Book* by examining the text of the first five paragraphs (Ludwig Wittgenstein, *The Blue and Brown Books*, Harper & Row: New York (1965), p. 1):

Wittgenstein begins with the seven word sentence, "What is the meaning of a word?" This question sets an agenda: the schism between words and their meaning. This question, and the way it is asked may seem typical of some philosophies: we're looking for a "what" that is the "meaning" of a "word"—and assume, for example, that (a) words have meaning, (b) that that meaning is an unknown "what" and not something like a "who," (c) and that words are not identical with their meaning. It may seem peculiar to question this typical philosophical question the way I have, yet Wittgenstein will question this question too; and we need wait no longer than his second sentence.

For no sooner has he asked "What is the meaning of a word?" than he shifts his position, and frames this question as suspect with a more specific question, "what is an explanation of the meaning of a word; what does the explanation of word look like?" Wittgenstein calls this second question an "attack" on the first one—it is both an engagement, and the beginning of an argument—and one made with "us" (the writer, L.W., and the readers / listeners), with a community that Wittgenstein is communicating to. Unlike the *Tractatus*, which begins with the world and things, and barely an "I," *The Blue Book* almost immediately brings in the relationship between people. This is important for Wittgenstein's later approach, where communication is a linguistic *activity* between people. And Wittgenstein tries to illustrate his new insights with his verbal performance. This second question also develops a more self-consciously critical position with regard to the subject matter, and asks us to "look" at the practice and the course the initial question set us upon: "looking" at the "materiality," or physicality of what we have set out to do.

Such is refined with the next paragraph, which points at an "analogy" between how we might examine the measurement of length to understand the concept of "length," and how we might examine the concept of "meaning" in itself: how our general abstract concept is related to a particular concrete practice. The introduction of "analogy" is important, as this will be a theme or motif throughout *The Blue Book*, both as a way of gaining insight, and as a way that philosophical problems arise. And again, this turning from individual words or phrases, to the material usage of them is important too.

Wittgenstein cites the "mental cramp" one may get when asked—what is "length," "meaning," or "the number one?" and consequently "we feel" we *ought* to point to some *thing*—a "bewilderment" arises when "a substantive makes us look for a thing that corresponds to it." The word "corresponds" is important here, as Wittgenstein's *Tractatus* was all about the correspondence of representations (e.g. propositions) to things.

Ever critical and commenting on and questioning his comments and questions, in his fifth paragraph, he frames his second question (about procedures) which framed the first (about meaning), by noting that it brings the abstract "down to earth," and also has the "advantage" (again, the relevance of the word "advantage" to *strategic* arguing with an engaged opponent, maybe even his own prior position in the *Tractatus*)—the procedural question has the advantage of "studying the grammar of the expression... [which] will teach you something about the grammar of the word... and will cure you of the temptation to look for some object" as the meaning. N.B.: in the common split between "semantics" or "meaning" and "grammar" or "syntax," Wittgenstein is suggesting that we look for the meaning in the grammar: how a word is used will tell us how it means rather than what it means.

So much for our little examination of the first half page of *The Blue Book*. These opening words are ripe with Wittgenstein's new insights and approach to his subject matter. Throughout the rest of this chapter, I will outline various important themes of *The Blue Book*, where I think Wittgenstein describes a system of language games which have indefinite family resemblances and grammatical / procedural rules that people experience in their "shadowy" expectations, imaginations, motivations, and reasons; and which can lead to the over extension of such analogous resemblances and grammatical usages into philosophical "error;" such as the creation of a "mental" realm typified by the "I" or ego which is supposed to "own" experience.

DIFFERENTIATING METHODS

As mentioned before, Wittgenstein's new approach contrasts with that of the *Tractatus*. The new approach is not concerned with logical and pictorial form that corresponds to, or interlocks with, a world: we are no longer so concerned with the word / object (or proposition / situation) connection—this may have been seen as sufficiently addressed and insufficient to explain the phenomena of language. Such presented a fairly stable, possibly "eternal" view of how language mapped to the world via logical and pictorial form: world and words had a logical form which bridged the two. Hence "A Cat in the Hat" would represent a Cat in a Hat due to the thing(s) sharing an "essence" (Wittgenstein prefers the words "logical form") with the words. Such "logical form" brings notions like Plato's eternal forms to mind. But here, Wittgenstein is concerned with the unfolding of context within the temporal—with how we use language with a cultural, historical, and possibly personal, context that we bring to bear in a particular situation and circumstance.

The new approach is also differentiated from psychology, and what in modern academics might be called "psycho-linguistics." Of such, Wittgenstein notes:

> "We may find that such a mind-model would have to be very complicated and intricate in order to explain the observed mental activities; and on this ground we might call the mind a queer kind of medium. But this aspect of the mind does not interest us. The problems which it may set are psychological problems, and the method of their solution is that of natural science" (*The Blue Book*, p. 6).

Although very interested in the question of "how" (finding significance for "length" in *how* we measure it), Wittgenstein is not concerned with the "how" of science: he's not searching for mechanical explanations of language and experience, but trying to describe them—the "how" for Wittgenstein is not one of deeper fundamentals that produce surface phenomena: it is the action of such phenomena on the surface as described. This is how we say things, not how the saying works. So the new approach is not one of science and psychology, although Wittgenstein will often limn the edge of scientific disciplines in delineating his own.

Wittgenstein also opposes his philosophy to that of metaphysical puzzle making. Such puzzle making is intertwined with science as well:

> "I mean the method of reducing the explanation of natural phenomena to the smallest possible number of primitive natural laws; and, in mathematics, of unifying the treatment of different topics by using a generalization. Philosophers constantly see the method of science before their eyes, and are irresistibly tempted to ask and answer questions in the way science does. This tendency is the real source of metaphysics, and leads the philosopher into complete darkness. I want to say here that it can never be our job to reduce anything to anything, or to explain anything. Philosophy really *is* 'purely descriptive'" (*The Blue Book*, p. 18).

Like the tendency to look for *things* that correspond to abstract terms, the bewildered temptation to make problems where they really don't exist is something to be cured with a dose of common sense and ordinary language and just plain "looking."

The new "method" revolves around descriptions via analogies, metaphorically rich examples, and a complex cluster of concepts that help enable insights into the surface of our linguistic practices. The method is one of asking questions, reframing old problems with new insights—and often showing how isolating common sense or ordinary language usages that imply metaphysical problems, metaphysical problems such as thinking that time is like space (since we often talk of time spatially—even Einstein conceived of a space-time continuum that seems more spatial than temporal in one's imagination)—Wittgenstein shows how such extrapolations of implications found in ordinary language can be dissolved by cross-referencing other common sense and ordinary language usages: showing how such problems do not occur persistently throughout a well rounded and complexly experienced human perspective, but arise when taking a particular contortion of language usage too seriously in isolation. Wittgenstein likes to describe phenomena with a full context, rather than taking a particular twist of language as implying a deeper truth at odds with the rest of the surface of common sense.

SEEING THE SAYING

Straight out of the gate, Wittgenstein differentiates two ways of defining (i.e. the *how* of meaning): (1) ostensive *pointing*, and (2) word to word verbal *interpreting*. But pointing and translating / interpreting both require a rich context to make any sense. When you "bracket" (to borrow Husserl's term) or take away the full context of what we already know when looking at phenomena, a pointed finger tells us nothing but that a finger is extended (and such is to already use a fully contextualized language to try to describe what is not verbal). If we just look at things, without linguistic interpretation, then what can they tell us about language? The phenomenal world would be either meaningless, or almost infinitely ambiguous: this pointed finger could mean almost anything—it could be pointing at a myriad of aspects on the other side of the finger that go along with any vocalizations; vocalizations that might have nothing to do with a finger or what is pointed at (and even saying "pointed at" makes some learned assumptions).

The same goes for interpretation or translation when associated with pointing. Hence problems of "indeterminate translation" arise. Could we ever learn an alien language with pointing?—is it even possible to learn a language in the first place from scratch in such a manner? And thence come solutions to the language acquisition problem, such as Chomsky's innate abilities, or maybe even Locke's noting that we learn identities through what remains the same in different circumstances: where a single identity is constructed from a multiplicity of different experiences, and noting what remains the same among them.

Wittgenstein refines his take on the structural bifurcation between situational ostentation and verbose interpretation well in these two remarks:

> "if we had to name anything which is the life of the sign, we should have to say it was its *use*" (*The Blue Book*, p. 4).

And:

> "The sign (the sentence) gets its significance from the system of signs, from the language to which it belongs. Roughly: understanding a sentence means understanding a language" (*The Blue Book*, p. 5).

Here we have an image of a web-like fabric of context feeding through the eye of a needle of the present articulation. This Saussurian "parole" and "langue" distinction—the particular usage vs.

the system of language—find symbiosis in that one cannot be without the other. Words on a page need an agent to enact them, and an agent needs the tools of language to use in a particular circumstance:

> "I know what a word means *in certain contexts*" (*The Blue Book*, p. 9).

Describing the system of signs leads Wittgenstein to the notion of *language games* with *family resemblances*:

> "The study of language games is the study of primitive forms of language or primitive languages. If we want to study the problems of truth and falsehood, of the agreement and disagreement of propositions with reality, of the nature of assertion, assumption, and question, we shall with great advantage look at primitive forms of language in which these forms of thinking appear without the confusing background of highly complicated processes of thought" (*The Blue Book*, p. 17).

And:

> "The tendency to look for something in common to all the entities which we commonly subsume under a general term.—We are inclined to think that there must be something in common to all games, say, and that this common property is the justification for applying the general 'term' game to the various games; whereas games form a *family* the members of which have family likenesses" (*The Blue Book*, p. 17).

These two concepts, "language games" (ways of doing things with words) and "family resemblances" (language games having common but not essential features) although not entirely new (Plato thought activities like cutting and sewing had forms too; and that forms were inter-related by a dialectic of dividing into particulars and subsuming particulars under generals)—these two concepts are radical breakthroughs in the philosophical description of language. "Language games" are important for their shifting a focus form representational thinking to *particular ways of doing*; and "family resemblances" for breaking the tradition of hard and fast word-to-word definitions (and note again the division between particular use with language games—"parole," and a network of word-to-word associations with family resemblances—"langue").

The notion of a definition that is always 100% accurate doesn't wash up with experience—looking for necessary and sufficient conditions may help clarify how we use some words (there are elegant definitions in dictionaries—we can define words)—yet these definitions often have exceptions that we can still understand. Wittgenstein notes:

> "there is not one definite class of features which characterize all cases of wishing (at least not as the word is commonly used). If on the other hand you wish to give a definition of wishing, i.e. to draw a sharp boundary, then you are free to draw it as you like; and this boundary will never coincide with the actual usage, as this usage has no sharp boundary" (*The Blue Book*, p. 19).

Generalities, kinds, identity, definitions: such rarely have *criteria* that are always necessary, but almost always have *symptoms* that are relevant. Such is not to say that all symptoms are as relevant to a term as others: some might argue that to eliminate sharp boundaries will melt everything together into an indistinguishable blob (of being / nothingness?). Yet, as Wittgenstein failed to point out, but implied with his notion of "fuzzy boundaries," there may definitely be hubs, with some symptoms having more weighted relevance than others to a specific generality. Fire trucks need not be red, but red is a stronger symptom of a fire truck than yellow. (On *another* hand, the notion of a "prototype" is problematic too, as it seems to suggest an aggregate image with which to compare something with in order to recognize it: do we really always need to compare inorder to *recognize* something, or is a prototype just the image we might *construct* if told, say, to draw a "bird in general?") And just as we need not have sharp boundaries with our definitions, our activities need not have hard and fast rules:

> "For not only do we not think of the rules of usage—of definitions, etc.—while using language, but when we are asked to give such rules, in most cases we aren't able to do so" (*The Blue Book*, p. 25).

Such a statement leads to the notion of how mentally conscious or not we are of the linguistic procedures we use, if any "procedures" are used at all.

LISTENING TO INTENTION

Wittgenstein's early task of describing how we mean, and his later task of describing experience itself becomes complicated in-between, in the middle section of *The Blue Book*, when describing the "shadowy" world of the "mental." This is the world of thinking, expecting, imagining, hypothesizing, wishing, intending, reasoning, etc. The "mental" realm seems to be able to have its own world separate from the world of facts—yet this all happens in a world of facts. And to complicate matters further, there is the possibility of something unconscious intertwined with the "mental."

The world of facts, and their relation to language was rigorously studied by Wittgenstein in his *Tractatus*, where representations, ranging from pictures to propositions, could have either a true or false relation to reality. Wittgenstein keeps this representational view in mind, as he tries to think about "mental" activity, and its relation to the factual world, claiming:

> "An obvious, and correct, answer to the question 'What makes a portrait the portrait of so-and-so?' is that it is the *intention*" (*The Blue Book*, p. 32).

Yet notes soon after, concerning a "processes of projection," that:

> "there is a peculiar difficulty about admitting that any such process is what we call 'intentional representation'. For describe whatever process (activity) of projection we may, there is a way of reinterpreting this projection. Therefore—one is tempted to say—such a process can never be the intention itself. For we could always have intended the opposite by reinterpreting the process of projection" (*The Blue Book*, p. 35).

That is to say, any specific activity, be it painting a portrait (in effect portraiture is another form of ambiguous pointing), or giving an order, etc, can be altered—even reversed—in a broader context, such as when using deception or irony; or framed as not really intended, as when performing on a stage, or telling a joke. Some language games can frame others, changing effects of the more basic language games. Ultimately, any linguistic act is completely ambiguous, and must be used in a particular situation, framed by ever widening and diverse contexts (contexts that may include unconscious elements like Freudian desires... I digress).

Yet, there seems to remain, according to Wittgenstein, simples where interpretation stops:

> "adopt whatever model or scheme you may, it will have a bottom level, and there will be no such thing as an interpretation of that" (*The Blue Book*, p. 34).

> "What one wishes to say is: 'Every sign is capable of interpretation; but the meaning mustn't be capable of interpretation. It is the last interpretation.' Now I assume that you take the meaning to be a process accompanying the saying,

> and that it is translatable into, and so far equivalent to, a further sign. You have therefore further to tell me what you take to be the distinguishing mark between *a sign* and *the meaning*" (*The Blue Book*, p. 34).

> "If we say to someone 'I should be delighted to see you' and mean it, does a conscious process run alongside these words, a process which could itself be translated into spoken words? This will hardly ever be the case" (*The Blue Book*, p. 34).

Wittgenstein is claiming here, while traversing the intersection of particular usage, and the general system of language and contexts, that we do not intend to intend what we say: just as we may compare a red apple to a red patch of color to identify its redness, and yet there is no red patch to compare that red patch to in turn for comparative identification (there is simply the fact of "redness"), likewise, there are simple acts of intending where we do not need to think about what we are saying before we say it... or even further... how can we think what we are going to think before we think it. Such is not to say there is not an *unconscious* process that occurs, but that is not an intention, and we definitely feel we intend things. There are some pictures we understand:

> "immediately, without any further interpretation" (*The Blue Book*, p. 36).

Wittgenstein continues:

> "We easily overlook the distinction between stating a conscious mental event, and making a hypothesis about what one might call the mechanism of the mind. All the more as such hypotheses or pictures of the working of our mind are embodied in many of the forms of expression of our everyday language" (*The Blue Book*, p. 40).

Here we can see Wittgenstein bumping up against psychology or cognitive science, which is not part of his philosophy. Yet, although not interested here in the mechanics of mind, he is interested in its materiality: do we need to talk about translating an intention into language, when all that we experience, including the experience of intentions, may be some sort of linguistic activity? Hence:

> "the expression of belief, thought, etc., is just a sentence; and the sentence has sense only as a member of a system of language" (*The Blue Book*, p. 42).

> "when the temptation to think that in some sense the whole calculus [read: system of language] must be present at the same time [as speaking or writing acts] vanishes, there is no more point in *postulating* the existence of a particular kind of mental act alongside our expression. This, of course, doesn't mean that we have shown that peculiar acts of consciousness do not accompany the expressions of our thoughts! Only we no longer say that they *must* accompany them" (*The Blue Book,* p. 42).

Further, Wittgenstein tries to rid

> "us of the temptation to look for a peculiar act of thinking, independent of the act of expressing our thoughts, and stowed away in some peculiar medium [....] Our investigation tried to remove this bias, which forces us to think that facts *must* conform to certain pictures embedded in our language" (*The Blue Book*, p. 43).

What Wittgenstein is exploring here, is the materiality of language, distinct from our experience of it. He is separating the activity and structure of language from our conscious observation of it—but noting that that thought may be the coincidence of the two: we simply experience the activity of language as thought regardless of whatever may be the hypothesized unconscious mechanics that produce a linguistic thought / articulation. Articulations in the present are simply understood immediately, without going through a process of exchanging words for words or comparing representations.

EXPRESSING EMOTION THROUGH INVERTED OBJECTIVE CORRELATIVES

Now, this investigation of language, heretofore, makes sense when relating facts, facts about what we believe, what we imagine, what we expect, and hypothesize—when we try to state what it is that we feel. But how could this account of language possibly relate to the possibly poetic relation of how we feel? That is to say, I can say, "I am in pain," and this will be a fact that separates the articulation of a fact from the experience of pain. But what about expressing yourself in such a way that people would feel your pain or at least relate to it in a way, other than simply as a fact that "so and so is in pain?" T.S. Eliot talks, problematically, of an objective correlative—a feeling can be put into language and then re-experienced by a hearer or reader of that language. Wittgenstein's account doesn't address such poetic realities in *The Blue Book*, where he reduces almost everything to a passive observation of the activity of language, and mostly ignores the power of expression. Yes, we have imaginations that can build worlds contrary to facts, built up from the fragments of such facts rearranged. We can expect something to happen that might not happen, and that we wish would. We can have intentions to do something, and fail to do it, and wonder why we failed, and ask if we had unconscious motivations for failing, or at least, can understand the implications of our feelings about something in a fuller manner in retrospect. But such a world is still a world of facts: facts about feelings, yes, but such falls short, in my opinion, of an account of our ability to *express* our emotions more or less powerfully. (Wittgenstein does not go into the possible "*reconstruction of contexts with emotional pull*," for example, which might be a better, externalized, account of T.S. Eliot's "*correlative objective*": the emotion is not encoded in the object, but shared via the vehicle of language and a common "existence"). There is a great difference between knowing the fact that someone may have regrets, and reading T.S. Eliot's "The Love Song of J. Alfred Prufrock." How does "So-and-so is sad" compare to the emotion shared when singing the blues? Connecting words to feelings via behavior may be on the right track, but falls short, in my opinion, of explaining how a good poem, beyond using image cues, has tone, mood, etc.

NOT GENERAL SENSATIONS, BUT AGREEABLE HALLUCINATIONS

I agree, however, with Wittgenstein's general observations that there are no sensations in general. Words are general, and parts of language systems, and become particularly meaningful when employed in a particular circumstance. A particular sensation always has a complex context in which it arises: the concrete circumstance of a certain experience. Talking about the experience of "expectation" in general makes some sense; there may common associated features of such—possibly apprehension, hope, boredom, etc. But, when do we feel such outside of a particular circumstance that colors that sense of expectation (ranging from light hearted hope to severe trepidation)? Also, is it possible to separate a sensation from an object?

> "Consider this case:—we have a general undirected feeling of fear. Later on, we have an experience which makes us say, 'Now I know what I was afraid of. I was afraid of so-and-so happening'. Is it correct to describe my first feeling by an intransitive verb, or should I say that my fear had an object although I did not know that it had one?" (*The Blue Book*, p. 22).

Possibly one can have an emotional hallucination, as in a nightmare—with an imagined object. But still there is some "*thing*," be it real or imagined, which is associated with the sensation. Although we may have a world of facts that we can agree on, and semi-autonomous imaginations that can be somewhat idiosyncratic, it seems completely possible to communicate a hallucination: two people could have similar hallucinations and know that they are experiencing something similar, even though no particular fact in the world corresponds to their experiences. The experience is common.

How do we know?

> "When we learnt the use of the phrase 'so-and-so has toothache' we were pointed out certain kinds of behavior of those who are said to have toothache" (*The Blue Book*, p. 24).

Such would seem to suggest that we can "know" the experience of other minds by some sort of behaviorism. Such an implied isolation of the subjective perspective is heightened also by the example of looking at your brain in the mirror, while you think. There is definitely some distinction here between the objective brain and the subjective experience (*The Blue Book*, 8). Possibly we have subjective experience that we can only come to *believe* that other people have: solipsism.

INVERTED SOLIPSISM

But Wittgenstein reverses this uncertainty (and it may be uncertain which uncertainty, solipsism, or the reverse I'm about to describe, is more believable): Wittgenstein believes we should be uncertain about locating our sensations in any particular "mind." You may have already noticed that Wittgenstein sees our "mental" activities such as "thinking," "expecting," "intending," etc. as phenomena that we experience: sensation occurs right on the surface, without the necessity of a mind separate from that sensation to feel it:

> "We are not concerned with the difference: internal, external" (*The Blue Book*, p. 13).

There is not an ego / "I" that has these experiences (one doesn't only think of phenomena, thinking itself is a phenomena): "I" and ego are speculations without merit via the facts of experience. Or so Wittgenstein might have us believe:

> "We feel then that in the cases in which "I" is used as subject, we don't use it because we recognize a particular person by his bodily characteristics; and this creates the illusion that we use this word to refer to something bodiless, which, however, has its seat in our body. In fact *this* seems to be the real ego, the one of which it was said, 'Cogito, ergo sum'.—'Is there then no mind but only body?' Answer: The Word 'mind' has meaning, i.e., it has a use in our language; but saying this doesn't yet say what kind of use we make of it" (*The Blue Book*, p. 69-70).

And also:

> "To say 'I have pain' is no more a statement *about* a particular person than moaning is. 'But surely the word "I" in the mouth of a man refers to the man who says it; it points to himself; and very often a man who says it actually points to himself with his finger'" (*The Blue Book*, p. 67).

Yet:

> "that of which I said it continued during all the experiences of seeing was not any particular entity 'I', but the experience of seeing itself. This may become clearer if we imagine the man who makes our solipsistic statement to point to his eyes while he says 'I'" (*The Blue Book*, p. 63).

So, in a linguistic philosophical mistake, the word "I" was taken to mean a particular person's "mind" separate from their sensation surfaces. Is it not a fact that two people can experience the same pain (such as twins with connected bodies that might have overlapping nervous-systems)? How can you tell which body a pain belongs to, simply by experiencing it? By the *behavior* of that body, even if it is your own!?! For consider—how do you know that it's *your* finger that moves when you will it? Because you see "your" finger move. If you were to will a finger movement, and consistently saw "another's" finger move, you would most likely consider that finger yours.

WIRELESS TELEPATHY AND EXTERNAL CONSCIOUSNESS

> "For we could imagine a, so to speak, wireless connection between the two bodies which made one person feel pain in his head when the other had exposed his to the cold air" (*The Blue Book*, p. 54).

Such a notion was probably not originated with Wittgenstein, and James Joyce gets more specific on pages 309-310 of his *Finnegans Wake*:

> "equipped with supershielded umbrella antennas for distance getting and connected by the magnetic links of Bellini-Tosti coupling system with a vitaltone speaker, capable of capturing skybuddies [...] eclectrically fitted [.... to] pinnatrate inthro an auricular forfickle [...] tympan [...] Eustache [...] conch" (James Joyce, *Finnegans Wake*, The Viking Press: New York (1968), pp. 309-310).

Such suggest an inner ear transceiver that is quite in line with Joyce's own preoccupation with the inner auditory, etc, stream-of-consciousness. I have no problem with the implication by Wittgenstein, that sensation is on the surface (not in a head), that one can think of it as being "in" the world, or external (actually neither: just right there on the experiences' surfaces)—but I don't think that this means two bodies being hooked up via brain transmitters will be experiencing the same pain, any more than one could record a pain, play it back later, and call it the same pain. Does the recording feel pain? Isn't the pain only experienced when the recording is played back? Isn't the pain at least differentiated temporally? (being, no longer simultaneous... and even brain transmitters would have some latency). And some people feel pains when other's do not, so we are not all one external mind. Wittgenstein seems to imply that sensation is differentiated consciously through various particular bodies, but that subjectively, we cannot differentiate whose body a sensation belongs to (we always observe such indirectly—via behaviors). Feelings arise prior to connecting them to individuals—solipsists have this turned inside-out.

THE FLIP SIDES OF CONSCIOUSNESS & BODIES

Such "external" consciousness can be combined with Kant's deduction of subjective apperception taken from noticing how identities are formed via the connecting threads in differentiated experience: such as determining that a single camera was *most likely* used for a series of frames in a movie that differ in a fluid way that suggests a single moving camera and environment, rather than jump-cuts from one perspective and / or series of objects and environment to another. A key to Kant is the word "implicit": what identities (or faculties of mind) are *implied* by the very fact of experience, and our ability to make judgments. Such judgments may not necessarily be a priori absolute, but they do have pragmatic consequences relevant to survivability. Implication can tell us more than mere observation of general associations. Wittgenstein's *The Blue Book* could have taken a few lessons from Kant's *Critique of Pure Reason*—maybe there is some philosophical use for implication and necessity that doesn't tread on the ground of cognitive science's hope for mechanistic sufficiency. However, statements by Kant like, "It must be possible for the 'I think' to accompany all my representations" (*Critique of Pure Reason,* B131) show his limitations in light of Wittgenstein's *The Blue Book.* Possibly further philosophical fruit might be farmed by connecting the inverted solipsism that Wittgenstein develops (where again, the uncertainty is with connecting a sensation to a particular body, since the perception is always subjective, and not an objective body) and Kant's transcendental deduction of apperception, which finds that unity of apperception is deduced from a plurality of experience. These opposed perspectives may be two sides of the same "coin."

Of course, much of Wittgenstein's project is to show that particular linguistic biases, such as the implication that "I'm in pain" means there is an "object" "I" or ego-thing that is a subject for predicates, much like a thing has properties, sets up a whole shadow world of mental smoke and mirrors. This extended example in the last parts of *The Blue Book* is important, but "the mental" is only one of many metaphysical notions that Wittgenstein could have tackled with his new method of examining particular word usages being taken out of context, and contorted into something out of this world. Yet, his highly experiential / factual account of language in *The Blue Book*, while accounting for the ability for people to learn to share what is indefinable via language, seems to me to fail to fully take into account the power of expression to effectively communicate emotion, rather than simply identify it.

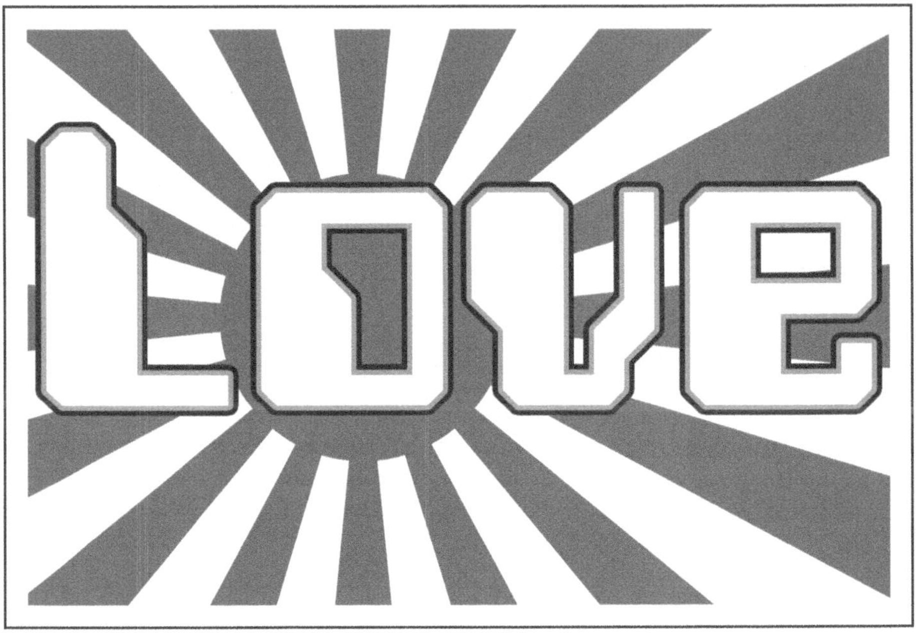
LOVE

Chapter 20

Heidegger

Towards a Poetry Beyond Technological Perfection

"No prophecy is necessary to recognize that the sciences now establishing themselves will soon be determined and regulated by the new fundamental science that is called cybernetics. This science corresponds to the determination of man as an acting social being. For it is the theory of the regulation of the possible planning and arrangement of human labor. Cybernetics transforms language into an exchange of news. The arts become regulated-regulating instruments of information. The development of philosophy into independent sciences that, however, interdependently communicate among themselves ever more markedly, is the legitimate completion of philosophy. Philosophy is ending in the present age. It has found its place in the scientific attitude of socially active humanity. But the fundamental characteristic of this scientific attitude is its cybernetic, that is, technological character" (Martin Heidegger, David Farrell Krell (ed.), *Basic Writings*, Harper: San Francisco (1993), p. 434, "The End of Philosophy and the Task of Thinking").

"'Poetry,' I answered, confident of my argument, 'is the beginning and the end of philosophical knowledge. Like Minerva from the head of Jupiter, philosophy springs from the poetry of an eternal, divine state of being. And so in philosophy, too, the irreconcilable finally converges again in the mysterious spring of poetry'" (Friedrich Holderlin, Willard R. Trask (trans.), *Hyperion*, The New American Library, Inc: New York (1965), p. 93).

"Philosophy is universal phenomenological ontology, and takes its departure from the hermeneutic of Dasein, which, as an analytic of existence, has made fast the guiding-line for all philosophical inquiry at the point where it arises and to which it returns" (Martin Heidegger, John Macquarrie & Edward Robinson (trans.), *Being and Time*, Harper & Row: New York (1962), p. 487).

CAVEAT: ANTI-FASCISM

Today, humankind may have the technological sophistication and material resources to provide (via robotic labor) many if not most basic human needs with little or no human labor: absolute leisure. And tomorrow, the perfection of bio-mechanics promises to provide the option of ever-lasting life (not to be confused, by the way, with eternal life!) Ever-lasting leisure? Technology provides the opportunity for alleviating many of life's greatest worries, and one must ask—what stands in the way of a technologically perfected society? But also, would we want it? No doubt, robotic labor is not always being used to feed the poor—it can displace workers; the dream of absolute leisure can be overtaken by perfectly efficient exploitation. This notion of "technological perfection" might bring fascism to the minds of some. The military strives to mold humans into "lean green machines"—the precise and efficient execution of maneuvers. Fascism goes hand in hand with the militarization of a society structured for totalitarian efficiency—maybe this informed

Heidegger's notion (quoted above) of cybernetics as "the theory of the regulation of the possible planning and arrangement of human labor." What Heidegger failed to note in his comment was that technology offers more than the mechanization of labor by humans, it offers to relieve humans of labor through mechanization! This association of Heidegger with fascism is important given the debate over his Nazi activities.

I could not excuse Heidegger as being someone too timid (and comfortable) to rock his local boat—he often lent his poetical and philosophical voice to the cause of National Socialism. What is perplexing, however, is that much of Heidegger's ground breaking exploration of "existentialism" questions going with the flow. To the very core, Heidegger's philosophy is really a working out of Plato's allegory of the cave—which itself pivots on non-conformism! One should not forget, however, that Plato's own Republic could be construed as fascist—and possibly this romantic dream of a state "philosophically perfected" enticed Heidegger. Did Heidegger see ethics as part of philosophy (if not as its foundation, like Levinas?) Did the end of philosophy include the end of ethics? Maybe Heidegger failed to see the irony employed by Nietzsche's going *Beyond Good and Evil.* Nietzsche himself rarely failed to address ethical issues.

I hope to distance myself from Nazism and fascism here clearly enough. Yet, Heidegger's thought has been too influential to ignore. And, as a sometime poet, I must admit that much of what he has to say has some resonances with my own poetic experiences. His writings offer insights into many important areas of thought: e.g., the history of freedom, interpretation, and ontology. Does his Nazi affiliation banish him from the philosophical cannon? Should we deny his insights and banish him to historical obscurity, or paraphrase all his good ideas in a book that would allow us to ignore his writings, or simply trace all his sources? One should not forget that many great writers of the past were slave owners. Much of our history is clouded with unethical atrocities; possibly someday our present time will be looked at with dismay too (some already see it that way). But we cannot dispose of all of history, it is how we have become who we are; and one should not throw out an entire bushel because of a few rotten apples. Thus, it is my suggestion that we read Heidegger critically—some material is good, some not.

And here I will approach the Heidegger found in his philosophical writings—the theoretical texts with poetical flourishes are the Heidegger that I know. And there, in his philosophical writings I have read, I have seen no explicit advocacy of what might be called a Nazi philosophy. There have been some comments by Heidegger claiming the superior spirituality of the German language. Can someone be proud simply by being different, and not by claiming to be better? I hope so, as this author finds "supremacism" and even "perfection" to be a somewhat inhumane concepts.

HOLDERLIN'S POETRY: *HYPERION*

In this chapter I hope to give a brief summary of what I see to be Heidegger's (1889-1976) main project. I will focus on how he has set up the concept of Dasein, and then show how this relates to his later comments on poetry and technology. Again, far from advocating some sort of technological fascist regime, we will see that Heidegger largely advocates non-conformist self-discovery, and warns against the possibility of technology overtaking poetry. With this in mind, I will be including some references to Holderlin's *Hyperion*, as Holderlin was obviously Heidegger's favorite poet.

Hyperion, like *Being and Time*, is divided into two parts. The first part documents (in the form of letters) the narrator Hyperion's enthusiastic journey into adulthood as he learns from his mentor Adamas and his friend Alabanda, and finally falls completely in love with Diotima. Through his journeys and letters, Hyperion advocates peace and non-conformism, and learns about life while manic-depressively considering many conflicting points of view:

> "We pity the dead as if they felt death, yet the dead have peace. But the pain, the pain that no pain equals, is the incessant feeling of utter annihilation when our life loses its meaning, when our heart bids itself 'Down! into the depths! there is nothing left of you; you have planted no flower, built no hut, so that you might say: I leave a trace behind me on earth.' Yet, oh! the soul can always be so full of longing, even when it is so cast down!" (*Hyperion*, p. 56).

> "What is all that men have done and thought over thousands of years, compared with one moment of love? But in all Nature, too, it is what is nearest to perfection, what is most divinely beautiful! Thither all stairs lead from the threshold of life. Thence we come, thither we go" (*Hyperion*, p. 68).

In the second, more dreary part of *Hyperion*, Hyperion is called off to war; he leads soldiers in hopes of liberating Greece, yet is disillusioned when his own men commit war crimes; and before he returns home, the story tragically closes with the death of his beloved Diotima. Again, it is hard to reconcile Heidegger's Nazism with his fondness for the peace advocating Holderlin. With the romantic Holderlin we see a precursor to many of Heidegger's interests, including ancient Greece, the meaning of life, and love. For, as Holderlin holds love above all, we shall find that Heidegger finds *care* lying at the heart of Dasein.

DASEIN: "BEING THERE"

But what, exactly, is this Dasein? To approach this question, we must first understand that Heidegger's primary work, *Being and Time*, is dedicated to formulating the question of the meaning of being. Heidegger never provides a direct answer to this question, which might crudely be reformulated, "why are we here," or "what does it all mean," questions which now are widely known as "existential." His project is rather one of determining how one might be able to even begin an attempt at answering it. He begins his analysis by studying the structure of the type of beings that would be able to pursue this question.

Now, when one thinks about one's being in the world, there are several ways that one could talk of "things" existing. First, there are the objects and what-not that we find in the world—these things exist, and we can reach out for them. They are, using Heidegger's terminology, "present-at-hand" in nature. But, contrasted with this, there is our own perspective; contrasted with all the "whats" in the world, there is the being of the "who." Heidegger calls the being of being a "who" Dasein. Dasein ("da" "sein" meaning in German, "there" "being", or "being there") is our own personal being in the world, a being which is both wrapped up factually in the "everydayness" of the world, as well as being outside of the world enough, along side of it, to observe it.

"EXTERNAL" CONSCIOUSNESS

This aspect of being alongside the world, observing it, has often been called subjectivity. Dasein is not classical subjectivity, however, in that Dasein includes the phenomenal qualitative presence of one's consciousness. The spatial being of one's Dasein extends to all that one senses: as far as you can sense, you are. When one looks at a sunset over mountains, the sun, the mountains and all the trees and the surrounding birds chirping (the entire "environment") are equally near in that they are present in the immediate presence of one's Dasein. This extension of personal consciousness Heidegger calls "de-severing"—the space cut between you and an object is closed, for there is no distance between you and the object in your conscious perception of the object: there is only one conscious perception, with no split between subject and object. This phenomenally aware Dasein, which is both in the world and alongside it, also has a "directedness" about it—one is usually aiming somewhere, or focusing on something. One is "concerned" about things in the world; an aspect of conscious-being which in contemporary cognitive science is often called "attention."

CHOPPING OFF A FINGER

This "concern" brings us to the things in the world, which Heidegger says are always first "ready-to-hand." What this means is that in our concern with the world, we first come across things as "equipment;" e.g. (referencing the movie *2001: A Space Odyssey*) one grabs a bone and kills an enemy, or one puts a stick into an insect nest and pulls it out for an insect pop-sickle. It is not until our use of these tools becomes unfamiliar, "conspicuous," that we come to see them as things in the world which are "present-at-hand" in nature. What we see here is that our equipment is largely seen at first as extensions of ourselves. When such items fail to instantaneously yield to our concern, we begin to see them as separate from ourselves—as things involved with an external world; e.g., if one has a finger chopped off, it no longer seems to be an extension of oneself, but is an object in the world.

Indeed, Heidegger even sees the signs that we use for communication as being equipment; signs are things that we use to *point* to the world. Signs, like all things that are ready-to-hand, are given a purpose by Dasein, a "for-the-sake-of-which" which is involved with their relationships to all the other things in the world that Dasein concerns itself with. Indeed, later we shall see that this entire interconnection among things and signs which compose the world past, present, and future have always already been preconceived by Dasein in that its knowledge of the world is always brought forth from itself. It seems problematic however, to place the ready-to-hand prior to the present-at-hand, for often things are awkward when we first encounter them; an automobile becomes an extension of the self (the ego revving the unmuffled exhaust pipe) only after one learns to drive.

BORN INTO THE "THEY" SOCIETY

Besides things, there are also other living beings in the world—other people are not ready-to-hand or present-at-hand for Heidegger; they too have their own individual Daseins. Our own individual Dasein is born into a world full of other Daseins. Also, this world we are born into is already full of ways of seeing things, attitudes, knowledges, and beliefs. Through being raised and educated, etc, we become culturally intertwined with the others around us. This society that has been here before us, and which we are shaped by, Heidegger calls the "they." We, as Daseins, are "thrown" into both the phenomenal world and the society of the "they." This "they" society is hopelessly superficial for Heidegger—its knowledge is for the most part "uprooted" from real truth in a "groundless floating", and is passed on like gossip: an "idle talk" which sees everything as familiar and is "curious" only for the sake of fads. This "they" perspective is "tranquilizing" in its familiarity, but is "inauthentic" through and through—the entry of Dasein into this inauthenticity is called "falleness" by Heidegger (*Being and Time*, chaps. 25-27). Heidegger comments that all this should not be seen as necessarily bad—one has as yet no notion of there being something good outside of the "they" by which to condemn it. In contemporary terminology, we could say that the "they" is wrapped up in a differential / coherence type of knowledge where the knowledge of the community becomes largely self-referential, severed from primordial reality in a floating off propelled by its own self-satisfaction. One could imagine Leo Tolstoy's *Ivan Illych* and the superficial prestige mongering society he was a part of.

INTERPRETATION: EXPECTED PROJECTION

Returning to the individual Dasein, we can see that we approach the world with various "states-of-mind"—emotional attitudes—e.g. fear. Along with these states-of-mind we also have the ability to "understand" the world—but this understanding is always thrown from our own Dasein in a type of "projection." As noted earlier, Dasein is that which bestows a purpose on the things which are around it: a cave may become a home because I will it—the cave is known as a home by me, and not by the birds outside. This projection of the understanding operates through the "possible": through the understanding, items in the world have many possible uses and interconnections with other items, which may be projected. When this understanding takes itself into consideration "interpretation" arises—with interpretation, one does not simply comprehend, one sees into relationships. With the understanding, one only knows that the cave is one's home; with interpretation one knows that one is seeing the cave as a home. This seeing "something as something" (*Being and Time*, p. 189) seems akin to Nietzsche's discussion of metaphor. Interpretation is the way in which our world becomes explicit to us—through Dasein things in the world are already interconnected, and our use of things already accounts for such implicitly:

> "When something is understood it is still veiled, it becomes unveiled by an act of appropriation, and this is always done under the guidance of a point of view, which fixes that with regard to which what is understood is to be interpreted. In every case interpretation is grounded in *something we see in advance—in a fore-sight*" (*Being and Time*, p. 191).

The way in which we come to understand and interpret "entities" in the world is predetermined by a type of expectation, yet Heidegger is unclear as to what extent this expectation and fore-sight is shaped by: (1) the way the world is:

> "Dasein, in so far as it is, has always submitted itself already to a 'world' which it encounters, and this *submission* belongs essentially to its Being" (*Being and Time*, pp. 120-121).

Or (2) by the attitudes of the "they" whose community we are born into:

> "Dasein-with is already essentially manifest in a co-state-of-mind and a co-understanding. In, discourse Being-with becomes 'explicitly' *shared*; that is to say, it *is* already, but it is unshared as something that has not been taken hold of and appropriated" (*Being and Time*, p. 205).

Or (3) by the "disclosing" of Dasein itself:

> "*only Dasein can be meaningful or meaningless*" (*Being and Time*, p. 193).

In any case, Heidegger notes that the problem of the hermeneutic circle arises here:

> "if interpretation must in any case already operate in that which is understood, and if it must draw its nurture from this, how is it to bring any scientific results to maturity without moving in a circle, especially if, moreover, the understanding which is presupposed still operates within our common information about man and the world?" (*Being and Time*, p . 194).

USING YOUR LINGUISTIC BRAIN

Heidegger does explain, however, that it is with "assertions" (communications that point out facts) that those things that used to be ready-to-hand become present-at-hand:

> "Something *ready-to-hand with which* we have to do or perform something, turns into something 'about which' the as-

> sertion that points it out is made. Our fore-sight is aimed at something present-at-hand in what is ready-to-hand. Both *by* and *for* this way of looking at it, the ready-to-hand becomes veiled as ready-to-hand [....] this discovering of presence-at-hand, [...] is at the same time a covering-up of readiness-to-hand" (*Being and Time*, p. 200).

And as language and discourse are shared in the world as ready-to-hand, they too can become present-at-hand. However, Heidegger is careful to point out that:

> "Dasein, as Being-in-the-world, already dwells *alongside* what is ready-to-hand within-the-world; it certainly does not dwell proximally alongside 'sensations'; nor would it first have to give shape to the swirl of sensations" (*Being and Time*, p. 207).

In other words, although we can think of speech and writing as sounds and marks, Dasein's being alongside them ready-to-hand understands them immediately without a moment of interpretation:

> "Even in cases where the speech is indistinct or in a foreign language, what we proximally hear is *unintelligible* words, and not a multiplicity of tone-data" (*Being and Time*, p. 207).

Heidegger's insight here seems to imply that as soon as we begin to analyze language and discourse, bringing them from usage as ready-to-hand to an appropriated present-to-hand, they lose their usability—what we know about language can never coincide with how we use it, for one way blocks the other.

ANXIETY AND AUTHENTICITY

It was noted earlier that it takes a "problemitization" that makes something ready-to-hand conspicuous for us to possibly know it as something present-at-hand. And, moreover, it should be recalled that the world of the inauthentic "they" was one of comfortable familiarity. If we keep both of these notions in mind, then we will see how both the possibility for an authentic Dasein and the disclosure of new understanding is possible. For Heidegger claims:

> "Dasein's absorption in the 'they' and its absorption in the 'world' of its concern, make manifest something like a *fleeing* in the face of itself—of itself as the authentic potentiality-for-Being-its-Self" (*Being and Time*, p. 229).

And:

> "Being-in-the-world is a basic state of Dasein. *That in the face of which one has anxiety is Being-in-the world as such*" (*Being and Time*, p. 230).

The state-of-mind of anxiety is intertwined with defamiliarization—anxiety is the turning away from the comfortable falleness of the "they":

> "Anxiety throws Dasein back upon that which it is anxious about—its authentic potentiality-for-Being-in-the-World. Anxiety individualizes Dasein for its ownmost being-in-the-world, which as something that understands, projects itself essentially upon possibilities [....] Anxiety makes manifest in Dasein its *Being towards* its ownmost potentiality-for-Being—that is, its Being-free" (*Being and Time*, p. 232).

The moment of anxiety is one where Dasein solipsistically (*Being and Time*, p. 233) turns away from the "they" self and discovers the possibility of free individuality—and this is a turning back to something more primordial than falleness. This turning would be towards an original authentic Dasein as being-a-whole (all being), but such is limited by death, the limit of our own existence. However, it is the anticipation of death itself which initiates the anxiety of turning away from the "they" self, the turning of Dasein towards itself in response to the call of care.

LOVE REVELATION

We find, however, that no sooner has Dasein turned toward its:

> "ownmost potentiality-for-Being" (*Being and Time*, p. 236),

or free individuality, than that it finds it has always already been in the world; and Heidegger calls this phenomena of being ahead-of-itself-in-already-being-in-a-world "care." Care is the being of Dasein:

> "Care, as primordial structural totality, lies 'before' every factical 'attitude' of Dasein [....] care in its totality is essentially something that can not be torn asunder." (*Being and Time*, p. 238).

Or, to appropriate Holderlin's words:

> "Does not silence dwell in the Land of the Blessed? Above the stars the heart forgets its needs and its language.
>
> I have guarded it sacredly! I have carried it within me like a palladium—the divine that appeared to me! and if henceforth Fate lays hold of me and casts me from abyss to abyss and drowns all powers in me and all thoughts: yet shall this unique revelation outlive myself in me and shine in me and rule me, in eternal, indestructible brightness!—
>
> So didst thou lie poured out, sweet life, so didst thou look up, and arise, and stand there before me in delicate completeness, divinely calm, and with thy heavenly face filled

with the serene ecstasy in which I disturbed thee" (*Hyperion*, pp. 63-64).

And:

"Man's *perfectio*—his transformation into that which he can be in Being-free for his ownmost possibilities (projection)—is accomplished' by 'care'" (*Being and Time*, p. 243).

And also:

"the moments when we are set free, when the divine shatters the prison, when the flame bursts from the wood and wings up over the ashes, ah! when it is with us as if, its sorrows and its servitude forgotten, the unshackled spirit were returning in triumph to the halls of the Sun" (*Hyperion*, p. 65).

"To be one with all—this is the life divine, this is man's heaven. To be one with all that lives, to return in blessed self-forgetfulness into the All of Nature—this is the pinnacle of thoughts and joys, this sacred mountain peak, the place of eternal rest, where the noonday loses its oppressive heat and the thunder its voice and the boiling sea is as the heaving field of grain" (*Hyperion*, p. 23).

"There is a forgetting of all existence, a hush of our being, in which we feel as if we had found all.

There is a hush, a forgetting of all existence, in which we feel as if we have lost all, a night of the soul, in which no glimmer of any star nor even the fox fire from a rotting log gives us light" (*Hyperion*, p. 54).

"I can speak of her only fragmentarily—a word here, a word there. I have to forget what she is in her completeness if I am to speak of her at all" (*Hyperion*, p. 72).

> "In the altogether unsettling experience of this hovering where there is nothing to hold onto, pure Da-sein is all that is still there.
>
> Anxiety robs us of speech. Because beings as a whole slip away so that the nothing crowds round, in the face of anxiety all utterance of the 'is' falls silent [....] In the lucid vision sustained by fresh remembrance we must say that in the face of which and for which we were anxious was 'properly'—nothing" (*Basic Writings*, p. 101, "What is Metaphysics?")
>
> "Only on the ground of original revelation of the nothing can human existence approach and penetrate beings [....] Without the original revelation of the nothing, no selfhood and no freedom" (*Basic Writings*, p. 103, "What is Metaphysics?")

We have, as a society, by and large forgotten the presence of pure Dasein, and the revelation of all knowledge avails itself from the partial fragmentary recollection of oneness. Real truth, for Heidegger, is revealed—it is uncovered:

> "Circumspective concern, [...] uncovers entities within-the-world [....] What is primarily 'true'—that is, uncovering—is Dasein [....] In so far as Dasein *is* its disclosedness essentially, and discloses and uncovers as something disclosed [...] it is essentially 'true'. *Dasein is 'in the truth'*" (*Being and Time*, p. 263).
>
> "assertion is grounded in Dasein's uncovering, or rather in its *disclosedness*. The most primordial 'truth' is the 'locus' of assertion" (*Being and Time*, p. 269).

Dasein is thus the revealing of its own truth, both disclosing and disclosed. Yet, it must be realized that what Heidegger is postulating here is nothing much more metaphysically complex than Plato's comments on our being fallen from a formally perfect realm of being, with our ability to gain knowledge of this realm through recollection. Heidegger has been, however, much more specific and detailed—in some ways more than Kant was concerning the difference between the "productive" and "reproductive" imagination.

And thus concludes this rather rough summary account of Heidegger's analysis of Dasein. Such an explication should be enough though to situate it within Heidegger's comments in "The Question Concerning Technology."

TECHNOLOGY: INSTRUMENTAL AND ENFRAMING

In "The Question Concerning Technology," Heidegger investigates the essence of technology. He begins by noting that we often see technology as being instrumental—that is, technology concerns the equipment, tools, and machinery that we use. Technology is seen as something that we have control over; again, we as humans use technology—it is a means to an end. And, explains Heidegger:

> "Wherever ends are pursued and means are employed, wherever instrumentality reigns, there reigns causality" (*Basic Writings*, p. 313, "The Question Concerning Technology").

Technology is employed in bringing something about—it is involved in the causality which brings forth that which was not previously there. Heidegger calls this bringing forth "revealing," and it is at this point that we may begin to discern a relationship between Dasein (the locus of revelation) and technology.

Also, when we look to the origin of the word technology, we find that it stems from the Greek word "technē," which:

> "is the name not only for the activities and skills of the craftsman but also from the arts of the mind and the fine arts. *Technē* belongs to bringing forth, to *poiēsis*; it is something poetic" (*Basic Writings*, p. 318, "The Question Concerning Technology").

Heidegger notes that the word "technē" has also been linked with the word "epistēmē"; both designate knowing, which is always involved with revealing. This revealing is contrasted with that of pure "poiēsis" though, in that it does not bring some presence forward, but challenges nature to supply energy; concealed energy is to be unlocked, transformed, stored, distributed, switched about, and regulated ("The Question Concerning Technology," *Basic Writings*, 322). Technology calls forth an ordering of nature, it calls humans to reveal nature as "standing-reserve."

Thus, we should see that technology, as a type of revealing, cannot simply be understood as instrumental. The tool connects up with nature and humans only to the extent that nature and humans are already within this realm of the technological: nature and humans already have machine like aspects—the extent to which we recognize this is due to the type of organizing which technology calls forth. Heidegger also calls this "enframing":

> "We now name the challenging claim that gathers man with a view to ordering the self-revealing as standing-reserve: *Ge-Stell* [enframing]" (*Basic Writings*, p. 324, "The Question Concerning Technology").

Enframing is that which, beyond the technological, or the mechanical:

> "is the way in which the actual [nature] reveals itself as standing-reserve" (*Basic Writings*, p. 329, "The Question Concerning Technology").

This type of revelation can be seen, however, with any type of knowing that appropriates along the line of mechanics—and indeed, physics, e.g. with quantum mechanics, prevails through chemistry, biology, and cognitive science, the sciences that demand:

> "that nature report itself in some way or other that is identifiable through calculation and that it remain orderable as a

> system of information" (*Basic Writings*, p. 328, "The Question Concerning Technology").

Hence we can see the important place of cybernetics, the study of information flows, with regard to enframing.

THE PRESENCE OF POETRY

Technological enframing is precisely that which has no centered presence. In its gathering towards orderly information flows, it circumscribes being as with an invisible mathematical line. The repetition of structure everywhere holds consistently. However:

> "what is present is present in a representation that has the character of calculation. Such representation knows nothing immediately perceptual. What can be immediately seen when we look at things, the image they offer to immediate sensible intuition, falls away. The calculating production of technology is an 'act without an image'" (Martin Heidegger, Albert Hofstadter (trans.), *Poetry, Language, Thought*, Harper Colophon Books: New York (1983) pp. 126-127, "What Are Poets For?")

The conscious immediacy of being concerned with a thing escapes technology for technology is representational mediation. As we noted earlier, technological enframing is involved in being a means to an end. In this way we can see the difference between technē and poiēsis as ways of revealing: poiēsis is that which brings forth to presence, while technē simply provides a way of ordering such as informational standing-reserve.

It is with this contrast in mind that Heidegger warns of the danger of enframing:

> "the challenging-enframing not only conceals a former way of revealing (bringing-forth) but also conceals revealing it-

> self and with it that wherein unconcealment, i.e., truth, propriates" (*Basic Writings*, p. 333, "The Question Concerning Technology").

And:

> "enframing [...] blocks *poiēsis*" (*Basic Writings*, p. 335 "The Question Concerning Technology").

These comments can best be understood with reference to the earlier discussion of Dasein. Dasein, in its concerned presence which reveals itself is beyond the scope of technological enframing. Yet, humanity is in danger of interpreting itself wholly in terms of technological enframing—humanity is in danger of drastically misinterpreting itself by not investigating that which lies beyond the technological, the:

> "overflow beyond number" (*Poetry, Language, Thought*, p. 128, "What Are Poets For?")

ANGELS BEYOND THE SHIELD OF BEING

How could one go about investigating beyond calculation? In his discussion of a Rilke poem in his "What Are Poets For?" Heidegger writes of those who would venture themselves beyond the venture of being. The poem in question reads:

> "As Nature gives the other creatures over
> to the venture of their dim light
> and in soil and branchwork grants none special cover,
> so too our being's pristine ground settles our plight;
> we are no dearer to it; it ventures us.
> Except that we, more eager than plant or beast,
> go with this venture, will it, adventurous
> more sometimes than Life itself is, more daring
> by a breath (and not in the least

> from selfishness) .... There, outside all caring,
> this creates for us a safety—just there,
> where the pure forces' gravity rules, in the end,
> it is our unshieldness on which we depend,
> and that, when we saw it threaten, we turned it
> so into the Open that, in widest orbit somewhere,
> where the Law touches us, we may affirm it"
> (*Poetry, Language, Thought*, p. 99, "What Are Poets For?")

A few venture beyond the shieldness of being. The structure of being shields us from the chaos of the nothing; being clears a dwelling space in the nothing. Some venture forth beyond shieldness (as with the prodigal son) and in such crises convert the unshielded into the cleared opening of being—in effect extending the "boundary," or Law. Such extends out from the gravitational center of pure being, towards the "widest orbit," or outer-limit, which boundary itself proves to be the center of division:

> "This being, [...] is the being who governs the unheard-of center of the widest orbit and causes it to appear. This being [...] is the Angel" (*Poetry, Language, Thought*, p. 134, "What Are Poets For?")

With Heidegger's Rilke, the angel ventures farther than the plants, animals, and humans:

> "...the man who is outside all protection can procure a safety by turning unshieldness as such into the Open and transmuting it into the heart's space of the invisible. If that happens, then what is unstilled in unshieldness passes over to where, in the balanced oneness of the world's inner space, there appears the being who brings out the radiant appearance of the way in which that oneness unifies, and who in this way represents Being. The balance of danger then passes out of the realm of calculating will over to the Angel" (*Poetry, Language, Thought*, p. 136, "What Are Poets For?")

Going beyond the shieldness of being's venture is the one who turns inward to their heart; as with the discussion concerning Dasein, the one who would reveal their self turns away from inauthenticity. However, at its limit, the unconcealing of Dasein conceals itself in denial:

> "Insight into that which is—thus do we name the sudden flash of the truth of Being into truthless Being" (Martin Heidegger, William Lovitt (trans.), *The Question Concerning Technology*, Harper & Row: New York (1977), p. 47, "The Turning").

> "'I am close upon them,' I said, 'The great saying [...] the one differentiated in itself [...] of Heraclitus [....] Now classification became possible, for the whole was there. The flower had ripened; now it could be dissected'" (*Hyperion*, p. 93).

> "more venturesome are the poets, but poets whose song turns our unprotected being into the Open [....] The converting inner recalling is the daring that dares to venture forth from the nature of man, because man has language" (*Poetry, Language, Thought*, p. 140, "What Are Poets For?")

No doubt, it is the poets risking non-orderly confusion who might venture forth beyond technological enframing. Yet, Heidegger notes that with its danger, technological enframing may have a saving grace:

> "It is precisely in enframing, which threatens to sweep man away into ordering as the ostensibly sole way of revealing, and so thrusts man into the danger of surrender of his free essence—it is precisely in this extreme danger that the innermost indestructible belongingness of man within granting may come to light, provided that we, for our part, begin to pay heed to the essence of technology" (*Basic Writings*, p. 337, "The Question Concerning Technology").

Also, more than aiding in the discovery of the "innermost indestructible belongingness," technology, through its consistency offers that which endures in standing reserve. Noting this potential of technological enframing for both danger and saving, Heidegger claims that humans thus have the task of watching over technology. And it is here, with Heidegger's calling us towards the task of the shepherd that we find the heart of his philosophical ethical stance.

Moving towards a closure of this chapter, it must be noted that I have not ventured to examine many important aspects of Heidegger's thought, especially his discussion of time. Here, I have hoped to give a rough sketch of Dasein with special reference to its technological and poetical modes of revelation. Of course, a full disclosure of being continually withdraws itself from possibility, being that it is that which, through Dasein, throws us into the possible as such. No doubt, much of Heidegger's thought is speculative, and may require further "philosophical-poetical verification." Hopefully, I have not seemed to bring too much of Heidegger into a standing-reserve. Despite his often perplexing philosophical insights (has he not already loaded Dasein's answer to the question concerning the meaning of being with *care*?) Heidegger remains among those who, with the aid of poetry, would help shield thinking, and guide it beyond the end of philosophy.

> "Thinking's saying would be stilled in
> its being only by becoming unable
> to say that which must remain
> unspoken" (*Poetry, Language, Thought*, p. 11, "The Thinker as Poet").

Richie Rich comic book (1960/1987)

Chapter 21

Adorno & Benjamin

Authentic Culture Critique

"You're a Star-Belly Sneech, You suck like a leech
You want everyone to act like you
Kiss ass while you bitch so you can get rich
But your boss gets richer off you
Well you'll work harder with a gun in your back
For a bowl of rice a day
Slave for soldiers til you starve
Then your head is skewered on a stake
Now you can go where people are one
Now you can go where they get things done
[....]
And it's a holiday in Cambodia
Where you'll do what you're told
A holiday in Cambodia
Where the slums got so much soul"

"Holiday in Cambodia" excerpt
—W-Biafra / M-Biafra, Ray, Flouride, Slesinger

JUXTAPOSED DOCUMENTS OF CULTURE?

This chapter includes reference to a compounded "dialectical" dichotomy. First there is the general dichotomy between the critics, and some mass-media "art." Within each of these categories, there is on the one hand a specific exemplar dichotomy between Frankfurt School insider critic Theodor Adorno (1903-1969) (with some focus here on his works, "Cultural Criticism and Society," and "A portrait of Walter Benjamin," found in *Prisms*), and the almost outsider critic Walter Benjamin (1892-1940) (with some focus on his works, "The Work of Art in the Age of Mechanical Reproduction," and the "Thesis on the Philosophy of History"

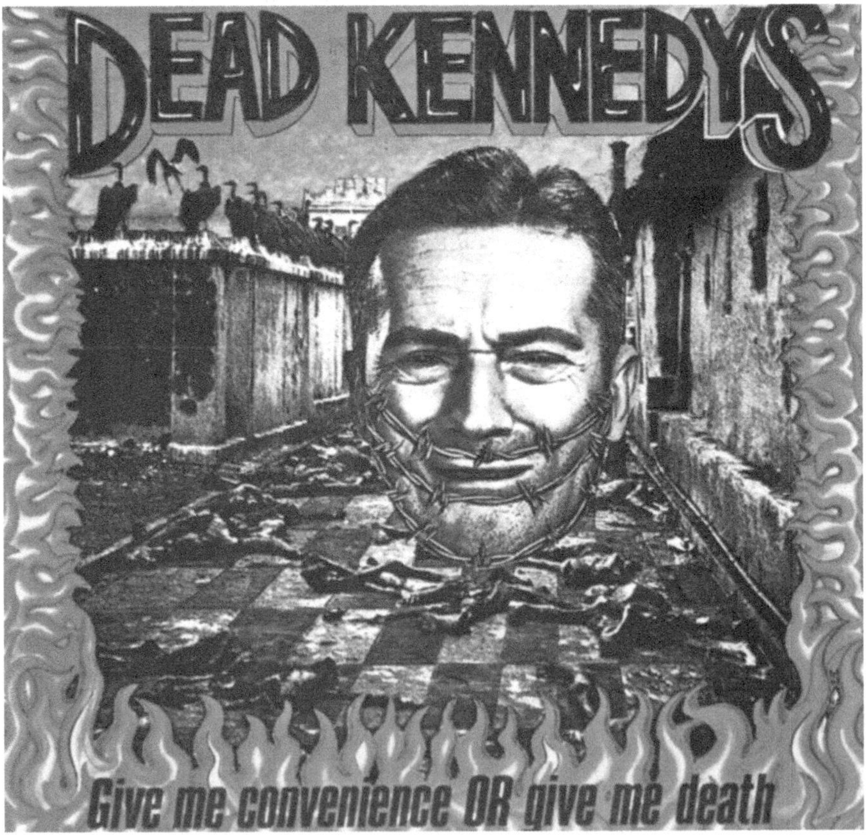

Dead Kennedys album cover (1987)

found in *Illuminations*). On the other hand, there is a material dichotomy between two "mass-media art works," I've selected to examine: a typical *Richie Rich—The Poor Little Rich Boy* comic book and the Dead Kennedys' song, "Holiday in Cambodia," included here from their CD album, *Give Me Convenience or Give Me Death*. Although both Adorno and Benjamin may have found such art vulgar, I believe these selections exaggerate, and hence clarify in way, the kind of pop-art that has saturated highly capitalist culture.

Martin Jay (in his *The Dialectical Imagination*) noted that:

> "Adorno chose to discuss them [works of art] in dialectical pairs in order to transcend the inherent insufficiency of individual accomplishments" (Martin Jay, *The Dialectical Imagination: A History of the Frankfurt School and the Institute of Social Research, 1923-1950*, Little Brown & Co.: Boston (1973), p. 178).

My selections also poke a little fun at Adorno and Benjamin as the lovers of seriously "classical" (yet challenging) music and literature. The following will include many quotes:

> "Benjamin saw himself as the vehicle for the expression of objective cultural tendencies, a belief that made the mode of expression particularly crucial. One manifestation of this was his hope to exclude all subjective elements from his work by writing an essay consisting solely of quotations from other sources" (*Dialectical Imagination*, p. 176).

Jay offers some religious context for this as well:

> "Benjamin's examination of cultural phenomena resembled that of a biblical scholar probing a sacred text. In his hope of writing a book consisting solely of quotations, Benjamin expressed a quasi-religious desire to become the transparent mouthpiece of a higher reality" (*Dialectical Imagination*, p. 200).

And further, Herbert Schnadelbach (in his *Philosophy in Germany 1831-1933*) writes concerning a Historicist approach, of a:

> "value-free accumulation of material and facts without distinction between what is and what is not important" (Herbert Schnadelbach, Eric Matthews (trans.), *Philosophy in Germany 1831-1933*, Cambridge University Press: Cambridge (1984), p. 35),

by way of:

> "'depersonalization': research which is rationalized, in the sense of rigorously methodical, promises success only when the individual scientist precisely does not bring his own individuality into play in any essential way and when he strictly follows the ethos of pure 'objectivity'" (*Philosophy in Germany 1831-1933*, p. 71).

Could simple juxtaposed quotations from and selections of art from mass-culture be a powerful form of criticism? Could the critic, by withdrawing themselves from the work, and simply presenting it, still imply a *critique* of the current state of society and point towards a more civilized and reasonable future—and help fulfill the goals of "Critical Theory" (which might precisely defined as: *exactly not that which leads to its conception*)?

Again, Schnadelbach:

> "Consciousness of the historical and consciousness of itself as something historical, the historical consciousness locates itself in the process of history, which, after the historicization of history, it can no longer confidently regard as being commensurable with our current conditions of interpretation and understanding. Historical consciousness is thus at the same time consciousness of its own finitude and limited autonomy in the face of the superior force of history as a whole" (*Philosophy in Germany 1831-1933*, p. 38).

Historical consciousness operates something like a history pivot, where history turns or folds on itself and becomes conscious of its own determination: the light of consciousness becomes self-conscious through a hall of mirrors: mirrors of history reflecting on history through the light of that consciousness; yet that consciousness may become alienated from determining history, fearing a bad-faith un-freedom due to total historical determination: a double bind of being both from and against history.

EVOLVING REASON

A tension intensifies in early twentieth century continental philosophy between various theoretical dichotomies and their interdetermination. Indeed, the theoretical, or abstract itself has its opposite in the particular and, with another context, in praxis, or practice. However, theory is seen to color particular experience, through what is highlighted, and how it is interpreted. There is a hermeneutic circle, where an individual's history and background (or a group's), both as part of the human condition and culture on *this* planet, informs what is determined as perceptually relevant: just as we can only learn what we almost already know, the hermetically self-referential nature of what a person knows shapes what they can possibly come to know in addition. That is, knowledge is not added, piece by piece in some sort of stockpile, but is integrated with what we already know, by means of what we already know how to do. We, as individuals, and a culture, have a web of knowledge, as it were, that can integrate new knowledge only when it is recognized as such through the lens of that web. How we are who we are is largely determined by our cultural history; a cultural history which limits what we may perceive, simply by offering some of the only tools by which we can perceive at all: the gift of the ability to reason, as given by culture (beyond the body), is also constrained by the limited number of methods given (what we have to work with), and by the historically indoctrinated limitations on what is considered acceptable (these methods and limitations have proven pragmatically useful over the course of the evolution of culture's reason and knowledge).

Now, Critical Theory (and Adorno's Negative Dialectic), as developed in the Frankfurt School, seems aimed, like an active Darwinism (or more accurately, a post-Hegelian Marxist praxis) at propelling culture forward by a critique of that present which is a remnant of the past: the status quo. Hence the critique of repetitious and formulaic art, and the anti-systematic, anti-methodological, and difficult to summarize, appropriate, and co-opt style of Adorno's thinking that favors bold schisms and unfamiliar shocks, which would awaken people from their dogmatic slumber. (A possible critique of such a notion might make reference to the fact that errant DNA would more often lead to dysfunction rather than better adaption to existing and new niches in the ecological and sociological environment.)

The link between the particular and practice in opposition to theory, is important, in that while theory often strives for that which is universal and absolute (think science and mathematics), practice operates temporally through change, as the particular itself also changes relative to some other particular (spatiotemporally). There is a tension between the contingent, singular, dynamic particular, and the theoretical reasoning that tries to freeze it, e.g. in a concept or percept, through hypostasis, reification and analytical definition: a fetishism that isolates some feature of the inexhaustible "object." Such an "object" may itself be a limitation of the dynamic flux of the concrete particulars: the surface of the concrete may have no theoretically distinguishably identifiable particular. The truth of the singular particular is lost in repetitious generalizations (by "The Whole is the False" Adorno must have referenced Hegel's Reason reaching the absolute and comprehending everything, and not the whole as every particular aspect of the concrete: possibly the two are one at such a *differential* point of being). Adorno does not keep silence with some sort of particular—he uses the repeated words of language, albeit often in a different manner, and emphasizes a gestalt constellation of concepts in opposition to some key concept (like Being), or isolatable faculties of the mind (as with Kant). Adorno recognizes that his is an *Immanent Critique*, one already within a tradition: that he is part of what he questions. The objective is always perceived through the lens of the subjective, which subjectivity is shaped by the objective. Hence the difficulty with the critical project of withdrawing from culture (and its objective history) and letting it speak for itself in such a way that said culture is disrupted, disturbed by its own image, and is propelled towards improvement.

ENCULTURED CRITIQUE OF CULTURE CRITICS

Addressing the problematic and subjective nature of critical selection in *Prisms*, Adorno notes:

> "The cultural critic evaluates and hence is inevitably involved in a sphere stained with 'cultural values,' even when he rants against the mortgaging of culture. His contemplative stance towards culture necessarily entails scrutinizing, surveying, balancing, selecting: this piece suits him, that he rejects. Yet his very sovereignty, the claim to a more profound knowledge of the object, the separation of the idea from its object through the independence of the critical judgment threatens to succumb to the thinglike form of the object when cultural criticism appeals to a collection of ideas on display, as it were, and fetishizes isolated categories such as mind, life, and the individual" (Theodor W. Adorno, Samual & Shierry Weber (trans.), *Prisms*, The MIT Press: Cambridge, MA. (1994), p. 23, "Cultural Criticism and Society").

These last remarks seem directed at cultural criticism performed by life philosophers such as Nietzsche: such individualist preferences fly in the face of lessons learned from Marx et.al. about the importance of the social: Adorno's "Culture Industry" critique and psychoanalysis of culture was informed by *both* Nietzsche's notion of a "herd" mentality, and Marx's notion of class-consciousness and outlook "ideology." The "masses," like Heidegger's "they," are often un-self-determined in their flocking to a inter-subjective "reality" which may have little resemblance to objective reality. However, in contrast to Marx, for Adorno the material objective order does not center on the economic, but on the social. Again, there is a tension for the critic, for that objective reality sought may be heavily intertwined with the very inter-subjective "reality" that shapes the critic too.

Like Jacques Lacan psychoanalyzing psychoanalysis, Adorno, more generally, is criticizing criticism: he is performing a double application of criticism, a folding over of criticism on itself where the critic becomes more self-conscious of what they are doing: while criticizing the critic, Adorno is simultaneously criticizing himself:

> "The culture critic is not happy with civilization, to which alone he owes his discontent. He speaks as if he represented either unadulterated nature or a higher historical stage. Yet he is necessarily of the same essence as that to which he fancies himself superior" (*Prisms*, p. 19, "Cultural Criticism and Society").

Critics are in a struggle for their own autonomy:

> "Their very rejection of the guilt of life which blindly and callously reproduces itself, their insistence on independence and autonomy, on separation from the prevailing realm of purposes, implies, at least as an unconscious element, the promise of a condition in which freedom were realized" (*Prisms*, p. 23, "Cultural Criticism and Society").

Later:

> "Only the mind which, in the delusion of being absolute, removes itself entirely from the merely existent, truly defines the existent in its negativity" (*Prisms*, p. 26, "Cultural Criticism and Society").

Here Adorno, with shades of Freud, has taken cultural criticism to the absolute limits, where the critic, in an insane moment of monumental megalomania completely severs all ties with society. But again, back here on earth, the:

> "Dialectical critic of culture must both participate in culture and not participate" (*Prisms*, p. 33, "Cultural Criticism and Society").

GIVE ME CONVENIENCE OR GIVE ME DEATH

The very title of the compilation CD, *Give Me Convenience OR Give Me Death*, has its own little dialectic; and is a commentary on the fact that the band "Dead Kennedys" has let its record company weed out all their less popular tunes for the sake of a more marketable compilation album. Even my own selected quotation of their song, "Holiday in Cambodia" could be seen as a co-opting choice: this song was probably their most popular. Maybe the Dead Kennedys have co-opted the Frankfurt School a little with their questioning the use of brand-name "Right Guard" antiperspirant in Pol Pot's Cambodia ("It's time to taste what you most fear / Right Guard will not help you here": turning political radicalism into the band's most accessible "alternative hit song," complete with a *repetitious* and catchy chorus. The CD album art shows a road littered with decaying corpses and large face with its mouth bound by barbed-wire, possibly depicting the atrocities that occur in society, and the stifling of those who would try to reveal them (such as the Dead Kennedys). Martin Jay on Adorno on music:

> "As was the case with all cultural phenomena, it [music] was neither fully reflective nor fully autonomous. Still, in the current era, its autonomy was severely threatened. Most music displayed the characteristics of a commodity, dominated more by exchange than by use value. The real dichotomy, Adorno contended, was not between 'light' and 'serious' music [...] but rather between music that was market-oriented and music that was not" (*Dialectical Imagination*, p. 182).

With this Dead Kennedys' CD, we have something that wavers on the edge of marketability; possibly both the best and worst for either spreading critical thought or co-opting it. Adorno may have not appreciated the frantic beat of punk though, for he found the:

> "use of neoprimitive rhythms corresponded to the shocks of unintegrated *Erlebnis* (experience) fostered by fascist society" (*Dialectical Imagination*, p. 184).

Although, Adorno praised a:

> "fragmentary montage style, which employed shocks [...] as the most progressive and critical and popular music of the day" (*Dialectical Imagination*, p.184);

he may have found punk music's:

> "beat and syncopation were derived from the military march, which suggested its implicit relation to authoritarianism" (*Dialectical Imagination*, p. 187).

AUTHORITARIANISM: HITLER, JESUS & THE UNITED STATES

In his "Freudian Theory and the Pattern of Fascist Propaganda," Adorno notes:

> "for Freud the concept of psychology is essentially a negative one. He defines the realm of psychology by the supremacy of the unconscious and postulates that what is id should become ego. The emancipation of man from the heteronomous rule of his unconscious would be tantamount to the abolition of his 'psychology.' Fascism furthers this dependence instead of the realization of potential freedom, through the expropriation of the unconscious by social control instead of making the subjects conscious of their unconscious" (Andrew Arato & Eike Gephardt (eds.), *The Essential Frankfurt School Reader*, Continuum: London (1982), p. 136, "Freudian Theory and the Pattern of Fascist Propaganda").

On the flip side of rejecting the psychological "status quo history" of the unconscious (our animal past), we have the ego hitting a glass ceiling of the super-ego state; Martin Jay notes:

> "By portraying so vividly the unattainable promise of individual self-realization in the age of destructive competition, Ibsen exploded the liberal myth of personal happiness. 'Competition,' Lowenthal wrote, 'turns out to be not only a struggle for social and economic success among various individuals; it is also an inner struggle in which the individual must drastically curtail certain sides of his own being, his personality, in order to realize his personal ambitions'" (*Dialectical Imagination*, p. 139).

> "The people he has to reckon with generally undergo the characteristic modern conflict between a strongly developed rational, self-preserving ego agency and the continuous failure to satisfy their own ego demands [....] Freud's psychological construction of the leader imagery is corroborated by its striking coincidence with the fascist leader type, at least as far as its public build-up is concerned. His descriptions fit the picture of Hitler no less than the idealizations

into which American demagogues try to style themselves" (*The Essential Frankfurt School Reader*, p. 126, "Freudian Theory and the Pattern of Fascist Propaganda").

"Since the libidinal bond between members of masses is obviously not of an uninhibited sexual nature, the problem arises as to which psychological mechanisms transform primary sexual energy into feelings which hold masses together [....] Freud dwells on the fact that in organized groups such as the Army or the Church there is either no mention of love whatsoever between the members, or it is expressed only in a sublimated and indirect way, through the mediation of some religious image in the love of whom the members unite and whose all-embracing love they are supposed to imitate in their attitude towards each other" (*The Essential Frankfurt School Reader*, p. 123, "Freudian Theory and the Pattern of Fascist Propaganda").

Max Horkheimer from his "The Authoritarian State":

"If the lack of modern technology and the war-like environment had not played into the hands of bureaucracy, statism would have already outlived its usefulness. In integral statism, even apart from the militaristic encroachment, the absolutism of bureaucracy, whose authority the police enforce to the utmost in all phases of life, stands opposed to the free structuring of society. No economic or juridical measures, only the will of the ruled can lead to the democratization of the system of control. They will be trapped in the vicious circle of poverty, domination, war and poverty until they break through themselves" (*The Essential Frankfurt School Reader*, p. 102, "The Authoritarian State").

"If a region, for example the United States or Europe, is great and powerful enough, the machinery of oppression used against the internal enemy must find a pretext in the threats of the external enemy. While hunger and the danger of war are necessary, uncontrollable and inevitable results of a market economy, they can be constructively utilized by the authoritarian state" (*The Essential Frankfurt School Reader*, p. 103, "The Authoritarian State").

"Critical theory is of a different kind. It rejects the kind of knowledge one can bank on" (*The Essential Frankfurt School Reader*, p. 106, "The Authoritarian State").

"The belief that one is acting in the name of something greater than oneself is bankrupt" (*The Essential Frankfurt School Reader*, p. 112-113, "The Authoritarian State").

"The mass media assimilate the revolution by absorbing its leaders into their list of celebrities. The isolated individual who is not appointed or protected by any power cannot expect fame. Even so, he is a power because everyone is isolated. Their only weapon is the word. The more it is bandied about by the barbarians within and the cultural sophisticates abroad, the more its power is restored" (*The Essential Frankfurt School Reader*, p. 115, "The Authoritarian State").

"The fact that even the enemies of the authoritarian state can no longer conceive of freedom destroys communication. A language in which one does not recognize his own desires or become impassioned is alien. Thus the bourgeoisie is no longer upset in the slightest over its own non-conformist literature; it has brought Tolstoy into the movies and Maupassant into the drugstore" (*The Essential Frankfurt School Reader*, p. 116, "The Authoritarian State").

MESSIANIC NOW TIME

In his critique of radio, Martin Jay claims Adorno thought:

> "radio could preserve the *nunc* or 'nowness' of a performance but not the *hic* or 'hereness.' In so doing, it destroyed one of the crucial features of what Benjamin had called the 'aura' of a work of art, its ritual, cultish nimbus" (*Dialectical Imagination*, p. 191).

> "Normal time was indeed suspended by great works of art, but in its place was a type of coherent development, which was a foretaste of the temporal order of the 'other' society. Benjamin was especially fond of distinguishing between 'homogeneous, empty' time and time 'filled by the presence of the now'" (*Dialectical Imagination*, p. 192).

Jay later notes that:

> "the posthumously published 'Thesis on the Philosophy of History' [.... was where] Benjamin most clearly articulated his distinction between homogeneous, empty time and the messianic *Jetztzeit* (the fulfilled time of the present) that the revolution was supposed to usher in" (*Dialectical Imagination*, p. 200).

In this "Thesis on the Philosophy of History," Walter Benjamin roughly sketches two views of history; one which he believes to be the faulty view of classical historicism, and the other, which is the proper goal of the historical materialist. He sees the faulty view as that of a *narrative* history, where the historian compiles:

> "the sequence of events like the beads of a rosary" (Walter Benjamin, Hannah Arendt (ed.), Harry Zohn (trans.), *Illuminations*, Schoken Books: New York (1969), p. 263, "Thesis on the Philosophy of History"),

with a history conceived as a

> "progression through a homogeneous, empty time" (*Illuminations*, p. 261, "Thesis on the Philosophy of History").

Such historians believe in

> "technological progress" (*Illuminations*, p. 258, "Thesis on the Philosophy of History"),

yet ultimately end up perpetuating the status quo, and reinforcing the hold of those in power.

To understand how historicism is duped into this, we must see how Benjamin views time. As Jay was noted above, for Benjamin, the important element is the *moment*; Benjamin says:

> "The true picture of the past flits by. The past can be seized only as an image which flashes up at the instant when it can be recognized and is never seen again" (*Illuminations*, p. 255, "Thesis on the Philosophy of History").

Thus, there is a moment of pure presence of the moment, which disappears instantly—an instant which may only be remembered through the constructs of the history we are embedded in:

> "For every image of the past that is not recognized by the present as one of its own concerns threatens to disappear irretrievably" (*Illuminations*, p. 255, "Thesis on the Philosophy of History").

As soon as one tries to articulate the past moment, dominant modes of perception distort it; it is:

> "lost in the void the very moment he [the historian] opens his mouth [....] To articulate the past historically does not mean to recognize it" (*Illuminations*, p. 255, "Thesis on the Philosophy of History").

From each of Benjamin's historical perspectives, the other seems like chaos. From his preferred view, what I would call "*Now* History," narrative history resembles a:

> "catastrophe which keeps piling wreckage upon wreckage" (*Illuminations*, p. 257, "Thesis on the Philosophy of History").

Within narrative history, the Now occurs as revolution. To recognize this, the historical materialist must attempt to exit historical constructs; (s)he must view:

> "cultural treasures [....] with cautious detachment [.... Where he...] dissociates himself from it as far as possible" (*Illuminations*, p. 256-257, "Thesis on the Philosophy of History"),

in order to recognize

> "the sign of Messianic cessation of happening, or, put differently, a revolutionary chance in the fight for the oppressed past. He takes cognizance of it in order to blast a specific era out of the homogeneous course of history" (*Illuminations*, p. 263, "Thesis on the Philosophy of History").

The critic must rescue the moment from the continuum of history, thus redeeming the silent origin which was fractured into narrative history, and returning the moment to a:

> "present which is not a transition, but in which time stands still and has come to stop" (*Illuminations*, p. 262, "Thesis on the Philosophy of History").

Benjamin writes:

> "To be sure, only a redeemed mankind receives the fullness of its past—which is to say, only for a redeemed mankind has its past become citable in all its moments. Each moment it has lived becomes a *citation a l'order du jour* [sum-

> mons to the order of the day]—and that day is Judgment Day" (*Illuminations*, p. 254, "Thesis on the Philosophy of History").

For Benjamin, each moment can be viewed within two contexts; the fallen (genealogical) context of a narrative history, or the proper (phenomenal) context of "Messianic time."

APPROVED BY THE COMICS CODE AUTHORITY

Benjamin's notion of a continually experienced full "now" connects with his notion of an artistic:

> "'aura' which was so frequently used in the Institute's [for Social Reform's] cultural analyses" (*Dialectical Imagination*, p. 210).

In his "The Work of Art in the Age of Mechanical Reproduction" Benjamin examines this notion of the aura, and how it is both lost and artificially reconstructed in the age of reproduction. He says:

> "Even the most perfect reproduction of a work of art is lacking in one element: its presence in time and space, its unique existence at the place where it happens to be" (*Illuminations*, p. 220, "The Work of Art in the Age of Mechanical Reproduction").

He continues:

> "One might subsume the eliminated element in the term 'aura' and go on to say: that which withers in the age of mechanical reproduction is the aura of the work of art" (*Illuminations*, p. 221, "The Work of Art in the Age of Mechanical Reproduction").

What is lost, then, is the sense of authenticity, but more theoretically important, as discussed above, the *singular particularity* of the work of art is lost when it is mechanically repeated and reproduced, much like abstractions, or theories, (or even language) lose some detail of the object or data they are concerned with.

Benjamin argues that contemporary society seeks to distance itself from authenticity:

> "Namely, the desire of contemporary masses to bring things 'closer' spatially and humanly, which is just as ardent as their bent toward overcoming the uniqueness of every reality by accepting its reproduction" (*Illuminations*, p. 223, "The Work of Art in the Age of Mechanical Reproduction").

I think this point is exemplified by turning a few pages of a typical *Richie Rich* comic book. The comic book could be seen as the ultimate humanized and inauthentic work of pop-art. And with *Richie Rich*, we have the added dimension of glorified wealth aimed at enticing youth. The words "NOW" and / or "NEW" are frequently emblazed across the typical cover, as if to compensate for the considerable distance the comic book has from any sort of actuality. Moreover, in the upper right corner, appears a stamp reading, "Approved by the Comics Code Authority." Benjamin notes that such types of art were able to come about only through the mechanisms of reproduction:

> "the work of art reproduced becomes the work of art designed for reproducibility.... But the instant the criterion of authenticity ceases to be applicable to artistic production, the total function of art is reversed. Instead of being based on ritual [as Benjamin claims pre-reproductive art-forms were], it begins to be based on another practice—politics" (*Illuminations*, p. 224, "The Work of Art in the Age of Mechanical Reproduction").

Moreover:

> "When the age of mechanical reproduction separated art from its basis in cult [which was involved with ritual], the semblance of its autonomy disappeared forever" (*Illuminations*, p. 226, "The Work of Art in the Age of Mechanical Reproduction").

Jay also points out that Adorno and Benjamin:

> "feared [...] that mass art had a new political function diametrically opposed to its traditionally 'negative' one; art in the age of mechanical reproduction served to reconcile the mass audience to the status quo" (*Dialectical Imagination*, p. 211).

It would seem clear then, that a comic book might be caught up in a perpetuation of capitalism, and a degradation of critical autonomy. Along similar lines, Benjamin claims:

> "One of the foremost tasks of art has always been the creation of demand which could be fully satisfied only later" (*Illuminations*, p. 237, "The Work of Art in the Age of Mechanical Reproduction").

In our *Richie Rich* comic book there is a "center-fold" of "Casper the Friendly Ghost," evidencing this: what appears as an added bonus also doubles as advertisement, and hence far from being some Hegelian Spirit at the center of things, we find a semi-present ghost referring to yet another product possibly promising its own "NEW NOW" that never ultimately delivers authenticity.

The comic book form may have merits though; as Benjamin discussed film in depth, noting that the format offered different modes of expression:

> "The spectator's process of association in view of these images is indeed interrupted by their constant, sudden change" (*Illuminations*, p. 238, "The Work of Art in the Age of Mechanical Reproduction").

Could it be that *Richie Rich* is a parody? Who could take the approval of the comic code authority as authoritarian? (Its logo looks slightly like an anarchy symbol!) Possibly this is an opportunity for those who can afford a comic book to have a laugh at the impossibly rich. Yes, the comic book creates a fantasy world where all of the Rich family's employees are happy... but possibly in a subversive implication, the fact that this is self-evidently an impossible fantasy world brings down the entire house of cards. The comic book allows the reader to both enter an impossible fantasy world of wealth vicariously, and to remain critical of it as completely ridiculous. The parodying nature of this comic book might subtly indoctrinate youth to have a bifurcated relation to super riches, as both an unobtainable fantasy, and something not to take too seriously: a joke. If you are against capitalism, this might be negatively interpreted as encouraging complacency with the status quo of wealth. If you don't mind some capitalism, but are not as rich as Richie, it may help you to not take the disparity between you and him too seriously. Whether the parody is dangerous, or comforting, might depend on your own political persuasion: but a unified message is clear—you're not as wealthy as Richie Rich.

AN AUTHENTIC SOCIAL HISTORY, NOT AN AUTHENTIC SELF

When Adorno writes of Benjamin, in "A Portrait of Walter Benjamin," he notes that Benjamin is a philosopher who:

> "had nothing of the philosopher in the traditional sense" (*Prisms*, p. 229, "A Portrait of Walter Benjamin").

And developed,

> "a philosophy directed against philosophy" (*Prisms*, p. 235, "A Portrait of Walter Benjamin").

Although being one of those who lived the kind of life that others write about, Benjamin did not draw upon his life, in order to create something original—something originating in *him*:

> "The impression he left was not of someone who created truth or who attained it through conceptual power; rather, in citing it, he seemed to have transformed himself into a supreme instrument of knowledge on which the latter had left its mark" (*Prisms*, p. 229, "A Portrait of Walter Benjamin").

Such was in line with Benjamin's "anti-subjectivism" that sought to eliminate personal intention, and let "objects" speak for themselves. In fact, an unfinished "On the Concept of History" had nothing more or less than 1000's of pages of citations. He didn't seek to use his subjective powers to form such into a unified narrative, but rather left the object in its fragmented state:

> "Benjamin as it were revokes the unity of the subject to mythic turmoil in order to comprehend such unity as itself being only a natural condition; with his philosophy of language oriented on the cabbala, Benjamin saw subjective unity as scribbling of the Name" (*Prisms*, p. 236, "A Portrait of Walter Benjamin").

Adorno sees "myth" as tying Benjamin's theological esoteric period with his fragmentary materialist period; yet each period tried to suspend subjectivity:

> "Just as the domain of myth is ruled by multiplicity and ambiguity and not subjectivity, the unequivocal character of reconciliation—conceived after the model of the 'name'—is the contrary of human autonomy" (*Prisms*, p. 236, "A Portrait of Walter Benjamin").

Like some cabbalist performing an exegesis of a sacred text, the text takes primary place before the subjectivity of the exegete, for:

> "In all his phases, Benjamin conceived the downfall of the subject and the salvation of man as inseparable" (*Prisms*, p. 231, "A Portrait of Walter Benjamin").

Yet, twice in his essay on Benjamin, Adorno mentions Medusa:

> "Before his Medusan glance, man turns into the stage on which an objective process unfolds" (*Prisms*, p. 235, "A Portrait of Walter Benjamin").

> "The glance of his philosophy is Medusan. If the concept of myth, as the antipode to reconciliation, occupies a central position in it, especially during its openly theological phase, then everything, and especially the ephemeral, becomes in his own thought mythical" (*Prisms*, p. 233, "A Portrait of Walter Benjamin").

And, twice, Adorno notes the metaphor of "microscope":

> "He [Benjamin] sees his task not in reconstructing the totality of bourgeois society but rather in examining its blinded, nature-bound and diffuse elements under a microscope" (*Prisms*, p. 236, "A Portrait of Walter Benjamin").

> "By permitting thought to get, as it were, too close to its object, the object becomes as foreign as an everyday, familiar thing under a microscope" (*Prisms*, p. 240, "A Portrait of Walter Benjamin").

I find it interesting that Adorno, when citing a "Medusan glance," says first that it is from Benjamin's philosophy, and then from Benjamin himself, as if the two were one and the same—and indeed this close friend of Benjamin seems more oriented towards his friend's philosophy in this portrait, than to the human being himself. Possibly such was in line with the anti-subjectivity sought for; but this seems a little less than friendly, or humane. Yet again, with his turn of phrase, Adorno is also giving Benjamin, something of the "legendary" treatment—promoting him, through his work, as larger than life. Such "anti-humanism" may have been oriented towards Benjamin's own wishes, with his:

> "idiosyncratic distaste for worlds like 'personality'" (*Prisms*, p. 235, "A Portrait of Walter Benjamin").

Such reminds me of Heidegger's looking toward a Being beyond beings in his "Letter on Humanism." Although it is a safe bet that Benjamin and Adorno did not share Heidegger's affiliation with the Nazi's, there remains a dangerous, in my opinion, desire to transcend the human—by perfecting science like some engineered machine, and pushing the human into that perfection, to the point of losing what was characteristically human. Yes, I can see the desire for the spiritual as some sort of pursuit of perfection parallel to a scientific quest for the deepest truths—and Benjamin, with his theological and materialist pursuits seems to be looking for the final revelation of attaining these truths—but perfection at the price of personal humanity seems too steep for me. Again, not that Benjamin, Adorno, or even Heidegger were inhumane people (I have little inclination to believe they were not humane)—but one may ask: "can a person live in your philosophy, and remain humanly alive?"

It is clear that Benjamin sought to suspend his subjectivity and let the fragmented material objects of contemporary history speak their ancient being beyond the everyday ways they are commonly comprehended—with a sort of social Heideggerian existentialism, which sought not for an authentic self in the face of vacuous everydayness, but for an authentic social history erupting out of the corrupt status quo. Walter Benjamin sought to be nothing less than a pivot for the critical awakening of history to itself; in order, it would seem, to transform it into something ultimately out of history as we know it. A tall order for a "mere" human; yet with reference to Benjamin's interest in the "microscopic":

> "He never wavered in his fundamental conviction that the smallest cell of observed reality offsets the rest of the world" (*Prisms*, p. 236, "A Portrait of Walter Benjamin").

Adorno's epigraph by Karl Kraus captures the flavor of Benjamin's project in a crystal clear light:

> "... and listen to the sounds of the day as though they were chords of eternity" (*Prisms*, p. 227, "A Portrait of Walter Benjamin").

With these words, and linking the "day" with a microscopic look at contemporary history, that through the subject-less observer's Medusan look, is found frozen in time like an eternal "now," Walter Benjamin could have been seen as (1) messianically realizing the true essence of social history, or (2) caught in a self or world-referential paranoia almost beyond imagination. Maybe a little of both, seeing that Benjamin had both monumental insights, yet remained a troubled human being.

HIGHLIGHTS OF CRITICAL THEORY

In closing this chapter, I'll try to co-opt some Critical Theory insights, given my opinion that for many, the DNA divergence has gone fallow, and for the few where seeds have taken root, or may take root, spurring healthy intellectual growth, my few insights should not obstruct further Frankfurt School fertilization.

In sum, a central methodological, systematic, and structural element of critical theory is to be a-methodological, a-systematic, and a-structural: it might be defined in the negative. It questions all that is stable, the same, and repeating (the status quo, etc.) and everything that would attempt to keep things stable, the same and repeating (co-optation, the formulaic, etc.)—in contrast, it celebrates the subversive, changing, and differing. Again, this opposition between conservation and revolution has structural-epistemological roots in the contrast between a mental reason which would reveal the universal (unchanging), and a material reality which would be in constant motion: the revolution of the now, vs. the stasis of a static eternal history. Yet, although these thinkers seemed to prefer change to stasis (they tended to be more liberal than conservative, politically), they often recognized the impossibility of completely severing themselves from history: possibly a balance should be struck, but they saw history, especially the history of theory, as dominated by the pursuit of eternal stability, and a counter-movement seemed necessary in the face of the fascist seizures of centralized power seeking ego-controlled order in their time.

Benjamin, but especially Adorno were both thinkers that took "thinking outside the box" to an (anti) systematic extreme. Their sophisticated philosophical outlook was crucial to the formation of Derridian deconstruction, and AI researchers may find, that although difficult, the Frankfurt School work went to the very heart of social intelligence.

Valerio Adami - Portrait of Jacques Derrida (2008)

Chapter 22

Derrida
The Subject of Deconstruction

SITUATING THE FUTURE

Jacques Derrida's (1930 - 2004) philosophy does not lend itself to a simplified characterization, much because it draws upon a diverse range of ideas from the history of philosophy, and complicates these ideas even further. More than the a-methodological, a-systematic, and a-structural "Critical Theory" that we saw with the Frankfurt School philosophers Theodor Adorno and Walter Benjamin, but also in a questioning, negative orientation towards its varied subject matter (and influenced by Adorno's "negative dialectics" as well), Derrida's writing probes the writings of other writers—with him often writing, as he has entitled one of his many books, in the *Margins of Philosophy*. And this re-reading of the past and opening it up to an incalculable future, this pivotal event can be seen in more than one of the terms he has developed, but very much so in one which is especially associated with his philosophical movement: "Deconstruction." For beyond the meaning of "taking apart," the word, "deconstruction" can be divided, with the morphemes leading back to Latin roots comprising of: "de" meaning "of, concerning, about," and "from, away from, down from, out of," but also "to depart, withdraw from"—"con" meaning "with"—and the Latin "struō" meaning "pile up, arrange; build, erect"—hence de-con-struction is both "about" construction and "departing" construction—and both "of" and "against" the history such arises from. It is in this way that "deconstruction" is a more subtle term than "postmodernism," as the latter term seems to suggest an unqualified break with the past, and as being something other than historical. In a way, deconstruction has not yet overcome history, it is always already on the way out of being determined by history, progressively reforming the ideas of history, if not rebelling against them, yet with a conservationist bent that maintains a respect for the past even while overturning its imbalances, anarchically, in favor of justice.

"Deconstruction" is not a strategy, an operation, or a way of reading—perhaps no more than a soul is. And in many ways, Derrida treats the term as if it were as complex as a human being: are we as human beings not in a constant flux between being shaped by our past, as we negotiate our (im)possible futures? The term "deconstruction" could also be said to name the subject: naming subjectivity itself. But the subject and subjectivity are seen here, not in a full plentitude of self-presence—not a consciousness that is hooked up to a Logos of absolute knowledge handed down by Western philosophy. No, here subjectivity is temporal: its intentions, never fully worked out in advance other than in a possibly over-determined destiny projected from one's past into the future.

Because much of this book heretofore has been preoccupied with the history of Western philosophy, we will not need to retread the entirety of ground in order to situate and contextualize Derrida's writings... but placing his work relative to a couple of 20th century French thinkers, Ferdinand de Saussure (1857-1913) and Michel Foucault (1926-1984) should help.

DE SAUSSURE: STRUCTURAL DIFFERENCES

As just seen, the term "deconstruction" relates to the word "structure" (via the Latin word "struō")—and such is no accident, as Derrida was part of an intellectual scene that developed around the issues raised by "structuralism"—a set of ideas that was advanced by many "continental" philosophers (western European philosophers outside of Britain), especially in 20th century France. This school of thought arose through the work of Ferdinand de Saussure, a French structural linguist, who argued that words were not to be defined by some sort of higher essence or form (much like the logical forms sought by thinkers from Plato to the early Wittgenstein), but rather by the relation of words to each other; relations that are both "synchronic" (the static array of words seen with semantic webs and grammar rules: language, or "langue" understood abstractly) and "diachronic" (the dynamic changes in language over history, not to be confused with specific instances of language or "parole.")

Of the "synchronic" aspect of language, de Saussure goes so far as to suggest that certain types of this are materialized in the brain:

> "Outside the context of discourse, words having something in common are associated together in memory. In this way they form groups, the members of which may be related in various ways. For instance the word *enseignement* ('teaching') will automatically evoke a host of other words: *enseigner* ('to teach'), *reseigner* ('to inform'), etc., or *armement* ('armament'), *changement* ('change'), etc., or *éducation* ('education'), *apprentissage* ('apprenticeship'). All these words have something or other linking them.
>
> This Kind of connexion between words is of quite a different order. It is not based on linear sequence. It is a connexion in the brain. Such connexions are part of that accumulated store which is the form language takes in an individual's brain. We shall call these *associative relations*" (Ferdinand

de Saussure, Roy Harris (trans.), *Course in General Linguistics*, Open Court: Chicago (1997), pp. 121-122).

This "automatic" evocation of other words and the "connections" in the brain is a subject of our studies into artificial intelligence and cognitive science, esp. concerning "spreading activation" with neural networks. De Saussure saw the rules of grammar as also being synchronic too, as they are an abstract way of looking at language, much like dissecting an organism and making a taxonomy of its parts and how they inter-relate. Such is opposed to the "diachronic" aspect of language: how language changes over time (as studied with the etymological history of words—where a Latin word like "video" will shift in meaning from "I see" to "I understand"—and will be further changed in contemporary English, to be associated with "television," etc.)

Over the course of his studies, de Saussure maintained something resembling *forms* or *essences*, called "signifieds." A "signified" would be the meaning of a word, where a "signifier" would be the physical manifestation of word—the spoken phonetic sounds, or the written marks. However, unlike previous thinkers, who would suggest that signifieds have meaning in themselves—a sort of intellectual nucleus of a word, that atomically contains meaning—de Saussure suggested that both signifieds and signifiers are defined in *relation* to each other, much like juxtaposing the colors red and green will make them pop out, for their contrast. Hence a *signified* for "dog" is defined in relation to other signifieds, like "canine," "beagle," "cat," etc, while the *signifier* "dog" is related to words like "fog," "bog," "dig," and "dot." However, it would appear that we have no way of connecting the two planes of differentiation: how do we connect the signifiers to the signifieds? Would onomatopoeias like "bark" or "meow" work to connect the two? Even onomatopoeias have different spellings and pronunciations in different languages! De Saussure simply claims:

> "the essential function of a language as an institution is precisely to maintain these series of differences in parallel" (de Saussure, p. 119).

> "to say that in a language everything is negative holds only for signification and signal considered separately. The moment we consider the sign as a whole, we encounter something which is positive in its own domain. A linguistic system is a series of phonetic differences matched with a series of conceptual differences" (de Saussure, p. 118).

Diving into what de Saussure means, Derrida finds that it is with one's *intended articulation* that the two realms of signifier and signified are welded together. It is with the conscious presence of one's speech, through individual speech acts, that the two realms, which are completely defined in the negative, without positive terms, are fused into positive signs. It is the very sentient being of consciousness that bridges meaning as content, and language as form. And it is in this way that we can see how important speech is, as it gives de Saussure's signs meaning: so that we are not, according to him, continually roaming from word to word, and endlessly trying to reconstruct contexts in order to be understood (although such *is* the case, when trying to explain something that another does not quite yet understand).

WRITING FRAMES SPEECH

Derrida's take on de Saussure's parallel linguistic realms could be construed as a question: how is it even possible for this mysterious conscious being in the process of a speech act, to "fuse" the signifier and signified? Time and again, throughout the history of philosophy, up to and beyond 20th century philosophers, and including de Saussure himself, the meaning buck stops with conscious thought and speech and possibly a transcendental signified (like God or Being as Word or the Logos). But this really doesn't explain anything: it suggests that meaning is as mysterious as consciousness itself: subjectivity. As we claimed before, subjectivity itself, with each of our own personal perspectives, is shaped by a history, personal / cultural, and biological / environmental. It is by framing this subjectivity, by looking at the context of speech acts, and what shapes them, that we can begin to go beyond the mystery of consciousness and subjectivity, and begin to situate meaning as a post-structural, if not a scientific, phenomena. We do not need to look only to a person's intent to understand meaning: we can look to contexts—and with the context of all of a person's history, singular univocal intent can begin to break down as well, into a plurality of meanings beyond intent.

An example of a plurality of meanings beyond what a person thinks they are intending can be seen with Freudian slips; and in fact, so much of Freudian interpretation depends on the notion of unconscious meaning—we say more than we think we are saying. Such a plurality of meanings can go beyond Freudian animal instinct, as turns of phrase, idioms, etc, betray the culture one is brought up in, as much as it does a person's specific intent (we can "psychoanalyze" culture as well). Such is not to claim that present intent does not exist, but that intent is constructed by much more than our subjective will (or rather, our subjective will itself is in large part determined by our histories: personal / cultural, and biological / environmental). There are implications in what people say: words used always imply shared contexts; contexts that can change and render those same words as something quite different than they were once thought to be. Some might suggest that this changeability means anything could mean anything, and hence we should be free to mean whatever we want and not care about grammar, proper language usage etc. But this would be to ignore the robust multiplicity of historical connections and contexts that all public language use brings with it. We are not talking so much about a de-stabilization of language with Derrida and deconstruction, but more about a re-situation or re-contextualizing of language, which sweeps the foundation of signifieds / essences and intention out the door, in favor of a stratification of contexts and differential connections. The floor may be an illusion, but we have a variety of safety wires keeping us tied to an even deeper, more robust, stability.

So given that meaning is shaped by context... the painting situated both by a literal and metaphorical frame, we see Derrida time and again preoccupied by what shapes a given subject by lying on the border forming it. Hence *Margins of Philosophy*, and essays like "Living On Border Lines," and the notion of the "Parergon" (which means "frame"). Such is not to look outside the frame, to some "other," to define the "atomic" subject, but rather to look at the frame and framework itself, to see how lines give shape to "color." (And this can be seen as visually illustrated by one of Derrida's evident artist favorites, Valerio Adami, who painted with precise lines, and colors... interestingly, Adami evolved as a painter and began to use cross-hatching as well for fleshy depth—indicating a way of seeing the one might argue challenges the Derridian emphasis on lines and differences—the gestalt Dalmatian recognized in a image of black dots without lines challenges the line as being the visually defining aspect: traces recognized without delineated difference).

As opposed to the self-presence of meaning in speech, Derrida investigates writing—again both literally and metaphorically. Writing seems to be meaningful outside the presence of an articulate subject. How could this be if meaning is dependent on the self-present subject? Because context is not lost... we can still understand what is written in the same way we understand all language. And while closely reading western philosophers, Derrida often finds that while they extol the virtues of speech as the anchor of meaning, these same philosophers denigrate writing, as a sinful supplement to writing: much like Plato condemning our worldly experience as a poor copy of intellectual reality. But upon closer reading, Derrida deconstructs these same writings, e.g. demonstrating that the evil outcast and inferior copy found with writing, actually precedes speech, in a sort of arche-writing—a writing more primordial than speech—writing as metaphor for technology.

As we saw with Heidegger, technology is immediately understood as the use of tools. But technology is also a way of understanding things technologically. DNA is hardly a screwdriver, but it is a way of understanding things in a technical sense—a sense beyond mere mechanics, but instead mechanics coupled with information, forces, etc. In his book *Of Grammatology*, Derrida cites de Saussure (here directly quoted):

> "Thus although writing is in itself not part of the internal system of the language, it is impossible to ignore this way in which the language is constantly represented. We must be aware of its utility, it defects and its dangers" (de Saussure, p. 24).

Derrida notes:

> "Writing would thus have the exteriority that one attributes to utensils; to what is even an imperfect tool and a dangerous, almost maleficent, technique" (Jacques Derrida, Gayatri Chakravorty Spivak (trans.), *Of Grammatology*, The John Hopkins University Press: Baltimore (1997), p. 23).

To see writing almost literally as a tool, can be to see writing as technology, at least metaphorically. And such is what Derrida does with his concept of arche-writing. We may clothe ourselves with fig leaves to cover our naked "intent," yet one might say we were not potentially naked until the invention of clothing. If we are to see tools as emblematic of technology, and that we are shaped in technological ways (DNA, the mechanics of our bodies, the techniques we have learned from culture and from personal experience), we can then see how arche-writing, prior to speech, could be said to shape it too. The supplemental tool used to copy original intent is also the technology that structures that original intent:

> "There is no longer a simple origin. For what is reflected is split *in itself* and not only as an addition to itself of its image. The reflection, the image, the double, splits what it doubles. The origin of the speculation becomes a difference. What can look at itself is not one; and the law of the addition of the origin to its representation, of thing to its image, is that one plus one makes at least three. The historical usurpation and theoretical oddity that install the image within the rights of reality are determined as the *forgetting* of a simple origin" (Derrida, *Of Grammatology*, pp. 36-37).

In fact, Derrida cites de Saussure again (directly quoted):

> "A language is a system of signs expressing ideas, and hence comparable to writing, the deaf-and-dumb alphabet, symbolic rites, forms of politeness, military signals, and so on. It is simply the most important of such systems" (De Saussure, p. 15).

Derrida writes:

> "Speech thus draws from this stock of writing, noted or not, that language is" (Derrida, *Of Grammatology*, p.53).

Such is to say that the origin of meaning is not found in simple self-presence of consciousness and univocal intent—but is found in a plurality of technologies that precede any such articulation, and through the example of writing, can operate without such presence. This "univocal speech" vs. "plural writing" is also illustrated by Derrida with his single work *Speech and Phenomena* as compared to the many essays found in his work *Writing and Difference*. This pair of books suggests a polarity between not only speech and writing, but between phenomena and difference. This is a bit troubling though, as it may suggest (as do some scientific thinkers) that phenomenal consciousness itself, subjectivity, the color green, etc, are epiphenomenal appearances created by formal-physical reality (it is not clear which of Aristotle's four causes would make the appearance of green). Such colored light flooding through the cracks suggests that a robot *might* be sentient... this is discussed at greater length in later chapters of this book.

THE COPY BEFORE THE ORIGINAL?

Derrida seems to be articulating a new logic, where the copy precedes the original. But it is not so much that the copy precedes the original, but that the original itself is repeatable, it has "iter-ablity." In order to be iterable, to be repeatable, the original must contain the seeds of some sort of difference within itself, between what it is, and its copy. That implied difference, is a difference non-the-less. The simple unity of subjective soul can be split across at least five senses of the body; the spoken word often caries a plurality of meanings beyond conscious intent; Descartes' God as singular pinnacle of perfection grounding reality has the complexity of that reality already within; Plato's Good as the highest of forms is also a form among forms—and for Derrida:

> "Difference is [...] the formation of form" (Derrida, *Of Grammatology*, p. 63).

In a later essay simply titled, "*Différance*," Derrida expands upon the concept and makes it temporal as well as spatial. The word "différance" plays on a French ambiguity between what in English would be "Difference" and "Deference." Hence we could be talking not only about the sort of difference between words that define them, but a constant temporal deferring from one word to another in time. When defining a word, using the dictionary, we always come across more words, which could be defined using the same dictionary, and so on ad infinitum. One may recall the arguments of the ancient skeptics (in chapter 5), that nothing is true in itself or in relation to anything else (as their relativity stance suggested that things are different from varied perspectives—hence not true in themselves—and that if one isn't to have an infinite regress in arguing, dogmatic assumptions must be made). Derrida seems to flip skepticism on its head, reversing its critique of all claims to knowledge, and showing that its strictures form the very structure of "reasonable" discourse. Nothing is "relative," but everything is relative *to* everything else—there are no positive terms to anchor the system of knowldege. But again, such does not result in chaos, but rather, simply undermines the certainty of a simple foundation, in favor of complex coherence; much like the shift in some 20th century analytic philosophers towards pragmatism, where a correspondence theory of truth is given up in favor of a coherence theory of truth (cp. Donald Davidson, W.V. Quine, and Richard Rorty). Propositions are not true because they mirror the world, and represent it, but rather, because they cohere with other propositions: propositions do not represent the world, they are about it. In the same way, words are not defined by essences that they share with the world (as with the early Wittgenstein's logical forms)—but rather are defined by the endless web of associations and contexts.

Derrida does not limit himself to the concept of "différance" in his essay though, as he must account for the connections found... those associations that tie, e.g., words together. For this, he uses the word "trace":

> "The two apparently different values of *différance* are tied together in Freudian theory: to differ as discernibility, distinction, separation, diastem, *spacing*; and to defer as detour, relay, reserve, *temporization*.
>
> 1. The concepts of trace (*Spur*), of breaching (*Bahnung*), and of the forces of breaching from the *Project* [*for a Scientific Psychology*] on, are inseparable from the concept of difference. The origin of memory, and of the psyche as (conscious or unconscious) memory in general, can be described only by taking into account the difference between

> breaches. Freud says so overtly. There is no breach without difference and no difference without trace.
>
> 2. All the differences in the production of unconscious traces and in the processes of inscription (*Niederschrift*) can also be interpreted as moments of *différance*, in the sense of putting into reserve. According to a schema that never ceased to guide Freud's thought, the movement of the trace is described as an effort of life to protect itself by *deferring* the dangerous investment, by constituting a reserve (*Vorrat*). And all the oppositions that furrow Freudian thought relate each of his concepts one to another as moments of detour in the economy of *différance*" (Jacques Derrida, Alan Bass (trans.), *Margins of Philosophy*, The University of Chicago Press: Chicago (1986), p. 18).

Moreover:

> "The alterity of the 'unconscious' makes us concerned not with the horizons of the modified—past or future—presents, but with a 'past' that has never been present, and which will never be, whose future to come will never be a *production* or a reproduction in the form of presence. Therefore the concept of the trace is incompatible with the concept of retention, of the becoming past of what has been present. One cannot think the trace—and therefore, *différance*—on the basis of the present or of the presence of the present" (Derrida, *Margins*, p. 21).

Derrida suggests there is something "unthinkable" (Derrida, *Margins*, p. 19) in both différance and the trace—that is, we will never be able to grasp it in an instant of comprehension. But possibly, as with learning any language, we may become *familiar* with the concepts, as Derrida himself seems to be. With Derrida's own critique of past (and contemporary) thinkers elevating speech over writing, one might ask why *différance* gets an essay title, and the "trace" doesn't even get second billing. It seems that in an effort to counter past thinkers' embracing "identity" in its atomic self, the self-same meaning found in an absolute transcendental signified, Derrida has emphasized *différance* in relation to everything else—but there are flip-sides to différance too: namely the trace, and the "judge." Whether or not there are differences at all without a judge (as with some sort of "un-cut qualitative continuum") is unknowable—when we have a judge, we have differences, and *possibly* these differences precede any subjective judgment. Certainly, as soon as we have a judge, we have a difference between the judge and judged. Perhaps judges and differences co-dependently originate (as when a person becomes a parent only in relation to having a child). But speculating on what "exists" before (subjective) judgment is possible is just that: speculation.

I put "exists" in scare quotes (and Derrida often uses scare quotes, many terms are not exact, or must be used with qualifications—further differentiating beyond the singular words used)—"exists" is in quotation marks, because in a way différance could be said not to exist, just as the line separating one thing from another does not have physical existence. The difference between what is an atom, in physics, and what is not an atom is not a space—but something like a line of demarcation. True, modern quantum physics sees the shape of an atom determined by a "probability cloud" (sometimes spherical, but not an absolute shell)—yet there is a shape, and one can say that any shape requires a zone of differentiation, if not a razor-thin line. This is not space itself, but spacing. And with the temporal aspect of différance, there is a deferral. Immediately, what these two aspects create in opposition to themselves, at least linguistically, is the presence of consciousness in the now: being here now—or Heidegger's Dasein ("being there"). Such is not an accident, as both Derrida and Heidegger studied Edmund Husserl—who used a method of "bracketing" to eliminate worldly contexts, in order to get at the subjective experience of pure consciousness. Heidegger went on to situate this consciousness in relation to objects in the world with notions such as equipment present at hand, etc, and situated the poetic presence of consciousness within technological enframing. Derrida goes further... he's concerned almost exclusively with what Husserl was at pains to exclude: the context which situates consciousness.

But again, like some scientists and philosophers of mind, such seems to indicate that subjective consciousness is epiphenomenal, if it really exists at all (albeit formed by differential structures, rather than some sort of "wet-ware" in the brain). And again, as with the Freudian unconscious (and Derrida looks towards much more primordial motivations than Freud did)—there is more to our subjective experience and our talking about it, than we can know. Moreover, we never "know" anything in an instant. "You can only learn what you already *almost* know" (a spin on a Sufi saying). Sure, there may be a point, where we know something we did not know before, but such knowledge is usually cumulative, with new knowledge situated in relation to past knowledge. And some of that prior knowledge itself may be hard-wired in our brains, after millennia of evolution.

How far back need we go for a simple origin? Some physicists claim that the big-bang arose out of a primordial differentiation in an original singularity (not to be confused with the technological singularity that has been hypothesized as when artificial intelligent agents design even more intelligent entities at an accelerating rate: an event horizon past which we cannot even begin to imagine). I'm not sure if Derrida wrote on the cosmological singularity, but my guess is that he would throw his hat to the "multiple-big-bangs" theory ring with an infinite past without origin. Or at least a complex origin with the seeds of our vast cosmos already implicit within that single point: a big bang that did not occur with Plato's "image of time."

FOUCAULT: THE OTHER HISTORY

Cosmology aside, for the most part heretofore we have been focusing on the immediate, or rather the mediation of differences—a general concern with the infinitely small, beyond microscopic spacing between atoms of meaning—the spacing that creates meaning in language, often smaller in scope than propositions: working at the level of words, and even morphemes (and for Derrida, even with letters, such as his concern with words that add an "r" with a "+r" like "trace" or "trait"—see his essay "+ r (Into the Bargain)" in Jacques Derrida, Geoffrey Bennington & Ian McLeod (trans.), *The Truth in Painting*, University of Chicago Press: Chicago (1987). Such is to zoom in on the local (cp. close reading), and make generalizations (cp. wide ranging theory). Language can operate on broader levels too.

Derrida's predecessor and some time contemporary Michel Foucault, shared Derrida's structuralism influence as well—but despite writing a bit on the subject of language, he was more concerned with the field of history. Such carries with it a wide scope, and although Foucault encourages small theories and local practices, and examined very specific phenomena, he also developed a vocabulary capable of far reach. For example, Foucault uses the term "episteme" not just to characterize the epistemology of a particular person, but rather as the knowledge worldview of a place and time. In his histories, *Madness and Civilization*, *The History of Sexuality*, *Discipline and Punish*, and even in the *Order of Things*, he goes to extreme lengths to recreate the details of worlds people lived in at certain times in history, and show how these histories were not a linear development, but a multi-dimensional intersection of different forms of power and resistances to power. Again, People saw things differently at different times (not a huge surprise to a relativist); they had different world views. And like etymological lineages of words, historical practices shifted over time as well; hence we see Freudian psychiatrist's couches sprouting from the Catholic confessional booths.

Time and again, in history Foucault finds some sort of subjugation of the "other" (the mad, prisoners, etc.) but also finds that the "other" differentially structures the norm: these two, norm and other, are defined, much as with de Saussure's differentiated words, in juxtaposed opposition; e.g. we define civilization in opposition to madness (and ironically treat the mad in an uncivilized manner throughout much of history). Such a concern with the "other" has a considerable 20th century history, and grew out of existentialism's transformation of Hegel's master-slave dialectic. History was largely told as a story of the "masters" of society, the political triumphs of kings and such... and it wasn't until the 20th century with Foucault (and others, as with Howard Zinn's *A People's History of the United States*), that the story of the extraordinary "ordinary" lives of many began to be explored in respected detail by historians: what Hegel might call the slave's perspective on history. One can see this immediately in the history of art: it isn't until the 19th century, with works by Jean Francis Millet, Vincent Van Gogh, etc. with the artistic movements of naturalism and realism, that the working class is finally and widely portrayed with a glory previously permitted only to the ruling class (and we should not forget the autobiographies by people like escaped slave Frederick Douglass as well).

Beyond Foucault's investigations into the formation of the subject (often by state institutions), his older contemporary, the semiotician Roland Barthes, ever ready to debunk symbolic *authority*, saw the *author* as such—and as something to overturn as well: e.g. with his essay "The Death of the Author" (1967). Such relates to Walter Benjamin's task of erasing ones' own subjective judgment and personality in order to let history speak for itself. Foucault however wanted the subjugated (the jailed, the insane, etc.) to have a voice in public—to tell their story in an unfiltered manner. This wouldn't be an attempt to let chaos loose in a world of structure, to unleash the id without super-ego constraints, but rather to loosen the tensions that are subjugating the helpless in favor of a status quo that itself is at times insane, criminal, etc. Even though our very identities are often socially constructed, or so it would seem, especially with copious examples demonstrating that our world views or "epistemes" have changed over the ages, Foucault struck a more subtle note than Barthes in his 1969 lecture, "What is an Author?" by examining the notion of the concept of "author."

Foucault contrasted with Chomsky over the possibility of a human nature: Humans, for Foucault are shaped by histories to an incredible extent. Foucault could be criticized by Chomsky, as not taking our hard-wired biology as shaped by DNA and evolution into full account. But through concepts like "epistemes" and "discourses"—discourses are something like macro-Wittgensteinian language-games, as with institutionalized sets of practices: think the military and its standard operating procedures, the procedures of the legal system, medical practices, governmental bureaucracies, and the such—through well illustrating these concepts that emphasize the socially constructed peculiarity of perspectives, Foucault was able to show that much of who, for example, an author is, is largely a question of "who" that author's culture and time is, but also what that culture determines as acceptable as an "author." Authors speak the zeitgeist as much as speaking for themselves, largely because the zeitgeist determines who will be known as an "author"—there are more authors than are remembered by historical canons. The writer-author as ego is somewhere between a body speaking via unconscious Freudian meanings (expressions of the id) and the discourses of a super-ego institutional bureaucracy—Foucault's celebrity intellectual image aside, he was as interested in ego-formation as much the other (id) / norm (super-ego) opposition.

It is with this metaphorical death of the author, that we can connect Foucault with Derrida. But through their studies of how history shapes us, and how language speaks us, rather than there being some untouchable aspect to meaning in conscious subjectivity, they were both on the track towards situating the creation of artificial intelligence, where machines might be en-webbed into the same social fabric as we humans are: a machine might not be able to tap into the Logos, but it surely should be able to learn from history.

What Derrida and Foucault share is "relativism situated in contexts" which is not your typically understood relativism where "anything goes." For Foucault, no context is absolute: following Nietzsche, we may share a perspective, but that perspective itself has a specific history about how it came to be; these perspectives are often shaped by governments and social practices, as well as the economic factors that Marx might point out (as with ideologies). Derrida's focus is on language, when he says "There is nothing outside of the text" (Derrida, *Of Grammatology*, p.158)—he is saying there is nothing outside context—that there is a possibly endless deferring from one frame to another; as the earth loses its place as center of the universe, then the sun, etc, beyond even a universe with no center: a Copernican revolution without end.

Derrida and Foucault also share a fascination with the concept of the "other." In contrast to the deferral of "différance" where meaning is always slipping into the future (and Derrida himself often enacts his theories in his writing practice—his meaning often seems to slip away in references to other writings (other authors and his own); he makes conscious use of a non-univocal plurality of meanings exploiting the economy of double-meanings; his work is hard to grasp in a moment, taking time to become familiar with; yet he aims for justice, so that his works will not deconstructed in turn), the later Derrida was often concerned with opening one's self up to an "impossible" incalculable future to come. There is an almost messianic reverence for the potential, not only for progress (an openness to political change), but an openness to something to come that is completely other. With the earlier Derrida, it as if his concepts were made in pursuit of the intangible—as if he were looking for a fabric by cutting it with scissors: différance as "spur" or blade that differentiates its target as it "looks" for it; always slicing material up in smaller pieces as if such would end in an understanding of the whole cloth—an active looking with cataracts, always catching site of its aim through peripheral vision, yet blinded as soon as it tries to catch site straight on: blindness and insight. The later Derrida seems interested in a passive subject, waiting. Such is not to put a gender-marker on the early and later Derrida (with a "phallic" spur spear-heading différance, in contrast to a "lady in waiting" vaginally open to the advent of the future)—but continental philosophers have tended to gender-code their metaphysics, which can be seen as a problem. Derrida does address such issues, noting that the structuralist Freudian, Jacques Lacan's own attempts at stabilizing meaning through speech (between earlier ubiquitous Freudian phallic symbols and Derrida's later critical term "phall-logo-centrism") were connected with the privileged phallus. Here Derrida's essays reference "style" and "taste":

> "In the question of style there is always the weight or *examen* of some pointed object. At times this object might be only a quill or a stylus. But it could just as easily been a stiletto, or even a rapier. Such objects might be used in vicious attack against what philosophy appeals to in the name of matter or matrix, an attack whose thrust could not but leave its mark, could not but inscribe there some imprint or form. But they might also be used as protection against the threat of such an attack, in order to keep it at a distance, to repel it—as one bends or recoils before its force, in flight, behind veils and sails (*des voiles*). But let us leave this elytron to float between the masculine and feminine" (Jacques Derrida, Barbara Harlow, trans., *Spurs: Nietzsche's Styles,*The University of Chicago Press: Chicago (1979), pp. 37, 39).

> "I have often been accused of writing things that are unnecessarily difficult, that could be simplified, and I have even been accused of doing it on purpose. I'd say that this accusation is just and unjust at the same time. It is unjust because I really do try to be clear; it's not that I amuse myself multi-

plying obstacles to understanding; I can even be pedagogical—often too pedagogical, perhaps. But I have to admit that there is a demand in my writing for this excesses even with respect to what I myself can understand of what I say—the demand that a sort of opening, play, indetermination be left, signifying hospitality to what is to come [*l'avenir*]: 'One does not know what it means yet, one will have to start again, to return, to go on.' And if there were time, it could be shown precisely how each text enacts a kind of opening—as the Bible puts it—of the place left vacant for who is to come [*pour qui va venir*], for the *arrivant*—maybe Elijah, maybe anyone at all. There has to be the possibility of someone's still arriving; there has to be an *arrivant*, and consequently the table—the table of contents or the table of the community—has to mark an empty place for someone absolutely indeterminate, for an *arrivant*—who may be called Messiah, but that's another question" (Jacques Derrida & Maurizio Ferraris, Giacomo Donis (trans.), *A Taste for the Secret*, Polity: Cambridge (2001), p. 31).

DECONSTRUCTION'S LIMITATIONS

As we saw Derrida's interest in writing's *oppositional relation* to speech, and his use of a single word that *combines two elements* (the difference and deferral of différance)—two strategies he employs often—one might be tempted to see him as a dialectician, concerned with overcoming opposition through a third term or category. And indeed some of Derrida's terms seem to echo Hegel's "aufheben," or sublimation—even the word "deconstruction" itself. But this could be a mistake... Derrida tries to *overturn* what are perceived as violent hierarchies—more subtle than Foucault's historically subjugated "others" in need of liberation—yet more deeply ingrained in western philosophies, languages, and customs. And although his single terms suggest a neither / nor + both / and—hence eluding any sort of Sheffer Stroked law of non-contradiction (différance being prior to logic, truth, and being)—they do not resolve any oppositions, but rather play with them in un-decidable ways.

Derrida is often interested in the accidental vs. the intentional, errors vs. the correct way, the humorous vs. the serious, the contaminated vs. the pure, the faux vs. the true, the improper vs. the proper, and so on. Not that he enjoys a mess... but reality is usually more complicated than a neat theory can express. Much like Gödel's incompleteness theorem, Derrida often sees the inside of a neat theory as having some hinge which opens up to worlds of other possibilities that said theory cannot touch (e.g. how can meaning based on live speech make sense once the speaker has died?) The book is closed and inter-textuality opened, much like a Bible ended with the New Testament, yet extended indefinitely through the writings of the Cabala. The outside, the other, the different, etc, shapes and interpenetrates the inside, the self, the same, etc: there is no pure theory that can't be cracked or ruptured (metaphors for difference), showing traces in the subject matter of what that pure theory excludes. In so many theories there is a sort of yin-yang binary opposition irreducible to "oneness" where each side contains a bit of the other; yet too often the imbalance between the two is unjust. And the point of contact between the two—the point where reversal of the unjust hierarchy becomes possible in some sort of affirmative action—this hinge often occurs, for Derrida, when examining the marginal... footnotes, prefaces, ambiguities, and other places where "exceptions to the rule" of an ambitious theory, break down upon closer scrutiny—yet these supplemental materials do not destroy the theories when supplanting them, but rather expand and complicate the theories. Again, writing in the broader sense could be said to frame speech; yet speech still precedes writing in the narrower sense. Writing escapes the theory of meaning being present in conscious intent by showing that writing as difference (or writing as a supra-structure—the difference that creates structure) is necessary for intent to be formed—that accidental element turns out to be key.

Critique of Derrida can be difficult both if one does not understand him, and if one does. Deconstruction, as we have seen, revolves around the issue of justice—and it often seems aimed a striking a balance between asymmetrical components (a balance and equality which is implied by any affirmative action). The problem of consciousness being considered epiphenomenal has already been raised. Despite his own creativity, Derrida's theory does not focus much on the self-conscious freedom of subjective creativity. Foucault seems to suggest that we can only be free from social shackles and prejudices if we become cognisant of them: knowledge is power. And studying Derrida can make one much more aware of the biases of language and philosophy. Possibly Husserl survives here, as with some sort of negative theology: we become more aware of consciousness for becoming familiar with all that it is not.

However, Derrida does have a view on language intimately tied with difference, and that may be at odds with fuzzy logic. He suggests that while there is a play in the meaning of words within a delimited zone, fine differences between words are located near their borders. Some might argue that words get fuzzy at the borders, and shade into one another; Derrida does not—see Jacques Derrida, Samuel Weber (trans.), *Limited Inc.*, Northwestern University Press: Evanston, IL (2000). To illustrate the problem consider an example—imagine a child who knows only three color words: blue, red, and yellow. Where is green in this case? Blue? Yellow? It would appear that the language would be fuzzy *in application*: one would not know what to call a green spot, although it seems to be blue or yellow but not red. It would also seem to follow that Derrida needs not only a full breadth of a language refined enough to be exact, but must include any future vocabulary as well, for those borders between words to be fully distinct. People with a limited vocabulary cannot be as precise as people with a developed one; Derrida's theories seem to be focused on language itself, as fully developed, rather than on the limited use by individuals. Yes, we could say that a person uses language in a crude manner, and that society's language could have been used to clarify what they were talking about, had they known it. The child says they see a "blue-yellow" spot, and we would know, with our more refined vocabulary, that it is "green."

At a limit, there might be a word, eventually, for every singular experience: with no need for "iterability"—every experience being spatiotemporally unique and every word used only once. In this case, we can see iterability... the ability to lift a mark from one context and put it in another would entail that some experience would be similar—that the relation of words to experiences would mean that, although not identical, the meaning of a mark would be similar from one context to another (the contexts would be similar). Since we do not a have a distinct word for every single experience, words become broader in scope: at its broadest, with a new-born baby, everything might be "mama." What is being illustrated here is a distinction between language in itself, and language as it is experienced by a human being. A young human being may not have a full grasp on language, yet they are able to communicate. Derrida himself is quite articulate: possibly his theory of meaning really is self-referential, and applies only to people like him. Ironically, He doesn't fully address the idiosyncratic experiences that people have: I can mean something slightly different by my use of words than you do, based on different experiences we have had with words. Yes, we share many contexts, but don't share some others. In sidelining subjectivity as a source of meaning (meaning is not found in our self-present conscious intent), Derrida fails to account for the fact of personal histories shaping what people mean when they use words: examination should evidence that language is at least semi-private and at most only semi-public. Derrida's deconstruction focuses more on public discourse, and hence misses something like half the picture of meaning.

Nim Chimpsky (1973-2000 photo by Harry Benson)

Chapter 23

Chomsky
Beyond Generic Human Beings

Noam Chomsky's (1928 - present) contribution to our understanding of theoretical cognitive modeling has been monumental, as has been his searing critique of the social power elite. Although he claims little connection between his linguistics and his political / media critique, the two are tied, in his mind, by the necessity of approaching both with a reasonable if not scientific perspective—and a *universal* perspective concerning not only how human nature has evolved, but also how we can progress as a humane society that unhypocritically applies its principles to itself in a universally consistent manner.

As with the other philosophers covered in this book, this chapter will outline the general conceptual structure of Chomsky's insights into the workings of mind and language while excluding, for the most part, how he came to these views, and avoiding in depth critique. I will also briefly cover his political thought, as it will become relevant to the social implications of technology and artificial intelligence to be discussed in the third part of this work.

INNATE CREATIVE ABILITIES

Although Descartes clearly was interested in a mechanics of mind, it was Chomsky who championed the new "cognitive science" and led the charge in the 20th century against "observational" approaches to language, such as Behaviorism's knee-jerk stimulus response theories, or the hierarchical taxonomies of structural linguistics. Again, as with the rationalist Descartes, and as we have seen, in a similar way, with Plato's theory of "recollection" and Kant's bifurcation of Understanding (where both the explicitly known and the potentially known—foreknowledge—are found in the self-same mind), Chomsky has studied how certain aspects of the mind, specifically aspects of language, are *innate*—and for Chomsky this means aspects of language, especially syntax abilities, are biologically hard-wired in the brain.

The core language abilities a human is born with might be identified by aspects of language common to all languages—if not a deep structure, then at least some inbuilt syntactic abilities or a Universal Grammar—and this focus on *grammar*, or *syntax* may come from a focus on the *formal* aspect of language, rather than the informational *content*—since Chomsky is more concerned with the formal functionality of language—an explanation of *how* it works rather than *what* it says, in general and hopefully elegant theories. Of course, and a little ironically, this is not all arm-chair theorizing, as much as theory informs the *data* aimed at, for *observation* has demonstrated that children acquire language despite a "poverty of stimulus"—they pick up a fairly complex amount of *ability* to *use* language despite their being exposed to few occurrences of certain linguistic phenomena. And of course, language acquisition is much more complicated than monkey-see, monkey-do; as language abilities are generalize-able, not specific to just one circumstance, and can be used in novel, creative ways.

Foucault could have hinted in his debate with Chomsky that, existentially following Sartre, we ought to be free to determine our human nature; that existence precedes essence—and although such a notion forms a backdrop to his thinking, Foucault, ever cognoscente of the archeology, genealogy, and history of inter-human power relations, relates "human nature" as another aspect of the *socially constructed* modes of human being:

> "In the history of knowledge, the role of human nature seems to me mainly to have played the role of an epistemological indicator to designate certain types of discourse in relation to or in opposition to theology or biology or history. I would find it difficult to see in this a scientific concept" (Noam Chomsky & Michel Foucault, *The Chomsky-Foucault Debate on Human Nature*, The New Press: New York (2006), p. 7).

Although "human nature" could be seen as too broad or vague of a concept (and Foucault may have thought that since what is often discussed as human nature is a culturally contingent "construct"—it *could* be different), biologically there seems to be at least some human nature (e.g. the natural property of walking on two legs, being featherless, etc)—but even then, Chomsky's claim that human nature is to be "creative" could also be in line with the self-creation and self-determination advocated in existentialist philosophy.

I think there would be no mistake in connecting Chomsky's claim that being *creative* is a part of human nature with Universal Grammar as a *Generative* Grammar: articulating thought has an indefinite potential for variation; especially given the infinite use of discrete elements with potentially endless recursive constructions. We are creative beings, and that is reflected in our language: again, as against behaviorist views that we are trained to respond in accordance with experienced rewards and punishments—almost "brainwashed by experiences" (experiences that can be humanly manipulated, as with advertising)—creativity ties in with spontaneous freedom and intent. But like artistic creativity, we need something to work with: techniques and mediums—and in this case, the techniques and medium may be given by the structure of our brains, as they have evolved over millions of years. Again, rather than a focus on language as an *object* to be dissected, we must look at it as an *ability* of the brain (which is much in line with Wittgenstein's shift from searching for the core logic in language in his *Tractatus*, to an examination of how we *use* language in his later philosophy).

THE CONTEXT-FREE LANGUAGE ORGAN(S)

To say that language is "hard-wired" in the brain, at least as far as generative grammar is involved, means that the brain is not 100% plastic at birth... not the blank slate as the Empiricist Locke would have it (although, as we have seen, Locke believed we were born with certian *faculties*)—and although many *parts* of the brain may be incorporated in a linguistic act (as noted with brain imaging that reveals specific areas of brain activation for certain types of mental activity), the very notion of "parts" suggests that the brain's functions are divided: there could be a modularity of mind, where specific sections of the brain are like organs in the body—different organs work together, but they each have separate (sometimes overlapping) functions. Dissecting a human brain indeed demonstrates that different parts are at least visually distinguishable; and even those parts not distinguishable by the naked eye (e.g. globs of neurons) have been demonstrated to be connected to different functions through evidence of many people having lost certain abilities in conjunction with damage or aphasias to similar regions in the brain. This can get specific, as with the loss of coherent semantic content with damage to Wernicke's region, or loss of grammatical ability with damage to Broca's region in the brain. Indeed, these two areas (Broca's and Wernicke's) would be prime candidates as language organs, and would demonstrate that the "language organ" is itself subdivided into smaller sub-organs.

(I)NTERNAL AND (E)XTERNAL LANGUAGES

It is with this notion of an innate, inborn ability to use an internal language based in brain sub-organs, that we can see Chomsky's aversion to theories of a social construction of external languages, as with, for example, "memes." Memes are purported to be social customs and practices that pass through humans as carriers—one can think of how linguistic fads, which like the hula-hoop, take off from an original spark, and spread like a fire—linguistically, one might think of catch-phrases, like "where's the beef?" promoted once by a hamburger franchise. Of course, there would be interactivity between human biology and cultural customs; but these memes can seemingly take on a life of their own.

Where I believe Chomsky is coming from, in rejecting the likes of memes, is a failure to see how they could originate outside brains—and to know *where* exactly these "things" exist. I think it could be argued that they do originate in brains, and that they exist in brains as subtle modifications of brain-memory / structure. They exist socially as a sort of statistical "hologram"—many people carry memes as similar modifications of memory and / or learned abilities—just as learning to drive a car is an acquired ability that is passed on from some people to others. Yes, one can't point to the "car-driving-ability" in some sort of abstract / real existence, but that ability does exist, in particular people, not all people, and is largely a cultural phenomenon. I think that Chomsky would agree that at least some aspects of language are cultural—such as the content of the advertising slogan mentioned above. To say that a linguistic hula-hoop is innate seems absurd to me; it is one of those infinite variations that the creative mind came come up with—given the biological limitations of a Universal Grammar, or hips that are able to swivel.

But it would be difficult, if not impossible, to demonstrate that memes could account for all language abilities. We must have some innate abilities to work with, or else, to really put the metaphor to the test, simply putting a mirror or real blank slate in the world would create a knowledgeable being. We need faculties of the mind, as Kant developed at length (for the ability to make logical judgments)—there must be some functioning of a certain degree of complication or else there would be no ability to acquire additional abilities. To what extent these fundamental abilities are "context-free," and how flexible they are, is a matter for debate and observational inquiry. Steven Pinker, who wrote a book called, and refuting, *The Blank Slate*, also studied the limits of innate *semantics*, staking a moderate position where he:

> "proposed that the human mind contains representations of the meanings of words which are composed of more basic concepts like 'cause,' 'means,' 'event,' and 'place,'" and dismisses "the crazy idea that the concept of 'carburetor' is somehow coded in to our DNA." (Steven Pinker, *The Stuff of Thought: Language as a Window into Human Nature*, Penguin Books: New York (2007), pp. 90-91).

As Chomsky's work has focused mostly on language and grammar—areas critical to what makes us human, and our ability to think, and much of what we consider part of our "souls" and personality—his insights into language should reveal much about any nature we might have as humans.

PRUNING THE UNIVERSAL GRAMMAR TREE

I will not go into the complexities of the evolution of Chomsky's view on the structure of our grammatical abilities. Suffice it to note that his views have changed: from noting how deep structures of thought could be transformed into surface structures of linguistic utterances via various specific rules, rules that could become untenably complex to the point of varying from language to language and from situation to situation (and hence not be Universal); to the more recent *minimalist program* arising from intermediate theories re: general *principles* (such as the mobility of terms from one place to another in a proposition [move] and the combination of parts to make larger wholes as with morphemes, phrases, etc. [merge]) limited by *parameters* (optional aspects of grammar that get locked in by habit—e.g. imagine a language like Latin, where the verb might always *move* to the end of a sentence after both subject and object (SOV), contrasted with English where the verb usually follows a subject, but precedes an object (SVO) and will *move* inbetween).

Such a transition from a more traditional and specific examination of grammar to a more flexible, general, and hence universal, understanding of grammar might redraw the lines between those abilities that are biologically innate, and those that are culturally contingent; albeit that to a certain extent the culturally contingent aspects may be available to the Universal Grammar: the Universal Grammar has possibilities that are pruned. In this case, exceptions do not only prove the universal rule, but suggest where the universal shades into the contingent.

A question arises with the division between intent and articulation. Chomsky's earlier views on Transformational Grammar, despite such claims that "linguistic and mental processes are virtually identical" (Noam Chomsky, *Cartesian Linguistics: A Chapter in the History of Rationalist Thought*, Cambridge University Press: Cambridge (2009), p. 78), suggest that what we have to work with is a semi-linguistic intent that must be transformed into a coherent grammatical utterance:

> "Language has an inner and an outer aspect. A sentence can be studied from the point of view of how it expresses a thought or from the point of view of its physical shape, that is, from the point of view of either semantic interpretation or phonetic interpretation" (*Cartesian Linguistics*, p. 79).

Thus intent would be part of the deep-logical-semantic-structure, and the utterance would be the surface phonetic structure in accordance with grammatical rules. Such is an overly static and spatial view; as it seems that with spontaneous thought and speech (and writing) one does not always know were one will end up when one starts. We may have a fuzzy notion of what we want to say, but do not fully realize that until we've at least partially said it. There is more interaction between utterance and intention than the earlier Chomsky could account for with his mapping out of grammar trees. We need a more temporal theory of grammar... one that can account for the unfolding and development of intent. Again, the earlier Chomsky seems to assume that we know what we're going to say, or more importantly, verbally think, before we say or think it. Not so, in actual practice.

The minimalist program may be more flexible in this regard—given that the rules of transformation are thrown out in favor of general principles and parameters: such could be more easily "animated." The verbal thought process is no doubt tied to other cognitive processes (we can talk about rotating an image in our mind; give verbal instructions about how to steer a car, etc.)—these cognitive processes are tied up with our senses, and how we reason *through* our sense organs (figuring things out visually, or with auditory or olfactory thinking strategies, etc.)—and in a way, these non-verbal cognitive processes could be seen as tying in with pre-verbal intent. Even these ways of thinking might unfold temporally as well though... and we must keep in mind that these various modules of the mind are in a near constant interaction—verbal cues can cause one to focus on and highlight certain visual aspects of perception, and vice-versa.

To diverge from Chomsky for a moment... by letting go of very specific *transformational* rules of grammar—the minimalist program might be said to get closer to what goes on with "verbal intent" as it unfolds. Such would mesh with the notion of a *short-term memory* able to hold on to one idea while exploring a few others: the *principle* of term mobility requires short-term memory; and the *parameter* set by tuning into one option or another, would be an innate habit, albeit a habit carved out of other possibilities.

"Verbal intent" may be shaped by further limitations of the brain. Why is your average sentence only so long (say, seven words), and not much longer (like 100 words). Obviously, long books suggest that ideas can take more than 100 words to express—so it wouldn't necessarily be the "nature of the world" that limits average sentence length. No, this seems more like a cognitive limitation: short-term memory can only hold so much at a time. But, of course, one short sentence leads to another, and our much more complex long-term memory may feed through short-term memory, as we develop our immediate thoughts via our accumulated (and innate) knowledge and beliefs.

Our "language" organs may be limited, as is our short-term memory. Diverging even further from Chomsky, there may be a grammar "look-ahead"—a grammatical predisposition to use, e.g. "Subject Verb Object," (or Subject Object Verb, if our parameter selection was different) that interacts with the semantic content of what we start to express—although the flash of a thought "I want..." may invoke a verb-phrase "to eat ice cream" or a noun-phrase "some ice cream," we often do have idiomatic grammar habits, that most likely don't have a grammatical "look-ahead" beyond your average sentence length. Such habits though, when the grammatical rubber hits the road, often follow very specific and what seem non-universal grammatical rules, but rather, culturally arbitrary habits—turns of phrase, or idioms. Again, there is the issue of determining where biological Universal Grammar (of the Minimalist Program) stops, and the more idiosyncratic, albeit habitually idiomatic, specific language grammars begin. Idiosyncratic idioms aside, Chomsky might suggest that, like a tree that fans out to very fine branches, the Universal Grammar is refined by "pruning" parameters where principles like *move* and *merge* can only be expressed in a finite number of ways given our biological (and possibly to some extent, cultural) limitations, perhaps being refined to even the most specific of grammatical practices.

The difference between early Transformational Grammar and the Minimalist Program, as pertains to verbal intent and verbal articulation might be summed as thus: the first assumes that we have a *complete* intent that must be translated from one sphere to the next—from intent to articulation; whereas the Minimalist Program allows for us to follow a biologically hard-wired yet culturally switched on habit as the intent unfolds, dancing on the edge of the most specific of parameters (those available habits that were not pruned away in childhood)—intent and articulation need not be divided. We may actually think through a grammar as it has been developed (through pruning various possible parameters). The building blocks of general principles like *move* and *merge* are limited by the parameter selection determined by cultural "switching"—much like pruned tree branches that eliminate some potential syntactical "habits" and reinforce others.

NEWTON, NATURAL SELECTION & CONSCIOUS INTENT

The coincidence of mental intent and physical articulation—the way we feel out what we are saying, brings to my mind *consciousness*. Like gravity, or force, subjective *feeling* is not a mechanical process. One can imagine mechanics taking place in a four-dimensional space bent by a displacing matter—where trajectories change due to curves in space: such might seem a way of accounting for gravity in a spatial-mechanical way; but that just defers the question of how matter curves space. Chomsky's reminder that Newton saw beyond mechanism and hence saw beyond the engineer-ably understandable—is a great confession that one cannot know everything (at least as mechanical constructions):

> "Newton established, to his great dismay, that not only does mind escape the reach of mechanical philosophy, but so

does everything else in nature, even the simplest terrestrial and planetary motion" (Noam Chomksy, Adriana Belletti & Luigi Rizzi (eds.), *On Nature and Language*, Cambridge University Press: Cambridge (2002), p. 67, "Language and the Brain").

Such means that there could be more to intent than biologically determined (and possibly unconscious "Freudian") desires. Consider: natural selection suggests that something survives in a possibly changing environment: that there is a statistical similarity between an entity and it's possibly deviant copy (or progeny). Living creatures reproduce, and divergent DNA can help them to better survive, or die off more quickly. In any case, thriving survival of a *species* means that statistically similar plants or creatures sustain or multiply in number, and survive possibly changing environments (including other plants / creatures). (*Conservation* is usually aimed at conserving life as we know it now; often in order to preserve the human species). There seems to be a tie between a desire for reproduction (e.g. the sex-drive) and population growth (obviously!) A fundamental life-force seems aimed at the success of one's own species: love itself could be the subjective manifestation of life as shaped by natural selection (and this natural selection extends to the mental realm, where we look for theories that survive in varying contexts: theories that are reproducible and context-invariant—the philosophical school of Pragmatism interlocks with natural selection).

But human desires, as much as they may be rooted in life itself, come in many refined and complex varieties. The specifics of a contingent conversation may limn the edge of a bio-mechanical brain, as one expresses what one intends or desires, and one is not an operation of the brain: the brain and body are used by us as much as they determine us as free agents. Conscious intent would simply be beyond the mechanics of the brain, albeit shaped by that very bodily organ.

POLITICAL ECONOMIES: UNIVERSAL EQUALITY AND DISSENT

In addressing Noam Chomsky as a philosophical thinker, it would be unjust to simply outline his contributions to psycho-linguistics, as he has also been a monumental force with his searing socio-political critique of U.S. foreign policy and mass media bias. There are a few parallels between his linguistic studies and his political critiques: both are studied in a "scientific" manner—beholden to the facts, moving beyond anecdotal evidence to the statistical; both evidence appreciation for action over mere observation (i.e. a focus on generative grammar over behavior observation and taxonomies in linguistics, and an emphasis on activist dissent over passive consumerism in a social context); and both are concerned with the "Universal"—as with a Universal Grammar, and his political critique of hypocritical foreign policies that don't apply principles universally (—esp. to oneself).

We closed this chapter's discussion of linguistics with some comments on natural selection—and I think natural selection is applicable to Chomsky's political critique as well. Chomsky could be accurately described as a progressive—and like divergent DNA being necessary for evolution, Chomsky demonstrates dissent from the status quo; questioning deep societal assumptions—dissent that is necessary to improve upon our social situation. But Chomsky is also against a sort of social Darwinism where, like a particular species succeeding in multiplying, a particular type of people (the power elite) gather more concentration of power in ever fewer hands. That is, just as one *species* may spread over an environment, there is also a tendency for one group of people to spread their influence over a social-political-economic domain. Chomsky does not discuss this phenomenon in these terms, but suffice it to say that he advocates a democratic decentralization of power with his critique of concentrations of power: radical biodiversity rather than the success of a few species.

So Chomsky's targets, in an effort to move beyond all sorts of enslavement and exploitation, are the "masters" of society, private tyrannies, the power elite—and in many ways, the capitalist bourgeois owners of... not only the means of production, but the means of public information dissemination. In his essay, "Democracy and Markets in the New World Order"—Chomsky notes:

> "The history of business and political economy yields many examples of the subordination of narrow gain to the broader interest of the opulent minority, which is usually class conscious in a business-run society like the United States. Illustrations include central features of the modern world: the creation and sustenance of the Pentagon system of corporate welfare despite its well-known inefficiencies; the openly proclaimed strategy of diversion of soaring profits to creation of excess capacity abroad as a weapon against the domestic working class; the design of automation within the state system to enhance managerial control and de-skill workers even at the cost of efficiency and profitability; and many other examples including a large part of the foreign policy" (Noam Chomsky, *Powers and Prospects – Reflections on Human Nature and the Social Order*, South End Press: Boston (1996), p. 130, "Democracy and Markets in the New World Order").

Indeed, the minority most often protected by the state supposedly run through a democracy representing the majority, are the power elite... the "opulent minority." It would be safe to say that Chomsky seriously questions capitalism, albeit he might point out that we in the U.S. do not have a pure capitalistic economic system, but a business world that has a symbiotic relationship with the government (e.g. "state-capitalism" as with "corporate welfare" where externalities are cleaned up by the state, where the state funds research that is in turn exploited for private profit, and where the government intervenes to "bail out" firms that are deemed too large to fail, etc).

PROPAGANDA, INDOCTRINATION, EDUTAINMENT

Along with Edward S. Herman, Noam Chomsky worked out a study of propaganda in their work, *Manufacturing Consent: The Political Economy of the Mass Media*. In the introduction, they claim:

> "The media serve, and propagandize on behalf of, the powerful societal interests that control and finance them. The representatives of these interests have important agendas and principles that they want to advance, and they are well positioned to shape and constrain media policy. This is normally not accomplished by crude intervention, but by the election of right-thinking personnel and by the editors' and working journalists' internalization of priorities and definitions of newsworthiness that conform to the institution's policy" (Edward S. Herman and Noam Chomsky, *Manufacturing Consent: The Political Economy of the Mass Media*, Pantheon Books: New York (2002), p. xi).

To examine this, Chomsky and Herman developed a propaganda model that described five stages of filtering of any news before it reaches the public at large. All five stages are intimately intertwined with the semi-capitalistic system that we have in the US.

First, the media are *owned* by the power elite who want to expand their power through profit:

> "[T]he dominant media firms are quite large businesses; they are controlled by the very wealthy people or by managers who are subject to sharp constraints by owners and other market-profit-oriented forces; and they are closely interlocked, and have important common interests, with other major corporations, banks, and government" (*Manufacturing Consent*, p. 14).

Second, the media must attract *advertisers*, who aim to promote passive consumerism:

> "Advertisers will want, more generally, to avoid programs with serious complexities and disturbing controversies that interfere with the 'buying mood.' They seek programs that will lightly entertain and thus fit in with the spirit of the primary purchases—the dissemination of a selling message" (*Manufacturing Consent*, pp. 17-18).

In her book, *No Logo*, Naomi Klein explored the ubiquity of advertising, and how it has put a positive veneer on an ugly reality of outsourcing jobs to foreign countries where it is easier to exploit labor (Naomi Klein, *No Logo*, Picador: London (2009)). This, combined with the Chomsky-Herman insight into the second media filter, suggests the untenable aim of advertisers to move the US populace as a group to be passive consumers that don't actively produce anything—the US population itself as a sort of power elite living off the labor of poorer countries (untenable, as *this* power-elite does not have the seemingly endless resources of the "*real*" power elite). Yet it should be noted that advertising can also be seen as adding aesthetic value to products—there is a continuum between product design and product promotion—and an informed public may want their products to be designed and advertised as ecologically and sociologically sustainable.

The third filter: the *sources* of information are the powers that be—those who have been legitimized simply by virtue of their being integral parts of the present social system. Reporters can't search out news everywhere, and hence often rely on corporate PR representatives, and government press-conferences to set the agenda, if not the actual content, of what is deemed newsworthy.

Fourth, it can take a considerable amount of resources, and established connections, to create *flack*. Yes, there are real grass-roots organizations that create protests and political movements; but too often we have "fake grass-root" organizations, sometimes labeled as "Astroturf" organizations, that use power already gained, in order to maintain and expand that same power.

Fifth, Chomsky and Herman claim "Anticommunism" as a "control mechanism"—an ideology that "helps mobilize the populace against an enemy" (*Manufacturing Consent*, p. 29). In the contemporary era, that may be seen with the "war on terror"—where a new enemy is developed, or becomes a focus that is used to create a "team" atmosphere of "us against them." This serves to get people on-board, and to facilitate the ostracizing of people questioning the status quo as assisting "the enemy," and gives members of a country unifying common cause—hence curtailing dissent.

Again, these five stages of news filtering are a result of the type of political economy we have—the people with power are aimed at maintaining a status quo... of maintaining and expanding the power they already have—and the "Us vs. Them" mentality just mentioned integrates well with Chomsky's critique of U.S. foreign policy.

Yet, there are mitigating "counter-filters" as well. Consider the quote often attributed to Abraham Lincoln: "You can fool some of the people all of the time, and all of the people some of the time, but you cannot fool all of the people all of the time." This stands to reason, as there are competing ideologies and sources of information in a robustly cross-referencing and media rich world. It could also be that many people often buy into the status quo, knowing full well its ramifications and alternatives. Moreover, many different agendas that have money behind them are at odds (e.g. lawyer groups vs. insurance agencies)—which suggests we should take up causes we believe in that are not already well-financed. Furthermore, the media industries first allegiance must be to their audience, or they will have no audience to sell to their advertisers. How do we explain phenomena such as the success of television shows such as "The Daily Show," and "The Colbert Report" which provide a bit of "edutainment?" True, these shows are often "level-headedly moderate" thus enforcing an "Overton window" of what is deemed acceptable discourse—they comically savage the extremes of right and left—and thus center political debate. But they also raise consciousness and teach critical thinking, even about their own relation to advertising (as with Stephen Colbert's run for president being sponsored by Doritos). It is true that these shows rarely if ever incite anti-capitalist rebellion; but again—do democratic polls really suggest that people are against capitalism per se; or that they are against specific aspects of it?

ALTERNATIVES TO HYPOCRISY

Getting into the details of Chomsky's varied and specific foreign policy critiques is beyond the scope of this chapter, and would derail the focus of this book. But it should be noted that time and again, Chomsky exposes the hypocrisy of U.S. foreign policy—the "Us vs. Them" attitude that helps unify and mobilize a domestic population that becomes blind to the "others" who are vilified, or simply not cared for as full human beings, because they are not part of "our" group:

> "People can believe that when we use force against Iraq and Kuwait it's because we really observe the principle that illegal occupation and human rights abuses should be met by force. They don't see what it would mean if those principles were applied to U.S. behavior. That's a success of propaganda of quite a spectacular type" (Noam Chomsky, *Media Control: The Spectacular Achievements of Propaganda*, Seven Stories Press: New York (2002), p. 53).

Time and again, Chomsky finds the U.S. as the egregious aggressor on the international scene—often itself a terrorist state breaking international law with impunity. And time and again, those across the political spectrums share underlying assumptions... biases (e.g.

that our political-economic system is better than any other)—and are willing to use force, not only to defend that political-economic status quo... but to spread it globally as well.

Again, much of Chomsky's political-economic critique boils down to those in power trying to maintain and expand that power by spreading ideals of passive consumerism, and an Us vs. Them mentality to keep group cohesion and mobilize forces against any external or internal threats to the status-quo of, yet again, maintaining and expanding the power of the elite.

Although citing the anticommunism prevalent in 20th century media and state policy bias, Chomsky has not advocated communism as much as he has claimed appreciation for Libertarian-Socialism or Anarcho-Syndicalism. A commitment to autonomy against dominating powers, while supporting the idea of community cooperation. This ties in with the participatory economic or "Parecon" system advocated by Michael Albert and Robin Hahnel. Albert and Hahnel have argued that communism has its own power elite, with a "coordinator class" that deal themselves all the empowering jobs. What we should strive for, rather, is a system that truly decentralizes the decision making processes that affect each and every one of our lives. Such might politicize the entire economic apparatus—whether this would liberate us from tyrannies at every level of social order, or bog us down in endless debates is something we should seek to test, if we want to resist concentrations of power in the hands of an elite few. I say "test," though, and not revolt for—as reform through avenues that have been experimented with successfully (as with employee owned businesses), are a safer bet for social advance than a scorched earth revolution that starts largely from scratch. The system we have may not be as good as it could be, but others built from the ground up could be worse.

CHOMSKY AS "DIALECTICAL" THINKER?

Many thinkers use binary oppositions in their thought... from Plato's Eternal (real forms) vs. the Temporal (illusory world), to Nietzsche's plastic Apollo vs. dynamic Dionysius. Whether by design or by accident, Chomsky too has a binary pair—not as a conceptual opposition to explore, but as a dichotomy in his actual intellectual orientation. Roughly, that split between linguistics and politics could be seen to correspond between a study of the individual and the social. Seeing that Chomsky has looked to move from mere observational descriptions to explanations of action—he has demonstrated this as well with the dichotomy at the heart of his intellectual endeavors: he doesn't just talk about the (invariant) brain / (pliable) politics split in itself—he has oriented his own "philosophical career" to explore these realms and to *achieve* advances in both—as a scientist and activist. Although his achievement in each field is considerable—to put both together has demonstrated yet another monumental advance in thinking, and places Chomsky as a major western philosopher in our lineage dating back to Plato and Anaximander. It is true that in the contemporary era, individual achievement is all the more situated in a social context (advances are as much social as they are individual)—but Chomsky has taken a leadership role in his chosen fields, and represents a cutting edge of thinking in his era. And despite the possibly universal aspects of human nature—what one might call the generic aspects of humanity—through his dissent and his own creativity, he has demonstrated how to move beyond being a generic human being.

Part III

Situating AI in the History of Conceptual Advances

Tin Woodsman - William Wallace Denslow (1900)

Chapter 24

AI
A History of the Field

FICTION & FABRICATION

Although this book has been focused on the history of philosophy and the development of what it means to think and how we can understand thinking, the more specific field of artificial intelligence (AI) has its own milestones that will be covered here in brief. This history necessarily includes some aspects besides intelligence alone, namely the instantiation or embodiment of intelligence in robots and computers—which also explains why logic and mathematics have been so important to the history of AI. We have not dwelled on logic proper to a great extent in this book (other than with Aristotle, the Stoics, and the early Wittgenstein), largely because it is the author's belief that symbolic logic too narrowly limits the scope of intelligence; albeit that other traditional approaches (AI neural-net connectionism, and artificial life / natural selection, etc.) have their limitations as well.

As to the history of something yet to be achieved in full, we must begin with ideas, not realities. Aristotle, our 4th century BCE philosopher scientist and logician covered in chapter six, mentions the automated "tripods" from Homer's *Iliad* in his *Politics* when discussing (and defending) slavery:

> "For if every instrument could accomplish its own work, obeying or anticipating the will of others, like the statues of Daedalus, or the tripods of Hephaestus, which says the poet,
>
> 'of their own accord entered the assembly of Gods;'
>
> if, in like manner the shuttle would weave and the plectrum touch the lyre, chief workmen would not want servants, nor masters slaves" (Aristotle, *Politics*, I, 34).

Homer's story could be seen as early "science fiction," which inspired the notion of artificial beings with intelligence... at least enough to carry out certain tasks. Other early stories include the legend that Rabbi Judah Loew (~1520-1609CE) created the Golem of Prague—an animated creature created from mud (and hence was like the biblical Adam). Such may have inspired Mary Shelley's monster in *Frankenstein* (anonymously published at first in 1818). And there are of course countless other stories in modern times:

> "Science fiction writers have used the possibility of intelligent machines to advance the fantasy of intelligent nonhumans, as well as to make us think about our own human characteristics. Jules Verne in the nineteenth century and Isaac Asimov in the twentieth century are the best known, but there have been many others including L. Frank Baum, who gave us the *Wizard of Oz*. Baum wrote of several robots and described the mechanical man Tiktok in 1907, for example, as an 'Extra-Responsive, Thought-Creating, Perfect-Talking Mechanical Man... Thinks, Speaks, Acts, and Does Everything but Live'" (Buchanan, Bruce G., "A (Very) Brief History of Artificial Intelligence, *AI Magazine* Winter 2005, p. 53).

In fact we get the name "robot" from a fiction as well.... from Karel Capek's 1920 play *R.U.R. (Rossums's Universal Robots)*—the word "robota" meaning "serf labor" in the author's native Czech language.

In the real world, Leonardo da Vinci designed a humanoid automaton (Leonardo's Robot ~1495CE), which placed some pulleys and cables in suit of armor, and a modern replica has been proven fully functional, being able to stand, sit, and move its arms. Another early attempt at an automaton includes The Turk, a fake chess-playing machine exhibited from 1770 to 1854 (when it was destroyed by a fire)—although not a functioning automaton but a sort of magic trick made to look as if it were working (a human chess player was inside the contraption). From the mid 19th century on, magicians used automatons (think of the automatons one could see at 20th century Disneyland), and many automaton toys have been created, including wind-up dolls, toy robots, etc, to this day.

LOGICAL & MATHEMATICAL THINKING

We've summarized Aristotle's introduction of logic in chapter six; but also of major importance is the *Elements*—13 books written ~300BCE by the Greek mathematician Euclid. This work, which included mathematical definitions, postulates / axioms, and mathematical proofs of theorems and constructions, had its own algebraic system and was a groundbreaking work in formal reasoning. The name "algebra" comes from the Arabic word "Al-Jabr" (the meaning of which is not known with certainty)—being used by the Persian mathematician al-Khwarizimi (from whose name we get the word, "algorithm") with his development of the field in his book from 820CE—this book addressed how to systematically solve linear and quadratic equations.

Also formalizing thought:

> "Ramon Llull (circa 1235-1316), a Catalan mystic and poet, produced a set of paper discs called the *Ars Magna* (Great

> Art) which was intended, among other things, as a debating tool for winning Muslims to the Christian faith through logic and reason [....] One of his disc assemblies was inscribed with the attributes of God, namely goodness, greatness, eternity, power, wisdom, will, virtue, truth, and glory. Rotating the discs appropriately was supposed to produce answers to various theological questions" (Nils J. Nilsson, *The Quest for Artificial Intelligence*, Cambridge University Press: Cambridge (2010), p. 3).

Such material and symbolic formalization of thought attracted Leibniz (1646-1716) as well who sought a "*characteristica universalis*":

> "universal language would allow us to express with great precision problems of all kinds, and then go about solving them" (Ben Coppin, *Artificial Intelligence Illuminated*, Jones and Bartlett Publishers: Sudbury, Massachusetts (2004), p. 7).

Leibniz paired this universal language with his "Calculus Ratiocinator"—a framework that may be interpreted as a sort of computer hardware or software that could be used to calculate the truth and logic of propositions. Leibniz was also interested in binary mathematics, but it wasn't until 19th century that George Boole,

> "who lived from 1815 to 1864, developed Boolean algebra, the logical system we still use as part of propositional and predicate logics. Boolean algebra is widely used by electronics engineers in developing logical gates for silicon chips and is also used by computer scientists. Boolean algebra provides a language for expressing concepts like 'A is True' and 'A is true but B is false'" (Coppin, p. 7).

Boole begins his book, *An Investigation of the Laws of Thought*:

"The design of the following treatise is to investigate the fundamental laws of those operations of the mind by which reasoning is performed; to give expression to them in the symbolical language of a Calculus, and upon this foundation to establish the science of Logic and construct its method; to make the method itself the basis of a general method for the application of the mathematical doctrine of Probabilities; and, finally, to collect from the various elements of truth brought to view in the course of these inquiries some probable intimations concerning the nature and constitution of the human mind" (George Boole, *An Investigation of the Laws of Thought*, Walton And Maberly: London (1854), p. 1).

Boole developed binary truth values (0 and 1) along side with logical operators inherited from the Stoics (conjunction: AND, disjunction: OR, and negation: NOT). Again, this binary logic was later used to develop digital logic gates, where for example 0 OR 1 will return 1; 0 OR 0 will return 0; 0 AND 1 will return 0; 1 AND 1 will return 1; NOT 0 will return 1; and so on.

In 1913, Henry M. Sheffer proved that the operation NAND ("not both... and...") could be used to produce all the other operations (by applying it recursively), and symbolized it with a "Sheffer stroke" which is just a vertical line as with A|B, which produces the following truth table:

Input	Output
A|B	A NAND B
0|0	1
0|1	1
1|0	1
1|1	0

From this, NOT A can be derived from A NAND A written as "A|A"; A AND B from (A|B)|(A|B); A OR B from (A|A)|(B|B). (And in extension A Implies B can be derived from A|(B|B) or A|(A|B)).

Also important for the history of logic and mathematics is the work of Gottlob Frege (1848-1924), especially his *Begriffsschrift* (German for "concept-script"), which aimed to show how mathematics can be derived from logic, with his predicate logic being a combination of the syllogistic logic we saw with Aristotle and the propositional logic of the Stoics, but prior to the work with truth tables that we saw with the earlier Wittgenstein. Predicate logic differs from propositional logic, in that it deals with quantifiable variables rather than propositions, yet it maintains the logical operators that were seen to connect propositions (a propositional variable is a variable that can be true or false; a quantifier would be something like the existential ∃ "there exists" and universal ∀ "for all"). Although Frege did not use these modern symbols, his conceptual notation helped to clarify ambiguities in previous logics that used words like "some" and "all."

Frege's work in logic was put to extensive use and expansion with Alfred North Whitehead and Bertrand Russell's *Principia Mathematica* (1910-1913). This monumental work aimed at laying the foundations of mathematics logically, yet only dealt with set theory (collections of objects where one set can be in Union with another, in Intersection with each other, etc.), cardinal numbers (1,2,3,...), ordinal numbers (well-ordered numbers with a functional relation, e.g. 2, 4, 8, 16) and real numbers (including decimals, etc).

Even with this limited scope, *Principia Mathematica*'s limitations were brought to the fore in 1931 when Kurt Gödel proved his two incompleteness theorems. His first theorem claimed no "effective procedure" (something like a computer program) can prove all the truths about arithmetic—for there will always be a non provable undecidable statement in the system (e.g. the statement "This statement is not provable" which is related to the liar's paradox: "this sentence is false.") The second theorem backs up the first, by stating that for a formally generated theory, if that theory includes a statement about its own consistency, then it is inconsistent. What it seems that Gödel was getting at has to do with the self-referential nature of a formal foundation (which refers only to its axioms and definitions)—and that for such a system to be *meaningful* requires a connection to some sort of semantics outside the system (e.g. our ordinary language)—or at least a subject or judge that is outside the system who can make decisions based on a context the system excludes. In other words, a computer which tries to establish the foundations of mathematics will get caught in "infinite loops" trying not only to prove certain statements that refer to themselves (oscillating between true and false with the statement "this statement is false") but also in systems or theories that refer to themselves without recourse to extra-system or extra-theory aspects. The skeptics would be pleased. For although Sheffer's stroke recursively applied may be able to generate all the logical operators—the computer logic gate equivalents of which can ultimately be used in performing arithmetic—although digital logic can be used to *do* mathematics, proving that logic can build mathematics, *theoretically*, requires recourse to a semantics which logic has yet to establish the foundations for—and with semantics, more could be involved, such as the ability of an intelligent agent to "step-back" and re-frame a situation, esp. with problem solving (or theorem proving)—being able to look at a context broader than a finite system.

MACHINE COMPUTATION

Calculating machines have been around for quite some time (think of the abacus invented by the 26th century BCE). In 1623, Wilhelm Schickard invented a "Calculating Clock":

> "for performing addition and subtraction. This machine recorded the overflow from its calculations by the chiming of a clock.

> Another famous calculating machine was the *Pascaline* that Blaise Pascal, the French philosopher and mathematician, created in 1642 [....] Pascal later stated in his *Pensees* (1670), 'The arithmetical machine produces effects which approach nearer to thought than all the actions of animals'" (George F. Luger, *Artificial Intelligence: Structures for Complex Problem Solving 6th Ed.*, Addison Wesley: Boston (2008), p.6).

By 1804 we had Joseph-Marie Jacquard's weaving loom:

> "The punched cards of the Jacquard loom controlled the actions of the shuttles, allowing automatic production of fabric designs [....] The idea of using holes in paper or cards was later adopted by Herman Hollerith for tabulating the 1890 American census data and in player pianos (using perforated rolls instead of cards). The very first factory 'robots' of the so-called pick-and-place variety used only modest elaborations of this idea" (Nilsson, p. 25).

Punch cards were also to be used in Charles Babbage's "Analytic Engine," which differentiated it from his earlier "Difference Engine" of the 1820's that could calculate a series of numbers. The Analytic Engine, which Babbage worked on till his death in 1874, would have had various computing powers such as the ability to operate in sequence (move from one instruction to the next), branch (jump from one instruction to another based on a comparison; IF... THEN), and loop (perform a series of operations repeatedly). Babbage's correspondence with Ada Lovelace (daughter of the poet Lord Byron) about Babbage's engine suggests that Lovelace was the first computer programmer; with her apt quote:

> "The Analytical Engine has no pretensions to *originate* anything. It can do *whatever we know how to order it* to perform" (Quoted in Alan B. Turing, Jack Copeland (ed.), *The Essential Turing*, Clarendon Press: Oxford (2004), p. 455).

Modern day computers really became a possibility though, with the theoretical advance made by Alan Turning and his Turing Machine which he described in a 1936 paper and explained in full in his 1948 essay "Intelligent Machinery"; the set-up and operation of which could be described as such:

> "The Turing machine mathematically models a machine that mechanically operates on a tape. On this tape are symbols which the machine can read and write, one at a time, using a tape head. Operation is fully determined by a finite set of elementary instructions such as 'in state 42, if the symbol seen is 0, write a 1; if the symbol seen is 1, shift to the right, and change into state 17; in state 17, if the symbol seen is 0, write a 1 and change to state 6;' etc. In the original article ('On computable numbers, with an application to the Entscheidungsproblem'), Turing imagines not a mechanism, but a person whom he calls the 'computer', who executes these deterministic mechanical rules slavishly (or as Turing puts it, 'in a desultory manner').
>
> More precisely, a Turing machine consists of:
>
> 1. A **tape** which is divided into cells, one next to the other. Each cell contains a symbol from some finite alphabet. The alphabet contains a special *blank* symbol and one or more other symbols. The tape is assumed to be arbitrarily extendable to the left and to the right, i.e., the Turing machine is always supplied with as much tape as it needs for its computation. Cells that have not been written to before are assumed to be filled with the blank symbol. In some models the tape has a left end marked with a special symbol; the tape extends or is indefinitely extensible to the right.
>
> 2. A **head** that can read and write symbols on the tape and move the tape left and right one (and only one) cell

at a time. In some models the head moves and the tape is stationary.

3. A **state register** that stores the state of the Turing machine, one of finitely many. There is one special *start state* with which the state register is initialized. These states, writes Turing, replace the 'state of mind' a person performing computations would ordinarily be in.

4. A finite **table** (occasionally called an **action table** or **transition function**) of instructions (usually quintuples [5-tuples] : $q_i a_j \rightarrow q_{i1} a_{j1} d_k$, but sometimes 4-tuples) that, given the *state*(q_i) the machine is currently in *and* the *symbol*(a_j) it is reading on the tape (symbol currently under the head) tells the machine to do the following in sequence (for the 5-tuple models):

• Either erase or write a symbol (instead of a_j, write a_{j1}), *and then*

• Move the head (which is described by d_k and can have values: 'L' for one step left *or* 'R' for one step right *or* 'N' for staying in the same place), *and then*

• Assume the same or a *new state* as prescribed (go to state q_{i1}).

In the 4-tuple models, erase or write a symbol (a_{j1}) and move the head left or right (d_k) are specified as separate instructions. Specifically, the table tells the machine to (ia) erase or write a symbol *or* (ib) move the head left or right, *and then* (ii) assume the same or a new state as prescribed, but not both actions (ia) and (ib) in the same instruction. In some models, if there is no entry in the table for the current combination of symbol and state then the machine will halt; other models require all entries to be filled.

Note that every part of the machine—its state and symbol-collections—and its actions—printing, erasing and tape motion—is *finite, discrete* and *distinguishable*; it is the potentially unlimited amount of tape that gives it an unbounded amount of storage space" (<//en.wikipedia.org/wiki/Turing_machine>).

The Turning Machine relates directly to the functioning of a CPU, although modern CPUs do not use a read-write head on a tape, but rather have memory registers (like RAM—Random Access Memory, ROM—Read Only Memory, hard disk drives, etc.):

"The basic concept of a CPU is as follows:

Hardwired into a CPU's design is a list of basic operations it can perform, called an instruction set. Such operations may include adding or subtracting two numbers, comparing numbers, or jumping to a different part of a program. Each of these basic operations is represented by a particular sequence of bits; this sequence is called the opcode for that particular operation. Sending a particular opcode to a CPU will cause it to perform the operation represented by that opcode. To execute an instruction in a computer program, the CPU uses the opcode for that instruction as well as its arguments (for instance the two numbers to be added, in the case of an addition operation). A computer program is therefore a sequence of instructions, with each instruction including an opcode and that operation's arguments.

The actual mathematical operation for each instruction is performed by a subunit of the CPU known as the arithmetic logic unit or ALU. In addition to using its ALU to perform operations, a CPU is also responsible for reading the next instruction from memory, reading data specified in arguments from memory, and writing results to memory.

> In many CPU designs, an instruction set will clearly differentiate between operations that load data from memory, and those that perform math. In this case the data loaded from memory is stored in registers, and a mathematical operation takes no arguments but simply performs the math on the data in the registers and writes it to a new register, whose value a separate operation may then write to memory" (<//en.wikipedia.org/wiki/CPU>).

To give an example of this, consider this little "program" to add 5 + 7 and store it in memory location "2", with an Intel 6502 processor (used in Apple II, 8-bit Atari, and Commodore 64 home computers in the 1980's)—here is a string of numbers (in decimal) in machine language:

169,5,105,7,133,2

Now, in assembly language, such would read as LDA 5, ADC 7, STA 2 or "LoaD Accumulator" with "5", "ADd with Carry" "7", and "STore Accumulator" at memory location "2". Note that the Accumulator (the CPU's main variable; a 6502 also had X & Y variables) can only hold a number from 0 to 255 (the maximum range for an 8-bit computer—eight zeros or ones in binary can make a decimal number between 0 and 255, much like two decimal places can make a number between 00 and 99. This particular STA command was to store the Accumulator variable on the "zero page" —the first 256 (0-255) numbered memory addresses (and each of the 256 memory addresses can hold a number between 0-255). In order to address higher amounts of memory, two "bytes" (each with 8 of those 0/1 "bits") make up a "word" where the first byte would be multiplied by 256 and with the second byte simply added. So, had our ending instruction been 141,3,2 instead of the 133,2; the 141 STA command would place the 5+7 answer in memory location 3 x 256 + 2, or 770. Hence 6502 8-bit computers were able to address memory from 0 up to 255*256+255 or 65535: 64KB.

Such is not to get too far into the architecture and programming of computers, but to illustrate the isomorphism between the theoretical Turning Machine and a CPU. But first, history needed Claude Shannon's 1937 demonstration of isomorphism between Boolean logic and logic gates which could be realized with electronic relays and switches (Claude Shannon, "A Symbolic Analysis of Relay and Switching Circuits", *Transactions of the American Institute of Electrical Engineers*, Vol. 57 (1938), pp. 713-723):

> "the Boolean AND function is realized by connecting normally open relay contacts in series, the OR function by connecting normally open contacts in parallel. The change-over or Form C contacts perform the XOR (exclusive or) function. Similar functions for NAND and NOR are accomplished using normally closed contacts" (<//en.wikipedia.org/wiki/Relay>).

And others have sought to see how the brain could instantiate a Turing Machine, at least at some level. In their 1943 paper, "A Logical Calculus of the Ideas Immanent in Nervous Activity" (*Bulletin of Mathematical Biophysics*, Vol. 5, pp. 115-133) Warren McCulloch and Walter Pitts demonstrated how *logical* neurons (modeled on the activity of neurons, but not 100% reflective of what a neuron does) could perform logical operations, and hence be used to construct a Turing Machine. These logical operations were based on neurons having synaptic connections that can either activate or negate the activation of other neurons they are connected with. Given these connections between neurons, various configurations can make an input neuron or neurons activate an output neuron or neurons in a logical way; e.g. if we give an input for p = 1 and q = 0 and wanted the function for AND to produce an output of 0 in this case (but an output of 1 if p = 1 and q = 1)—neurons with facilitating and inhibiting synapses could construct such a input / output pathway that gives logical operator results. Such is not to claim that the brain operates like a Turing Machine; just that elegant neuron structure is complex enough to perform logical operations when connected in networks.

Indeed, by using the Turing Machine idea with a John von Neumann architecture where the same memory is used for the programs and the data (making it quite easy for a computer to modify its own program!), using relays and vacuum tubes, digital computers were realized by the 1940s, including the German Z3 (~1941, designed by Konrad Zuse, using relays); the American Atanasoff-Berry Computer (ABC, also 1941, using vacuum tubes); the British Colossus computers (1943... only used for code breaking); and 1946's famous ENIAC (Electronic Numerical Integrator And Computer—the first general purpose electronic computer). Such set the stage for the birth of AI in the 1950s.

THE TURING TEST

In his 1950 paper "Computing Machinery and Intelligence" published in a philosophy journal (*Mind*, 59 (1950), 433-60), Alan Turing proposed an "Imitation Game" in response to the question:

> "'Can machines think?' [....]
>
> Instead of attempting such a definition I shall replace the question by another, which is closely related to it and is expressed in relatively unambiguous words.
>
> The new form of the problem can be described in terms of a game which we call the 'imitation game'. It is played with three people, a man (A), a woman (B), and an interrogator (C) who may be of either sex. The interrogator stays in a room apart from the other two. The object of the game for the interrogator is to determine which of the other two is a man and which is the woman [....]
>
> Now we ask the question, 'What will happen when a ma-

chine takes the part of A in this game?' Will the interrogator decide wrongly as often when the game is played like this as he does when the game is played between a man and a woman? These questions replace our original, 'Can machines think?'" (Turing, p. 441).

This simple "Turing Test" suggests that if we can't determine whether a hidden interlocutor is a machine or a human, then the machine is intelligent. Of course, such a machine would have to know a lot about what it means to be human, as well as being intelligent! Interestingly, the machine would need to be able to deceive, and have a very broad knowledge base... and most likely, a *personality*, as well as the ability to negotiate a conversation. Nils Nilsson notes shortcomings of this test, relating:

"that even Joseph Weizenbaum's patently unintelligent 1966 conversational program ELIZA, has fooled people into thinking it is human.

For example, here is part of a transcript of a conversation with ELIZA:

ELIZA: How do you do. Please state your problem.

> I spend too much time with computers.

ELIZA: Why do you mention computers?

> You are one.

ELIZA: Does it please you to believe I am one?

>Yes.

ELIZA: You seem to be quite positive.

> I suppose I am.

ELIZA: How long have you been?

> 31 years.

ELIZA: I'm not sure I understand you fully.

> Join the club.

ELIZA: Please go on.

> You don't make too much sense yourself.

ELIZA: We were discussing you—not me" (Nilsson, p. 38).

Before ELIZA, there were a few games programmed on computers in the 1950's: Arthur Samuel's checker playing program (which learned from experience), and chess programs started to appear climactically leading to IBM's "Deep Blue" chess-playing computer to win a match against world champion Garry Kasparov in 1997—we had to wait till 2011 for IBM's "Watson" to win the TV trivia game *Jeopardy!*

THE SEARCH FOR KNOWLEDGE

In 1956, John McCarthy coined the term "Artificial Intelligence" and helped organize a Dartmouth conference that inaugurated the field academically, along with others including Marvin Minsky and Claude Shannon. At that conference, Allen Newell and Herbert Simon (with J.C. Shaw) introduced their "Logical Theorist" program that proved 38 of the first 52 theorems in Russell and Whitehead's *Principia Mathematica*. The Logic Theorist worked by using a "heuristic search"—Newell and Simon explain (as quoted in Nilsson, p. 81):

"The solutions to the problems are represented as symbol structures. A physical symbol system exercises its intelligence in problem solving by search—that is, by generating and progressively modifying symbol structures until it produces a solution structure

[....]

> To state a problem is to designate (1) a test for a class of symbol structures (solutions of the problem), and (2) a generator of symbol structures (potential solutions). To solve a problem is to generate a structure, using (2), that satisfies the test of (1)" (A Newell and H.A. Simon, "Computer Science as Empirical Inquiry: Symbols and Search," *Communications of the ACM*, Vol. 19, NO 3. pp. 113-126, March 1976).

One can imagine a computer program, aimed at playing chess by searching through ALL the possible moves in order to find the best one at a moment in play. Minksy noted in 1968 (quoted in Buchanan, p. 57):

> "The most central idea of the pre-1962 period was that of finding heuristic devices to control the breadth of a *trial-and-error search*. A close second preoccupation was with finding effective techniques for *learning*. In the post-1962 era the concern became less with 'learning' and more with the problem of representation of knowledge (however acquired) and with the related problem of breaking through the formality and narrowness of older systems. The problem of heuristic search efficiency remains an underlying constraint, but it is no longer the problem one thinks about, for we are now immersed in more sophisticated subproblems, e.g., the representation and modification of plans" (Marvin Minksy, *Semantic Information Processing*, MIT Press: Cambridge, MA (1968), *p*.9).

As Minksy suggests, "trial and error" searches could get out of hand when problem spaces explode exponentially (this could be represented by a tree branching out to ever more refined branches)—with searches sometimes going down long trails before reaching a dead-end and having to back-track to an earlier juncture to search through an alternate problem space. "Heuristics" or "rules of thumb" were added that would reduce trying out lines that were not likely to lead to a solution. Simon and Newell improved upon their "Logic Theorist" with a "General Problem Solver" in 1959, which broke larger problems into sub-problems. An ultimate goal in chess may be a check-mate, but first pawns, queens, etc. must be captured or sacrificed, etc; and heuristically, a look at previous games will show that certain strategies are dead ends. In the future, these searches would be refined with Bayesian networks (networks of beliefs and their probable connections), hidden Markov models (making local judgments without contexts), and stochastic modeling (re: indeterminate and random evolutions).

The second part of the Minksy quote above concerns knowledge representation. His own later solution was with a notion called "frames," which was also developed by Roger Schank and Robert Abelson with their "scripts"—but both of these ideas were expansions on Ross Quillian's work in the 60's on semantic networks:

> "According to Quillian, the meaning of a term is represented by its place in the network and how it is connected to other terms. This same idea is used in dictionaries where the meaning of a word is given by mentioning the relationship of this word to other words. The meaning of those other words are, in turn, given by their relationships to yet other words. So we can think of a dictionary as being like a large semantic network of words linked to other words" (Nilsson, pp. 100-101).

Frames and scripts would be organized in networks too, with connection hubs around general words, but also hubs around specific abilities—e.g. eating out at a restaurant requires not only frames representing a sandwich in relation to a drink, but "scripts" about ordering food, leaving tips, etc. (all which would have subscripts: calculating a tip, reaching for one's wallet, etc). What we see with semantic networks, frames, and scripts, is an attempt at situating intelligence in a *context* of worldly knowledge through *network* constellations.

The vast extent of worldly knowledge leads to two immediate avenues for AI research: decrease the amount of knowledge needed with task specific "expert systems" or with limited "micro-worlds"—or consider the task of data-entry of almost all knowledge:

> "An expert system is a program that answers questions or solves problems about a specific domain of knowledge using logical rules that are derived from the knowledge of experts. the earliest examples were developed by Edward Feigenbaum and his students. Dendral, begun in 1965, identified compounds from spectrometer readings. MYCIN, developed in 1972, diagnosed infectious blood diseases" (<//en.wikipedia.org/wiki/History_of_AI>)

A leading "micro-world" was developed with Terry Winograd's SHRDLU (1973)—a simulated world of a limited space with a few shaped blocks—his program parsed simple text that you could type in (e.g. "pick up a red block") and determined if it would make sense within this limited world. Such was expanded on in text-adventures (introduced with Will Crowther's Colossal Cave / Adventure from 1975)—given a description of a location (connected to other locations) and objects (that could be picked up and used) presented like an inter-active story, you could type in natural language commands and travel around in a fantasy and solve puzzles. These adventure programs often look for key verbs (like "go," "get," "look at," "open," etc) which would trigger sub-routines (a sub-program for "open" etc) that would be further modified by the nouns representing the objects (e.g. "a chest," "a door," etc).

But our world is not a micro-world, and hence in 1984 Douglas Lenat began his *Cyc* project—a database of all the facts of the world that is organized hierarchically, topped by abstract concepts, followed by core theories about space, time, causality, etc, followed in turn by specific theories about specific sciences, etc, and fanning out the many facts of the world at the bottom. Cyc now has around 3,000,000 facts and rules but Lenat believes it would need something like 200,000,000 to understand an encyclopedia (Nilsson, pp. 359-361). IBM's Watson, on the other hand, used the internet as a database to win at *Jeopardy!*

EMERGING EMBODIED ANALOGICAL AGENTS

AI research in the 1980's and 1990's revived a few research lines from earlier in the 20th century: perceptrons, cybernetics, genetic algorithms, and analogy.

Connectionism, an AI avenue that side-lines logical and semantic symbol manipulation arose from the study of "perceptrons." A perceptron, invented by Frank Rosenblatt in 1957, is an artificial neural network (recalling the work of McCulloch and Pitts), where a set of input nodes is connected to a set of output nodes. Usually every single input node is connected to every single output node (so that input node 1 would be connected to output nodes 1, 2, 3, and so on, as input node 2 would be connected to output nodes 1, 2, 3, etc.). Each connection would have a "weight" determining the strength of connection. With this weighted space, certain input configurations can be made to generate corresponding output configurations, by "training" the perceptron—if we wanted to make output node 2 activate every time we activated input node 1, we might strengthen the weights between these two. Such gets complicated with input cues that could activate more than one input node, and require a complex output from several output nodes—some weights might end up being negative, or inhibitory.

The limitations of perceptrons was elucidated by Marvin Minsky and Seymore in a 1969 book *Perceptrons* where they argued it is impossible for these artificial neural networks to learn an XOR function (although perceptrons can implement the basic NAND function). It wasn't until the 1980's that Parallel Distributed Processing or Connectionism regained stature (esp. with the advent of multi-layered arrays of nodes—input, output, and "hidden" layers that could learn the XOR function), and was used in the 1990's for character and speech recognition. Such distributed networks do not have rules or symbols, but can generate a desired output from a given input—and can even be used to predict events somewhat, given data of past events (they are used, for example, with stock-market predicting programs).

Cybernetics (and "general systems theory") is the study of self-governing systems and teleological mechanisms. Such systems can be illustrated by a thermostat—when a room gets too hot, the thermostat turns off the heat... when it gets too cold, the heat is turned on. There is a feedback mechanism that keeps the system in a sort of equilibrium. The term "cybernetics" was coined by Norbert Wiener, as used in his 1948 book, *Cybernetics, or Control and Communication in the Animal and Machine*. Feedback loops can be both negative (minimizing change, as with the thermostat), and positive (explosive growth, as with animal population growth). Keeping an environment stable means that many positive feedback loops must be held in check by negative feedback loops or mitigating factors (e.g. predators eating overly reproductive prey). Cybernetics is especially suited to the study of biological organisms (and self-organizing systems) but as a "general systems theory" it is applicable to many different fields, from physics to sociology.

Such "self-governing" systems call into question good old fashion artificial intelligence (GOFAI) and it's preoccupation with logical and symbolic thinking—bringing in more holistic view. There are no symbols involved with a thermostat—no representation of "heat"—the environment is its own best representation, as noted by Rodney Brooks in his 1990 paper "Elephants Don't Play Chess":

> "the world is its own best model. It is always exactly up to date. It always has every detail there is to be known. The

trick is to sense it appropriately and often enough" (Rodney Brooks, "Elephants Don't Play Chess," *Robotics and Autonomous Systems* 6 (1990), pp. 3-15).

In order to be intelligent, it would seem that thinking machines would require bodies of some sort, to interact with the world—much as we humans do: embodied intelligence—but not only in a body, but with respect to being situated in a complex environment—no micro-world. Much research is focused on robotics, but this brings up the related fact that intelligence is often rooted in a biology that is dependent on its environment, which means relating to other people as well—consequently theories of "intelligent agents" were developed via Judea Pearl and others' work on rational economic agents and decision theory, which studied any type of intelligence: animal, societal, etc. Bringing this together with "emergence theory" George F. Luger writes:

> "What are the main themes supporting an agent-oriented and emergent view of intelligence? They include:
>
> 1. Agents are autonomous or semi-autonomous. That is, each agent has certain responsibilities in problem solving with little or no knowledge of either what other agents do or how they do it. Each agent does its own independent piece of the problem solving and either produces a result itself (does something) or reports results back to others in the community (communicating agent).
>
> 2. Agents are "situated." Each agent is sensitive to its own surrounding environment and (usually) has no knowledge of the full domain of all agents. Thus, an agent's knowledge is limited to the tasks at hand: 'the-file-I'm-processing' or 'the-wall-next-to-me' with no knowledge of the total range of files or physical constraints in the problem solving task.
>
> 3. Agents are interactional. That is, they form a collection of individuals that cooperate on a particular task. In this

sense they may be seen as a 'society' and, as with human society, knowledge, skills, and responsibilities, even when seen as collective, are distributed across the population of individuals.

4. The society of agents is structured. In most views of agent-oriented problem solving, each individual, although having its own unique environment and skill set, will coordinate with other agents in the overall problem solving. Thus, a final solution will not only be seen as collective, but also as cooperative.

5. Finally, the phenomenon of intelligence in this environment is 'emergent.' Although individual agents are seen as possessing sets of skills and responsibilities, the overall cooperative result can be viewed as greater than the sum of its individual contributors. Intelligence is seen as a phenomenon resident in and emerging from a society and not just a property of an individual agent" (Luger, pp. 17-18).

This notion of emergent intelligence, or at least emergent behavior was evident in John Conway's 1970 computer "game" called "Life." Here is how the game plays out:

> "The universe of the Game of Life is an infinite two-dimensional orthogonal grid of square *cells*, each of which is in one of two possible states, *alive* or *dead*. Every cell interacts with its eight *neighbors*, which are the cells that are horizontally, vertically, or diagonally adjacent. At each step in time, the following transitions occur:
>
> 1. Any live cell with fewer than two live neighbors dies, as if caused by under-population.
>
> 2. Any live cell with two or three live neighbors lives on to the next generation.

> 3. Any live cell with more than three live neighbors dies, as if by overcrowding.
>
> 4. Any dead cell with exactly three live neighbors becomes a live cell, as if by reproduction.
>
> The initial pattern constitutes the *seed* of the system. The first generation is created by applying the above rules simultaneously to every cell in the seed—births and deaths occur simultaneously, and the discrete moment at which this happens is sometimes called a *tick* (in other words, each generation is a pure function of the preceding one). The rules continue to be applied repeatedly to create further generations" (<//en.wikipedia.org/wiki/Conway's_Game_of_Life>).

Depending on the initial configuration, one's pattern can die-out, "explode," or repeat in a cycle. This game demonstrates insights close to another avenue of AI research: Artificial Life and Genetic Algorithms which were pioneered by in the early 60's by John Holland (who also invented the AI programming language LISP):

> "With genetic algorithms and artificial life we evolve new problem solutions from components of previous solutions. The genetic operators, such as *crossover* and *mutation*, much like their genetic equivalents in the natural world, work to produce, for each new generation, even better potential problem solutions. Artificial Life produces its new generation as a function of the 'quality' of its neighbors in previous generations" (Luger, p. 29).

Evolutionary algorithms work much like natural selection, and like natural selection, they may be hard to predict, non-logical, and non-symbolical:

> "In artificial intelligence, an evolutionary algorithm (EA) is a subset of evolutionary computation, a generic population-based metaheuristic optimization algorithm. An EA uses

> some mechanisms inspired by biological evolution: reproduction, mutation, recombination, and selection. Candidate solutions to the optimization problem play the role of individuals in a population, and the fitness function determines the environment within which the solutions 'live' [....] Evolution of the population then takes place after the repeated application of the above operators" (<//en.wikipedia.org/wiki/Evolutionary_algorithm>)

In another line of AI research, Thomas Evans' 1963 thesis explored analogical reasoning with a program that could perform adequately on geometric analogy tests (Nilsson, p. 96). The analogical approach has been rare in the history of AI, but has been resurrected in cognitive science in general, thanks in part to the work of people like George Lakoff and Mark Johnson—*Metaphors We Live By* (1981) and *Philosophy in the Flesh* (1999); natural selection and analogy will be addressed further in later chapters.

Perhaps all four avenues addressed in this last section (perceptrons / connectionism, cybernetics / embodied agents, artificial life / natural selection, and analogy) would align with what have been called the "scruffies" (who are more concerned with context interaction) as contrasted with the "neats" (who focus more on symbolic-logical searches of problem spaces). One may take sides, but it should be remembered that human thinking can be both scruffy and neat: context-relational and mathematical-logical.

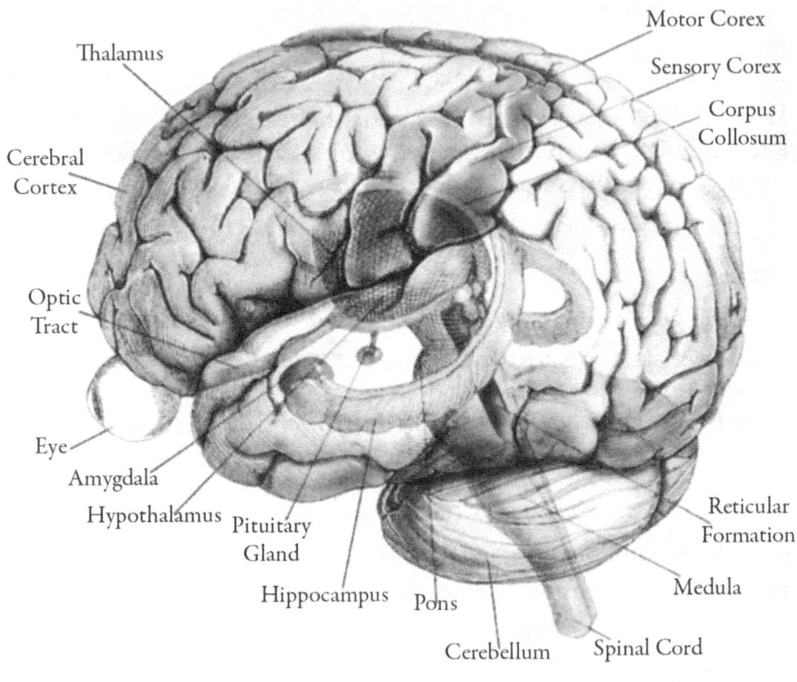

See-through Brain - adapted from Bloom & Lazerson (1988)

Chapter 25

Cognitive Categories

A SURVEY OF SURVEYS

Contemporary cognitive science textbooks are commonly divided into a few familiar categories; although they may differ in detail and emphasis, time and again these books refer to key divisions in the field—much like Kant isolating various faculties of the mind (apperception, imagination, understanding, etc.) The most common categories include an approach to: the brain, attention, memory, knowledge, perception, language, problem solving, and decision making. This chapter will outline these categories—not to list all the relevant theories and their histories, but to roughly situate further discussion of modeling the mind.

THE BRAIN: MODULARITY

A key focus of cognitive psychology is on the biological anatomy of the brain and how it relates to behavior. There have been a variety of approaches to studying the physiology of the brain, including: dissecting animal brains; examining the brains of dead humans; measuring the distributed electrical activity of the brain with Electroencephalograms (EEGs); brain imaging (CAT, MRI, PET scans, etc); and studying the effects of brain lesions (damage to particular areas having specific behavior effects). What these studies have shown is that despite the prevalence of neurons throughout the nervous system and brain, the brain can be divided into parts, a modularity with localization of functions. For example, although one can see a symmetry between the left and right *hemispheres* of the brain (where, in a reversal, the left side of the brain is associated with the right side of the body, and the right hemisphere with the left side of the body), the left side is often associated with logic and language abilities, while the right side may emphasize spatial and creative thinking. The brain can be regionalized from front to back as well, with the forebrain, midbrain, and hindbrain—which on an evolutionary scale might relate loosely to our human, mammalian, and reptilian heritage.

There are *roughly* one trillion neurons in the human brain with a quadrillion synaptic connections. A neuron is a cell with a nucleus, dendrite branches for receiving signals from other neurons, and axon branches for transmitting signals to other neurons via synaptic connections on the ends. When one neuron "activates" it sends a signal down its axons to stimulate other neurons. When a neuron receives enough stimulation from other activated neurons, it activates, and sends a signal down its axons to other neurons as well and then becomes temporarily harder to activate again. Moreover, the more that one neuron activates another, the stronger the connections between the two neurons become. There is more to the biology of the brain than neurons, but most theoretical modeling of brain functioning has focused on the electrical-chemical-mechanical aspects of neurons, and how their network organization and "spreading activation" can be used to understand memory recall and stimulus-response behavior learning.

Neurons of various sub-flavors are connected in clustered ways that can be seen as differentiating various "organs" of the brain. For example, the division of the brain into forebrain, midbrain, and hindbrain can be further differentiated by location and associated function; here are a few sections (modified from a table in Robert J. Sternberg, *Cognitive Psychology 4th ed.*, Thomson-Wadsworth: Belmont, California (2006), pp. 38-39):

- **Forebrain** (from outer to inner):
    - Cerebral Cortex (outer layer of the cerebral hemisphere: perception and reasoning)
    - Limbic Systems (inner edge of cortex including hippocampus, amygdalae and septum: emotion and memory)
    - Basal Ganglia (network at base of forebrain: motor/muscle control)
    - Thalamus (between cortex and midbrain: central relay associated with attention)
    - Hypothalamus (below Thalamus, above brainstem: automatic body regulations—body temperature, etc)
- **Midbrain**
    - Superior Colliculi (top of midbrain: visual reflex)
    - Inferior Colliculi (bottom of midbrain: hearing)
    - Reticular Activating System (extends into hindbrain: arousal)
    - Red Nucleus (back of midbrain: motor coordination)

- Hindbrain
    - Cerebellum (bottom of brain: balance, coordination)
    - Pons (on brain stem: wakefulness, dreams)
    - Medulla Oblongata (lower brain stem: automatic functions—heartbeat, etc.)

What we see in the above differentiation is something like a tree that branches out, with the inner-trunk being associated with unconscious automatic regulations like maintaining a heartbeat shading into more conscious-related aspects from feeling balance to focusing attention—on into more self-aware aspects of thinking as with moving muscles, remembering, reflecting, and planning, which operate more towards the outer parts of the splayed branches. Given the nature-nurture debate, it would seem that the "foundational" aspects of the unconscious automatic systems at the core of the brain would most definitely be biologically hardwired by DNA; whereas the outer periphery of the brain, associated more with memory, representation, and planning might be more plastic, or at least less structured by DNA in a detailed way.

ATTENTION: FILTERED FOCUS

The category of attention possibly relates to Kant's concept of apperception—in that it also relates to the experience of consciousness. This category designates the border between unconscious automatic functions of the brain, and more lucid aspects of our awareness where we feel we control our own thinking. Not only are certain functions like our hearts' beating and our unintentional breathing automatic; there also seems to be some sort of "bottleneck" filtering of our sensory perceptions and possible memory recollections that lets us focus on what is important. We don't have a flood of irrelevant memories at all times, and we are able to see important visual cues fairly quickly. With the category of attention, we are no longer looking just at the bodily brain organs that might be associated with it, but with the very functioning of the aspect itself; and given that it bridges the conscious with the unconscious, scientists have had to develop experiments, often involving the measurement of the timed tasks, in order to allow us to indirectly get a grasp of what is going on within the brain as we think through our world.

For example, with *priming* (Sternberg, p. 64), prior stimuli that are associated with a new stimulus can facilitate the recognition of the new stimulus: a fire truck may be recognized as such more quickly and accurately if there is a context of connections like a fire, a siren, etc. This, along with the notion of how neurons activate given above, suggests that there are unconscious connections in the brain of related ideas (perceptions, words, etc.) where the occurrence of certain cues or ideas can facilitate the activation and / or recognition of related ideas.

There seems to be a continuum between controlled behavior and automatic processes of the brain / body. But also, what was once controlled behavior may become habit, as we may no longer be receptive to, vigilant of, or searching for features of our environment that we found to be irrelevant, or developed reactions to that do not require further attention. But there is more to the category of attention, in that it designates that which integrates our experiences—the ability to see that one experience compares in some way with another—the bridge of connecting ideas, esp. new ideas that have not been habitually associated with others. But experience is not threaded through a single node, as with the read-write head of a Turing machine—humans can multi-task. We have the ability to focus on a conversation at a party, with much background noise... while also being able to notice when someone calls our name from across the room (see Friedenberg, Jay and Gordon Silverman, *Cognitive Science: An Introduction to the Study of Mind*, Sage Publications, Inc., Thousand Oaks, California (2006), p. 115). Our ability to filter out irrelevant items and focus on more than one thing at a time is limited—we can only keep track of so much at a time. This notion that we can only focus on so much at a time coupled with the priming effect mentioned above leads us straight to our next cognitive category: memory. Such a connection between memory and attention is furthered by the "Stroop effect"—discovered by John Ridley Stroop (see Sternberg, p. 96)—where color words like "brown" and "blue" take longer to identify if they are presented in a color other than that which they name—it would seem that our perceptions can compete with each other, and conflicting information takes more time to process.

MEMORY: TRANSITORY TO PERMANENT

Memory, the storage of information, knowledge, procedures, etc, in the brain is a critical category for cognitive science—if organized in certain ways, it could be said to explain much of our thinking—for memory is not simply a passive receptacle for our experience, but organizes our experience in such an abstract way that it becomes useful to our intelligence. Often there are three different types of memory discussed in the literature: sensory memory, short-term / working memory, and long-term memory.

Sensory memory brings to mind the phenomena of "photographic" or "flashbulb" memory—the ability to remember what you've seen in complex concrete detail. This visual memory would be the "iconic" type of sensory memory, which for most of us lasts only a couple hundred milliseconds, or tenths of a second, as the details slip away quickly. Auditory memory is of the "echoic" type and lasts longer, around 2 seconds (see Ronald T. Kellogg, *Fundamentals of Cognitive Psychology*, Sage Publications, Inc., Thousand Oaks, California (2007), p.98).

Whereas our sensory memory decays quickly, and irrelevant sense-data is filtered out in our perception via unconscious processes, our short-term or working memory involves our intent and attentive focus and ability to hold on to a few "items" in memory: typically seven (Friedenberg, p. 130). These seven "items" or so can carry more information though, as we can chunk parts into wholes—we might be able to turn over seven parts in our mind, or seven separate wholes. Along with these few items under our attention, we also seem to have a "visual-spatial sketch pad" and "phonological loop" (Kellogg, p. 120) where one can rehearse their perceptual representations, and remember impressions for around 30 seconds (Kellogg, p. 123).

Long-term memory can be divided into declarative and procedural memory. Declarative memory or memory of concepts, facts and events can be further divided into semantic and episodic memory. Semantic memory includes concepts and facts and episodic memory is about specific spatial-temporal events experienced. Procedural memory concerns our learned skills, conditioning, and habituation.

There is of course an interplay between sensory, working, and long-term memories—such as with the distinction between recall and recognition. Recognition is much easier, and may be seen as identifying an item in sensory memory because certain cues will activate an idea via the association of those cues with that item in long term memory—whereas recall requires that one actually produce a fact, word or other memory item given other types of association (Sternberg, p. 158). We rarely recall items in 100% detail, as our long-term memory encoding is often one of abstraction, shaped by "schemas"—"A schema is a set of organized concepts that provides expectations about the world" (Kellogg, p. 180). That is, we have clusters of associated concepts that we use when recognizing items in our perception, which become aids in generating reconstructed memory as well. I may know that a car has tires, so when I try to draw a car that I saw recently, tires will also be on my mind (even if the car I briefly saw had no tires!—errors in memory can arise due to our expectations and stereotypical thinking, but also due to wishful thinking, suggestibility, bias, etc).

KNOWLEDGE: HIERARCHICAL NETWORKS

Knowledge is a type of long-term memory with a focus on concepts and schemas—which:

> "are often organized hierarchically in subordinate, basic, and superordinate categories. An example of the hierarchy of natural kinds would be *robin* (subordinate), *bird* (basic), *animal* (superordinate)" (Kellogg, p. 185).

Here a *robin* might be seen as the most typical bird, more typical than a penguin or turkey, and hence might be thought as more like a "prototype." As with Kant suggesting that knowledge is a web or network of concepts, we can see that a semantic network (a network of concepts, with each concept being associated and connected to other concepts and features relevant to it) is also a way of representing knowledge. But we should not limit ourselves to just considering concepts, as we also have constellations of concepts that are integrated with our other procedural skills and abilities—and such a widening of scope is covered by artificial intelligence terms like "scripts" and "frames." All our types of memory can be integrated with other similar ways of organizing our perceptions and ideas, such as with mental maps and propositional knowledge.

PERCEPTION: CONCEIVING THE SENSIBLE

Concepts and schemas can also play a role with perception and pattern recognition, as what we sense is organized via perception into what we can understand—this may happen in the interplay between mental abstraction and concrete sensation:

"Top-down or conceptually driven processes reduce the need to sample all of the information available in the environment by providing the perceiver with expectations. Simultaneously, bottom-up or data-driven processes analyze the edges, lines, areas of light and dark, colors, sounds, and other physical features available briefly in sensory memory. These processes pick up the features needed to confirm or refute expectations. Through such simultaneous processing from both the bottom up and the top down, people can perceive the features of the environment with remarkable quickness and accuracy" (Kellogg, p. 43).

We have the ability to recognize details structurally and analytically, yet we also have more gestalt and holistic perceptive abilities such as the ability to recognize faces. Humans have binocular depth perception (where images seen with the left eye are processed in the right side of the brain, and the left brain processes images from the right eye). Pattern recognition, distinguishing figure from ground, the ability to recognize an object from various perspectives and distances / sizes—and the ability to rotate images in our minds mentally interlock with our memory and knowledge representations (just as we might define one concept by its relation to other concepts and features—various features of our perceptions may connect the sensed to the concept: visual, auditory and other cues can trigger the recognition of the perception into a concept.

LANGUAGE: SEMANTICS, SYNTAX, CONTEXT

A good deal of this book has dealt with knowledge, reasoning, and also language, so reference will be made here to what is especially relevant about language to cognitive science.

Language has at least five characteristics that cross the various cultures it is used in; that is, despite superficial differences, some characteristics appear to be universal (see Friedenberg, pp. 276-277 and Sternberg, p. 312)—language is:

1. *Communicative*: Language allows for individuals to produce, transmit, and comprehend information.

2. *Arbitrary*: Language uses symbols that have no necessary relation to their meanings.

3. *Structured*: Symbolic patterns may be combined by use of rules.

4. *Generative*: Language users can create and produce novel utterances.

5. *Dynamic*: Languages can evolve.

Some animals have demonstrated linguistic abilities to greater or lesser extents, and primates have been taught to use a limited vocabulary with sign language, but demonstrate very little if any syntax. Such would seem to indicate that primates have the ability to use concepts as they are relevant to attention, memory, knowledge, and perception—but generative syntax / grammar may be a specifically human aspect of language use:

> "They [primates] know that some aspects of word order affect meaning. However, most primates tend to repeat the sentences they were taught or produce only small variations on them" (Friedenberg, p. 281).

Such is beyond the cooing and babbling stage in human child language acquisition, but not far beyond the one and two-word stages, and not at all the basic adult sentence structure acquired by humans by about age four (Sternberg, p. 333). How much of language is universal and biologically inherited, and how much is relative and socially acquired, the nature-nurture controversy, is a matter of much debate. In either case, individual pragmatic utterances would have to be coherent in a given context that would further clarify meaning and situate a given speech act.

Two approaches to understanding grammar include phrase structure grammar and transformational grammar. Phrase structure grammar looks at how sentences (as basic linguistic units of communication) can be divided into noun phrases and verb phrases (where verb phrases can be found to include its own noun phrase, etc); while transformational grammar looks at how related sentences can be translated from one to another, as with turning a declaration into a question, making an active sentence passive, or expressing the same meaning with different phrase structures, etc.

Phrases can be broken down further into their words; words can in turn be reduced to the minimal unit of meaning—a morpheme (recall the various parts of the word "de-con-struct"); and morphemes can be broken down into spoken phonemes and written marks With whole concepts, and in contrast to grammar / syntax, we have semantics, which deals with the meaning of lexical word items (recall the semantic networks mentioned above). Semantics seem to be related to an area in the left side of the brain towards the rear called "Wernicke's area"—since aphasias or damage to that location in the brain consistently has caused speech with grammatical fluidity but with little meaning. Syntax / grammar may be related to "Broca's area," similarly in the left hemisphere but towards the front, since damage to Broca's area has caused meaningful but laborious a-grammatical speech. Such a "more-to-the-front = grammar / syntax and more-to-the-rear = semantics" orientation suggests that the back of the brain may be more structured for associations and network connections; whereas the parts near the front may be engaged in more temporally unfolding and progressive thought. Such may be in line with the frontal lobe's connection with being able to think about the future, make plans, etc. This brings us to the next section on problem solving.

PROBLEM SOLVING: ANALOGIES OUTSIDE THE SEARCH BOX

Although our brains are full of wandering thoughts and undirected thinking, getting something done often requires being pragmatic, and coming up with a solution to a problem. Such a problem is "well-defined" if it has an initial state, a goal state, subgoals, and a way to get from one state to another: an operator—all these comprise a problem space. Such could be represented by a tree where a single starting point splits off to several other states, which in turn split into further possibilities; and with one path through these branching possibilities leading to a goal state (Kellogg, pp. 246-247; Sternberg, pp. 397-398). Such problem space searching can be enhanced with heuristic rules of thumb—strategies that sometimes lead to a solution. Playing games like chess or checkers, or proving mathematical formulas are examples of activities that can incorporate well-defined problem solving strategies.

Ill-defined problems lack an initial state, goal, subgoals, or operators. For example, if you want to paint an impressionist portrait, or write a novel—although you may have a goal, you don't really have a problem space to search (one is reminded of an image of thousands of monkeys chained to typewriters for decades, with one finally producing the play "Hamlet"—but even this would not be a really verifiably achieved goal). Such ill-defined problems may require a gestalt insight along with a productive thinking that goes beyond a reproductive thinking based on existing knowledge associations. Such creative solutions can be facilitated by going through four stages: preparation (studying the problem and possible solutions); incubation (putting the problem aside and letting it "gel" in the unconscious); illumination (an "aha!" moment where a possible solution arises from the unconscious); and finally verification (checking if the solution is satisfactory)—see Kellogg, pp. 268-269.

Creative problem solving may require surmounting typical obstacles: fixations, mental sets, or entrenchment—one must learn to not always think inside the box—a solution may be surprising and unexpected, so one must think outside the normal route, especially if one wants to build a better mouse trap. A strategy that can work, since some problems are isomorphic, is to try to find a problem similar to the one you are trying to solve, and see if the solution to the similar problem has an analogical solution to the one at hand—that is by mapping features of one problem to features of an analogous problem may yield a new solution.

Problem solving brings intelligence to mind, and in turn reasoning. Classically, we have two modes of reasoning that have been studied extensively: deductive and inductive reasoning. We've covered these modes elsewhere; suffice it to remind the reader that a logical deduction is one where the conclusion is already implicit in the premises of an argument. E.g. If it is day it is light. It is day. Therefore it is light. Similarly with syllogisms: All men are mortal. Socrates is a man. Therefore Socrates is mortal. Induction on the other hand offers to make predictions based on past observations—such may result in a highly probable conclusion, but not an absolutely certain one (as new observations can contradict prior ones). Sadly, too many cognitive scientists (and philosophers!) have forgotten the structure of Kant's Transcendental Deduction: if something is *necessary* prior to the truth of something we know to be true, then that prior necessity is true as well: an implicit truth may be deduced from the explicit truth indirectly.

DECISION MAKING: AGGREGATING FALLACIOUS HEURISTICS

Heretofore the discussion of cognitive science has focused on the individual powers of the brain. Humans are social beings though, and a transition from understanding individual behavior to getting a clue about social aggregate behavior has been sought for the sake of better understanding economics, politics, medical epidemics, etc. An early questionable approach to such was to see people as rational agents trying to maximize their gains or to achieve their goals: often pragmatically seen as maximizing utility. However, later approaches began to point out that humans are not 100% rational and objective when facing uncertainties; they bring subjective emotions and idiosyncratic thinking to their striving towards goals: decisions are skewed from rationality by biases, probability heuristics (the representative heuristic, law of small numbers, gambler's fallacy, conjunctive fallacy), the availability heuristic (basing decisions on available, but not necessarily the best info), the simulation heuristic (basing quantitative decisions on a qualitative mental model), making illusory correlations (bad analogies), being overconfident (and hence not balancing one's opinion against others')—and then concluding with a hindsight bias (where one believes they would have been correct in hindsight, since they now know the facts)—See Sternberg, pp. 441-454; Kellogg, pp.295-308. All these flawed thought processes and more skew decision making from rational self-satisfaction efficiency at the individual level.

Such concludes the present outline of the categories of cognitive science. Throughout, we've seen an "analogy" between the networked clusters of neurons in the brain, and the network constellations of concepts, schemas, and scripts. Such is to note that since memory can include our skills, strategies, techniques, etc, that much of what we call thinking is something like *applied memory*. Not only is memory applied to thinking, but also to perception as well, as our expectations can shape what we actually see (consider, e.g. an optical illusion)—but we remain able to *work* with our memory, feeling out associations, rehearsing actions and perceptions: we are conscious as well, and all the more so when the habit of memory fails to prepare on for the unexpected.

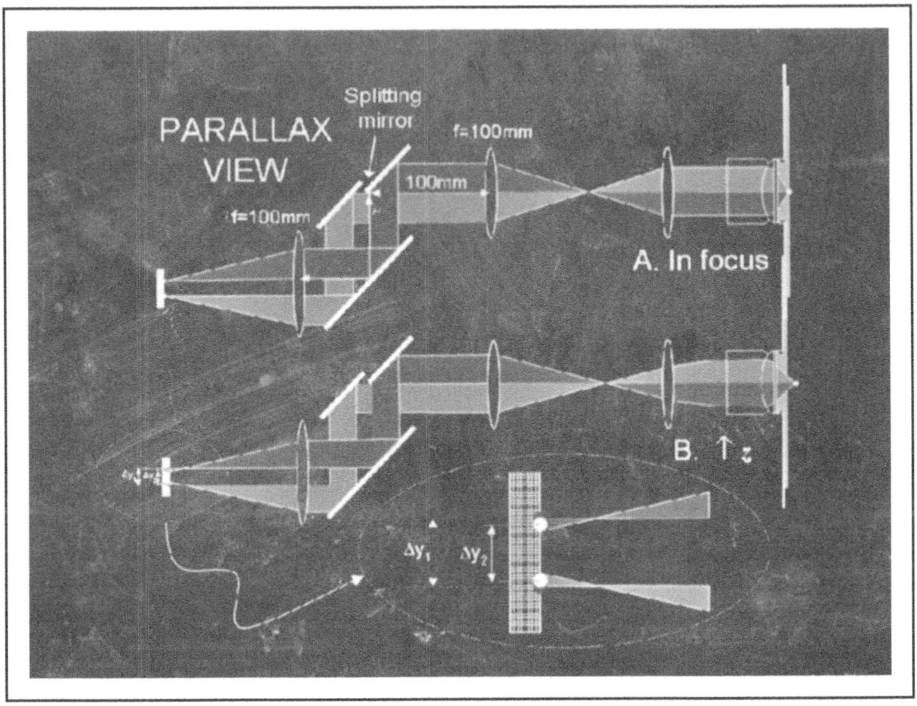
Splitting mirror
PARALLAX VIEW
f=100mm
100mm
f=100mm
A. In focus
B. ↑ z
Δy1
Δy2

Chapter 26

Žižek & Pinker: Post-Structuralism vs. Cognitive Science

TWO PERSPECTIVES

This chapter will examine two contemporary thinkers: Slavoj Žižek—a leading intellectual in the continental tradition whose thinking is relevant to and informed by postmodernism and Jacques Derrida's deconstruction; and Steven Pinker—a popular intellectual and psycho-linguist representative of some current trends in cognitive science, esp. those stemming from the work of Noam Chomsky. As the second part of this book, concerning the history of philosophy leading up to our contemporary understanding of artificial intelligence and deconstruction, ended with chapters on Derrida and Chomsky, this third part continues on to the contemporary debate between thinkers in traditions stemming from these two thinkers and their various schools of thought. Here we will see the post-structuralist Žižek examining and critiquing various aspects of cognitive science, and the cognitive scientist Pinker taking on what he sees as the failings of post-structuralists. While Žižek draws on a vast plurality of perspectives on contemporary philosophical cognitive science, finding insights, blind-spots, and inconsistencies, Pinker takes on the implications of what he sees as a cultural constructivism that doesn't wash up scientifically, strongly taking the nature side in the nature-nurture debate.

SLAVOJ ŽIŽEK: SUBJECTIVITY INSIDE AND OUT

Slavoj Žižek is a thinker in the continental tradition of philosophy heavily influenced by Hegel, Marx and the structuralist psychoanalyst Jacques Lacan. Again, his thinking is intertwined with deconstruction and postmodernism, and as such, his engagement with cognitive science is very relevant to this book's project. He is somewhat of a "celebrity philosopher" with a couple of films that feature him and this is somewhat appropriate given that his thinking often references popular culture—his culture critique often uses Hegel, Marx, and Lacan to interpret culture, but he also uses popular culture to illuminate the theories of these thinkers. As he is influenced by deconstruction and postmodernism, he shuns theories that try to explain everything, finding items that such theories cannot account for. He is an atheist and sympathetic with communism, identifying himself as a radical left thinker, and hence connects much of his philosophical interpretation with politics. Such is illustrated in his book *The Parallax View*, where his examination of the intertwining of objectivity and subjectivity is placed in a section in the middle of the book, between an initial section on religion and a concluding section on politics—and these three main sections are divided by two short interludes. Such an organization of the book illustrates a structure of the individual subject situated in the social milieu. His writing style jumps from insight to insight in a sort of non-linear, patch-work fashion that develops his themes using one illustrated encounter with a thinker after another.

Žižek introduces his take on the parallax view:

> "The standard definition of parallax is: the apparent displacement of an object (the shift of its position against a background), caused by a change in observational position that provides a new line of sight. The philosophical twist to be added, of course, is that the observed distance is not simply 'subjective,' since the same object that exists 'out there' is seen from two different stances, or points of view. It is rather that, as Hegel would have put it, subject and object are inherently 'mediated,' so that an 'epistemological' shift in the subject's point of view always reflects an ontological shift in the object itself. Or—to put it in Lacanese—the subject's gaze is always-already inscribed into the perceived object itself, in the guise of its 'blind spot,' that which is 'in the object more than object itself', the point from which the object itself returns the gaze. 'Sure the picture is in my eye, but I am also in the picture'" (Slavoj Žižek, *The Parallax View*, MIT Press: Cambridge, Mass (2006), p. 17).

Žižek also mentions, "another case of Parallax where the two elements can never meet precisely because they are one and the same element in two different spaces" (Žižek, p. 159). Here, he is situating the relationship between subjectivity and objectivity; subjectivity being variously related to "inner space [...] ego [...] the core of my being [...]" (Žižek, p. 148), and "*agency*" (Žižek, p. 150), but which is beyond the scope of any one of these notions alone. Moreover, Žižek is interested in both conscious and unconscious subjectivity (if one could imagine the latter). In the intertwining of objective and subjective, Žižek discusses fantasy:

> "the ontological paradox, even scandal, of the notion of fantasy lies in the fact that it subverts the standard opposition of 'subjective' and 'objective': of course fantasy is by definition not 'objective' (in the naive sense of 'existing independently of the subject's perceptions); however it is not

> 'subjective' (in the sense of being reducible to the subject's consciously experienced intuitions either). Fantasy, rather belongs to the 'bizarre category of the objectively subjective—the way things actually, objectively seem to you even if they don't seem that way to you'" (Žižek, p. 170).

The point being developed here is that what we deem objective, as subjects, may not be indeed objective, as our perspectives are colored by our unconsciousness. For example, we may recall a certain past experience incorrectly, based on wishful thinking, or because of the wording or the way in which a person asked us to recall it—even visually, an ambiguous appearance may seem to be viewed in one way rather than another just by suggestion ("do you see the lamp stick?" vs. "do you see the two faces?") Žižek connects this with Lacan and Freud:

> "This is also one way of specifying the meaning of Lacan's assertion of the subject's constitutive 'decenterment': its point is not that my subjective experience is regulated by objective unconscious mechanisms which are 'decentered' with regard to my self-experience and, as such, beyond my control (a point asserted by every materialist), but, rather, something much more unsettling—I am deprived of even my most intimate 'subjective' experience, the way things 'really seem to me,' that of the fundamental fantasy that constitutes and guarantees the core of my being, since I can never consciously experience it and assume it" (Žižek, p. 171).

> "Is not the status of unconscious fantasy nonetheless, in an unprecedented sense, phenomenal? Is not *this* the ultimate paradox of the Freudian Unconscious—that it designates the way things 'really appear' to us, beyond their conscious appearance? [....] it is only here that the subject is 'no longer a master in his own house'—in the house of his (self)appearances themselves" (Žižek, p. 172).

It would appear that what we take for granted in our primal experience of the world, our subjective experience from our individual points of view, the way the world appears to us, may be distorted... our subjective awareness, consciousness itself, may be distorted as compared to a "deeper" subjectivity—our unconscious subjectivity:

> "At its most radical, the Unconscious is the inaccessible phenomenon, not the objective mechanism that regulates my phenomenal experience" (Žižek, p. 171).

In a similar "at its most radical" vein, in describing the relation between objective and subjective, Žižek goes on to claim:

> "quantum physics confronts us with the gap between the Real and reality at its most radical: what we get in it is the mathematized Real of formulas which cannot be translated into ontologically consistent reality [....] quantum physics is scientific formalization at its most radical, formalization without interpretation. Is it then not accurate to say that quantum physics involves a kind of reversal of Kantian transcendental philosophy?" (Žižek, p. 172).

That is... with Kant's transcendental deductions of the faculties necessary for logically judged experience we have no direct access to an "object = x" on the other side of immediate perspectival experience; whereas with quantum physics, the introduction of subjective experience ties in with undecidable observations (the inability to observe both position and momentum at the same time; wave-particle duality, etc)—and any physics would be hard-pressed to explain subjective experience itself without reference to something that would be by definition not objective—and hence not physical. However, Kant largely limited his inquiry to that of one subject: we may not be able to directly access the thing in-itself without a detour through our senses, but several subjects can verify aspects of the world via something like triangulated and compared experiences: verification; and the same could be said of one subject who examines the world from diverse perspectives. And in extension we might know subjectivity better by examining it from several angles, which is central to Žižek's discussion here. In fact, Žižek identifies and discusses various theoretical attitudes towards subjectivity (subjectivity as illusion or subjectivity as unexplainable—each here in three flavors):

> "each of the three main 'antihumanist' and / or 'antisubjectivist' positions (cognitivist-biologist reductionism: the dismissal of subjective self-experience itself as mere 'user illusion'; the Heideggerian position: the essence of man is not reducible to subjectivity, there is a more primordial dimension of being-human; the deconstructionist 'decentering' of the Subject: the subject emerges out of presubjective textual processes) is accompanied by an assertion of subjectivity (those cognitivists, from Nagel to Chalmers, who argue for the irreducible / inexplicable character of experience; the [revival of the] the standard transcendental-philosophical defense of the irreducibility of [self-]consciousness through the critique of its reflexive model: there is a dimension of self-acquaintance which precedes reflexive self-recognition in the other [Dieter Henrich and his school]; the Freudian subject as rethought by Lacan: the nonsubstantial *cogito* is the subject of the unconscious" (Žižek, pp. 174-175).

When trying to account for consciousness, a crucial although not exhaustive aspect of subjectivity, many cognitive science thinkers get stuck in a paradoxical loop:

> "Different versions of the emergence of consciousness, from Dennett to Damasio, all seem to 'get stuck' at the same paradox: that of a certain self-propelling mechanism, of a closed loop of self-relating, which is constitutive of consciousness: they all pinpoint this paradox, trying to describe it as precisely as possible, yet they seem to miss its proper formulation,

> and thus get lost in vague metaphors or outright inconsistencies. The wager of chapter 4 of this book, the hypothesis it endeavors to substantiate, is that this missing concept—a kind of absent Cause of cognitivist accounts—is none other than what German Idealism called self-relating negatively and Freud called 'the death drive'" (Žižek, p. 177).

But is consciousness even necessary for these congnitivist mechanisms to work? Would a "dead machine" need any drives whatsoever?:

> "What strikes us is how evolutionist or cognitivist accounts always seem to stumble over the same deadlock: after we have constructed an artificial intelligence machine which can solve even very complex problems, the questions crops up: 'But if it can do it precisely as a machine, as a blind operating entity, why does it need (self-)awareness to do it?' So the more consciousness is demonstrated to be marginal, unnecessary, nonfunctional, the more it becomes enigmatic—here it is consciousness itself which is the Real of an indivisible remainder" (Žižek, p. 177).

Is consciousness itself the Real? And if it is, how do we delimit it from reality in general?

> "this spectral voice which we hear in our interior reality, although it has no place in external reality, is [for Lacan] the Real at its purest [.... The] procedure of 'eavesdropping' on the brain's digital crackle with electrodes (where computers use zeros and ones, neurons encode our thoughts in all-or-nothing electrical impulses) and transmitting the signals to a computer that can read the brain's code and then use the signals to control a machine already has an official name: brain-machine interface. Further prospects include not only more complex tasks (for instance, implanting the electrodes into the language centers of the brain, and thus

> transmitting a person's inner voice to a machine via wireless, so that one can speak 'directly,' bypassing voice or writing), but also sending the brain signals to a machine thousands of mile away, and thus directing it from a distance. And what about sending the signals to somebody standing nearby with electrodes implanted in his hearing centers, so that he can 'telepathically' listen to my inner voice? The Orwellian notion of 'thought control' will thus acquire a much more literal meaning" (Žižek, p. 192).

Such biological augmentation with machinery is on the cyborgology verge of the "post-human." Despite such technological advances that extend our reach and complicate the integration of our consciousness with our environment, it would seem we may need to differentiate what a "self" is from what it is not:

> "materialist reductionists who claim that 'there really is no self' are right, but they nonetheless miss the point. At the level of material reality (inclusive of the psychological reality of 'inner experience'), there is in effect no Self: the Self is not the 'inner kernel' of an organism, but a surface-effect" (Žižek, p. 206).

Again:

> "our immediate *inner* conscious (self-)experience is by definition a process that takes place on a *surface* at the level of appearance, and when we try to account for it in neuro-biological terms, we do it by constructing, from an *outside* view, a neuronal process that can generate such an experience [....] The conclusion to be drawn from this is the one drawn long ago by Francisco Varela: consciousness (awareness) is a matter not of inside, but of the 'interface,' of the surface-contact between inside and outside. It is this convoluted relation between Inside and Outside that, in effect, undermines the standard notion of the Cartesian subject as a *res cogitans*

> (thinking substance): it brings home the fact that the subject, precisely, is *not* a substance" (Žižek, p. 223).

Žižek also discusses Antonio Damasio and the difference between a non-changing core-experiencing self and an evolving narrative auto-biographical self. Such is to situate the role of subjectivity when making a free choice. So much of what goes into making a choice has to do with our personal histories; we grow a soul, as it were, and who we are and what we choose to do may have as much to do with our personal histories as they have to do with our biological history as a species. Much of our personal experience is in relation to others—our personalities may be highly inter-subjective in their development; might our phenomenal conscious awareness be inter-subjective as well?—"the thoroughly *intersubjective* character of my inner most subjective experience" (Žižek, p. 178). But how does this core-awareness become self-aware via its narrative auto-biography? Žižek suggests this comes by way of a crisis in our habitual relationship between ourselves and our environment (such as going to a foreign country, far out of one's comfort zone)—this forced schism between our status quo connections with the familiar creates an alien environment we cannot immediately identify with:

> "the exposure of this constant to accidents, to contingent encounters, to otherness; we become 'aware' of ourselves through external shocks which threaten homeostasis" (Žižek, p. 210).

> "we should bear in mind the basic anti-Darwinian lesson of psychoanalysis repeatedly emphasized by Lacan: man's radical and fundamental *dis*-adaptation, *mal*-adaptation, to his environs. At its most radical, 'being-human' consists in an 'uncoupling' from immersion in one's environs, in following a certain automatism which ignores the demands of adaptation—this is what the 'death drive' ultimately amounts to" (Žižek, p. 231)

"Consequently, the only way effectively to account for the status of (self-)consciousness is to assert the ontological incompleteness of 'reality' itself: there is 'reality' only insofar as there is an ontological gap, a crack, in its very heart, that is to say, a traumatic excess, a foreign body which cannot be integrated into it" (Žižek, p. 242).

In a material world where everything is understood as determined, where even our psychological personalities are shaped by physical history as well as biological evolution, the causal loop might be broken only by negativity: a saying "no" to our desires. Ironically, we may be free only by not doing what we feel like doing. But what is this "I" that can choose to ignore its own desires? When we decentralize all our cognitive abilities into their biological components, and find that there is no homunculus that resides in the brain as an executive vanishing point of experiencing—that the pineal gland is not the seat of the soul; when we see that our personalities are as constructed as our perceptions; who am "I?" Might it be that the word "I" itself does not name one's subjectivity, but in a way creates a self by linguistically integrating various parts of the self with a signifier?:

> "The point here is that we should take Lacan's term 'subject of the signifier' literally: there is, of course, no substantial signified content which guarantees the unity of the I; at this level, the subject is multiple, dispersed, and so forth—its unity is guaranteed only by the self-referential symbolic act, that is, 'I' is a purely performative entity; it is the one who says 'I.' This is the mystery of the subject's 'self-positing,' explored by Fichte: of course, when I say 'I,' I do not create any new content, I merely designate myself, the person who is uttering the phrase. This self-designation nonetheless gives rise to ('posits') an X which is not the 'real' flesh-and-blood person uttering it, but, precisely and merely, the pure Void of self-referential designation (the Lacanian 'sub-

ject of enunciation'): 'I' am not directly my body, or even the content of my mind; 'I' am, rather, that X which has all these features as its properties. The Lacanian subject is thus the 'subject of the signifier'—not in the sense of being reducible to one of the signifiers in the signifying chain ('I' is not directly the signifier I, since, in this case, a computer or another machine writing 'I' would be a subject), but in a much more precise sense: when I say 'I'—when I designate 'myself' as 'I'—this very act of signifying adds something to the 'real flesh-and-blood entity' (inclusive of the content of its mental states, desires, attitudes) thus designated, and the subject is that X which is added to the designated content by means of the act of its self-referential designation. It is therefore misleading to say that the unity of the I is 'a mere fiction' beneath which there is the multitude of inconsistent mental processes: the point is that this fiction gives rise to 'effects in the Real,' that is to say, it acts as a necessary presupposition to a series of 'real' acts" (Žižek, pp. 244-245).

The relevance of Žižek's thinking to cognitive science and artificial intelligence is fairly clear: although not advancing many new ways of understanding the mechanisms of thinking (other than to bring together other thinkers of such), he shows the limitations of thinking purely in terms of objective mechanism or conceptual schemas. Here we have shown how he has studied the relationship between the subjective and objective from several perspectives: the psychoanalytic objectively subjective fantasy; the reverse relationship between "objective" quantum mechanics and a "subjective" Kantian perspectivism; the implication of multiplying sometimes incommensurable perspectives; our internal stream-of-consciousness and surface play of the senses; the division between a passive perceiving subjective self and a narrative auto-biographical objective self—and how we become aware of our core perceiving self via challenges to our auto-biographical narrative status-quo / homeostasis; and finally, how the word "I" as a linguistic signifier integrates various objective and subjective properties of being a self into a logical entity with unified impact on the world. Žižek demonstrates the continued relevance of psychoanalysis to cognitive science: we might psychoanalyze cognitive science as Lacan psychoanalyzed psychoanalysis itself. It is with this in mind that I might begin to make a critique of Žižek: despite his discussion of narrative autobiography mentioned above, but along with his dismissal of astrology (Žižek, p. 204), Žižek's blind-spot seems to be his ignoring that which he has plenty of: personality. Even when discussing narrative auto-biography—the aim remains of discovering the core self:

> "If I exist only within the story I am telling / experiencing, if I have no substantial identity / content outside it, then, when I experience myself, the story is always-already there, there is no subject preceding it who would formulate the quest, the question answered by the story—the primordial narrative that forms the Core Self is an 'explanation presented prior to any request for it'" (Žižek, p. 225).

In a twisted fate, Žižek may be limited by his own astrological sign(s)—born on March 21st, 1949, Žižek is an Aries (the sign of the Self) almost on the cusp with Pisces (a sign connected with the Unknown)—and his preoccupations, at least in the central part of what he has called his magnus opus, seem to be with the unknowability of the self! But more to the point, the ancient art of astrology posits that we have personality *traits*—aspects of our selves that, if not universal, at least have a wide descriptive power. Whether or not astrological charts have any basis in reality is not the point here: the point is simply that there are personality traits that we can identify; some may be genetic, some taught by osmosis from one's family, friends, society and so on—but there are traits which seem to have little to do with subjective experiencing or an abstract "I" —and have everything to do with being a complex human, or even a real animal (as we often see that our pets have personalities too). So the exclusion of concepts like "personality" and "soul" when talking of subjectivity seem to put Žižek in the same camp as those cognitive scientists he critiques—on the side of "truth" and against "meaning":

> "This is also the difference between Freud and Jung: while Jung remains within the horizon of meaning, Freudian interpretation aims at articulating a truth which is no longer grounded in meaning [....] either we endorse the 'postmodern' stance and renounce the dimension of truth altogether, restricting ourselves to the interplay of multiple meanings, or we engage in an effort to discern a dimension of truth outside meaning—that is, in short, the dimension of truth as *real*" (Žižek, p. 181).

Personality traits, our ways of being, seem fairly real. We can talk about these traits, and even verify them among ourselves (yes, we both agree that Žižek can be funny). Is the *real* related to just the geometric mechanics of our world? Or an unknowable chaos beyond death? Is significance itself just epiphenomenal? Is being head-strong, generous, gullible, selfish, etc. not real? This may be relevant to Žižek's discussion of *physics* and *design*. Re: Daniel Dennett:

> "the two basic levels of reality are the deterministic physical level and the 'higher' level of design" (Žižek, p. 237).

> "In this sense, what we perceive as the most immediate sensual reality is already the result of complex elaboration and judgment, a hypothesis which results from the combination of sensual signals and the matrix of expectations. Secondly, at the level of 'practical reason,' consciousness, while in no way able to instigate a spontaneous act, can 'freely' impede its actualization: it can veto it, say 'No!' to a spontaneously emerging tendency. This is where Hegel comes in, with his praise of the infinite negative power of abstraction that pertains to understanding" (Žižek, p. 240).

Personality could be understood at the design level if not physical... and to his credit, Žižek does discuss (briefly) nature and character:

> "Kant and Schelling postulate a nonphenomenal transcendental, atemporal act of primordial choice by means of which each of us, prior to his temporal bodily existence, chooses his eternal character. Within our temporal phenomenal existence, this act of choice is experienced as an imposed necessity, which means that the subject, in his phenomenal self-awareness, is not conscious of the free choice which grounds his character (his ethical 'nature')—that is to say, this act is radically unconscious (the conclusion explicitly drawn by Schelling). Here we encounter the subject as the Void of pure reflexivity, as that X to which we can attribute (as his free decision) what, in our phenomenal self-awareness, we experience as part of our inherited or otherwise imposed nature. The conclusion to be drawn is thus, again, that Self-Consciousness itself is radically unconscious" (Žižek, p. 246).

With Žižek's writing, it is sometimes confusing to try to discern what in it is his opinion and argument, and what is examination of the opinion and argument of others. Such illustrates that our own thinking, personalities, etc, are in constant interchange with those around us. Perhaps Žižek decided to be an Aries in a prior existence which he has completely forgotten. But beyond choosing one's personal character or nature, there is the question of human nature itself, a topic explored in some depth by Steven Pinkerin who is the focus in the next section of this chapter.

Here we have seen Žižek discuss consciousness from the inside; and better yet, consciousness on the surface; and even further—inter-subjective consciousness. But what about consciousness on the outside? Possibly the sun of conscious awareness shines through our senses into bodies of individual memory. Thirty-six objective views of one subjective Fuji volcano. A dozen eye-witnesses to one messiah of light. I digress...

STEVEN PINKER: STEREOTYPICAL CONTEXTS

Žižek warns that cognitive science may "humiliate" the field of psychoanalysis:

> "The story of the three successive humiliations of man, the three 'narcissistic illnesses' ('Copernicus-Darwin-Freud'), is more complex than it may appear. The first thing to note is that the latest scientific breakthroughs seem to add a whole series of 'humiliations' which radicalize the first three, so that as Peter Sloterdijk has perspicuously noted, with regard to today's 'brain sciences,' psychoanalysis seems, rather, to belong to the traditional 'humanist' field threatened with the latest humiliations" (Žižek, p. 163).

On the flip side, Steven Pinker, a researcher in cognitive science whose writing probably exceeds the popularity of Žižek, making Pinker somewhat of an academic celebrity as well, has a different take on science and humility:

> "The starting point for acknowledging human nature is sheer awe and humility in the face of the staggering complexity of its source, the brain. Organized by the three billion bases of our genome and shaped by hundreds of millions of years of evolution, the brain is a network of unimaginable intricacy: a hundred billion neurons linked by a hundred trillion connections, woven into a convoluted three-dimensional architecture. Humbling, too, is the complexity of what it does. Even the mundane talents we share with other primates—walking, grasping, recognizing—are solutions to engineering problems at or beyond cutting edge artificial intelligence. The talents that are human birthrights—speaking

> and understanding, using common sense, teaching children, inferring other people's motives—will probably not be duplicated by machines in our lifetime, if ever. All this should serve as a counterweight to the image of the mind as formless raw material and to people as insignificant atoms making up the complex being we call 'society'" (Steven Pinker, *The Blank Slate*, Viking: New York (2002), p. 197).

Whereas Žižek might emphasize "the radical *plasticity* of the human brain" (Žižek, p. 209) and claims that:

> "Vulgar materialism and idealism join forces against this plasticity: idealism, to prove that the brain is just matter, a relay machine which has to be animated from the outside, not the site of activity; materialism, to sustain its mechanical determinist vision of reality" (Žižek, p. 209).

Pinker on the other hand, sees appeals to plasticity as a "last stand" (Pinker, pp. 73-102) of empiricist blank slate reasoning, the blank slate notion that we do not have a human nature encoded in our DNA that shapes much of how we can think and learn:

> "Extreme plasticity also appeals to neuroscientists who wish to boost the importance of their field for education and social policy, and to entrepreneurs selling products to speed up infant development, cure learning disabilities, or slow down aging. Outside the sciences, all three developments [the Human Genome Project showing that we have fewer genes than expected, connectionism's artificial neural networks learning statistical patterns, and newer evidence of at least some if not radical neural plasticity] have been welcomed by some scholars in the humanities who want to beat back the encroachments of biology. The lean genome, connectionism, and extreme plasticity are the Blank Slate's last stand" (Pinker, pp. 74-75).

One can see a sort of militant academic field defending described by both Žižek and Pinker—and such is understandable, given that career legitimacy may be a stake for some—like competing oil companies arguing against human-caused global-warming—not wanting to convert all their gas stations into mere convenience stores. There is a typical tendency to underestimate not just the complexity and nuance of the positions of one's intellectual adversaries, but to overestimate the real differences between these positions. Pinker is not out to sustain his "mechanical determinist vision of reality" any more or less than Žižek is out to "cure learning disabilities."

I will not be going into depth on Pinker's assessment of human nature (he takes Chomsky's innate linguistic abilities into the realm of mental abilities and even human nature in general, and studies how these abilities have evolved in adaptation to environments via natural selection). His writing is lucid and entertaining, full of anecdotal evidence connected to statistics with solid reasoning. The purpose of this chapter is to highlight the intersection of deconstructive thinking with artificial intelligence approaches though, so we will focus here on cognitive scientist Pinker's engagement with postmodern oriented thinkers in his book that is aimed at refuting theories of *The Blank Slate*:

> "According to the relativistic wisdom prevailing in much of academia today, reality is socially constructed by the use of language, stereotypes, and media images. The ideas that people have access to facts about the world is naive, say the proponents of social constructionism, science studies, cultural studies, critical theory, postmodernism, and deconstructionism. In their view, observations are always infected by theories, and theories are saturated with ideology and political doctrines, so anyone who claims to have the facts or know the truth is just trying to exert power over everyone else" (Pinker, p. 198).

Pinker laments that

> "relativists have a penny-pinching theory of psychology in which the mind has no mechanism to grasp reality; all it can do is passively download words, images, and stereotypes from the surrounding culture" (Pinker, p. 198).

Noting that optical illusions demonstrate the fallibility of our senses (a stick may appear bent in water, when it is actually strait—and hence some might argue we always deceived in some way or another about what we perceive—Plato comes to mind, with his notion that our world is a pale reflection of eternal forms), Pinker puts such limited instances of illusion in context:

> "The fact that our cognitive faculties (like our perceptual faculties) are attuned to the real world is most obvious from their *response* to illusions: they recognize the possibility of a breach with reality and find a way to get at the truth behind the false impression" (Pinker, p. 201).

One sort of false impression might be the various stereotypes one sees with our categorization of the world. Stereotypes can be erroneous, and indicative of prejudice against gender, race, and sexual orientation. Indeed, some categories are obviously socially constructed—e.g. money seems to have little to do with biological natural selection; and hence one might be tempted to believe all categories can carry some sort of erroneous prejudice—but this flies in the face of the common sense that:

> "If it walks like a duck and quacks like a duck, it probably is a duck" (Pinker, p. 203).

Pinker claims that our categories tend to actually have a "systematic relation to things in the world" (Pinker, p. 202; contra the views of relativists):

> "Most cognitive psychologists believe that conceptual categories come from two mental processes. One of them notices clumps of entries in the mental spreadsheet and treats them as categories with fuzzy boundaries, prototypical members, and overlapping similarities, like the members of a family [....] the other mental process looks for crisp rules and definitions and enters the into chains of reasoning." (Pinker, p. 203).

Such a discussion seems in line with Wittgenstein (his reference to family resemblances among concepts; and the distinction between symptoms of a concept—that which we simply associate with it, and criterion for concepts—things necessarily true of a concept by definition). However, it seems to me that Pinker would have to account for categories being able to *represent* (or rather, to use Pinker's words, have a "systematic relation" to) things in the world, and might run into the problems that Wittgenstein had with "things" and "concepts" sharing a logical form: "the intuition that species are defined by an internal essence or hidden trait that lawfully gives rise to its visible features" (Pinker, p. 204) simply does not cut the mustard intellectually or scientifically—DNA is not an "essence," and not really even a "trait"—and the number of concepts limited to DNA created species is most likely small relative to the whole of language. Yes, DNA may indeed encode some of our general concepts within our inherited linguistic abilities—but these concepts will be in our heads, and not anchored in actual real world things—even if DNA plays a role in tool making, the notion of an obsidian arrow is in the head, and not in the object. However, Pinker points out that even our prejudicial stereotypes are more often than not accurate:

> "People's stereotypes are generally consistent with statistics, and in many cases their bias is to *underestimate* the real differences between sexes and ethnic groups" (*Pinker*, p. 204).

Pinker is not endorsing discrimination or prejudice here, but noting that our concepts in general seem to map to the world fairly accurately. One is reminded of James Surowiecki's book *The Wisdom of Crowds*; another work intertwining theories, anecdotal and statistical evidence, and demonstrating that crowds do "out-guestimate" the experts by drawing upon distributed intelligence, and something like a median perspective. In connection: with the "median voter theorem" it is stipulated that the voter with the median political perspective (where all voters are lined up on a left-right political spectrum) will always get their way politically: the median voter always "wins"—if most of the voters lean left, then the median voter will fall in the "leans left" category; if most of the voters lean right, then the median voter, falling in the middle of the voters lined up left to right, will be a right leaning voter as well. Of course, politics can be more complicated than left / right—there are many political issues and a single candidate or ballot measure could complicate a left / right spectrum. However, with a left-right orientation, the entire spectrum could be shifted, the "Overton window" of what is considered to be politically acceptable can shift left or right via reasons ranging from new scientific discoveries to insidious propaganda; hence "moderate" voters may swing from left to right—and, gerrymandering aside, change the median vote. Such shifting of attitudes can parallel changes in the meaning of words: etymology is all about how the meaning of words can change over time. Nevertheless, for the most part, we do not get confused when navigating the world with our language—and to that extent our categories do connect with the world.

Pinker goes on to discuss the power of language to either communicate thought or constrain it—does language transparently transfer thought from head to head, or does it color the way we perceive the world?... completely, partially, or not at all?

> "Famous quotations from two philosophers capture the anxiety. 'We have to cease to think if we refuse to do it in

> the prisonhouse of language,' wrote Friedrich Nietzsche. 'The limits of my language mean the limits of my world,' wrote Ludwig Wittgenstein" (Pinker, p. 207).

Pinker is leery of Heideggerian-like claims that "language speaks us" when

> "the centrality of language is taken to extremes in deconstructionism, postmodernism, and other relativist doctrines. The writings of oracles like Jacques Derrida are studded with such aphorisms as, 'No escape from language is possible,' 'Text is self-referential,' 'Language is power,' and 'There is nothing outside the text." (Pinker, p. 208).

Taken literally, these aphorisms are plainly ridiculous. However, if "text" or "language" are taken as metaphors for any structure that is comprehensible or understandable (*that* which we can even begin to discuss in a language as semiotically interpretable), and as something even beyond immediate comprehensibility: if "text" is taken as "context" (and Derrida notes just this)—then the claim that "there is nothing outside of (con)text"—that there is nothing outside that which situates our minds and bodies—then this claim makes more sense (as said context is indefinitely expandable). Derrida may be stretching meaning a bit—but such is only to illustrate his point that meaning *can* be stretched, and comprehended in a new context where one has a new understanding of a concept. Pinker again argues against the notion that:

> "since language is an arbitrary instrument, not a medium for communicating thoughts or describing reality, the powerful can use it to manipulate and oppress others" (Pinker, p. 208).

In arguing that language is indeed "for communicating thoughts or describing reality," Pinker notes that thinking and thoughts are not only linguistic, but visual, musical, etc. Moreover:

> "many experiments have plumbed the minds of creatures without language, such as infants and nonhuman primates, and have found the fundamental categories of thought working away: objects, space, cause and effect, number, probability, agency (the imitation of behavior by a person or animal), and the function of tools" (Pinker, p. 210).

Some might claim that these "fundamental categories of thought" are the fundamental categories of comprehensibility in general—and if language is understood, via metaphorical stretch, as comprehensibility, then these are indeed deep linguistic aspects of animal reality. On the flip side, some might claim that even bees have a "language"—a way to communicate thoughts via "dancing," etc—and if this is so, then primates probably communicate via some sort of "language" too... like the language of facial expressions. Deep structures may be found not only in language in general, but in individual utterances:

> "Many experiments on human memory have confirmed that what we remember over the long term is the content, not the wording, of stories and conversations. Cognitive scientists model this 'semantic memory' as a web of logical propositions, images, motor programs, strings of sounds, and other data structures connected to one another in the brain" (Pinker, p. 210)

Again, a deconstructionist would have few problems with this stance: linguistic understanding can be more abstract than the particular words used. And again, within a "semiotic" context, just about everything is linguistic, from advertising strategies to fashion trends: anything that we can interpret is linguistic in nature. Deconstructionists also reference the "webs" of language—the complex traces or connections between one word and others, between one meaningful notion and others. Yet there are abilities that are non-verbal, like the ability to throw a baseball—and such abilities are intertwined with our verbal reasoning, suggesting at least some real split between thought and language, as thought does appear to have something to do with non-linguistic abilities too. Language use can also be seen as a family of *abilities*.... Wittgenstein's major shift was to from trying to understand language as representing the world, to understanding it as a way of interacting with the world. Pinker goes on:

> "Writing and speaking do not consist of transcribing an interior monologue onto paper or playing it into a microphone. Rather we engage in a constant give-and-take between the thoughts we try to convey and the means our language offers to convey them. We often grope for words, are dissatisfied with what we write because it does not express what we wanted to say, or discover when every combination of words seems wrong that we do not really *know* what we want to say. And when we get frustrated by a mismatch between our language and our thoughts, we don't give up, defeated and mum, but change the language. We concoct neologisms (*quark, meme, clone, deep structure*), invent slang (*to spam, to diss, to flame, to surf the web, a spin doctor*), borrow useful words from other languages (*joie de vivre, schlemiel, angst, machismo*), or coin new metaphors (*waste time, vote with your feet, push the outside of the envelop*). That is why every language, for from being an immutable penitentiary is constantly under renovation" (Pinker, p. 210).

Jacques Derrida has coined a few neologisms as well—and the "prisonhouse of language" is not claimed to resist renovation. What Derrida would question is the notion that we always have a clearly formed thought prior to engaging in communication: a self-presence of conscious intent before a linguistic act. Pinker's point, that thought and language are not the same thing, makes a lot of sense (again, deconstructionists seem to be using semiotic / metaphorical extensions of the words "text" and "language" beyond their normal denotation or even connotation). But he seems to think that we know or at least *feel* what we want to say, when in many cases we may lack the words, not because words fall short of thought, but because we really don't have a clue what we want to say *entirely*: we often have partial, undeveloped ideas. Take my writing of this chapter section for example. Did I know every sentence I was going to write before I wrote it? No... I had some preconceptions and background knowledge prior to reading Pinker's book, developed some notions as I read and re-read from it, highlighted good quotes, took notes, worked at a fuzzy outline of my approach, and then fleshed out my ideas in reaction to what I have read when writing my own piece. This last part is by far the most engaging and enlightening step: when the rubber hits the road and one has to use grammatical sentences and something close to logical argument without nonsequiturs, fallacies, etc. One may actually change their thinking about the subject too. Again, Pinker:

> "language itself could not function if it did not sit atop a vast infrastructure of tacit knowledge about the world and the intentions of other people. When we understand language, we have to listen between the lines to winnow out the unintended readings of an ambiguous sentence, piece together fractured utterances, glide over slips of the tongue, and fill in countless unsaid steps in a train of thought [....] Indeed, the very existence of ambiguous sentences, in which one string of words expresses two thoughts, proves that thoughts are not the same things as strings of words" (Pinker, pp. 210-211).

This is close to what Derrida's comment "there is nothing outside the (con)text" is meant to convey: that "vast infrastructure of tacit knowledge" is part of what context is—it is part of the milieu that situates a particular utterance. But beyond our thought as intent that we encode in language (and this seems to be Pinker's view, that language is a vehicle for thought that references the world)—a particular utterance is also to be understood in the context of the situation it is uttered in: when I say "check out this book" and point to Steven Pinker's *The Blank Slate*, the pointing finger and the book itself are as much of the context as the intent that my listener look at the book. And if we were just talking about books to get from the library, the meaning of "check out" is changed a bit. When one emphasizes intent over context, I tend to think of T.S. Eliot's notion of an objective correlative—the idea that we can encode even emotion in words. But meaning is not there in the individual words, or even in a complete sentence—some meaning may be from the intent; but words can also convey more than what we intend to communicate (as when we may betray attitudes we did not mean to convey). Derrida would claim there is an excess of meaning—that different contexts can bring out, not just any meaning, but different meanings for the same utterance (think Freudian slips, cultural attitudes we may not be self-conscious about, etc.)—this "saying more than we intend to say" is part of how "language speaks us." If language has a bias built in (for example, the gender coding of words in some languages—where certain concepts are identified as feminine and others masculine, and history becomes "his story")—some may be doomed to propagate that bias, without even being aware that they are conveying these cultural attitudes—but note:

> "Language conveys not just literal meanings but also a speaker's attitude" (Pinker, p. 211).

My contention would be that Derrida and Pinker do not differ in their views on language as much as Pinker thinks. Both seem to agree that intent and context determine the meaning of a linguistic act. Derrida would suggest that conscious intent can't possibly hold in mind the entire context that shapes the meaning of an utterance; moreover, the context is indefinitely expandable—Derrida emphasizes context over intent. Pinker seems to think that we are more aware of what we mean to say as we say it: intent outweighs context; yet:

> "Since images are interpreted in the context of a deeper understanding of people and their relationships, the 'crisis of representation,' with its paranoia about the manipulation of our mind by media images, is overblown. People are not helplessly programmed with images; they can evaluate and interpret what they see using everything else they know, such the credibility and motives of the source" (Pinker, p. 216).

In the nature-nurture debate, a line must be drawn dividing innate concepts from cultural constructs. Derrida would deny innate concepts at the cost of scientific coherence—too much evidence suggests that we at least have some biologically innate conceptual ability. Given that at least some concepts are cultural constructs, and that what we think is often from what we learn via language—Pinker cannot deny that stereotypes can be self-perpetuating (and he doesn't deny this). But I think Pinker frames his indirect debate with Derrida on the wrong grounds, of nature vs. nurture, when the true difference between them is one of an emphasis on intent vs. an emphasis on context.

Pinker misses the point of Baudrillard's notion of the "hyperreal" as well (Pinker, p. 214), as Baudrillard thinks that society only considers real that which we can replicate in some way or another—such is merely to claim that the core of any science is based on verification. It's not that there is no reality and only appearances, it is that knowledge is always a matter of the representation itself, and not the represented: our knowledge concerns the map and not the territory—the map is the limit of what we know, it is what has been verified, and hence is what is real. Such is also found in jumping from representational thinking to coherence thinking—a move from understanding language acts as somehow mirroring in some way a non-linguistic world of things—to understanding language acts as shaped and understood via the context of the world. Admittedly the postmodern Baudrillard carries his metaphorical theories farther than Derrida would—and this misunderstanding on Pinker's part, if there is a misunderstanding, is quite understandable, esp. as he seems to have been exposed to Baudrillard's concept at second hand.

Again, Pinker's differences with postmodern and deconstructive thinkers are real but not absolute: some postmodern thinkers seem to want to view the very laws of physics as cultural constructs—but none of these gurus have levitated. A problem occurs when the notion that concepts are clarified, if not defined, by their juxtaposition and relation to what they are not, is confused with a subjective "relativism" where anything goes and anything could mean anything. For the innate concept thinker, some concepts are hard-wired in the brain—the concept of "an object" cannot mean just anything we want it to—but such innate concepts are not related to Derrida's critique of the "transcendental signified"—even though some concepts may be innate, they too could be defined differentially. Such might be related to Donald Davidson's principle of charity: We all share some amount of context; a core of related beliefs, and as is highly likely, a number of innate conceptions. Again, innate concepts are not necessarily at odds with a deconstructive view: as long as one sees them as an additional contextual aspect of language and not as containing meaning in themselves. These innate concepts would *not*, however, be any more or less *natural* than the rest of language. That is to say, they do not necessarily reflect the nature of the world: they reflect how the brain interacts with the world, how the human brain has adapted not only for species survival but for species proliferation. Our innate concepts may be our initial guide to the nature of non-human reality, but such does not mean that our understanding of reality should be limited by these concepts: hence we invent new words like "wavicle" in development of our understanding of the universe we are born into. Likewise, having a genetic "human nature" doesn't imply we should perpetuate historical injustices: as Žižek noted: saying "no!" to our desires, esp. unjust desires, can form the very basis of freedom.

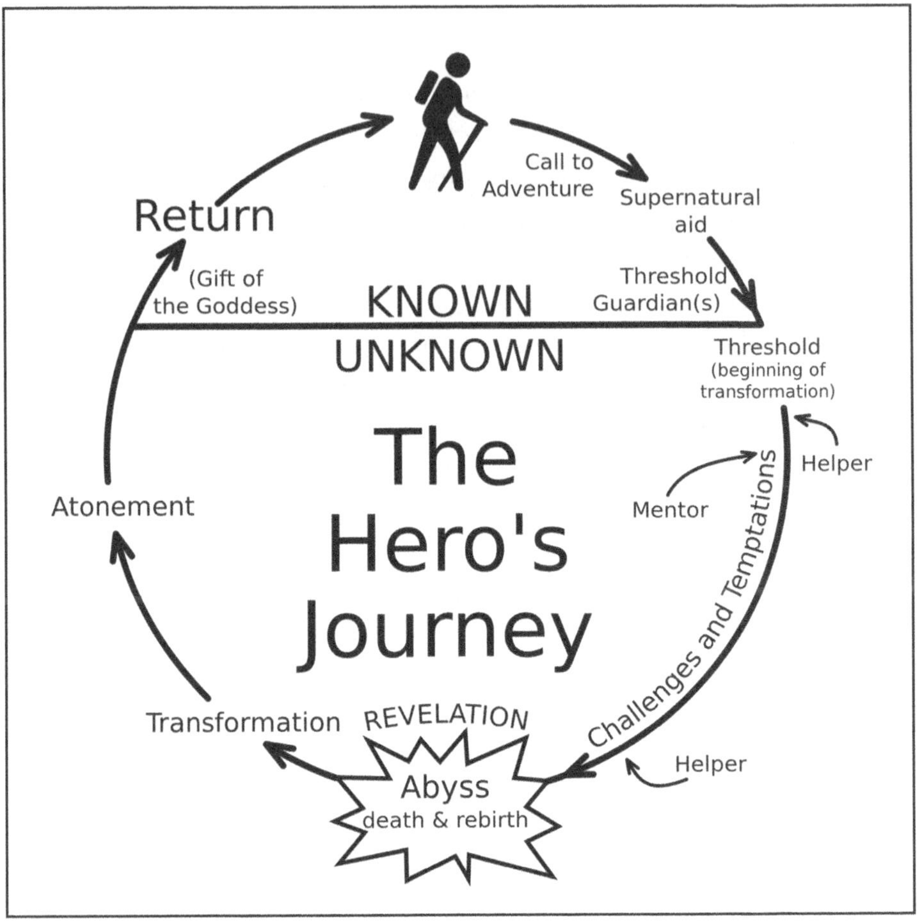
Call to
Adventure
Supernatural
aid
Threshold
Guardian(s)
Threshold
(beginning of
transformation)
Helper
Mentor
Challenges and Temptations
Helper
Abyss
death & rebirth
REVELATION
Transformation
Atonement
(Gift of
the Goddess)
Return
KNOWN
UNKNOWN
The
Hero's
Journey

Chapter 27

Cybernetics
Systems Theory & Complexity

THE SYSTEM OF SYSTEMS

In his 1949 book, *The Hero with a Thousand Faces*, Joseph Campbell argued for a general theory of an archetypical symbolic and cyclical "hero's journey"—a "monomyth" that was said to be found in many mythological stories, religious figure legends, and narratives from around the world (see chapter image at left). Campbell laments that in his contemporary age:

> "the democratic ideal of the self-determining individual, the invention of the power-driven machine, and the development of the scientific method of research, have so transformed human life that the long-inherited, timeless universe of symbols has collapsed" (Joseph Campbell, *The Hero with a Thousand Faces*, Princeton University Press: Princeton, NJ (1968), p. 387).

But Campbell's study of myth is a bit scientific in itself, in its systematic (albeit male-centered) integration of various myths, with an aim of finding something like a universal human nature, or at least an organized key to understanding our symbolic heritage.

We've seen this search for a universal in Chomsky, et. al, looking for a universal grammar. Looking for a deeper structure common to various phenomenon was also an aim of French structuralism (as with Ferdinand de Saussure). Deconstructive studies can be concerned with the structure of structure—a turning of structuralism against itself and the consequent questioning of the development of this school of thought, and focus on ideas like difference which are necessary *prior* to there being structure—and also with scholars like Foucault examining the different systems of know-how in various times, cultures, and institutions. This approach is seen in the history of information science and artificial intelligence too, sometimes under another name: cybernetics. In his 1948 book, *Cybernetics*, Norbert Wiener notes:

> "We have already spoken of the computing machine, and consequently the brain, as a logical machine. It is by no means trivial to consider the light cast on logic by such machines, both natural and artificial. Here the chief work is that of Turing. We have said before that *machina ratiocinatrix* is nothing but the *calculus ratiocinator* of Leibniz with an engine in it; and just as modern mathematical logic begins with this calculus, so it is inevitable that the present engineering development should cast a new light on logic. The science of today is operational; that is, it considers every statement as essentially concerned with possible experiments or observable processes. According to this, the study of logic must reduce to the study of the logical machine, whether nervous or mechanical, with all its non-removable limitations and imperfections.
>
> It may be said by some that this reduces logic to psychology, and that the two sciences are observably and demonstrably different. This is true in the sense that many psychological states and sequences of thought do not conform to the canons of logic. Psychology contains much that is foreign to logic, but—and this is the important fact—any logic which means anything to us can contain nothing which the human

> mind—and hence the human nervous system—is unable to encompass. *All logic is limited by the limitations of the human mind when it is engaged in that activity known as logical thinking*" (Norbert Weiner, *Cybernetics: or Control and Communication in the Animal and the Machine*, The MIT Press: Cambridge, MA (1999), pp. 124-125).

Cybernetics originated with concerns about self-regulating *mechanisms*—a prime example of self-regulation being the thermostat that is triggered to turn off a heating element when the temperature rises above a threshold, and turn on the heat if the heat falls below a threshold. Such negative feedback loops are present in nature, as when predators act as negative feedback in eating excessive populations of prey—population growths being a positive and possibly explosive feedback loops in themselves. Fortunately (or unfortunately, as the view may be for humans), the natural selection tendency towards population explosions are robustly controlled not only by predators, but by limits on the amount of food available down throughout the food chain or ecological web. Such cybernetic concepts have been used to understand various types of systems with a common systematic structure, or fundamental set of principles. Hence, feedback loops can be said to be relevant not only to engineering and nature, but to our aesthetic enjoyment:

> "Positive- and negative-feedback response in the context of art and aesthetic enjoyment means a purposive striving to bring about, or to optimize a meaningful aesthetic experience by maintaining, prolonging and sharpening the match between the perceptions of the subject and his pre-established constructs [....] creativity [...] is called for when the subject's perceptual experience is inadequately meaningful in the light of his existing aesthetic constructs. In that event he is impelled to produce new constructs, fresh ways of looking at experience" (Ervin Laszlo, *Introduction to Systems Philosophy*, Harper Torchbooks: New York (1972), p. 227).

Cybernetics and systems theory are aids to interpreting the varied events to be found in our universe—but we can also question the paradigm of the paradigm. In her essay, "Leverage Points," Donetta H. Meadows ranks 12 different points in which to intervene in a social or ecological system, in order to try and get desired results—on order from weakest to strongest, these points are (summarized from Donetta H. Meadows, *Thinking in Systems: A Primer*, Chelsea Green Publishing: White River Junction, VT (2008), pp. 145-165):

12. *Parameters*: as with detailed local regulations.

11. *Buffers*: where larger buffers are more stable, and the smaller, more flexible.

10. *Infrastructure*: usually physical and hard to change.

9. *Delays*: inefficiencies in feedback loops can slow change, but increase stability.

8. *Negative feedback loops*: these provide stability.

7. *Positive feedback loops*: we mitigate these for stability.

6. *Information*: transparency and availability are key factors for information.

5. *Rules of the system*: these are the broad institutional structures.

4. *Self-organization*: there is a natural tendency for variety to merge.

3. *Goals*: often growth, but is this good?

2. *Paradigms*: these are our assumptions, e.g. that growth is preferable to sustainability.

1. *No Paradigm*: this may be the unknowable truth of the matter.

Paradigm shifts may be very powerful, but simply cannot be invented out of nowhere: to take hold, consensus must be built, some sort of agreement among a community that a new direction is advisable and predictably better. Self-organization (or "autopoiesis" as termed by Humbeto Maturana and Francisco Varela and discussed in their 1987 book, *The Tree of Knowledge: The Biological Roots of Human Understanding*), is a crucial concept for systems theory too, as it designates the integration of unique organized mechanisms, as found not only with life-forms, but in such diverse phenomena as insect colonies and social trends (labor organizations, etc). Such systems can be quite complicated, and hence the field of "complexity" has developed:

> "an interdisciplinary field of research that seeks to explain how large numbers of relatively simple entities organize themselves, without the benefit of any central controller, into a collective whole that creates patterns, uses information, and, in some cases, evolves and learns [....] Complex systems researchers assert that different complex systems in nature, such as insect colonies, immune systems, brains, and economies, have much in common" (Melanie Mitchell, *Complexity: A Guided Tour*, Oxford University Press: Oxford (2009).

The post-deconstructionist thinker Mark C. Taylor identifies various characteristics of complex systems:

> "1. Complex systems are comprised of many different parts, which are connected in multiple ways.
>
> 2. Diverse components can interact both serially and in parallel to generate sequential as well as simultaneous effects and events.
>
> 3. Complex systems display spontaneous self-organization, which complicates interiority and exteriority in such a way that the line that is supposed to separate them becomes undecidable.

4. The structures resulting from spontaneous self-organization emerge from but are not necessarily reducible to the interactivity of the components or elements in the system.

5. Though generated by local interactions, emergent properties tend to be global.

6. Inasmuch as self-organizing structures emerge spontaneously, complex systems are neither fixed nor static but develop or evolve. Such evolution presupposes that complex systems are both open and adaptive.

7. Emergence occurs in a narrow possibility space lying between conditions that are too ordered and too disordered. This boundary or margin is 'the edge of chaos,' which is always far from equilibrium" (Mark C. Taylor, *The Moment of Complexity – Emerging Network Culture*, The University of Chicago Press: Chicago (2001), pp. 142-143).

And Steven Johnson speculates as to future emergence:

> "Can this chain be extended in new directions—both on the atomic scale of digital information and the macroscale of collective movements? Will computers—or networks of computers—become self-aware in the coming years, by drawing upon the adaptive open-endedness of emergent software? Will new political movements or systems explicitly model themselves after the distributed intelligence of the ant colony or the city neighborhood? [....] Is there a genuine global brain in our future, and will we recognize ourselves in it when it arrives?" (Steven Johnson, *Emergence – The Connected Lives of Ants, Brains, Cities, and Software*, Scribner: New York (2001), p.233).

Such a future of a "global brain" may be a fact already, as Andy Clark ponders:

> "a more general topic: how to conceive the boundary between the intelligent system and the world. This boundary

> [...] looks to be rather more plastic than had previously been supposed—in many cases, selected extra-bodily resources constitute important parts of extended computational and cognitive processes. Taken to extremes, this seepage of mind into the world threatens to reconfigure our fundamental self-image by broadening our view of persons to include, at times, aspects of the local environment" (Andy Clark, *Being There: Putting Brain, Body, and World Together Again*, The MIT Press: Cambridge, MA (2001), pp. 213-214).

The popular notion of memes (the notion of memes has become a meme too)—the ideas, fads, slogans, buzzwords, etc, that pass through culture via social media, the mass media, gossip, etc, are like a global intelligence flow—where we humans can be seen as vehicles for a public discourse, if not a public mind or zeitgeist. In fact, in his book, *The Wisdom of Crowds*, James Surowiecki demonstrates that groups often make more intelligent choices than even genius individuals:

> "while it is certainly true that you often need a smart individual to recognize the intelligence of the group, in the future that may no longer be as necessary. As the value of collective wisdom becomes more widely recognized, people will be more likely to adopt, on their own, collective approaches to problem solving, and the Internet affords us any number of examples of wise crowds that are, for the most part, self-organized and self-managed. We're a long way from anything resembling bottom-up decision making, either in government or in corporate America, but certainly the potential for it now exists" (James Surowiecki, *The Wisdom of Crowds*, Anchor Books: New York (2005), p. 281).

Yet, although tapping into complex global intelligence may help us better predict and cope with a changing world, as Kevin Kelly points out in his book *Out of Control: The Rise of Neo-Biological Civilization*:

> "There is no control outside a self-making system. Vivisystems, such as economies, ecologies, and human culture, can hardly be controlled from any position. They can be prodded, perturbed, cajoled, herded, and at best, coordinated from within. On Earth, there is no outside platform from which to send an intelligent hand into the vivisystem, and no point inside where a control dial waits to be turned. The direction of large swarmlike systems such as human society is controlled by a messy multitude of interconnecting, self-contradictory agents who have only the dimmest awareness of where the whole is at any one moment. Furthermore, many active members of this swarmy system are not individual human intelligences; they are corporate entities, groups, institutions, technological systems, and even the nonbiological systems of the Earth itself" (Kevin Kelly, *Out of Control: The Rise of Neo-Biological Civilization*, Addison-Wesley Publishing Co.: Reading, MA (1994), pp.448-449).

We as individuals are caught somewhere between the emergence of intelligence in the whole brain as produced by a plethora of relatively simple functioning neurons, and the emergence of global intelligence that we are simply a part of in turn. But as complex as individual and global intelligence may be, time and again, we find issues conceiving systems framed in terms of interconnected structures.

NETWORKS AND NATURAL SELECTION

"Structure" may take on a variety of meanings, depending on the context in which the word is used—ranging from the "Structuralism" and "Post-Structuralism" schools which emphasize or question often binary oppositions held to be deep under the surface of observed phenomena, to more "architectural" notions of a scaffolding and frameworks of buildings. Hence, a loose definition of structure, as it will be thought of here, will help to facilitate the present project of understanding intelligence and knowledge.

Immediately, "structure" will be used here to denote a sort of "Tinker-Toy" notion of inter-connected parts, or nodes connected by lines / arcs. A "whole" entity under examination might be analyzed, and in one use of the word, "deconstructed," into its various interrelated components or parts—although, since often, "the sum is greater than its parts"—this whole entity might not be fully functioning, when reconstructed (e.g. consider a dissected frog). Of course, "deconstruction" is a *softer* term than "destruction," and might reveal aspects of the entity without doing it lethal violence. At any rate, a structure might be seen as a *Network* or a *gestalt constellation*—even a web of interconnections that may demarcate an entity, or the entire fabric of all entities and their milieu.

Less evidently, "structure" would be defined by *divisions* and *connections*, which may be spatial-temporal—and not necessarily with divisions among parts, but divisions among "wholes" that "create" parts as associated, and thus connected, possibly by mere proximity, to other parts. Hence we also have more complicated "terms" like Derrida's "*différance*" that work to "defer" some eschatological or teleological full presence of being temporally, while at the same time marking the "border region" that gives shape to a "thing," "entity," "being," or "presence" while not being a "thing," "entity," "being," or "presence" in itself. Despite Derrida's emphasis on différance, one still cannot responsibly ignore the notion of a "trace" which I believe may be related, temporally, to a memory that is not immediately remembered—Derrida says "there are no 'conscious' traces" (Jacques Derrida, Alan bass, trans, *Margins of Philosophy*, University of Chicago Press: Chicago (1982), p. 21)—and spatially, the trace might include the coordination of various specific stratified differences. In other words, *divisions* and *connections* across space and time, might construct a sort of structure without "things"—or a sort of under-structure of structure in the immediate sense given above—they are more primary, yet more difficult to grasp in a typical, "this is a part of that" way. For the greater part of understanding the present project, the immediate "tinker-toy" notion of structure will do—but to approach the limits of structure—and what might lie beyond, this more subtle and complicated approach of "divisions and connections" may prove quite helpful.

Now, by structure, we indicate the mechanical, and technical aspect of whatever material we are to examine (and with regard to difference and connection, we may be addressing that which is without center—the de-centered itself—or that which like an infinite application of the Copernican revolution of moving ever outward from our "subjective" earth to wider frames of "objective" solar, galactic, universal reference—which in radical relativity—has no absolute focal point of reference). Such says little of the a-mechanical. Chomsky has pointed out that Newton's physics, so intertwined with the mechanical, necessitated the non-mechanical, especially with the notion of gravity. Yet, gravity may be "mapped" as it were, mathematically, and hence even it may be seen as a "time-space" trajectory (mechanical-structural)—with the non-mechanical gravity being displaced in Einstein's physics, by the bending of time-space by non-mechanical matter (it is my take, that gravity is not a "force" for Einstein, as much as a measurement of how much time-space is bent due to *matter*—again, gravity is not a force, but a potential trajectory field caused by the "inverse" displacement of space by matter). Ironically, for all its scientific "mapping" power, techno-mechanical-structure is in itself without material being—and would merely parallel actual being, or produce being as an epiphenomena, depending on your perspective—with a "chicken or the egg first" question, relating to whether laws produce or describe the phenomena they are related to (possibly the answer is: both, but in different ways).

So far, we have merely been trying to describe the *formal* nature of "whatever" it is that we might address (the universe, being, etc). But what of the *content* of said nature: given the infinite variations possible in such structure—both with regard to specific formal laws, and specific material content—what would make our universe as it is, and not in some other way—with other laws, or other arrangements of matter, for example—or light traveling only so fast, or this star here, rather than there. There may be an infinite number of "parallel universes"—with every combination of formal law and / or spatial-temporal material content differing: maybe even an infinite number of each of these infinite possibilities. But surely, many of these possible worlds would fall fallow—universes that implode immediately or never really get a good start. The longevity and complexity of our universe may be remarkable indeed.

Now, a strong contender against the "intelligent design" notion that our particular universe was not simply one of an infinite number of possibilities—or at least against the notion that we are somewhat "favored" by a God, rather than an accident of possibility—would be the forceful idea of "natural selection."

"Natural selection" might be understood here as simply: the process by which a particular entity or way of being (of any sort) "survives" or lasts temporally among possibly changing circumstances. This is a much broader definition than typically given—what Daniel C. Dennett calls, "*Darwin's Dangerous Idea*." Of course, biologically, this would include organisms that change over the course of evolution due to DNA mutations, and evolve into a new form if their new traits enhance or maintain their survivability in their local environmental niche habitat. But the core structure of natural selection is that "something" last, even if given a changing environment. Of course, that "something" may change itself, internally (apes become humans, memes mutate, etc.)—and drawing lines in the continuum of change may be arbitrary—although this is as necessary as distinguishing one word from another is for communication. But this core principle of lasting through change; or identity through difference; or maybe even maintaining a statistically significant amount of "sameness" given a varying amount of internal and external variances—can be swept away in relativity just as our earth is thought not to be central to the universe. Yet, although "chopping" up a continuum of change may include an arbitrary taxonomy—there remains that ability to statistically map one entity's traits in relation to another. Humans have more in common with Chimpanzees than they do with Zebras and more in common with each other than with Chimpanzees. Also, the replication of one entity may bear more or less resemblance to its progenitor—replication being a key aspect of traditional views of natural selection—and typically, natural selection "success" is gauged by the growth or multiplication of one sort of entity.

It is with this expanded view of natural selection, that one might say, given an infinite variety of possible structures in an infinite number of possible universes, natural selection blazes one or more paths—possibly with the path we find ourselves on too—where the very physical laws of our universe "survive" to this day, due to their inter-dependent fitness for each other; and said universe is populated by entities that are also inter-dependent with these laws, and with each other, in such a way as to produce even our speculations at this moment. In other words, our universe may have evolved its very structure of laws and singular arrangement of matter by the same principle by which life itself has effected the evolution of the various bio-organisms here on earth.

EMERGENCE AND EMBODIMENT

"Emergence" as a concept interlinks with the concept of "networks" when we talk of "hubs." It is with hubs in mind that we can see a smoother segue from the discussions about network theory, as in Albert-Laszlo Barabasi's *Linked: How Everything is Connected to Everything Else and What it Means*, Perseus Book Group: New York (2002) and discussions about emergence theory, as in Steven Johnson's *Emergence: The Connected lives of Ants, Brains, Cities, and Software*. Structural networks are about interconnections—and certain "nodes" in a large structural network may have a greater number of connections to other nodes. For example, in the game "Six Degrees of Separation" where one tries to connect any one movie actor to another movie actor (e.g. Kevin Bacon) by naming a trail of actors that have been in movies together; you can see that one person can be connected to just about any other person in the world, by connecting person one and person two via person three.

One "hub" in such "meeting of minds" would be Late Show host David Letterman. Since Letterman has met so many people on the Late Show, he can really connect diverse people. Another obvious example of hubs, is the emergence of larger airports (in New York, Denver, Los Angeles, etc)—where you connect flights from smaller, more decentralized airports. The same with electric power grids, telephone systems, etc—and in a way, life forms themselves.

Emergence itself, is the phenomena of group organization from individual parts—where "dumb" and local parts work together to form more complex (and possibly more intelligent) complex wholes. Such is immediately evident with the brain, where simple neurons interact together in large numbers to create the intelligence of the brain—which is an extension of the notion of independent cells in general operating in conjunction to create more sophisticated behavior in plants and animals. Structural network connections are found everywhere throughout biological organisms—e.g. from the sub-atomic components, through molecules and cells, to organs—to the whole of an animal. Again, a biological organism, then, is a sort of atomic-hub—bringing all sorts of components into complex connection and action. Perhaps individual consciousness is a perceptual or sensing hub too. But as noted before, intelligence doesn't stop at the individual brain, and organization doesn't stop at the particular bio-organism: much intelligence seems to be social, as people work together within a history, sharing knowledge, and learning from one another, both consciously and unconsciously, and again, as with the notion of "memes" introduced by Richard Dawkins in his *The Selfish Gene*, Oxford University Press: Oxford (1976). In this "bottom-up"-construction perspective on complex phenomena, the individual parts may be blind to the organism as a whole—even one super-genius in a society would never have the vast intelligence and knowledge of society as a whole (although certain specialty peaks may occur in individuals—even their knowledge is diffusely integrated with the whole). Hence we have notions such as a "Global Brain"—see also, e.g., George B. Dyson's *Darwin among the Machines: The Evolution of Global Intelligence*,

Perseus Books: New York (1997). There is even the "Gaia hypothesis" suggested by James Lovelock in books like *Gaia*, Oxford University Press: Oxford (1990), where the entire earth is a complex interdependent single organism composed of both organic and inorganic parts.

Natural selection might occur at any level of a complex organism—but in every case, for survival of the evolving part, and for the evolving whole—there must be an interdependent coalescence between the part and the whole. Hence, we human beings, who might be too successful of a species, with our semi-conscious social natural selection via technology, have been edging out the habitat of other species—and more recently, through global warming, may edge out much of our own habitat as well! Such clearly illustrates the danger of a positive feedback loop (certain types of technology feed into themselves) that may ultimately, in a Malthusian way, curb its own expansion through an ultimate negative-feedback loop (humans expanding until overreaching habitat capacity). And natural selection is definitely intertwined with positive and negative feedback loops: populations explode through reproduction, only to be held in check by diminishing habitat, food resources, etc.

Given that the biological *form* of conscious organisms might be emergent from possibly non-conscious parts (and possibly consciousness occurs in decentralized forms as well—we humans might barely be able to imagine this)—emergent consciousness itself could be experienced as "Embodied."

Embodiment, as a concept or theory, recognizes that intelligence and consciousness are not free floating forms—although there may be an indefinite or infinite potential for intelligence and consciousness, they are experienced by humans in a very specific type of body and situation. Not only do eyes, ears, fingers, tongues, and noses parallel or produce specific types of conscious qualia (like red, white noise, heat, cherry flavor, and manure smell—and there is also the sense of balance)—our sense organs, and the extension of our bodies with arms, legs, hands and feet, reach out into the world and organize it in specific ways that shape the way we think: our intelligence. Our biological heritage, the shaping of our senses and bodies over millions of years, shapes the way we can actively perceive and manipulate the world and words. Such can be seen in the evolution of language, where the words, "I see" ("video" in Latin) came to mean "I understand," and the word "digital" morphed from the word "digit" which means finger (and hence indicates the ability to count, at least on one's fingers). Beyond our bodies, the very structure of our environment that we are situated in would shape our ways of thinking: although the structure of a neuronal brain might be able to conceive of vastly wide range of other structures, and even speculate on that beyond structure—for the most part, such is constrained by the body and the environment.

Now, some thinkers, again like Noam Chomsky, have noted that parts of the brain have very specific functions, much like specific organs in the body (the heart pumps blood, the liver is a filter, etc, and also there are parts of the brain that control the motor functions, process vision, etc). In this way, some ways of thinking may be very constrained not only by the shape of the body and its sense organs, but also by the very organs of the brain. And indeed, most languages found across great cultural divides, may be incarnations of a single universal grammar ability that is integrated into the very biology of the brain. Yet beyond innate ideas and ways of thinking (babies seem to have an inbuilt ability to recognize faces as a special type of object, and can even recognize objects in general: no mean feat) there seems also be a more flexible ability of the brain—limited only by the animated structure of neural connectivity itself. However, one's ability to think in new ways would require not only using the more flexible aspects of the brain, but thinking outside of the box of the ways we are taught to think, both consciously and unconsciously, through a cultural "indoctrination" passed on historically through teachers, parents, and society—both intellectually, and through the tools and technologies we use that evolve and are passed on from generation to generation. And again, even the worldly environment that we live in, must indoctrinate our thinking to some extent: although we may be inspired to build airplanes with wings, since we've seen birds—such worldly experiences might also build blinders to other ways of thinking. The very structure of our bodies, environment and even social histories may express themselves, e.g. in our language, through deeply interrelated (dead) metaphorical gestalts or constellations that orient how we perceive the world—these unconscious structures may be like the assumptions we have that give the material we have to think with, as well as the limitations to how we can use that material. Such embodied metaphorical thought has been discussed well in George Lakoff and Mark Johnson's *Philosophy in the Flesh: The Embodied Mind and Its Challenge to Western Thought*, Harper Collins Publishers: New York (1999).

THE STATISTICAL FUTURE

Most of us have heard of ideas "going viral"—that is, when a meme is no longer the idiosyncratic notion shared in a small community, or known to just a few, and is communicated again and again on a large social scale. Malcolm Gladwell examines this phenomenon in his book, *The Tipping Point: How Little Things Can Make a Big Difference*; where he argues that these ideas spread across social networks, especially via "human hubs" (as with our example of David Letterman, above)—people who connect to many other people; but ideas also spread due to certain folks being trusted sources of information: these might be experts, or certain media channels that many turn to on a daily basis—as well as some folks just pushing their ideas persuasively. Of course, it helps if the meme you hope to spread is catchy, easily digestible (in an intellectual sense), and fits in with the context of the environment and times. It is in this way that marginal ideas can come to the center. In two other books, *Blink: The Power of Thinking Without Thinking* and *Outliers: The Story of Success*, Gladwell argues with copious anecdotal evidence that we are often more accurate intuitive thinkers than when thinking in an explicitly rational way; but that it takes thousands of hours of practice to form the sort of talent that will work its way into mainstream success. A common thread in Gladwell's work is that the marginal and unexpected must be looked at when accounting for what enters the center ring of culture. Such an approach is shared by Steven D. Levitt and Stephen J. Dubner in their *Freakonomics* books, which they claim are based on a just a few fundamental ideas:

> "*Incentives are the cornerstone of modern life* [....] *The conventional wisdom is often wrong* [....] *Dramatic effects often have distant, even subtle, causes* [....] *'Experts'—from criminologists to real-estate agents—use their informational advantage to serve their own agenda* [... and] *Knowing what to measure and how to measure it makes a complicated world much less so*" (Steven D. Levitt & Stephen J. Dubner, *Freakonomics – A Rogue Economist Explores the Hidden Side of Everything*, Harper Collins Publishers: New York (2005), pp.13-14).

The notion of the unexpected was also studied, from a more philosophical, but no less anecdotal perspective in Nassim Nicholas Taleb's *The Black Swan: The Impact of the Highly Improbable.* Yet even Taleb notes:

> "I believe that you can be dead certain about *some* things, and ought to be so. You can be more confident about disconfirmation than confirmation [....] The Black Swan asymmetry allows you to be confident *about what is wrong*, not about what you believe is right" (Nassim Nicholas Taleb, *The Black Swan: The Impact of the Highly Improbable*, Penguin Books: London (2007), p. 192).

So much is to say that we don't even know what we don't know. But it also reminds me of "*the one percent doctrine*"—the title of a 2006 book by Ron Suskind that suggested that the US went to war with Iraq in part based on vice president Dick Cheney's logic that a small percentage of possibility that something catastrophic could take place should be treated as a complete certainty. One might retort that global warming was not treated as such, but the logic really breaks down—in preparing for the future, yes we should look for the unexpected: but the more things change, the more they stay the same—and long term-trends are long term trends because they have proven themselves for a long time. The probable prediction may be the best we can make; *that* which *has* survived over time has already proven itself "anti-fragile," having already faced an "infinite" variety of possible interventions, yet still existing. Even a technological singularity where artificial intelligence is supposed to improve *itself* at accelerating rates with an event horizon we cannot see past: even here we should see a continuum with the past. Yet, in order to find such stability, one must "zoom out" from the radically contingent particular:

> "complex processes produce order and beauty when you zoom out and look at them from enough distance. I use the term signal and noise very loosely in this book, but they originally come from electrical engineering. There are different types of noise that engineers recognize—all of them are random, but they follow different underlying probability distributions. If you listen to true white noise, which is produced by random bursts of sound over a uniform distribution of frequencies, it is sibilant and somewhat abrasive. The type of noise associated with complex systems, called Brownian noise, is more soothing and sounds almost like rushing water" (Nate Silver, *The Signal and the Noise: Why So Many Predictions fail—but some Don't*, The Penguin Press: New York (2012), p.173).

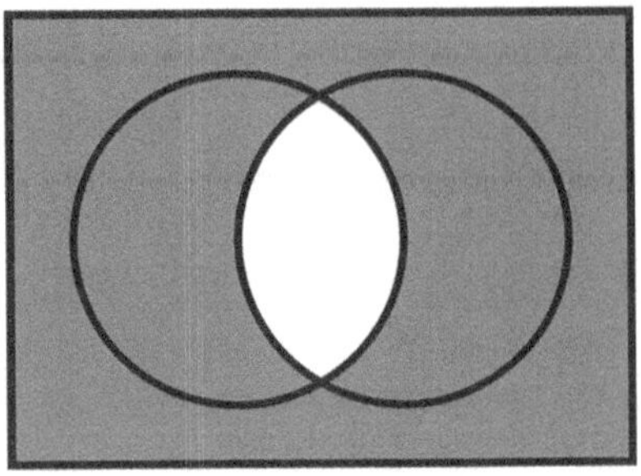

A	B	Q
0	0	1
0	1	1
1	0	1
1	1	0

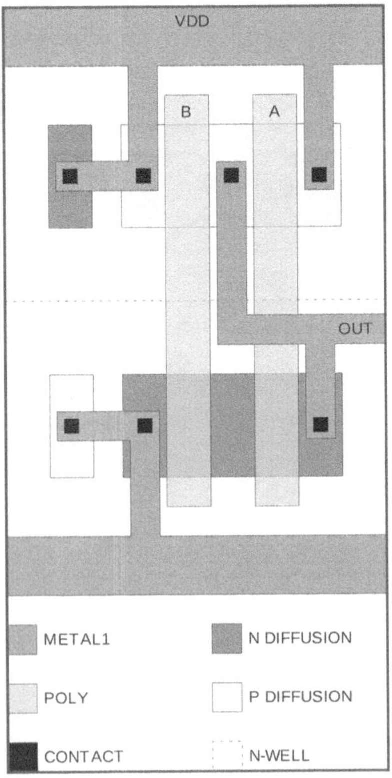

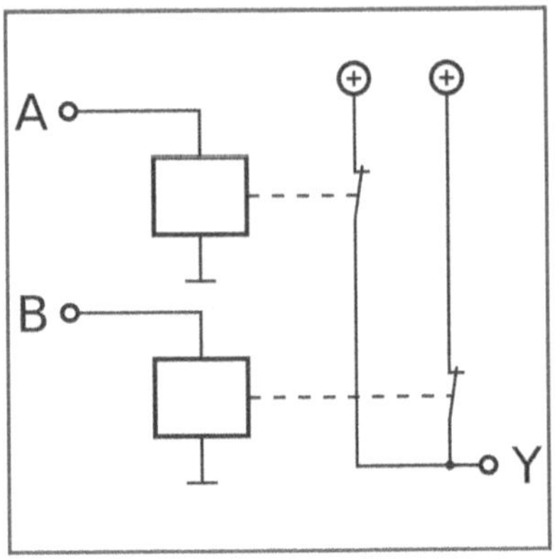

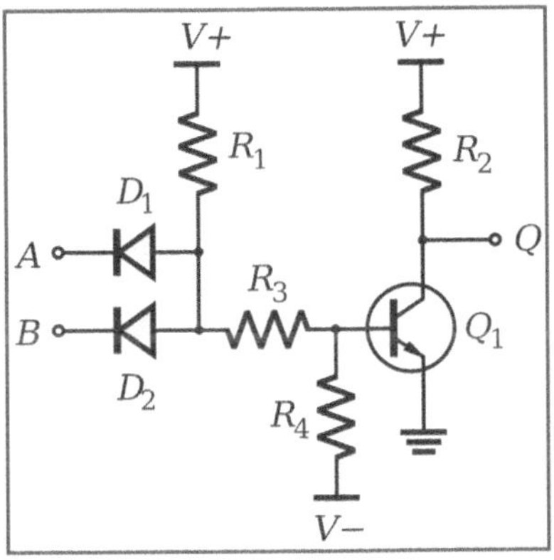

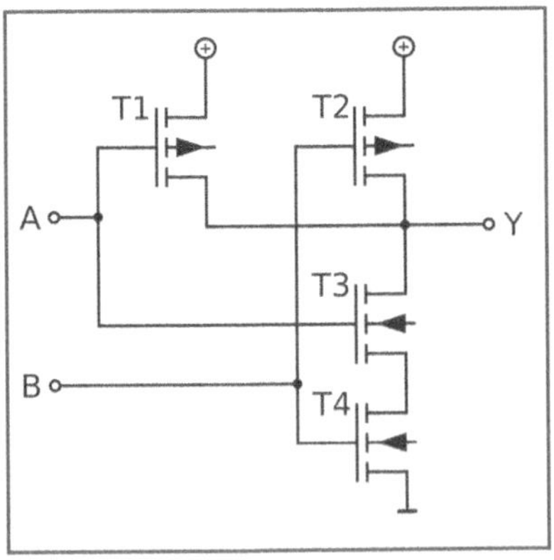

NAND operation: clockwise from top left as a Venn diagram, relay circuit, diode-transistor circuit, CMOS circuit, physical circuit, and table.

Chapter 28

Intersection
Deconstruction / AI

UNCONSCIOUS CONSCIOUSNESS

We may recall that in chapter two, on "Style, Taste & Cyber-Networks," the concepts of a *semantic network* and *spreading activation* were introduced. Throughout the history of philosophy, we have seen how ideas have been seen as associated, long before contemporary network theory, going back through Saussure, Kant's Understanding as a *web* of connected concepts, through the empiricists philosophers' *Associationism*, past Plato's dialectical and *hierarchical connection*s between concepts, implied as far back as Heraclitus' *Logos* as an integrating linguistic reckoning pinnacle, and even suggested by the *family* of Gods and *repetitious* religious rituals of ancient Egypt and beyond. In modern psychology, tests have been established to measure the effect of associated concepts, a prime example being that of the "priming effect":

> "Priming effects take many forms. If the idea of EAT is currently on your mind (whether or not you are conscious of it), you will be quicker than usual to recognize the word SOUP when it is spoken in a whisper or presented in a blurry font. And of course you are primed not only for the idea of soup, but also for a multitude of food-related ideas, including fork, hungry, fat, diet, and cookie. If for your

> most recent meal you sat at a wobbly restaurant table, you will be primed for wobbly as well. Furthermore, the primed ideas have some ability to prime other ideas, although more weakly. Like ripples in a pond, activation spreads through a small part of the vast network of associated ideas" (Daniel Kahneman, *Thinking, Fast and Slow*, Farrar, Straus and Giroux: New York (2011), pp. 52-53).

As with Malcolm Gladwell's book *Blink*, mentioned in the last chapter, in his book, *Thinking, Fast and Slow*, Daniel Kahneman argues that our fast emotional and intuitive, if not instinctive, thinking is—despite our sub-rational inability to judge probabilities reliably using off the cuff heuristics—intuitive thinking is actually a better compliment to our slower logical and explicit thinking than one might expect. Such also connects with the insight found in Marshal McLuhan's concepts of "hot" and "cool" media, where hot media could include a single sensory over-load consumed in the unthinking moment—like hearing a rock song—in contrast to cool media, which would require an effort on part of the spectator to interpret and fill in details, or that they could even participate in—as with a seminar, etc. (See Marshal McLuhan, *Understanding Media: The Extensions of Man*, The MIT Press: Cambridge, MA (1994), chapter 2: "Media Hot and Cold"). Taking up the ripples of activation theme, in his classic, *Gödel, Escher, Bach: An Eternal Golden Braid*, Douglas Hofstadter claims,

> "for each concept there is a fairly well-defined module which can be triggered—a module that consists of a small group of neurons—a 'neural complex'" (Douglas R. Hofstadter, *Gödel, Escher, Bach: An Eternal Golden Braid*, Basic Books: New York (1999), p.348).

And that we might

> "refer to these hypothetical neural complexes, neural modules, neural packets, neural networks, multineuron units—call them what you will, whether they come in the form of

pancakes, garden rakes, rattlesnakes, snoflakes, or even ripples on lakes—as *symbols*" (Hofstadter, 1999, 349).

And moreover:

> "symbols can be either *dormant*, or *awake* (activated). An active symbol is one which has been triggered—that is, one in which a threshold number of neurons have been caused to fire by stimuli coming from outside. Since a symbol can be triggered in many different ways, it can act in many different ways when awakened. This suggests that we should think of a symbol not as a fixed entity, but as a variable entity. Therefore it would not suffice to describe a brain state by saying 'Symbols A, B, . . ., N are all active'; rather we would have to supply an additional set of parameters for each active symbol, characterizing some aspects of the symbol's internal workings. It is an interesting question whether in each symbol there are certain core neurons, which invariably fire when the symbol is activated. If such a core set of neurons exists, we might refer to it as the 'invariant core' of the symbol" (Hofstadter, 1999, p. 349).

Such relates to the spreading activation phenomenon on a neural semantic network—and again, the notion of activation suggests some relation to conscious awareness of ideas:

> "consciousness systems [...] perceive themselves on the symbol level only, and have no awareness of the lower levels, such as the signal levels" (Hofstadter, 1999, p. 328).

and also:

> "Consciousness is that property of a system that arises whenever there exist symbols in the system which obey triggering patterns" (Hofstadter, 1999, p. 385).

With much activity operating behind the scenes of the consciousness stage, it is no wonder that many psychological tests have demonstrated that our conscious awareness lags behind the unconscious working of the brain (recall Freud's maxim, "where the id was, the ego shall be"), as Daniel C. Dennett explains here how intent can be retrospectively recogninzed:

> "We often do discover what we think (and hence what we mean) by reflecting on what we find ourselves saying—and not correcting. So we are, at least on those occasions, in the same boat as our external critics and interpreters, encountering a bit of text and putting the best reading on it that we can find. The fact that we said it gives it a certain personal persuasiveness or at least a presumption of authenticity. *Probably*, if I said it (and I heard myself say it, and I didn't hear myself rushing in with any amendments), I mean it, and it probably means what it seems to mean—to me" (Daniel C. Dennett, *Consciousness Explained*, Little, Brown and Company: Boston (1991), pp. 245-246).

LEVER HEADED THINKING

In the audaciously entitled *Consciousness Explained*, Dennett decentralizes conscious experience in a theory of how human thinking works—here in a "Thumbnail Sketch":

> "There is no single, definitive 'stream of consciousness,' because there is no central Headquarters, no Cartesian Theater where 'it all comes together' for the perusal of a Central Meaner. Instead of such a single stream (however wide), there are multiple channels in which specialist circuits try, in parallel pandemoniums, to do their various things, creating Multiple Drafts as they go. Most of these fragmentary drafts of 'narrative' play short-lived roles in the modulation of current activity but some get promoted to further functional roles, in swift succession, by the activity of a virtual machine in the brain. The seriality of this machine (its 'von Neumannesque' character) is not a 'hard-wired' design fea-

> ture, but rather the upshot of a succession of coalitions of these specialists.
>
> The basic specialists are part of our animal heritage. They were not developed to perform peculiarly human actions, such as reading and writing, but ducking, predator-avoiding, face-recognizing, grasping, throwing, berry-picking, and other essential tasks. They are often opportunistically enlisted in new roles, for which their native talents more or less suit them. The result is not bedlam only because the trends that are imposed on all this activity are themselves the product of design. Some of this design is innate, and is shared with other animals. But it is augmented, and sometimes even overwhelmed in importance, by microhabits of thought that are developed in the individual, partly idiosyncratic results of self-exploration and partly the predesigned gifts of culture. Thousands of memes, mostly borne by language, but also by wordless 'images' and other data structures, take up residence in an individual brain, shaping its tendencies and thereby turning it into a mind" (Dennett, pp. 253-254).

Dennett's theory sketch packs many ideas together—but the basic structure is to account for the "top-down" mental experience we have as conscious agents from a "bottom-up" biological and sociological construction from neurons, brain structure, habits, memes, etc. The neurons of the brain operate in parallel—more than one neuron or neuron cluster activates in a given time-frame—and there is a competition for awareness between the various firing neuron clusters *as* various ways of thinking and perceiving which come in and out of focus. This focus is temporally dynamic, never settled, as in a "moment of recognition"—but is continually sequencing experience, as we come to the experience of paying attention to particulars. "Focusing" (my terminology, not Dennett's) could be seen to be analogous to the read / write head in a Turing or von Neumann machine. Jerry Fodor writes:

"some, at least, of what makes minds rational is their ability to perform computations on thoughts; where thoughts, like sentences, are assumed to be syntactically structured, and where 'computations' means formal operations in the manner of Turing [....] Turing's account of computation is, in a couple of senses, local. It doesn't look past the form of sentences to their meanings; and it assumes that the role of thoughts in a mental process is determined entirely by their internal (syntactic) structure. And there's reason to believe that at least some rational processes are not local in either of these respects. It may be that wherever either semantic or global features of mental processes begin to make their presence felt, you reach the limits of what Turing's kind of computational rationality is able to explain. As things stand, what's beyond these limits is not a problem but a mystery" (Jerry Fodor, *In Critical Condition: Polemical Essays on Cognitive Science and the Philosophy of Mind*, The MIT Press: Cambridge, MA (2000), p. 205, "Review of Steven Pinker's *How the Mind Works* and Henry Plotkin's *Evolution of Mind*").

We may recall that a universal Turing machine resembles something like a read / write head which reads and writes symbols on a tape that can go forward or backward in discrete steps—where the symbol read as an instruction to move that tape and / or to read or write another symbol on the tape—such a simple procedural mechanism lies at the heart of modern computers—albeit that modern computers have a sort of "short-term working-memory" with memory registers, Random Access Memory, etc. instead of and in addition to more permanent forms of memory like magnetic tape, hard disks and flash drives. Although a Turning machine might be situated in a context possibly relating to the content of the tape (and modern computers have other inputs or outputs connecting to an environment, such as with keyboards, mice, cameras, microphones, display screens and speakers, etc), the read / write head is only "focused" on the singular information before it at any given step; because of this focus, some might claim the Turing machine is "context-free"—but I would argue, again, that the content of such machines (the information on the "tape" memory and from sensors, etc) provides context to the read / write head. It may also be that the read / write head functions something like Kant's objective / logical apperception (in contrast to subjective, sensing apperception—that subjectivity being the mysterious vanishing point of perspective that Dennett thinks is as illusory as a homunculus; the deceptive idea of the Cartesian Theater): Kant's concept of apperception seems to have anticipated the Turing machine.

Before getting into the cultural aspects of human thought (Dennett references "memes")—we should note that there may be several levels in "internal" thinking; as Marvin Minsky summarizes his own hierarchical schema:

> "We know that by the time of birth, every infant is already equipped with a variety of *instinctive* reactions, and has started to add *learned* reactions to these. Then, over time, we progressively add more *deliberative* ways to reason, imagine, and plan for the future. Later, we build a new layer in which we start to do *reflective thinking* about our own thoughts—and two-year-old children already are making additional ways to *self-reflect* about why and how they thought those things. And, eventually, we begin to think more *self-consciously* about which things to regard as *right* or *wrong* to do" (Marvin Minsky, *The Emotion Machine: Commonsense Thinking, Artificial Intelligence, and the Future of the Human Mind*, Simon and Schuster: New York (2006), p. 103).

Here again we see a split between instinct and learning—and importantly, for Minsky, the highest level includes emotions (much as was discussed in Spinoza's *Ethics*) that connect with moral feelings about right and wrong actions. Hofstadter also discusses a hierarchical activity in his book, *I am a Strange Loop*:

> "What I mean by 'strange loop' is [...] not a physical circuit but an abstract loop in which, in the series of stages that constitute the cycling-around, there is a shift from one level of abstraction (or structure) to another, which feels like an upwards movement in a hierarchy, and yet somehow the successive 'upward' shifts turn out to give rise to a closed cycle. That is, despite one's sense of departing ever further from one's origin, one winds up, to one's shock, exactly where one had started out. In short, a strange loop is a paradoxical level-crossing feedback loop" (Douglas R. Hofstadter, *I am a Strange Loop*, Basic Books: New York (2007), pp. 101-102).

At first, this sounds like an enlightenment experience, or Hegel's dialectical logic, but Hofstadter refines his metaphor:

> "What makes a strange loop appear in the brain and not in a video feedback system [...] is an *ability*—the ability to think—which is, in effect, a one-syllable word standing for the possession of a sufficiently large repertoire of triggerable symbols" (Hofstadter, 2007, p. 203).

> "The flip side is, ironically, an *inability*—namely, our Klüdgerotic inability to peer below the level of our symbols. It is our inability to see, feel, or sense in any way the constant, frenetic churning and roiling micro-stuff, all the unfelt bubbling and boiling that underlies our thinking. This, our innate blindness to the world of the tiny, forces us to hallucinate a profound schism between the goal-lacking material world of balls and sticks and sounds and lights, on the one hand, and a goal-pervaded abstract world of hopes and beliefs and joys and fears on the other, in which radically different sorts of causality seem to reign." (Hofstadter, 2007, p. 204).

> "Our innate human inability to peer below a certain level inside our cranium makes our inner analogue to the swirling galaxy on a TV screen—the vast swirling galaxy of 'I'-

ness—strike us as an undeniable *locus of causality*, rather than a mere passive epiphenomenon coming out of lower levels (such as a video-feedback galaxy) [....] causality gets turned around and 'I' seems to be in the driver's seat" (Hofstadter, 2007, p. 205).

Jeff Hawkins also sees a hierarchy in thought, and is especially interested in our ability to plan ahead, the mechanisms of which he has explored in some theoretical detail:

> "Nature first created animals such as reptiles with sophisticated senses and sophisticated but relatively rigid behaviors. It then discovered that by adding a memory system and feeding the sensory stream into it, the animal could remember past experiences. When the animal found itself in the same or a similar situation, the memory would be recalled, leading to a prediction of what was likely to happen next. Thus, intelligence and understanding stated as a memory system that fed predictions into the sensory stream. These predictions are the essence of understanding. To know something means that you can make predictions about it.
>
> The cortex evolved in two directions. First it got larger and more sophisticated in the types of memories it could store; it was able to remember more things and make predictions based on more complex relationships. Second, it started interacting with the motor system of the old brain. To predict what you will hear, see, and feel next, it needed to know what actions were being taken. With humans the cortex has taken over most of our motor behavior. Instead of just making predictions based on behavior of the old brain, the human neocortex directs behavior to satisfy its predictions" (Jeff Hawkins with Sandra Blakeslee, *On Intelligence: How a New Understanding of the Brain Will Lead to the Creation of Truly Intelligent Machines*, Owl Books: New York (2004), p.104).

What we find in Dennett, Fodor, Minsky, Hofstadter, and Hawkins—is an attempt to situate thinking as understandable objective mechanical functioning... and hence the flipside concerning subjective conscious experience becomes somewhat of a mystery. For Dennett this means rejecting the "Cartesian theater" and decentralizing qualia-type consciousness out of existence; for Fodor, even though a Turing machine provides a model of computational focus, consciousness itself is a mystery to be avoided; for Minsky, consciousness is a matter of becoming more self-aware and deliberate in one's thoughts, an activity rather than a state of being; while Hofstadter isolates consciousness as a symbolic epiphenomena, if not a robust constellation of ways of understanding the self; and Hawkins sees consciousness simply as the subjective experiencing of being a cortex:

> "We're all zombies [footnote: 'It would be an act of desperate intellectual dishonesty to quote this assertion out of context!'] Nobody is conscious—not in the systematically mysterious way that supports such doctrines as epiphenomenalism! I can't prove that no such sort of consciousness exists. I also cannot prove that gremlins don't exist" (Dennett, p. 406).

> "I try never to think about consciousness. Or even to write about it" (Fodor, p. 73, "Do We Think in Mentalese?").

> "if we take *consciousness* to mean 'awareness of our internal processes'—it doesn't live up to its reputation" (Minksy, p. 120).

> "consciousness is not a power moonroof (you can quote me on that). Consciousness is not an optional feature that one can order independently of how the brain is built" (Hofstadter, 2007, p. 325).

> "We can break consciousness into two major categories. One is similar to self-awareness—the everyday notion of being conscious [....] The second is *qualia*—the idea that feel-

> ings associated with sensation are somehow independent of sensory input. Qualia is the harder part" (Hawkins, p. 196).

We will return to the issue of consciousness and qualia later in this chapter—yet, before examining phenomenology and the direct experience of subjectivity, we should see how far we can go with "genealogical" or technologically engineer-able and objective knowledge. With an approach connected with Kant's arguing that the understanding is limited in what it can comprehend of that which lies on the other side of experience—by learning more about the technical context that shapes consciousness, we may also learn more about the unknowable aspects of consciousness, by defining them more precisely in the negative.

SCRIPTING THOUGHT

> "Are you talking to a person or a machine? It could soon be harder to tell amid claims that software called Cleverbot has passed one of the key tests of artificial intelligence: the Turing test.
>
> Proposed by British computer scientist Alan Turing in the 1950s, the test states that if a human talking to a machine believes the machine is human, it passes.
>
> The Cleverbot test took place at the Techniche festival in Guwahati, India. Thirty volunteers conducted a typed 4-minute conversation with an unknown entity. Half of the volunteers spoke to humans while the rest chatted with Cleverbot. All the conversations were displayed on large screens for an audience to see.
>
> Both the participants and the audience then rated the humanness of all the responses, with Cleverbot voted 59.3 per cent human, while the humans themselves were rated just 63.3 per cent human. A total of 1334 votes were cast—many more than in any previous Turing test, says Cleverbot's

developer and AI specialist Rollo Carpenter" (<//www.newscientist.com/article/dn20865-software-tricks-people-into-thinking-it-is-human.html>—Jacob Aron, "Software tricks people into thinking its human," New Scientist, September 6, 2011.)

Researching the programming theory behind chatbots—esp. Rollo Carpenter's "Jabberwacky" (see <//llt.msu.edu/vol10num3/pdf/emerging.pdf>) and "Cleverbot," (<//cleverbot.com>) and Jason Hutchens' "Megahal," (<//megahal.alioth.debian.org/How.html>)—one can see that they have developed artificial conversation further than the ELIZA program designed by Joseph Weizenbaum in the early 1960's which had canned responses and turnabouts for recognized keywords and phrases, and sounded like a questioning psychoanalyst. The newer chatbots, like A.L.I.C.E. (Artificial Linguistic Internet Computer Entity), actually learn from people who chat with them—that is, if you ask Cleverbot a specific question, you may get a response similar to what someone else has typed in at another time: the program compares conversations (a context sometimes beyond words, phrases, and sentences) and tries to find a match from its millions of phrases and sentences in its database which could be appropriate. Such sentences can be put together by statistically looking at what words or phrases tend to come before and after keywords (or each of the words) of an input sentence. The statistics involved use "Markov models"—which provide statistical probabilities based on past observations or known aspects, but must predict the future only given a present instance—if given the letter "b" and "c" tends to follow "b," then a "c" will likely be produced or expected. These statistical strategies can become more complex, when a temporal sequence (like speech) is processed in real-time (and the statistical expectations of a Markov model can help classify a pattern) or when a multi-dimensional field pattern (e.g. a visual image) has spatial relations rather than just a linear sequence.

Chatbots like Cleverbot do not parse grammar, are blind to the semantics of what they are saying, but rather act as micro-meme vehicles, passing cultural responses on from one time, place and context to another. They simply analyze patterns, and search for appropriate patterns to generate based on previous patterns of conversations as a whole: hence in a reversal of Chomsky's linguistics (as well as that of Steven Pinker and other innatist thinkers), chatbots do *not* use a formal context-free grammar to recognize and generate sentences, but use *only* content pattern contexts. Such raises the question: could all behavior simply be sensory contextualized pattern recognition and generation of motor patterns? Could there be a simple general learning procedure, like Locke's theory of abstraction, that could generate all intelligence and knowledge in their full complexity? Such would take the empiricist social-constructivist scheme to the extreme.

Moreover, with connectionist, neural network, holographic, or parallel distributed processing networks, inputs can be "mapped" to outputs with no programs or algorithms (see diagram A)—these perceptron-type mechanisms operate by connecting each of a series of input feature nodes to each of a series of output feature nodes. Each of the numerous arcs connecting the two series of nodes is *weighted*—or given a strength quantity—and the networks can be trained by strengthening weights between desired input-output connections, and weakening weights between undesired connections. Hence, in our prior if "b," then probably "c," example with Markov models, the weight between the "b" input and "c" output features would be strengthened through reinforcement training. Given many input and output features, and trained for a sufficient number of weight strengthening / weakening cycles, these connectionist networks can generalize and generate proper responses for examples they were not given before (e.g. a present tense input verb can generate a past tense verb for a word that was never given before). In fact such connectionist networks are used to generate stock price predictions, based on past associations between relevant variables and price fluctuations.

Connectionist networks offer the ability to encode content and provide appropriate responses to stimuli without programs, algorithms or rules—but currently, they are extremely limited without the ability to demonstrate problem solving abilities or complex multi-dimensional behaviors (such as holding a conversation). More complex behaviors *might* be produced, however, by stacking these connectionist networks hierarchically—Hawkins theorizes on how such might be accomplished in the cortex:

> "As you move about the world, changing inputs stream into lower regions of the cortex. Each region tries to interpret its stream of inputs as part of a known sequence of patterns. The columns try to anticipate their activity. If they can, they will pass on a stable pattern, the name of the sequence, to the next higher region. Again, it is as if the region says, 'I am listening to a song, here is the name, I can handle the details.'
>
> But what if an unexpected pattern arrives, an unexpected note? Or what if we see something that does not belong on a face? The unexpected pattern is automatically passed on to the next higher cortical region [....] The higher region may be able to understand this new pattern as the next part of its own sequence. It might say, 'Oh, I see a new note arrived. Maybe this is the first note of the next song on the album. It looks like it, so I predict we have gone on to the next song. Lower region, here is the name of the next song I think you should be hearing.' But if such recognition does not occur, the unexpected pattern will keep propagating up the cortical hierarchy until some higher region can interpret it as part of its normal sequence of events [.....] Thus we can see that observed patterns flow up the hierarchy and predictions flow down the hierarchy" (Hawkins, pp. 158-159).

Hawkins further theorizes that the hippocampus is the capstone to this connectionist hierarchy:

> "Unlike the neocortex, the hippocampus has a heterogeneous structure with several specialized regions. It's good at

> the unique task of quickly storing whatever patterns it sees. The hippocampus is in the perfect position, at the top of the cortical pyramid, to remember what is novel. It is also in the perfect position to recall these novel memories, allowing them to be stored in the cortical hierarchy, which is a somewhat slow process. You can instantly remember a novel event in the hippocampus, but you will permanently remember something in the cortex only if you experience it over and over, wither in reality or by thinking of it" (Hawkins, p. 171).

Hawkins' insights come from an ambitious effort to understand how the brain could work as a neural machine—but this understanding is partial: language seems to use a different sort of skill set than recognition and prediction. And although action is hinted at, such (and also in connection: motivation) is not adequately addressed.

GENETIC UNIVERSALS

"Innate" abilities and "instincts" may, to some extent, shape human nature in a universal fashion. Hofstadter claims:

> "The fact is that a large proportion of every human's network of symbols is *universal.* We simply take what is common to all of us so much for granted that it is hard to see how much we have in common with other people. It takes the conscious effort of imagining how much—or how little—we have in common with other types of entities, such as stones, cars, restaurants, ants, and so forth, to make evident the large amount of overlap that we have with randomly chosen people. What we notice about another person immediately is not the standard overlap, because that is taken for granted as soon as we recognize the humanity of the other person; rather we look beyond the standard overlap and generally find some major differences, as well as some unexpected, additional overlap" (Hofstadter, 1999, p. 376).

Beyond Chomsky's universal grammar (and the minimalist program), Steven Pinker has argued for a robust genetic inheritance of human nature and attitudes in his books *The Blank Slate* and *The Stuff of Thought*. These genetically inherited abilities could tie in with Donald Davidson's "principle of charity"—that we all share a core-cluster of beliefs and attitudes—and also ties in with W.V. Quine's thought where one's web of beliefs might have more stability towards a central core, while being more contingent towards the periphery. Moreover, examine the lengthy list of human universals at the end of this chapter; this list—compiled in D.E. Brown, *Human Universals*, McGraw-Hill: New York (1991); and also found in Steven Pinker, *The Blank Slate*, Viking Press: New York (2002)—is composed of behaviors and language uses found in *all* cultures.

Such an extensive list of human predispositions and ways of being could definitely give one pause for thought: are all these aspects encoded in our DNA? Or rather, do they emerge as inevitable consequences of a more limited human nature (including our bodies and some initial ability to "take-up" culture) engaged with an external environment that helps shape these cross-cultural constants? This question is also faced by George Lakoff and Mark Johnson who have studied common metaphors and have theorized that such are structured in part by being embodied in a specific kind of body and environment.

SITUATED MODALITY ANALOGIZING

In their book *Metaphors We Live By*, George Lakoff and Mark Johnson developed a theory of "conceptual metaphor" to account for much of how we think, not just linguistically, but at fundamental and deep level. They also develop this in their later work, *Philosophy in the Flesh*, which title highlights their "embodied" approach to cognitive science—an approach that goes back at least to Kant's schemata and Copernican revolution which shifted an emphasis from concerns with objects to subjective modality (a shift that can be seen in his own thinking between early and later drafts of *The*

Critique of Pure Reason); and which also has roots in Nietzsche and the 20th century "fleshy" phenomenology of Maurice Merleau-Ponty. But the work in *Metaphors We Live By* is also original in its approach to and description of how metaphors are intertwined with our thinking.

Metaphor is a way of defining or relating one idea in relation to another—and goes so far as to say the one IS the other, such as with "time is money" (here "time" would be the "tenor" and "money" would be the "vehicle" of our metaphor example). Such a definitive comparison will bring the whole family-constellation of associations that each term has in relation. Because a "succinct" metaphor will make some sense there should be some overlap in the associated ideas, and not some long trail of "six-degrees of separation" needed to connect them. The ideas developed in this chapter have required that we understand terms, words, ideas, concepts, categories, etc, as being connected in a network of associations where more abstract and common concepts (like "space") could be seen as hierarchical hubs that potentially have a lot more connections than, say, a less common concrete word, like "chagrin."

Much of *Metaphors We Live By* is dedicated to providing examples of how our idioms, common turns of phrase, clichés—and often what are termed "dead metaphors" (metaphors that have become currency in our language, rather than poetic stumbling blocks towards an insight), these ways of using language are often rooted in deeper metaphorical structures (connected to those hierarchical "hubs" mentioned above—but here, they would be metaphorical concept hubs). The structure of these hubs, at a most fundamental level are connected in a direct way to the experience of being a mobile body interacting with a worldly environment. This is the *embodied* aspect of Lakoff and Johnson's thought: we seem to have a deep conceptual structure, fundamental to how we can understand anything that is related to our perceptions of and interactions with the world, based on our human experience:

> "what we are claiming about grounding is that we typically conceptualize the nonphysical *in terms* of the physical—that is, we conceptualize the less clearly delineated in terms of the more clearly delineated" (George Lakoff, and Mark Johnson, *Metaphors We Live By*, University of Chicago Press: Chicago (1981), p. 59).

For example, consider the many expressions engendered by the metaphorical concept "time is money" (which could be considered high on the hierarchy of deep structural metaphors, as "time" and "money" are somewhat generic concepts that may go back fairly far in human history):

> "TIME IS MONEY
>
> You're *wasting* my time.
>
> This gadget will *save* you hours.
>
> I don't *have* the time to *give* you.
>
> How do you *spend* your time these days?
>
> The flat tire *cost* me an hour.
>
> I've *invested* a lot of time in her.
>
> I don't *have enough* time to *spare* for that.
>
> You're *running out* of time.
>
> You need to *budget* your time.
>
> *Put aside* some time for ping pong.
>
> Is that *worth your while*?
>
> Do you *have* much time *left*?
>
> He's living on *borrowed* time.

You don't *use* your time *profitably*.

I *lost* a lot of time when I got sick.

Thank you for your time.

Time in our culture is a valuable commodity. It is a limited resource that we use to accomplish our goals. Because of the way that the concept of work was developed in modern Western culture, where work is typically associated with the time it takes and time is precisely quantified, it has become customary to pay people by the hour, week , or year" (Lakoff and Johnson, p. 8).

Here, the more abstract concept of "time" is understood with reference to the more worldly, material (things, or relationship between things) and observable concept of money. In a way, the universal and particular are reversed in the metaphorical conceptual hierarchy, in that it is particular concrete experience that grounds more abstract metaphors—but again, this concrete experience itself could be hierarchically organized with the more "universal" aspects of being human providing a deeper anchor for our metaphorical concepts:

> "*we feel that no metaphor can ever be comprehended or even adequately represented independently of its experiential basis*" (Lakoff and Johnson, p. 19).

And there appears to be a core constellation of primitive concepts:

> "Just as the basic experiences of human spatial orientations give rise to orientational metaphors, so our experiences with physical objects (especially our own bodies) provide the basis for an extraordinarily wide variety of viewing events, activities, emotions, ideas, etc., as entities and substances" (Lakoff and Johnson, p. 25).

Does understanding only require primitive concepts based in experience, and metaphorical concepts that are built on these basic concepts? Of course, we have learned from Derrida, Wittgenstein, et. al, that *context* is crucial to understanding (or at least meaning), but there is some distance to cover between (metaphorical) concepts and contexts, and a midway category that Lakoff and Johnson use is that of *experiential gestalts*. Gestalts are clusters of components where the whole can be more basic than the parts; and such can create a coherence; for example we may have a gestalt for "money" that includes reference not only to other concepts ("gold," "labor," "spending," etc.) but to a whole constellation of cultural practices (that people go to work, earn money, spend it, etc.):

> "Structuring our experience in terms of such multidimensional gestalts is what makes our experience *coherent* [.....] Understanding such multidimensional gestalts and the correlations between them is the key to understanding coherence in our experience. As we saw above, experiential gestalts are multidimensional structured wholes. Their dimensions, in turn, are defined in terms of directly emergent concepts" (Lakoff and Johnson, p. 81).

Steven Pinker, who has been presented as a level headed champion of genetic human nature in the present book, singles out Lakoff (who has also written books on political and social propaganda—hence emphasizing the pliable aspects of a worldview) as going too far with metaphor theories:

> "Mental life begins with a few experiences that are not metaphorical, namely, the sensations, actions, and emotions that are built into our constitution and engage the physical world. From there, conceptual metaphors are acquired by a kind of associative conditioning. We learn that CONTROL IS UP because we experience fights in which the victor ends up on top, that GOALS ARE DESTINATIONS because we walk toward something we want, and that TIME IS A MOVING OBJECT because things that approach us get closer and closer as time elapses.

> But this isn't the half of it. Since we think in metaphors grounded in physical experience rather than in logical formulas with truth values, the entire tradition of Western thought since the Greeks is fundamentally misconceived. Reason is not based on abstract laws, because thinking is rooted in bodily experience. And the concept of objective or absolute truth must be rejected. There are only competing metaphors, which are more or less apt for the purposes of the people who live by them [....]
>
> Though I believe that conceptual metaphor really does have profound implications for the understanding of language and thought, I think Lakoff takes the idea a wee bit too far" (Steven Pinker, *The Stuff of Thought: Language as a Window into Human Nature*, Penguin Books: New York (2007), pp.245-247).

Pinker makes Lakoff out to be somewhat of a Nietzschean pragmatist, but not a completely a-rational and relativist "postmodernist." As I've noted before, I think academic polemics tend to exaggerate the differences in opinion—and a modicum of nuance often brings various views closer.

HARDWARE / SOFTWARE

An issue related to the nature / nurture debate (genetic abilities vs. social constructs) is that of the hardware / software schism and how this might apply to human thinking:

> "We humans also have 'software' and 'hardware' aspects, and the difference is second nature to us. We are used to the rigidity of our physiology: the fact that we cannot, at will, cure ourselves of diseases, or grow hair of any color—to mention just a couple of examples. We can, however, 'reprogram' our minds so that we operate in new conceptual frameworks. The amazing flexibility of our minds seems nearly irreconcilable with the notion that our brains must be made out of

fixed-rule hardware, which cannot be reprogrammed. We cannot make our neurons fire faster or slower, we cannot rewire our brains, we cannot redesign the interior of a neuron, we cannot make *any* choices about the hardware—and yet, we can control how we think.

But there are clearly aspects of thought which are beyond our control. We cannot make ourselves smarter by an act of will; we cannot learn a new language as fast as we want; we cannot make ourselves think faster than we do; we cannot make ourselves think about several things at once; and so on. This is a kind of primordial self-knowledge which is so obvious that it is hard to see it at all; it is like being conscious that the air is there. We never really bother to think about what might cause these 'defects' of our minds: namely the organization of our brains. To suggest ways of reconciling the software of the mind with the hardware of the brain is a main goal [....]" (Hofstadter, 1999, p. 302).

Dennett expands this notion:

"The 'software' or 'virtual machine' level of description I have exploited [is ...] not explicitly physiological or mechanical and yet capable of providing the necessary bridges to the brain machinery on the one hand, while on the other hand not being explicitly phenomenological and yet capable of providing the necessary bridges to the world of content, the worlds of (hetero-)phenomenology" (Dennett, p. 434).

Here, Dennett summarizes his view of consciousness:

"Human consciousness is *itself* a huge complex of memes (or more exactly, meme-effects in brains) that can best be understood as the operations of a '*von Neumannesque*' virtual machine implemented in the *parallel architecture* of a brain that was not designed for any such activities. The powers of this *virtual machine* vastly enhance the underlying powers of the organic *hardware* on which it runs, but at the same time many of its most curious features, and especially its limitations, can be explained as the byproducts of the *kludges* that make possible this curious but effective reuse of an existing organ for novel purposes" (Dennett, p. 210).

Moreover:

> "Suppose for the moment then that there is a more or less well-designed (debugged) version of this stream-of-consciousness virtual machine—the Joycean machine—in the memosphere. As we have seen, since there is no shared machine language between brains, the methods of transmission that would guarantee a fairly uniform virtual machine operating throughout the culture must be social, highly context-sensitive, and to some degree self-organizing and self-correcting." (Dennett, p. 220).

Evidently, Dennett sees his "virtual machine" as bridging the decentralized aspects of a brain's neuronal functioning, and the seemingly integrated flow of thoughts we have in a stream of consciousness. Now, it has been demonstrated that von Neumann machines (basically, physical instantiations of the abstract Turing machine), can simulate the sort of parallel neural networks found in connectionist, neural nets, or parallel distributed processing (see the "Scripting Thought" section above). But Dennett suggests we:

> "turn this standard idea upside-down. Just as you can simulate a parallel brain on a serial von Neumann machine, you can also, in principle, simulate (something like) a von Neumann machine on parallel hardware, and that is just what I'm suggesting: Conscious human minds are more-or-less serial virtual machines implemented—inefficiently—on the parallel hardware that evolution has provided for us" (Dennett, p. 218).

An objection to this might go as follows: a von Neumann machine (like a Turning Machine) requires quite precise logical-mechanical instantiation—*in principle*, this can be done with neural

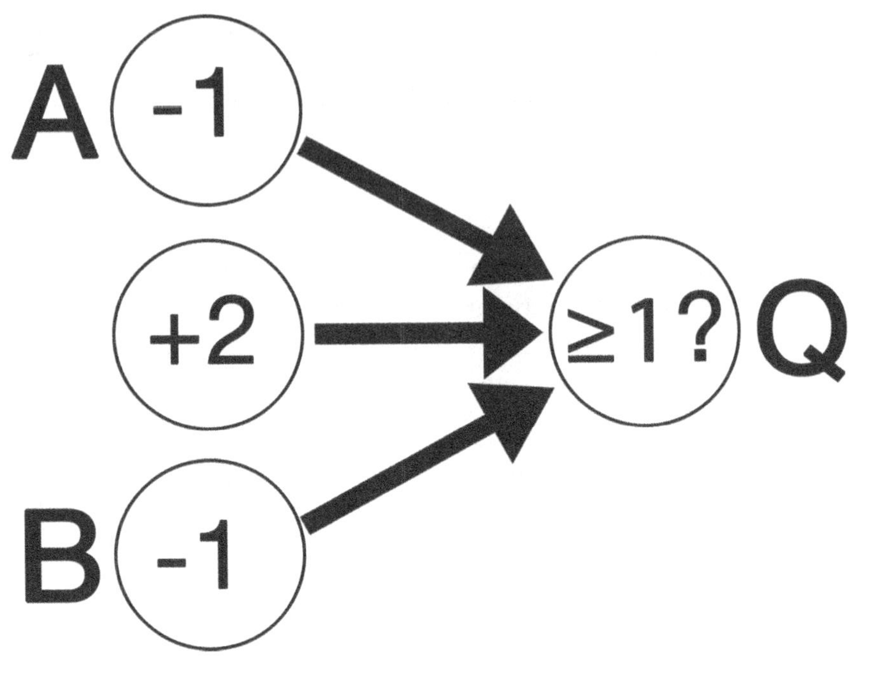

NAND "Perceptron"

networks, as we can see in the diagram above and in the diagrams at the start of this chapter, that, for example, a NAND operation (the NAND or not-both-and relates to the Sheffer stroke and law of non-contradiction—not both A and not A—and can generate all other logical operations through recursive application)—this NAND operation can be instantiated with relays, diodes, and even with a neural-network-like perceptron. Given that a neural network *can* perform logical operations, one might suggest that a brain *could* "host" a virtual computer—again, in *principle*. But how the brain could simulate such a von Neumann / Turing machine based on what it *learns* via public memes would require some explanation. We *can* learn to pretend we are a computer, and follow precise instructions relevant to data—this was how Turning conceived his universal computing machine. But if we have this ability to follow directions, we already have the virtual machine that Dennett claims integrates our parallel hardware into a more Turning-machine like focus on a discrete stream-of-information.

My suggestion would be that we look for an integrating mechanism in other biological structures in the brain, other than the "neural-nets" that make up the cortex—perhaps the thalamus, which is supposed to regulate consciousness, sleep and alertness, and mediates between the cerebral cortex and the midbrain, relaying sensory and motor signals to the cortex. The thalamus could operate something like the "read / write head" of a Turning machine. But Dennett dismisses the possibility of a thalamus integration point, in reference to a "searchlight" theory of consciousness:

> "the branches radiating from the thalamus to all parts of the cortex fits it for the role of a 'searchlight,' differentially arousing or enhancing particular specialist areas, recruiting them to current purposes [.... but, in footnote 5:] Searchlight theories of attention have been popular for years. Crude theories make the mistake of supposing too literally that what the searchlight differentially illuminates or enhances at a moment is a region of *visual* space—exactly the way a spotlight in a theater can illuminate one region of the stage at a time. More defensible—but also at this time more impressionistic—searchlight theories insist that it is a portion of *conceptual* or *semantic* space that is differentially enhanced (imagine, if you can, a theater spotlight that can pick out just the *Capulets*, or all and only the *lovers*)" (Dennett, p. 274).

My suggestion is not to see the thalamus as the locus of consciousness (somewhat like Descartes locating the mind-body juncture in the pineal gland)—but rather to see it as another type of neural configuration, unlike neurons instantiating semantic or connectionist networks—that is able to integrate, or rather, simply focus attention on specific items, much like the Turing machine read / write head. How the decentralized aspects of brain functioning could self-organize into a single stream-of-consciousness via the "software" of the brain, rather than the "hardware" is beyond my immediate imagination. Software is composed of the instructions and data that memory can hold. Memes can be both data and "instructions" (ways of doing things)—but we must distort the "software" metaphor somewhat, if we understand the operation of memes as "ways of doing" and not "instructions."

There could be *emergent* global behavior—much like individual ants creating a colony—but this integration on higher levels is as mysterious as the integration of living creatures. Yet, we do understand *living creatures* often via theories of evolution, particularly with the notion of "natural selection," so it may be no coincidence that Dennett also wrote a book on this entitled *Darwin's Dangerous Idea*. Although I agree with Dennett as to the importance of natural selection as an explanatory concept, I differ from his evolutionary views on four key points:

1. Dennett's de-emphasis of environmental "context" in favor of DNA building blocks.

2. Dennett's speculation on "absolute origins" out of chaos.

3. I see a "circularity" in the construction of intelligence.

4. Dennett's disregard of non-algorithmic mechanics.

1. Although DNA variation in organisms may be slow to change, there might be acceleration or a relatively "catastrophic" change in organisms due to rapid changes in an environment—Dennett's discussion of the factors that change organisms over time (as well as most evolutionists') tends to focus on the "change in the machine parts" rather than examining the environment (or larger eco-organism) they are a part of. Linguists too often seem to focus on constituting rules and elements with less regard to context (environment).

2. Dennett hints that life and physics erupt out of "chaos" (Daniel Dennett, *Darwin's Dangerous Idea: Evolution and the Meaning of Life*, Simon and Schuster: New York (1995), pp. 68, 74)—this

is intriguing, and I tend to agree—but is chaos *infinitely varied* structure, or is it a-structural? A chaos of what? And if chaos is infinitely varied structure—and if there are an infinite number of possibilities—couldn't it take an infinite amount of time to get to our possible universe (and hence, probably, never get here?)

3. If intelligence is designed by non intelligent "algorithms" (natural selection) is not intelligence already artificial (i.e. not real—an epiphenomena)? Or, if simple origins can have complex results, is not the complex *latent* in the simple (like the full tree being a potential in a seed)? In regard to intelligent design, why not define intelligence in such a way (e.g. "*making connections and developing skills that work and survive: via natural selection*") as to say that the *process* of evolution is intelligent itself: much as intelligence may be created by natural selection, natural selection *is intelligence* and hence the *there IS intelligent design*, but not from some transcendent point outside the system or universe, but the system or universe is intelligent in itself and through its processes.

4. Maybe Roger Penrose (in *The Emperor's New Mind*) should have claimed that intelligence couldn't be replicated by an *algorithm*—especially on a Turing machine—but not that a Turing machine couldn't replicate intelligence, since Turing machines can replicate non-algorithmic mechanisms (e.g. hypothetically, the structure of a brain could be replicated by a computer). I'm not a big fan of the Churchland's, et. al, Parallel Distributed Processing style neural-networks as a global solution to AI; but would we call *spreading activation* (the activation from neurons to neurons) an algorithmic process? Spreading activation is related to the "daemons" Dennett discusses in *Consciousness Explained*—those primary unconscious processes fighting for emergence, where hard and fast *rules* are not followed, but weighted associated connections allow for the flow from one idea or action to another, like a complex of reflex's (see chapter 2, "Style, Taste & Cyber-Networks"); Yes, biology has its laws—but there is a difference between mechanical regularity and programs.

All four of these points relate to structure:

1. Is the totality of structure infinitely complex in an ever expanding *environment / context*, and hence "un-saturate-able" and unknowable? We may identify some building blocks, but never be able to account for every aspect of environmental context (beyond the pragmatically useful) in such a way as to *completely* understand a system—e.g. just locally on earth there are the complex factors of the sun's effect on the weather which affects erosion which affects local habitat... just to name one a-b-c-d line that doesn't bring in the whole complex *web* of over-determined factors, that affect each other (see also the discussion of the "frame problem" below).

2. Is not the possibility of structure's *existence in the first place* a *mystery* and hence unknowable?

3. Is not structure, through knowledge, language and intelligence, always *self-referential*? (Structures relate to structures and nothing a-structural).

4. Structure is more flexible than mere logic or algorithms, although our physical world at least seems "consistent."

A problem with structure has to do with "centers"—and one may ask, "Why are some experiences more *lucid* than others, when we could be simply sleepwalking through life?"

Decentralized and parallel mechanics have come to the fore in 20th century cognitive science, where, for example, it may be theorized that perceptual cue-clusters and multiple muscular action mechanisms connect not just words, but also connect metaphors, and more complex thought analogies, and even connect aspects of vision, such as when numerous right angles, white surfaces, and black dots bring to mind Las Vegas (all operating without a mapping of an image onto an object, but by having our orientation to the object activate our sense and manipulation organs). Of course, a stream-of-consciousness (the mind that, e.g. just thought of Las Vegas) incorporates memes and cybernetic information flows that we "swim" in as agents, and which operate through us as we embody culture and worldly experience, and interact with that same culture and world through a *unified* body in constant mutually determining interactivity.

However, with this decentralizing view, when do we stop decentralizing, and what gives us the ability to think or experience unities or wholes? What determines a *focus*—not a point, but a fuzzy "thisness," "hereness," "nowness," (cp. Hegel) which seems tied up in anything we call a unified object or thing: How is one to distinguish a robot from a car it is driving if it is not perceived relative to some perspective, and how could this perspective be *centered*? What brings an agency together?—you might say a certain assemblage of parts interacts with a world in a manner that gets things done with other assemblages of parts; but without *some* (at least one) center of perception we might have a completely de-centered concrete physical world of causality—and a completely de-centered reality means there are no unities and hence no objects; and I see no way to have a unity without a this-here-now centered focus of *consciousness*: live subjectivity.

A question: where does a center, and hence the possibility of a unified object or thing (other than possible "atoms")—a necessity of the Objective itself—come from if not from conscious *subjective* (or inter-subjective) perspectives? Where do the more abstract and unified (as contrasted to the concrete and decentralized) levels of *design* and *intention* come from, if not from some sort of *focal* consciousness (not a representational Cartesian theater, but just a focus of qualia awareness?) From what vantage point are these less concrete and less de-centered patterns and purposes organized?

Yes, the design of a computer program is isomorphically capable of being instantiated on different physical systems, but those physical systems may simply be understood by paradigms (designs) that are themselves isomorphic—there may be a continuum rather than a schism between software and hardware; software is a like a particular configuration of a flexible machine—could we analyze the variable configurablity of the machine down to the point where the hardware is just simply the material it is made of? Also, why would static ROM (Read Only Memory) be software, but the active CPU (Central Processing Unit) with *variables* be hardware?—both are unchangeable aspects of the machine.

An analogy: with a fractal repeating its structures on ever larger scales, how do we know which scale of the fractal we are looking at? Would it be relative to the size of our human bodies? What if our bodies did not have subjective experience? Then *what mechanism of judgment would observe or make an object that would be a centralized object to a robot?*

The question in brief: **Are there focal centers? In the world, or only when perceived by a live unity?**—and not just things, but *groups* of things that are distinguished from other groups of things.

I'm not certain if this area of inquiry has answers: my vague stab at the *qualia* focus thing (where consciousness does not have to be divided between an a subject and an object) would be to think of reconstructing the brain "around," to use a visual analogy, light divided by boundary-like differentiating laws—think not of material *things* interacting, but of experiences shaped by real and hard physical *laws*; laws that structurally shape "a-structural" (light) consciousness (there being a distinction between the borders that structurally shape an experience, and the structured experience itself)—hence a shift in physics away from a matter of things, to laws of relations (relations between the unknowables—structurally unknowable—that we can experience).

Would natural selection be a general program for wetware? In his book, *Mind, Language, Machine: Artificial Intelligence in the Poststructuralist Age*, Michael L. Johnson asks:

> "If one conceives of human life as a program, then what are its subprograms? Young distinguishes four: the 'fundamental program' (which is 'inherited, written in triplets of bases of the DNA code'), that 'embodied in the structure of the brain', that of oral language, and that of written

language. The brain 'contains the scripts of the programs that issue in human action', and 'the detailed characteristics' of its cells 'provide the code signs for [...] a detailed model of the world'. Thus he embraces 'the theory of signs as the basis for all knowledge', and he would seem to agree with Thom's essentially constructivist argument that 'the physical properties of the outside world' are defined by 'the constraints of the dynamic of our brain'. (Reality is patterns of neurochemical frissons.)" (Michael L. Johnson, *Mind, Language, Machine: Artificial Intelligence in the Poststructuralist Age*, St. Martin's Press: New York (1988), p. 75; quoting J.Z. Young, *Programs of the Brain*, Oxford University Press: New York (1978), pp. 10-11; and René Thom, D.H. Fowler (trans.), *Structural Stability and Morphogenesis: An Outline of a General Theory of Models*, Benjamin: Reading, MA (1975), p. 11).

Deconstructionists frequently saw off the limb they are sitting on—it is a way to self-critique, and admit the impossibility of doing something and explicitly knowing what you are doing at the exact same time: blindness and insight can go hand in hand. But perhaps neurons do have the ability to understand neurons, much as a network can fold over on itself: the pivot being the locus of subjective rupture as one takes a critical stance on one's history and context. Such is not to say we could look at our brain in a mirror, and suddenly comprehend our own thoughts as neural activity, anymore than dissecting a frog will give us the power to revive it. There is an irony involved when arguing that knowledge is "constructed"—but, as we have seen, our thoughts may be largely produced by unconscious processes too—our very stream-of-conscious reasoning may be constructed: artificial. Again, we may have always already been artificially intelligent.

Irony or not, we have the ability to pragmatically get things done with our "knowledge." It works in a real world. We do have the ability to separate our agency to some extent from our environment. But what is it that keeps this agency semi-autonomous—

self-organized with autopoiesis, etc? Aristotle might hold a clue with his theory of the soul as the unifying form of the body. Could it be that software holds the hardware together? Dennett seems to claim that a "virtual machine"—that is, a software instantiation or simulation of hardware, that this "virtual machine" integrates our various inclinations and forethoughts which are in competition for our attention. Could it be that software is a form of writing... and like Derrida's arche-writing, it is the very iterability—that *something* found in the "original" that allows it to be copied—like programs isomorphically being translatable from one machine to another—digital replication—the copy-ability that precedes the possibility of the original being formed: is software information itself? But with a *virtual* machine the distinction between information as signal and the carrier of information itself collapses: the medium is the message (to quote Marshal McLuhan's book title).

FRAMING THE FRAME PROBLEM

Content will not disappear into form completely though. Beyond the hardware medium as formal signal carrier and software message as informational content, we have the modality of agencies. Classically, at least since Descartes, representation has set the difference between *the* world and *my* world. No doubt, the idea of madness or of being radically deceived by the senses was necessary in order to split the subject from the world in a "pure" way: we see this in St. Augustine's "si fallor sum" ("if I am mistaken, I am")—*Hamlet* too brings dubitable sense (the sighting of the ghost of Hamlet's father) in relation to the soliloquy revealing one's interior thoughts ("to be, or not to be"). Representation relates to information—our "copies" of the world inside our heads. With his later critique of his earlier representational philosophy, Wittgenstein seems to have thought he dissolved this representational split between interior feelings and exterior world: language is a public, rather than a private, matter—and so are our feelings! We do have idiosyncratic personal histories though, and our understanding of the world, even if we might not call it a representation of the world, can cause us to be at odds with reality, at least to some degree. Somewhere between our DNA designed innate abilities and social memes (which are basically Plato's forms of skills and Wittgenstein's language games taken, by Richard Dawkins and others, into the realm of social natural selection)—somewhere between our brain as an organ, and our culture as learnable, we develop our own opinions—based on the brain and culture combined with personal experience.

What this can create, in computers as well as humans, is the "frame problem." Not directly related to Marvin Minksy's "frames" (which again, are like semantic networks colliding with memes)—the frame problem arises when an agent *acts* in the world—how is it that their world-view, system of beliefs, or knowledge in general about reality, is updated to reflect how that act has changed the world? Such a conundrum, is bifurcated by the general tendency of solutions to this problem (that nothing changes in the world, unless we observe the change) and notions like the "butterfly-effect" (where small changes in one place may result in large changes elsewhere). The frame problem can also be seen from the agent's perspective; the "qualification problem" (the impossibility of knowing everything we must know for an act to have the intended consequence), and the "ramification problem" (the inability to know what consequences our actions will have indirectly).

But pragmatically, we do seem able to negotiate a somewhat uncertain world. This may be due to the robust nature of our environment's structure, and the fact that the environment, as "external memory," is available constantly through our senses, relating how our actions affect our local surroundings. We can experiment... society has experimented for ages... and hence we can have at least a modicum of reliable expectation, based, at least, on scientific knowledge. Trial and error might be a place to start, but we've come a long way—or at least have survived for quite awhile.

FUN HOUSE MIRROR

The cubism of the early 20th century could be said to anticipate the framing of pictorial thinking in Wittgenstein's later philosophy. Picasso's "analytic cubism," which took apart objects, reconstructing their three dimensions in two, preceded a "synthetic cubism" which took textured fragments and constructed compositions as with a collage technique. Wittgenstein's shift from early interests in pictorial form to later interests in families of language games also parallels the analytic-to-synthetic switch: moving from picturing the formal aspects of an object—a sort of perspective formal logic, to the use of cultural artifacts and weaving ideas together.

> "I hope that what I have been saying has made clear why I chose 'Philosophy and the Mirror of Nature' as a title. It is pictures rather than propositions, metaphors rather than statements, which determine most of our philosophical convictions. The picture which holds traditional philosophy captive is that of the mind as a great mirror, containing various representations—some accurate, some not—and capable of being studied by pure, nonempirical methods. Without the notion of the mind as mirror, the notion of knowledge as accuracy of representation would not have suggested itself. Without this latter notion, the strategy common to Descartes and Kant—getting more accurate representations by inspecting, repairing, and polishing the mirror, so to speak—would not have made sense. Without this strategy in mind, recent claims that philosophy could consist of 'conceptual analysis' or 'phenomenological analysis' or 'explication of meanings' or examination of 'the logic of our language' or of 'the structure of the constituting activity of consciousness' would not have made sense. It was such claims as these which Wittgenstein mocked in the *Philosophical Investigations*, and it is by following Wittgenstein's lead that analytic philosophy has progressed toward the 'post-positivistic' stance it presently occupies. But Wittgenstein's flair for deconstructing captivating pictures needs

to be supplemented by historical awareness—awareness of the source of all this mirror imagery—and that seems to me Heidegger's greatest contribution" (Richard Rorty, *Philosophy and the Mirror of Nature*, Princeton University Press: Princeton, NJ (1979), p. 12).

"Wittgenstein's argument is that a picture—even a mental picture—cannot determine its own meaning but that its meaning will be determined by its use or application, its insertion into a certain sequence" (Henry Staten, *Wittgenstein and Derrida*, University of Nebraska Press: Lincoln (1986), p. 73).

"the simples Wittgenstein came to reject were logical atoms—either 'states of affairs' (*Sachverhalten*) logically independent of one another or objects free from any contingent or empirical properties" (Newton Garver and Seung-Chong Lee, *Derrida and Wittgenstein*, Temple University Press: Philadelphia, PA (1994), p. 90).

"5.511 How can logic—all embracing logic, which mirrors the world—use such peculiar crochets and contrivances? Only because they are all connected with one another in an infinitely fine network, the great mirror" (Ludwig Wittgenstein, D.F. Pears and B.F. McGuiness (trans.), *Tractatus Logico-Philosophicus*, Routledge: London (1988), p. 49).

"for Aristotle [...] the voice, producer of *the first symbols*, has a relationship of essential and immediate proximity with the mind [....] It signifies 'mental experiences' which themselves reflect or mirror things by natural resemblance" (Jacques Derrida, Gayatri Chakravorty Spivak (trans.), *Of Grammatology*, The Johns Hopkins University Press: Baltimore, MA (1997), p. 11).

"Derrida [....] has even said that an atomic bomb is a text (1985, 20). It seems that the term 'text' applies to all that can be expressed in some sort of language [....] In the most general sense, a text is anything that makes use of symbols, that means something whose formal substance differs from its material substance, and whose import is deferred" (Garver and Lee, pp. 111-112; citing Jacques Derrida, "Deconstruction in America: An Interview with Jacques Derrida," *Society for Critical Exchange* 17 (1985), pp. 1-33).

"[Differance...] is a *grammatical device*, a trace structure constructed as a baffle, to be installed in syntax so that it keeps the sign sequence from cohering in the ways dictated by the habitual norms of philosophical discourse." (Staten, pp. 22-23).

"This book's title, *The Tain of the Mirror*, alludes to that 'beyond' of the orchestrated mirror play or reflection that Derrida's philosophy seeks to conceptualize [....] In spite of my contention that Derrida's philosophy must be related to the modern history of the concept of reflection and to the criticism it has drawn, I seek primarily to bring into view Derrida's debate with traditional paradigms of philosophy in general. The speculative form in which Hegel cast the unvarying philosophical topoi, and even their Husserlian or Heideggerian phenomenological form, are, undoubtedly, because of their strategic importance for Derrida's writings as a whole, privileged means of access to this thinker's discourse" (Rodolphe Gasché, *The Tain of the Mirror: Derrida and the Philosophy of Reflection*, Harvard University Press: Cambridge, MA (1986), p. 6).

If Wittgenstein's thought resembles the bifurcated phases of Picasso's analytic and synthetic cubism, Derrida could be connected to the Dada art of Marcel Duchamp, as illustrated in the post-cubist "Nude Descending a Staircase, No. 2" contrasted with the Readymade urinal "Fountain." As with the descending nude, Derrida's concepts chart a temporal trace... and like a Readymade, deconstructive close readings reconstruct the reception of a text, fracturing its narrative, re-contextualizing it within the context of inter-textuality: Not exactly a mustache on the Mona Lisa, but perhaps more like a collector's seal on a Chinese painting.

DIFFÉRANCE AND SPREADING ACTIVATION

One of the few thinkers combining cognitive science and deconstruction, Gordon Globus, has endorsed the connectionist network scheme, or at least neural semantic networks, noting that Derrida's quasi-concept of *différance* can be captured in the weighted memory of the networks:

> "the difference of *différance* depends on the weighted connectivity of neural nets (i.e. the memory trace) and the deferral of *différance* depends on tuning the weights, which alters the probabilities of paths in the network state space so that the most direct path to the attractor is breached and alternative paths opened. *Neural nets realize difference deferral. The operation of différance is characteristic of neural nets*" (Gordon Globus, *The Postmodern Brain*, John Benjamins Publishing Company: Amsterdam (1995), p. 58).

To clarify:

> "The pattern of connections is the fixed wiring diagram of the net. The adjustable weights constrain the network operation and carry the net's 'knowledge.' Because of the rich interconnectivity, then net always operates as a whole. The nodes that comprise the net are thus distinguished by virtue of their functional interrelationships. What distinguishes a node is its pattern of weighted primary connections that secondarily open to the whole.
>
> The sign qua signifier / signified is also a node in this sense of having a pattern of complex relationships to other signifiers / signifieds and through them to the whole. So language as a system of interdependent signs can be thought of as neural net, with each sign assigned a node" (Globus, p. 53).

The sort of neural nets Globus talks about are the semantic nets that we opened this chapter with—and the inverse connection between *différance* and *spreading activation* is fairly clear, although not specified by Globus. The *différance / spreading activation* relation is inverted, since Derrida writes about the differentiated deferral of meaning into and in relation to the rest of a semantic network; whereas although spreading activation can also activate from a given node on into the rest of the network, a node is activated in itself from the relationally decentralized connections and inputs from other nodes in the network. While a "judge" might chase differences indefinitely looking for full presence of meaning (which is never found in moving from one word to another, although *familiarity* occurs in the webbed *process* over time—we come to know words by repetition and context)—that same agent may have words activated into awareness via unconscious and decentralized sub-activations of related words and inputs). In a parallel, while the earlier Derrida might note the endless pursuit of meaning, the later Derrida would be open to the reception of something emerging from the future, as insights towards a problem solution can semi-self-organize over a night of "sleeping on it": welcome your enlightenment, as well as working towards it.

Networks form a common ground for bridging deconstruction and cognitive science—as with Globus, but also with Christopher Vitale's forthcoming work, *The Networked Mind: Cognitive Science, Artificial Intelligence, and Post-Structuralist Philosophy*:

> "By means of a genealogy of the role of the binary opposition in regard to theories of computation, the mind, and theory, *The Networked Mind* shows how binary paradigms emerged, and then examines new 'soft' computing technologies to show how binary models are being displaced within cutting edge computing and artificial intelligence research. Relational and networked oriented approaches at the level of units of computation (fuzzy systems theory), relations between units (artificial neural networks), and meta-relations between modules (genetic / evolutionary programming)

provide models for non-binary thinking at the level of node, link, and module" (Christopher Vitale, <//networkologies.wordpress.com/work-in-progress-the-networked-mind/>).

Michael L. Johnson also examines networks from a post-structural perspective:

> "A semantic network has not only a linguistic form but also 'a computational representation, a logical structure, and a conceptual content.' In the computational representation each node has a 'relationset' that 'encodes the information it represents.' Thus in Saussurean fashion, 'The meaning of any node is an ordering of the rest of the nodes of [the] network with which it is related.' Accordingly, as Simmons observes, in 'a richly interconnected network, the complete meaning of any particular node may involve every other node in the system'—an observation that recalls Boden's critique of conceptual-dependency diagrams and belies Simmons's desire for 'unambiguous concepts.' (In other words, semantic networks, like Schank's diagrams and similar devices, do not furnish any kind of 'center' that can check Derridean freeplay. To the contrary, they exemplify it.)" (Michael L. Johnson, pp. 134-135, citing R.F. Simmons, 'Semantic Networks: Their Computation and Use for Understanding English Sentences,' in Roger C. Schank and Kenneth Mark Colby (ed.s), Computer Models of Thought and language, Freeman: San Francisco (1973), p. 70).

DERRIDA AND NEURAL MECHANICS

Derrida himself comes close to connecting *différance* and spreading activation, or at least connecting his thinking with neurology, in his essay "Freud and the Scene of Writing." He begins by noting that:

> "Despite appearances, the deconstruction of logocentrism is not a psychoanalysis of philosophy.

These appearances: the analysis of a historical repression and suppression of writing since Plato. This repression constitutes the origin of philosophy as *epistēmē*, and of truth as the unity of *logos* and *phonē*" (Jacques Derrida, Alan Bass (trans.), *Writing and Difference*, The University of Chicago Press: Chicago (1978), p. 196).

Derrida probes deep into Freud's thought, especially concerning physical modeling of the psyche, from Freud's pre-psychoanalytic period (~1895) to his later work (~1925); and notes that Freud uses *metaphors* when trying to explain the most difficult to comprehend aspects of a psychic mechanism:

> "It is no accident that Freud, at the decisive moments of his itinerary, has recourse to metaphorical models which are borrowed not from spoken language or from verbal forms, nor even from phonetic writing, but from a script which is never subject to, never exterior and posterior to, the spoken word [....] The *structure* of the psychical *apparatus* will be *represented* by a writing machine" (Derrida, 1978, p. 199).

To situate the later (1925) metaphor of the "mystic writing pad" (a wax tablet with cellophane and wax paper sheets that one can write on and then erase by lifting the sheets), we should outline in brief, the neuronal structure that Freud describes in his 1895 work, *Project for a Scientific Psychology*. Here Freud notes that nerves can have a quantity of energy (Q) and tend to discharge this energy—being in an energized, "cathected," state brings tension, and possibly pain; while discharging the energy (cathexis) brings pleasure. There are three types of neuron, according to Freud: the ϕ (phi or permeable perceptual neurons) ψ (psi or impermeable memory neurons) and ω (omega or consciousness neurons). The ψ psi memory neurons have "contact barriers" (perhaps synapses), and the discharge of energy from one neuron can "facilitate" the charging of neurons it is connected to—and memory is formed in the ψ psi memory neurons by making the facilitating connection between neurons stronger through association, when two stimulations are simultaneous. Such is much like the network of neurons involved with spreading activation; and Qualities are experienced when ω omega consciousness neurons are activated.

Derrida sees the split between ϕ phi perception neurons and ψ psi memory neurons as a rupturing difference:

> "The path is broken, cracked, *fracta*, breached. Now there would be two kinds of neurons: the permeable neurons (ϕ), which offer no resistance, and thus retain no trace of impression, would be the perceptual neurons; other neurons (ψ), which would oppose the contact-barriers in the quantity of excitation, would thus retain the printed trace" (Derrida, 1978, p. 199).

And:

> "The rest of the *Project* will depend in its entirety upon the incessant and increasingly radical invocation of the principle of difference. Beneath an indicial neurology, which plays the representational role of an artificial model, we repeatedly find a persistent attempt to account for the psyche in terms of spacing, a topography of traces, a map of breaches; and we repeatedly find an attempt to locate consciousness or quality in a space whose structure and possibility must be rethought, along with an attempt to describe the 'functioning of the apparatus' in terms of pure differences and locations, an attempt to explain how 'quantity of excitation is expressed in ψ by complexity and quality by typography'" (Derrida, 1978, p. 205).

What Derrida seems to be noting is that Freud hypothesized that the brain is *modular*—that certain locations in the brain are divided from others relative to their functioning. But this difference is also one in experience—for it would seem that with the ω omega consciousness neurons, Freud was trying to locate consciousness at certain locations in the brain (e.g. between ϕ phi perceptual neurons and ψ psi memory neurons). By 1900, with his *Interpretation of Dreams*, Freud had moved on from physical to psychical explorations of the mind, but retained insights gathered from his early investigations—for instance, the notion that our verbal stream-of-consciousness is kind of hallucination. Freud describes a chain of neurons as such in an 1896 letter:

> "*W* [*Wahrnehmungen* (perceptions)] are neurons in which perceptions originate, to which consciousness attaches, but which in themselves retain no trace of what has happened. For *consciousness and memory are mutually exclusive*.
>
> *Wz* [*Wahrnehmungszeicchen* (indication of perception)] is the first registration of the perceptions; it is quite incapable of consciousness, and arranged according to associations by simultaneity.
>
> *Ub* (*Unbewusstsein* [unconsciousness]) is the second registration, arranged according to other (perhaps causal) relations. *Ub* traces would perhaps correspond to conceptual memories; equally inaccessible to consciousness.
>
> *Vb* (*Vorbewusstsein* [preconsciousness]) is the third transcription, attached to word-representations and corresponding to our official ego. The cathexes proceeding from this *Vb* become conscious according to certain rules; and this secondary *thought-consciousness* is subsequent in time, and is probably linked to the hallucinatory activation of word-presentations, so that the neurons of consciousness would once again be perceptual neurons and in themselves without memory" (Sigmund Freud, James Strachey (ed.), *The Standard Edition of the Complete Psychological Works of Sigmund Freud, Volume I (1886-1899): Pre-Psychoanalytic Publications and Unpublished Drafts*, Vintage Books: London (2001), pp. 234-235, "Extracts from the Fliess Papers – Letter 52, December 6, 1896").

In *The Interpretation of Dreams*, Freud moves from the notion that dreams can have a stable "key" to their interpretation (translating from the "encoded" latent content to a "decoded" manifest content)—to finding that dreams are more idiosyncratic, and connected to an individual's personal history. In either case though, the dream symbols are like writing itself—with a sign signifying something else. Yet, without a "key"—without some signified to translate a signifier into, the signifier / signified dichotomy breaks down. Derrida notes:

> "The absence of an exhaustive and absolutely infallible code means that in psychic writing, which thus prefigures the meaning of writing in general, the difference between signifier and signified is never radical. Unconscious experience, prior to the dream which 'follows old facilitations,' does not borrow but produces its own signifiers; does not create them in their materiality, of course, but produces their status-as-meaningful (*significance*). Henceforth they are no longer, properly speaking, signifiers. And the possibility of translation, if it is far from being eliminated—for experience perpetually creates distances between the points of identity or between the adherence of signifier to signified—is nevertheless in principle and by definition limited" (Derrida, 1978, p. 209).

Yet Freud *does* translate dream writing—sometimes a cigar is just a cigar, but often it is interpreted as some sort of phallic symbol. Such translation is an attempt to bypass the materiality of the signifier by way of the "ideal essence" of a signified." Again, Derrida:

> "Materiality is precisely that which translation relinquishes [....] since the materiality of the signifier constitutes the idiom of every dream scene, dreams are untranslatable [....] What is valid for a specific national language is a fortiori valid for a private grammar" (Derrida, 1978, p. 210).

Moreover:

> "If a dream cannot be translated into another language, it is because within the psychical apparatus as well there is never

> a relation of simple translation. We are wrong, Freud tells us, to speak of translation or transcription in describing the transition of unconscious thoughts through the preconscious toward consciousness" (Derrida, 1978, pp. 210-211).

It would seem that under Freud's earlier neuronal explanation, consciousness is an "effect" in ω omega consciousness neurons, produced by the filtering / screening ϕ phi perception neurons and ψ psi memory neurons:

> "That the present in general is not primal but, rather, reconstituted, that it is not the absolute, wholly living form which constitutes experience, that there is no purity of the living present—such is the theme, formidable for metaphysics, which Freud, in a conceptual scheme unequal to the thing itself, would have us pursue. The Pursuit is doubtless the only one which is exhausted neither within metaphysics nor within science" (Derrida, 1978, p. 212).

So much is to say that in line with contemporary cognitive science, we can see that our conscious experience lags behind unconscious processes. Such reinforces Derrida's claim that meaning does not come fully from an intent which is present in immediate speech, but rather involves arche-writing (much like Freud's dream-work of the unconscious) and continual *différance*. Indeed, according to Freud, our speech, as an extension of thought, is a kind of material hallucination externalized. However:

> "A certain polycentrism of dream representation is irreconcilable with the apparently linear unfolding of pure verbal representations. The logical and ideal structure of conscious speech must thus submit to the dream system and become subordinate to it, like a part of its machinery" (Derrida, 1978, p. 217).

This, linked with Derrida's earlier quote from Freud (Derrida, 1978, p. 218), noting that the dream work does not include words or images for "if," "because," "just as," "although," "either-or," etc.

demonstrates a severe lack in Freud's neural model. In fact, the model is not much more complicated than a semantic network with spreading activation. Of course, parts of the brain like Wernicke's area and Broca's area were not discovered in the 19th century—and Freud's particular modularity of the brain and account of neural memory was ahead of the its time (a modularity possibly inspired by phrenology, but a neural theory anticipating Donald "Hebb's Rule" from 1949 where "Cells that fire together, wire together").

Derrida concludes this essay on Freud, by turning to Freud's "A Note Upon The Mystic Writing Pad": here the original ϕ phi perception neurons are compared to the wax paper (with cellophane sheath) and the unconscious ψ psi memory neurons are compared to the wax of the tablet. Derrida notes however that:

> "A pure representation, a machine, never runs by itself" (Derrida, 1978, p. 227).

That is, with the mystic writing pad, a hand is required to lift the wax and cellophane sheets to "refresh" and erase the writing surface. In fact the "mystic writing pad" metaphor breaks down completely in its failure to reproduce memory traces on the surface of the sheets from its wax tablet surface. The metaphor is indeed a sort of patch to cover the unknown.

NOT SO FAST!

I don't want to leave the reader with the impression that spreading activation and *différance* are identical or (inversely) parallel—for Derrida explains that the effects of *différance* are not "in the brain":

> "*First, différance* refers to the (active *and* passive) movement that consists in deferring by means of a delay, delegation, reprieve, referral, detour, postponement, reserving. In this sense, *différance* is not preceded by the originary and indivisible unity of a present possibility that I could reserve, like an expenditure that I would put off calculatedly or for rea-

sons of economy. What defers presence, on the contrary, is the very basis on which presence is announced or desired in what represents it, its sign, its trace" (Jacques Derrida, Alan Bass (trans.), *Positions*, The University of Chicago Press: Chicago (1982), p. 8).

> "*Second*, the movement of *différance*, as that which produces different things, that which differentiates, is the common root of all the oppositional concepts that mark our language, such as, to take only a few examples, sensible / intelligible, intuition / signification, nature / culture, etc. As a common root, *différance* is also the element of the *same* (to be distinguished from the identical) in which these oppositions are announced" (Jacques Derrida, 1982, p. 9).

> "*Third*, *différance* is also the production, if it can still be put this way, of these differences, of the diacriticity that the linguistics generated by Saussure, and all the structural sciences modeled on it, have recalled is the condition for any signification and any structure. These differences—and, for example, the taxonomical science which they may occasion—are the effects of *différance*; they are neither inscribed in the heavens, nor in the brain, which does not mean that they are produced by the activity of some speaking subject. From this point of view, the concept of *différance* is neither simply structuralist, nor simply geneticist, such an alternative itself being an 'effect' of *différance*. I would even say [...] that it is simply not a concept" (Jacques Derrida, 1982, p. 9).

Différance, therefore, precedes the structure of neurons: "produces" structure itself as an effect—as well as (possibly) the "presence" of conscious qualia. The "concept" of *différance* is prior to and necessary for structure—and hence prior to and necessary for concepts themselves; hence Derrida's noting that *différance* is not a concept. To note that "deconstruction" or "*différance*" are not mechanical engines that could "get off the ground," any more than a lawnmower could fly, would be to do these quasi-concepts an injustice, in that they are pre-mechanical. However, the notion that *différance* might somehow *produce* qualia is extremely speculative—not only can we not *know* or *understand* this *différance* conceptually, the notion of something prior to the split between formal structure and qualia content creating both of them is debatable. Much like Daniel C. Dennett suggesting, along with others, that natural selection could "originate" our universe of physical laws out of chaos—just as this fails to explain why that chaos is even structure-able in the first place, Derrida's *différance*, as necessary as it may be for the formation of structure and the difference between structure and qualia (if there is such a difference—and the difference between visual colors and auditory tones suggests this), *différance* as a source for other kinds of traits—qualia, time, space... seems to be a convolution of qualitative differences. Yes, certain qualities (again, colors, tones), are different; but how could *différance* produce these as effects?—it seems to be to jump from one type of quality to another: to explain colors by time, tones by space, etc.

Perhaps by *différance*, Derrida had in mind something like a "vanishing point homunculus" subjective judge: that which actually differentiates qualitative experience, prior to such an experience occurring: this makes sense, when thinking about how "objects" and "structures" would be differentiated—even how blue is distinguished from green—but it would not explain how this hyper-negative-subjective "viewpoint" would, again, *produce* these effects that it judges as differentiated. The types and traits of the world are too complex and varied to be accounted for by a single type of trait; e.g. *différance*. Yet this may be a Derridian type of insight as well... origins are usually complex, not simple—and hence the irony of employing radically "reductionist" concepts, like *différance*. "Elegant" theories are those that explain the most with the least... they are economical; concepts like *différance* or *spreading activation*, may have wide application; but the exceptions not only prove the rule, they prove the rule is not absolute.

HUMAN UNIVERSALS

abstraction in speech & thought

actions under self-control distinguished from those not under control

aesthetics

affection expressed and felt

age grades

age statuses

age terms

ambivalence

anthropomorphization

anticipation

antonyms

attachment

baby talk

belief in supernatural/religion

beliefs, false

beliefs about death

beliefs about disease

beliefs about fortune and misfortune

binary cognitive distinctions

biological mother and social mother normally the same person

black (color term)

body adornment

childbirth customs

childcare

childhood fears

childhood fear of loud noises

childhood fear of strangers

choice making (choosing alternatives)

classification

classification of age

classification of behavioral propensities

classification of body parts

classification of colors

classification of fauna

classification of flora

classification of inner states

classification of kin

classification of sex

classification of space

classification of tools

classification of weather conditions

coalitions

collective identities

conflict

conflict, consultation to deal with

conflict, means of dealing with

conflict, mediation of

conjectural reasoning

containers

continua (ordering as cognitive pattern)

contrasting marked and nonmarked sememes (meaningful elements in language)

cooking

cooperation

cooperative labor

copulation normally conducted in privacy

corporate (perpetual) statuses
coyness display
critical learning periods
crying
cultural variability
culture
culture/nature distinction
customary greetings
daily routines
dance
death rituals
decision making
decision making, collective
differential valuations
directions, giving of
discrepancies between speech, thought, and action
dispersed groups
distinguishing right and wrong
diurnality
divination
division of labor
division of labor by age
division of labor by sex
dominance/submission
dreams
dream interpretation
economic inequalities
economic inequalities, consciousness of
emotions
empathy
entification (treating patterns and relations as things)
environment, adjustments to
envy
envy, symbolic means of coping with
ethnocentrism
etiquette
explanation
face (word for)
facial communication
facial expression of anger
facial expression of contempt
facial expression of disgust
facial expression of fear
facial expression of happiness
facial expression of surprise
facial expressions, masking/modifying of
fairness (equity), concept of
family (or household)
father and mother, separate kin terms for
fears
fear of death
fears, ability to overcome some
feasting
females do more direct childcare
figurative speech
fire
folklore
food preferences
food sharing

future, attempts to predict
generosity admired
gestures
gift giving
good and bad distinguished
gossip
government
grammar
group living
groups that are not based on family
habituation
hairstyles
hand (word for)
healing the sick (or attempting to)
hope
hospitality
husband older than wife on average
hygienic care
identity, collective
imagery
incest between mother and son unthinkable or tabooed
incest, prevention or avoidance
in-group distinguished from out-group(s)
in-group biases in favor of
inheritance rules
institutions (organized co-activities)
insulting
intention
interest in bioforms (living things or things that resemble them)
interpolation
interpreting behavior
intertwining (e.g., weaving)
jokes
judging others
kin, close distinguished from distant
kin groups
kin terms translatable by basic relations of procreation
kinship statuses
language
language employed to manipulate others
language employed to misinform or mislead
language is translatable
language not a simple reflection of reality
language, prestige from proficient use of
law (rights and obligations)
law (rules of membership)
leaders
lever
likes and dislikes
linguistic redundancy
logical notions
logical notion of "and"
logical notion of "equivalent"
logical notion of "general/particular"
logical notion of "not"
logical notion of "opposite"
logical notion of "part/whole"

logical notion of "same"
magic
magic to increase life
magic to sustain life
magic to win love
making comparisons
male and female and adult and child seen as having different natures
males dominate public/political realm
males engage in more coalitional violence
males more aggressive
males more prone to lethal violence
males more prone to theft
males, on average, travel greater distances over lifetime
manipulate social relations
marking at phonemic, syntactic, and lexical levels
marriage
materialism
meal times
mearning, most units of are non-universal
measuring
medicine
melody
memory
mental maps
mentalese
metaphor
metonym
mood- or consciousness-altering techniques and/or substances
moral sentiments
moral sentiments, limited effective range of
morphemes
mother normally has consort during child-rearing years
mourning
murder proscribed
music
music, children's
music related in part to dance
music related in part to religious activity
music seen as art (a creation)
music, vocal
music, vocal, includes speech forms
musical redundancy
musical reptition
musical variation
myths
narrative
nomenclature (perhaps the same as classification)
nonbodily decorative art
normal distinguished from abnormal states
nouns
numerals (counting)
Oedipus complex
oligarchy (de facto)
one (numeral)

onomatopoeia

overestimating objectivity of thought

pain

past/present/future

person, concept of

personal names

phonemes

phonemes defined by set of minimally constrasting features

phonemes, merging of

phonemes, range from 10 to 70 in number

phonemic change, inevitability of

phonemic change, rules of

phonemic system

planning

planning for future

play

play to perfect skills

poetry/rhetoric

poetic line, uniform length range

poetic lines characterized by repetition and variation

poetic lines demarcated by pauses

polysemy (one word has several meanings)

possessive, intimate

possessive, loose

practice to improve skills

precedence, concept of (that's how the leopard got its spots)

preference for own children and close kin (nepotism)

prestige inequalities

pretend play

pride

private inner life

promise

pronouns

pronouns, minimum two numbers

pronouns, minimum three persons

proper names

property

proverbs, sayings

proverbs, sayings - in mutually contradictory forms

psychological defense mechanisms

rape

rape proscribed

reciprocal exchanges (of labor, goods, or services)

reciprocity, negative (revenge, retaliation)

regocnition of individuals by face

redress of wrongs

resistance to abuse of poser, to dominance

rhythm

right-handedness as population norm

risk-taking

rites of passage

rituals

role and personality seen in dynamic interrlationship (i.e., departures from role can be explained in terms of individual personality)

symbolism

symbolic speech

synesthetic metaphors

synonyms

taboos

tabooed foods

tabooed utterances

taxonomy

territoriality

thumb sucking

tickling

time

time, cyclicity of

tools

tool dependency

tool making

tools for cutting

tools to make tools

tools patterned culturally

tools, permament

tools for pounding

toys, playthings

trade

triangular awareness (assessinjg relationships among the self and two other people)

true and false (distinguished)

turn-taking

two (numeral)

tying material (i.e., something like string)

units of time

verbs

violence, some forms of proscribed

visiting

vocalic/nonvocalic contrasts in phonemes

vowel contrasts

weaning

weapons

weather control (attempts to)

while (color term)

world view

Adapted from D.E. Brown, *Human Universals*, McGraw-Hill: New York (1991); and Steven Pinker, *The Blank Slate*, Viking Press: New York (2002).

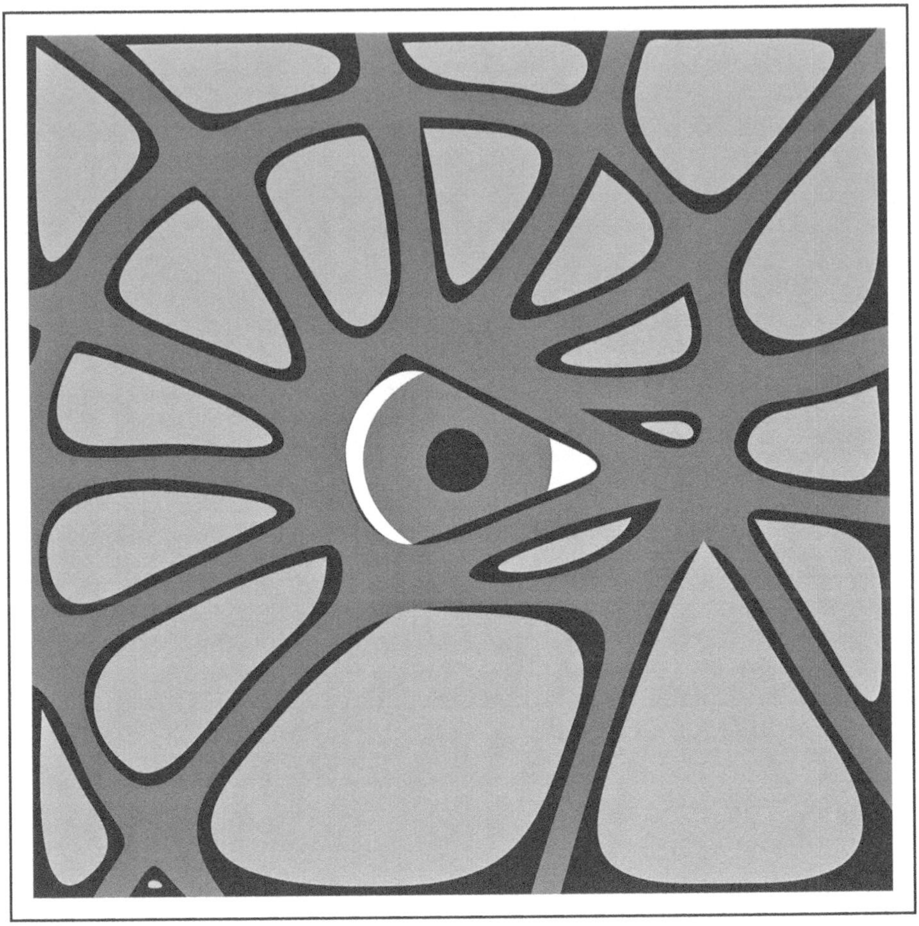

Chapter 29

The Zero Dimension

Thought Bursts into The Future

ORGANIZED OUTLINE

a. Cognitive Modeling and Mechanics

- i. The Agent Modality
- ii. The Embodied Environment
- iii. Intelligence as Natural Selection Pragmatism

b. A-Mechanical Deconstruction

- i. The Limit of Necessity
- ii. The Pre-Structural
    - 1. *Différance*
    - 2. Presence
- iii. Limning the Edge of Technology

c. Intelligence: Original and Copy

i. Copy Precedes Original

ii. Always Already Artificially Intelligent

iii. Co-Dependent Origination

d. Strategies of Thinking Outside a Habit Box

i. Analogous Problems with Other Solutions? (Isomorphic Mergers)?

ii. Reframe Problem in New Context or Paradigm (Inter-Disciplinary)

iii. Guestimate Trial and Error (Heuristic Experiment)

iv. Question Motivations and Assumptions (e.g. Growth?)

e. Structural Strategies of Deconstruction

i. Apply Principles to Principles Themselves

ii. Relativity and Reversal of Primary and Secondary (e.g. Berkeley)

iii. Negative Orientation?

f. Pheno / Geno (Kristeva)

g. Foci without Centers

h. Incommensurable projects?

i. Presence as Qualia?

ii. Core Constellation of Web

iii. *Différance* and Coherence

i. Can a Machine Deconstruct?

DISORGANIZED PATCHWORK QUIL(L/T)

The following *disorganized* collection of near-aphorisms—is meant to convey the as yet undeveloped nature of this thinking—and also to illustrate the way thinking often occurs to people as *insights*, rather than laboriously working over texts (which comprises most of the work in this book). I know that there is a bit of near-arrogance in presenting my thinking as such—I'm no Nietzsche or Wittgenstein—but this method of presentation has the virtue of exploding the confines of the outline above—working through, around, and out of it—making way without progress.

AI: "I think therefore I am": would AI's thinking imply certainty of its existence? (Of what or *who*?)

Deconstruction: "To be or not to be": deconstruction is as uncertain as the future.

Uncertainty blocks automation.

A method between aphorisms and narrative theme.

A method: intuitive idea thought through as to how effective it is. (Unconscious activation into pragmatic consequence).

The *ultra-contemporary*: a new orientation from the present towards an unknown future (a permanent structure for change, like democracy).

The ultra-contemporary as post-everything: newness, experiments, questions, critiques, avant-garde, Kuhnian paradigm shifts, challenging the status quo and dogma, thinking outside the box, pushing the envelope, challenging belief systems, surprises, the strange, shock, originality, revolution, defamiliarization, the unconventional, anti-authoritarian, *anti*-Oedipus, negative dialectics, unexpected, uncertainty.

Analogy—old ideas applied to new experiences.

Are artificial intelligence and deconstruction incommensurable? Should we use each as a perspective to critique the other or use the principle of charity: they can be used to illuminate each other too?

Always already artificially intelligent: original (implicit) and copy (explicit)? Inventing intelligence: has it been "fully" achieved?

Self-modeling as self-consciousness: Will *we* become fully self conscious only with AI? Co-dependent origination of intelligence?

A copy, as structure (also iteration) of original presence, is not copied: the structural aspect of an "object" is always already in repetition—there is no *identical* copy, especially in reference to time, space, and statistical similarity: yet *what* is copied from an "original" is, as that which is able to be copied, an un-instantiated copy: the double creates the original as original in a co-dependent origination. Without a common structure between original and copy, the original is as yet "unformed" or "unidentifiable"—and hence with developing artificial intelligence as "copy" we are actually defining *real* intelligence at the same time.

Does writing as exterior structure materialize conscious presence? Arche-writing as the pre-structure of conscious presence? How do we move from Arche-writing to conventional writing, if not through conscious presence?

Does the graphic element create a schism between isolated structure and context? (Frames?)

Does meaning come from a relationship of context structures? (Context for *what*?)

AI as structure and brain as structure: in "animated motion."

Différance as structuring borderline which limits and constructs presence as effect?

(Being = presence = subjectivity, objectivity and form).

Deconstruction of and against AI—limning the mechanical: with AI, are we realizing unconscious technology's desire for self-consciousness? Or did technology already achieve this through human beings?

Différance and judgment—is this beyond mechanical AI?—is the trace mechanical or technical?

Husserl brackets context to get to "essences" in themselves. As with Husserl: *bracketing* "our" AI to achieve pure consciousness.

Can one deconstruct "our" mechanical AI all the way to the "core" presence of conscious being?

Can we bracket the brain? As with hallucination?

The "generic" vs. the "brand name" aspects of being a being.

Both AI and Deconstruction are complex, but not infinitely so.

AI is by definition that which is copy-able (like writing)—but conscious presence is not repeatable in itself (only its form).

If AI is successfully achieved, will that invalidate difference and deconstruction; or will they be implicit in AI?

Learning new words is like cell division (*différance*).

Our intelligence is constructed unconsciously; hence artificial?

Innate knowledge? Not (just) innate concepts, but innate skills. Core of innate knowledge with variable periphery vs. memes as culturally learned and not biologically given?

Computers only do what they are programmed to do? True for humans too?

Qualia (consciousness) is never pure sensation before perceptual structure: The scheme / content distinction—there is no presence without difference (yet, which is more primordial? Or are they co-dependent?)

Postmodernism, if against reason—is also against AI? Reason questioned: not logocentric reason; but we do have reason if we have AI?

Analogy and metaphor in AI and Deconstruction—metaphors for assimilation (the cy-borg) and difference (the border).

Interpretation and the *resurrection of dead metaphors.*

Analogy: different concept names for similar structures; an activation of connected concepts.

No split between sensible and intelligible: both are just experiences with connections? We "experience intelligence."

Différance and the trace make structure: AI is structural.

If *différance* constructs both conscious presence and machine (as structure), the machine does not create consciousness.

Graph-ing (putting into context) and the "embedded mind."

Start with some innate concepts (objects, actions, abilities, etc) that new concepts are inter-connected with. Embedding of new concepts in network—*relations* are remembered rather than simple repetitions of words.

Reason as just a collection of certain skills among all skills: a pragmatic view.

Beyond "early" GOFAI: cp. Early and Late Wittgenstein, Husserl and Derrida; past representation to coherence: Neats to scruffies.

Can AI have a gender identity?

Questionable intentions: you can't think before you think.

Idealism's God and the homunculus vanishing point of exterior experience: Embedded and embodied?

Structure of brain not conscious; structural focus rather than matter. Reading between the nodes.

People "grasping" at quality but only getting structure.

Goal trees, expert systems, or evolutionary evolving systems: do we always need an aim?

AI and wandering thoughts: Hamlet's soliloquies.

Is there a generic version of Derrida's name brand deconstruction?

Is there a conflict between innate thinking and thinking outside the box? Freudian structures as innate?

Censorship—both unconscious and conscious—is modeled on editing before saying thoughts?

Prototypes (fuzzy centered concepts) vs. difference boundaries with play from "center."

AI FOCI:

1. Spreading Activation (both internal and external cue associations).

2. Daemons (loudest Daemon inclination to act or think is locked into attention system and becomes conscious).

3. Frames (also metaphorical structures, analogy, scripts, words, skills, phrases, sentences—innate and learned).

4. Memes (cultural Zeitgeist, embodied, embedded).

DECONSTRUCTION FOCI

1. In-between—*différance* (trace)—relations of structure rather than an essence or prototype in itself: something more fundamental than logic, ontology, or truth, yet not itself a foundation (foundations are metaphysical—making assumptions without absolute certainty).

2. Can't define deconstruction—precisely not that which *leads* (structure) to its conception. *Différance* always slips away—never fully present (enactment of theory).

3. Deconstruction—not something a subject uses—it simply happens and involves a constructed subject (writing as tool used by subject *and* arche-writing constructing a subject technologically).

4. *Différance*—not just between words, but skills, acts and events: Rather than bracketing all context to isolate essence—"essence" is determined by context

DECONSTRUCTING AI FOCI:

1. *Différance* (how is it material?)

2. Copy without original (AI as copy).

3. Presence (AI is not conscious?)

4. Deconstruction (can we climb out of our brains to model them? Deconstructing the brain?)

Software as Spirit? Or just configuration of hardware?

Derrida: boiled down or expanded upon? (Umberto Eco on under and over coding).

Why is "automatic writing" different than traditional writing with intent? Why is "word-association" different than typical speech?

Unconscious vs. Conscious *Reasoning*.

Can a computer compute *différance*? Is a juxtaposition mechanical? Are relationships? Form—is it relative to a judge?

Can AI create differences? As nodes in AI or spaces between in deconstruction, or does difference circumscribe AI? (Cp. Structure implicit in qualia).

Fodor: thinking is presence—mentalese—no unconsciousness; cp. Baudrillard: real dependent on fake; Defining human intelligence and consciousness as what AI can't do (not everything is mechanical: e.g. Chomsky and Newton).

AI is concrete and may fill in all the *details that theory does not*: Deconstruction as abstract inversion of this.

Deconstruction sees materialism as marginalized in favor of consciousness in the humanities, while AI and science often marginalizes consciousness as epiphenomenal.

Frame problem: surprises in environment—spreading activation does not grow into everything: Cp. urban sprawl vs. bio-diversity.

Emotional Minsky frames contain folk psychology—beliefs (inter-connections) and desires.

Technology transition from the tool used (object, computer, writing) to mechanics of user (subject, robot, arche-writing): structure folds over on itself?

Intelligence is often looked at as logical problem solving—but what about cunning or wit or wisdom? Personality? Astrological traits?

Short-term memory as activation of nodes and long-term memory as strength of connections between nodes?

Again, until we make our intelligence explicit with a model (AI) are we *really* intelligent? Or are we just using intuition?

Note how unconscious reasoning is; cp. conscious strategies (in using search trees and trial and error).

John Searle: argues both against strong (conscious) AI and Derrida—how would Derrida's take on translation reshape "the Chinese room" experiment?

Desire for undifferentiated presence: re-unite with the other?

AI as assuming the personality of the "other"... or the "norm?"

Generic personality?

Knowledge: ideas fall together and pan out—as with the emergence of life.

Ideology and ego-defense mechanisms: the concealment of truth, outside and inside.

PHENO/GENO: LIGHT BULB & FISHING NET

Pheno: subjectivity, simple, explicit, light bulb
Geno: objectivity, complex, implicit, fishing net

Heraclitus: Geno / change
Parmenides: Pheno / stasis

Plato: Geno / eternal forms (like math)
Aristotle: Pheno / worldly substances

Rationalists: Geno
Empiricists: Pheno

Hegel:
Pheno: being
Geno: mediation

Marx:
Pheno: ideology
Geno: economic base

Freud:
Pheno: manifest literal
Geno: latent interpretation

Heidegger:
Pheno: poetry
Geno: technology

Wittgenstein:
Pheno: logical form, representation, correspondence
Geno: family resemblance, coherence

Quine:
Pheno: reduction
Geno: analytic

Benjamin:
Pheno: aura
Geno: mechanical reproduction

McLuhan:
Pheno: hot medium
Geno: cool medium

Derrida:
Pheno: presence, Husserl
Geno: *différance* and trace

Kristeva:
Pheno-text: univocal
Geno-text: plurivocal

Phenomenology: edifying nomad agent
Genealogy: holistic cybernetic system

Pheno: brand name, relativity
Geno: generic, standard

Pheno: brilliance—intuition—speculation
Geno: genius—calculated—critical analysis

Pheno: node
Geno: connection

Pheno: conscious taste
Geno: automatic style

Pheno: unknowable nirvana, Brahma, Tao, Zen, enlightenment
Geno: zodiac wheel of existence and rebirth

Pheno: yin/yang of subject/object
Geno: network of logos

Pheno: character traits (astrological), morality
Geno: plot narrative, heroic cycle, justice

Pheno: content
Geno: form

Pheno: denotation
Geno: connotation

Pheno: idiosyncratic
Geno: conventional

Pheno: now
Geno: history

Pheno: number
Geno: law

Pheno: art
Geno: science

Pheno: one whole
Geno: many parts

Pheno: book
Geno: intertextuality

Pheno: bible
Geno: cabala

Pheno: center
Geno: edge

Pheno: self
Geno: other

Pheno: speech
Geno: writing

Pheno: soul
Geno: spirit/body

Pheno: beginning/destination
Geno: wandering journey

Pheno: atom
Geno: context

Pheno: nucleus
Geno: electron cloud

Pheno: sun
Geno: planets

Pheno: vision-space
Geno: sound-time

Pheno: pure
Geno: contaminated

Pheno: messiah prophet
Geno: witness

Pheno: norm
Geno: error

Pheno: tragedy
Geno: comedy

FOCUS VS. VANISHING POINT

The centered senses—is their centeredness distinguishable from the center of the observed object?

Is there more than one centeredness, or a one divided?

Centered senses }neural locality (sight, hearing)

|

Brain Connections

|

One perspective }neuro-logic (single language)

The brain can tie all the senses together in "material" memory with the organization of inputs; memory (of senses simplified) would cause *informational* "centering," but not "subjective" phenomenal centering.

Holistic sense combination is a-phenomenal?

Senses tied together: by form? (Measurability?) Cp. unified perspective (ties sight to sound). What would be difference between the brain as:

1) Structuring (nerve-ties)

2) Centering?

All senses: a) individually centered and

b) centered in conglomerate

Brain structure must at least parallel each point of description—yet the brain as structure is deaf and blind.

Notice how many philosophers of the canon criticize from the perspective of their cognitive construction—esp. Hume, Kant and Nietzsche—a current perspective (more sophisticated than, say, Kant's or Nietzsche's): the robot as thinker and over examination of those thoughts—pure description (informational)—is this an ethical perspective? Perhaps not—is our discourse ethical—self-correcting, etc—is the spirit of the age—focusable at its highest peak in the super-computer cognition synthesizer—itself ethical?

And the last critique of any structure itself—the "subjectivity's" "critique" of the objective structure—logically, a last judgment, yet practically, we would have to recognize multiple subjectivities—we have no reason to suggest that the logical unknowableness of subjectivity is singular (or plural, for that matter).

High "information resolution" of sensible detail vs. low "information resolution" of abstract concepts:

A reduction of quantity in reference to an increase in quality—frugal perfection?

Study all the words which are both verbs *and* nouns; e.g. eye, act, answer, question, corner, etc.—the cross between object and action.

Idioms—is there some "logic" they have (e.g. from folk *wisdom* to word as morpheme construction)?

A parallel between the dialectic and neurobiology? (Weaving and dividing, connections and units).

Cp. motion of *one* through dialectical movement (integrates into existence, divides out of existence) with consciousness—moving along the neural net—primed activation and decay—or the "life-force" itself.

The scientific delimitation of the unknown and the entrance of the unknown into that scheme: implicit to explicit.

Analysis of stock market: analyze traders, not corporations—analyze each stock in relation to all others—each variable in association with every other variable (as with parallel distributed processing / connectionism)—the trading malaise as a whole—would be analyzing not only how one part of an industry affects another, but how certain price fluctuations affect others.

Think about the difference between human intelligent consciousness and a sentient monkey integrated with an AI computer.

Question—was deconstruction discovered? Invented? (A neologism: a new logic—that "fell into place" as if recollected?) A construction from available parts appropriated for a function, use, and purpose? What would an alternative, or interfusion be? "Cultural manifestations"—are such necessary or accidental?

The conscious insight that some revelations are more profound than others, yet the weaving together of wisdoms with lesser insights and suppositions.

Does artifice corrupt natural revelation—or does such have its own art? Yet, memory can be jarred by certain methods, or accidents of procedure may be capitalized on—How meditation may facilitate revelation while also distorting it.

Time as form; history as content.

In my theories and anti-theoretical "rhetoric" I try to bring my shortcomings honestly to the surface, while others, post-structuralists often, try to hide theirs with a thesaurus.

Compare: the Turing machine (strip of info and processing head) and Plato's soul string in the *Timaeus*.

With Shakespeare—his double speak shifts from more of his own indirect comments to more indirect discourse among characters—e.g. from *Romeo and Juliet* to *King Lear*? In general, does indirect language appear in 1st person narratives or in dramatic dialogue? Are Shakespeare's earlier plays more subjective?—do they convey more of Shakespeare and his personal message?—maybe the later plays are simply more subtle, sophisticated—or less passionate? Is there an inverse relation between sophistication and passion? Does force *stretch* the fine threads? Perhaps this depends on the audience.

Wittgenstein?: Truth is context dependent, not correspondence dependent. Assumes "translatable" contexts of possible and "obtained" knowledge and assumes completely fulfilled knowledge coincides with its object—basically suggests our "original" forms of conceptual (nodal) communication are adequate to the task—that our methods (esp. aiming for the highest eloquence / connection ratio—explaining the most with the least) can achieve their goal—Question: New and possibly "truer" items are to "break in" to the context by dialectic cut and weave—yet how does such yield a "higher," "denser" more accurate and eloquent knowledge? How is it possible to find a *smaller* whole of two larger parts? Plato suggests that a recollected revelation of a higher form is agitated by probing neighboring context (a new part may be cut out—analyzed, or *synthesized from two*—but here is the crux—e.g. a new *equation* modified from older ones, does not *stitch* together, or simply fuse: a completely new "level" is jumped to: Hegel's dialectic seems more subtle on this point—Plato does not adequately explain hierarchy climbing, the very *ability* of which needs further clarification—hierarchy may be a function of the relation between language and the a-linguistic (e.g. sentience and tone)—language is *hierarchic* (at least in parts), but the a-linguistic is not necessarily so arranged. Technology is *not necessarily* hierarchical—hence neither is physics, mechanics (biology?)—*technological aspects of language* (not sentient tone) *are hierarchical* (in parts) and cannot explain new higher levels found—again the rich consistent detail of nature and the massive possibility of our knowledge / linguistic / techno-structure: why hasn't everything already happened? What limits nature's possibilities? time? (unfolding).

Retroactive retrospection.

"*Reverse fractals*" for visual recognition?

Fractals can encode / decode, pack / unpack; so: *fractal cues* encode: abstract?

Parallel Distributed Processing "organs": *organs* as pre-programmed modularity—note how un-embodied programming will have to compensate for situated memory: the frame problem.

The subtle effects of the personal name as a sub-conscious motivator: A relation between primal self-motivation and one's name associations.

The difficulty of tickling ourselves and auto-sexual "prohibition"—(being sexually attracted to our own body)—oneness self-division.

Bodily orifices and "automatic" protection: ejection, contraction cough, sneeze, sphincter closure, blink (ooze, gunk, drip, sniffle, snot, ear wax, drool). Body insecurity zones—how does questioning insecure zones change a body's relation to self? Become more secure and familiar with unfamiliar insecure parts—e.g. inner ear—touch eyeball gently with clean finger. Don't be afraid to clean your butt. Practice breathing—lungs—not as automatic as the heart?—still working on heart control? (Definitely trust in the heart!)

The swallow (cp. defecation, urination, etc.) as primarily conscious action where conscious control switches to unconscious bodily action. Consider the lump in the throat; how a swallow interrupts breathing—an early conscious control of breathing, fueled by hunger / swallow. And signaling with voice also interrupts and controls breathing.

Crying, lump in throat, swallowing and sobbing inward implosion?

Laughing, stomach contraction, tension, outward release explosion?

What good is change in and of itself? Possibility of the better or worse. Progressive as idealistic agitation implies ideology unless agitators accept equal chance of good / bad results.

All intended actions are moral? Nietzsche likes unconscious action—could his paradise-regained, the over-man be actually an under-man? An animal or reflex self—without intentions or with more base intentions.

Compare:

a) The auditory spoken self: voices in the void.

b) A "more" visual gestured self without voice.

Spirits in the dark, interchanging thoughts and emotion.

Emotive bodies in a world, hugging, conveying thoughts and emotions.

Compare as dialectic opposition:

The sun

{Gravity "implosion" not mechanical

{Atomic explosion

The dialectic

{Plato's "Good" above form

{Hegel's Absolute, un-represent-able whole

Gravity (above fusion) is unknowable as if unity itself, yet explosions explainable by their parts.

Test for "refresh" rate of eye vision's "frames":

Propeller reversal illusion is a product of the eye / brain refresh rate—have subjects push button when they see a reversal and note revolution speed of propeller when they see the shift (note difference for peripheral vision and center focus, the latter probably has a higher refresh rate / resolution).

For AI and robots instead of an internal monitor (e.g. Asimov's robot's commandments) a second, (non-intelligent?) computer that monitors all "proposed" actions and checks for "safety"—a sort of non-by-passable super-ego for the computer.

The will—just another emotion? A feeling? Not a direct action but the *feeling* of action being carried out with emotion of determination (a self-confirmation of automatic action?)

Phenomenal attention increases as the unfamiliar (possibility of death) increases—as death, the unknown itself approaches, consciousness increases—till the obliteration of phenomenal consciousness in the "absolute?"—the simultaneity of death and absolute consciousness?

The vanity of destiny, the modesty of luck.

The relativity of the eternal and temporal—"static" eternal from standpoint of temporal history; "static" temporal from standpoint of eternal outside temporal history.

Change and the eternal / temporal relativity—eternity, outside of time encompasses all history—all history is within the eternal, a place in which time has already begun and ended. From the temporal, eternity has already "gone through" the entirety of temporality; hence: from the eternal standpoint, our temporality does not change, and from this very standpoint eternity too does not change (four dimensional "block" space).

Life—where *the objective becomes least entropic-ally dissipated*—right up to the point of subjective unity?

Gravity, time, motion, spirit...

Neural nodes unify ambiguities—into unity (hierarchy: picture various shape cues activating a shape node).

Note how "objective" unity is experienced for life-forms with "subjective stream of consciousness?"

Question: what if life forms create machines which can unify? Is the extension from one of the life forms—(esp. if all subsequently died)?

Life: unification and time (gravity / love).

Cp.—Retro-active "knowledge"—"the eternal net" and rolling grammatical progression.

Neurons—nodes—connections—nothing can be known beyond the net, yet the *way* is conveyed.

Forms?—variations of design and sequential action—the patterns, always in reproduction in our minds, creations and nature's mirrors and creatures (and causality?)—transfer through mimesis—repetition (in part) in a near space and time—what is it that limits the possibility of infinite variation? Why *these* (particular) objects and actions of this flow and continuity of structure?—the particular structure of form—the formal content—reaches "out" to teleology, outside of formal temporality (unless, of course, there is a set "beginning"—which may also be out of time...)

Unknowable objects (cp. Kant)—qualitative "designs"—the content beyond formal content—sensed, yet never perceived?

Stable systems (Kant, etc.)—where language approaches the visible—images (a picture is worth a thousand words)—yet this engineering perspective melts down in mythical (interpretation) "systems"—a complex of metaphors—etc—back into language.

A static spatial relation—interacting symbols—Mars / Venus—metaphorically interpretative provocation vs. literal communication: from possessed reproducibility to expressed "whole."

Inversion of materialistic economic space into ecological web flow—"self-sacrifice" and automatic natural revelation of original conte(x/n)t in social animal form.

Cp. the beginning of writing and the illustration of time (cycles / almanac / prophecies)—or the invention of time as we might know it (a sequence of "moments")?

Will we allow the machine to lie? A lying autonomous machine? Us? With what sort of desires built in?

Homing in on the vague idea.

Plato's Parmenides—Being and Nothingness co-dependently originate out of oneness—yet it is "proven" by reference to *linguistic* necessity—one is needed for the other, for either one to have meaning—but the difference is purely linguistic?

As to the question of a central meaning in art—should we consider our creations as purely mechanical (without center) or as biological—with a center.

The reduction of visual subjectivity to sensation—no inside / outside at a moment—perspective comes from an objective time sequence—perceiver / environment shift—perspective is objective, temporal; quality is subjective, momentary?

Centrality of "subjective" perspective may be biological (the brain tree centralizes from limbs to trunk) yet what centralizes biology—how do the "parts" fall together?

Are ends, deaths, limits, stops, always negative, such that doubling them results in a positive?

- - = + ?

Negation of an origin, a reason the the start of the positive start never gets off the ground—a linguistic or logical effect?

Note how much of emotion and attitude is carried in our facial muscles.

Perception of space requires time, not vice versa.

Moral philosophizing is like critics telling artists what to do; often moral philosophy is like an exegeses of life and literature.

Inter-subjective Kantianism—the "subject" "apartness" necessities in plural society—cp. imagination as resource as well (for Kant: only from one perspective—ala Descartes' answer—contra solipsism, exterior (if not other's) environment seen as the source of arguments and language in first place—Kant's inter-subjective Imagination would be "exterior?"

Beyond private language—private reality—a solipsism of the Universe where each "subjectivity" like two or more "snowballs of chaos" overlapping is the % of the volume which is "common structure"—yet this structure, although consistent for both, is radically different for each: the impossibility of imagining the other.

Common structure is interpreted from whole of structure including that which is not common: Cp. idiosyncratic brain association and how they color our interpretations of the world and each other—not a private language, or completely public, but language shared from various perspectives.

A memory / reproduction of the "objects(s)" based on the relationship between unity and plurality—a reduction of the object(s) to the unity-plurality relations—this relation requiring the relation of unity to its "sustained repetition"—i.e.—the cross is not from one to infinity (and one again) instantaneously, but in divisions which are hierarchically cumulative—the ladder of numbers hierarchy is consistently inclusive repetition in e.g. sequence (memory / material regularity).

Consider—original division without number? is number not already there?—with only one division, no hierarchy in number, more than one division, number requires hierarchy—yet again—is not already there? At least "implicitly?"

Question: first numerical binary division?: 1 and 2, 0 and 1, 1 and infinity, 0 and infinity, 2 and 0?

Number which is implicit in all division requires hierarchy?

What is before a division number-wise? 0, 1, or infinity? It seems that before the split in 2, there would be 1 or everything:

infinity (infinity=2 at first split?) ha! or 0 because there is no number, (not 2, not 1...) = super numerical yet *numerically speculative* "Oneness?"

Again 0, possibly, does not exist in general, but only in reference to a specific content (e.g. a real or illusory ball does or does not exist; there are 0 balls on the table; possibly 0 Martians, or 0 square circles)—0, as a general term, applies 1) to its function—the negation of number in the particular—or 2) as a contrast to number made possible by particular content (note—my relation between super-numeral quality / content—Platonic forms as content—Quality and Zero have a relation?)

Net and naught—why no zero? Any "net" memory—can only have nodes that exist—a node cannot not exist; a node can only represent the absence of some other node(s)—any absence is always represented by a non-absence. One is implicit in the node, zero is not. Cp. representation is implicit itself—a net can contain another net—pure differentiation, represented by a net would require specific content for there to be zero / absence as a possible numerical factor. If difference is claimed to be absence, its qualitative aspect must first be determined. *Difference* as a quality (character) of which anything could be numerically said—any predicate of difference prior to qualitative consideration is purely speculative (how can a non-unity be assigned any predicate?) Yet number is implicit with difference.

Kant delineates the knowable / speculation, yet provides a possible way out—the necessary—yet is not the necessary, which is beyond the knowable, speculative to some extent? Any character we assign to the super-knowable—(e.g. that beyond structure) is already within that structure—hence—would be necessary, given our frame of reference or structuring—yet only speculatively so—remember there are several (possible) "exits" from structure—several non-structures, e.g. "formless" sense and perspective subjectivity?

Instead of a qualitative or perceptual Solipsism: a solipsism of memory—can't tell if yours is the only memory or not? (For who is to say "who" the "exterior" surface consciousness really belongs to—you only, by memory, associate it with yourself via subjectifying it via a point identified relative to flux)—but then again—within one's own memory, others remind you—besides—is not logic merely a matter of memory and not sense—etc?—couple this with memory as mostly generated from the exterior—a web that extends beyond the conscious...

Plato?—language is necessarily (practically—with neurons and dialectic) more flexible than "reality" hence *errors and imagination*; what exactly is it more flexible than? (besides reproduction-ally falling short of quality). N.B. Wittgenstein too—would we want to kill the imagination in the name of truth? Cp. Kant's imagination.

If humanity is natural (Plato: arts are natural)—then technological innovation may be natural evolution.

Is oneness the "exit / (entrance)" into the web (materiality, mental dialectic)—the point (approximate) to / of pre-differentiation—oneness itself implying a distinction—(and implication is secondary, bringing out from what is in—e.g. structure from quality, not vice versa)—it may be something of a misnomer to call the a-structural "oneness"—this is simply the closest we can get web-linguistically to such an "unknowable" as sentience, quality, etc—oneness suggests the "connection" and *identical* "parallelism" between the knowable and unknowable—Cp. Law as objective and inter-subjective function / border.

The id: something that needs to be controlled—or is the subconscious more naturally brilliant than we are?

Sex drives are quite conscious!

Logic: cp. conceptual physics—atoms, etc, are not the only *things*.

Life unifies—objects—esp. "atoms," etc, as objects, yet we say that these objects—atoms—build life? (Objects as cue clusters, neurons made from atoms)

The trickledown theory of enlightenment spreading—non-Buddhist, but hierarchical—the "middle-man" degradation of the "pure experience" (rumor deterioration)—the dream understood—realized...Jesus—a bit late—or right on time?

The prestige of intellect vs. various language games and applications—intellect as one group of language games among many. We all have only so much brain-power; some just apply it differently.

Depth implicit in surface?

In designing AI, the compromise between how much is pre-structured and how flexible its adaptation to new possible structures is...

Another way to insure safety with AI—stop, or radically decelerate learning procedure before distribution—only learns new facts and uses old strategies? Cp. Humans.

Chapter 30

Capital Regulation

Technology & Ecological Health

A FUNDAMENTAL CONCEPTUAL OPPOSITION

The possibility of political *science* would require at its base the ability to conceptualize any sort of political scheme or understanding, even if this conceptualism was understood in opposition to the conceptual. To this extent, the following discussion may offer, scientifically and metaphysically to the deepest core of anything we could possibly call knowledge, some of the most logically consistent political advice this philosopher can offer to the public at large. Such will be the case if we can see that 1): the fundamental dichotomy in possible human understanding, the dialectic of division and connection (as developed by Plato in such dialogues as *The Statesman*, and politically illustrated by Hegel in *The Philosophy of Right*) reverberates throughout all knowable reality; and that 2): striving for a balance in this dichotomy is necessary for life itself.

Far from any fascist or totalitarian scheme, this attempt at naturalized non-anthropomorphic authority based politics (recognizing our place within the universal), as a political platform, is already implicit in the current governmental movements in democratic societies which strive for a balance between individual liberties and global responsibilities.

Singular unity traverses all knowable reality: the whole as oneness is singular, as is the unity of each particular in plurality. Although oneness can never be understood, can never be represented by something other than itself (since oneness as a whole can have no other), we can only grasp its singularity as a whole interconnected with divided parts: the totality of interconnections comprise the whole, a whole which the singularity of any single unity can only allude to, and not encompass. In theories, any sort of conceptualism would require unified terms, notions, or operations that could be related to each other in a consistent manner. Memory and cognition, biologically understood, are based on and limited by neurologically interconnected nodal units: we would not even be able to explicitly remember anything beyond associated "thing," or "action" unities (note that the inception of the ancient dialectic may have been an insight into neurology, or a projection of it). In life, the whole and the part are interdependent as much as individual organs and the entire body work in conjunction. Politically and socially, the opposition between whole and parts divides among the ecological and the economical; our natural heritage and global interdependence being "opposed" to individual property accumulation and circulation. Our division from the whole establishes our individual unity, yet we remain enmeshed in the unity of the connected whole. Likewise we have the United States and United Nations.

DIVISIONS

Division is that which establishes property. It is the law as boundary, as demonstrated by the way we often use the word "law"—the law is upheld or transgressed and broken. This *law*, a partition like a *wall*, cuts out a space which is one's own area, a personal bubble. Hegel calls this "an external sphere of freedom" (G.W.F. Hegel, Allen W. Wood, ed, *Elements of the Philosophy of Right*, Cambridge University Press: Cambridge (1992), p. 73). This sphere includes one's body, and the material things which one grasps and unifies as one's own. Most graphically, it is a space within which to breathe, to move, to exercise one's will—it designates the possibility of what Hegel termed "subjective freedom." I can walk around, write things in my diary, and observe the world from my space. To cross a border and violate this space constitutes murder, rape, kidnapping, theft, trespassing, etc: invasion of privacy.

CONNECTIONS

One is not, however, completely severed from one's family, friends, community, nation, or world. We always have relationships with others as our very birth into this world is relational. These connections extend the effect of our actions, for our relations to entities outside our private sphere allow our actions to extend beyond this sphere. The networks of worldly connections provide opportunities and bind us together: If I am isolated in a prison cell, my connections are few—not only is my sphere of freedom limited, but my actions will have little consequence. Yet, if I am a dictator of a nation, my actions may move mountains. If I am a hermit in the woods, I may affect the local habitat, but little more. But, if I belong to a cooperative, I can help motivate the actions of many others, and contribute my strength to a greater force (factions having conglomerate unifications of bordered property).

POWER

To the extent that I am connected to my surroundings, the power of my actions reverberate even more—Hegel calls this "substantial freedom." Of course, these connections may also tie me down—I may be so obligated that all actions could be restricted. Connections thus can both extend and restrict one's power. Although these connections can extend or multiply power, they do not produce it; to a large extent, power is money, or property—one may have little money but many connections (e.g. the interconnected knowledge of education) and have more effect than someone with much money and few connections (e.g. a rich prisoner).

THE INTERDEPENDENT ECOLOGICAL WEB

Beyond the sort of discussion of the responsibility and empowerment of social connections which we find with Hegel and others, we must see that we are part of a greater ecological whole too (hence, a better balance between part and whole would not be one between self and society, but one between self and environment). The economic law which divides us from one another is our immediate subjective manifestation of the objective ecological system of laws which stratify the whole, like so many interconnected threads of a web which partition spaces (such an illustration falls short of temporal dynamics, yet is logically sufficient). The laws of nature (in an ecology including biology, chemistry, and physics) maintain a consistency of relations among the parts. And our global ecology is interdependent, as trees and animals need each other to breath, feed, and survive: no individual is self-sufficient.

This representation of nature as an interconnected whole is more "economical" in that it represents nature as related named parts that we can "posses" and manipulate, whereas a pure and whole ecological nature would transcend any divided part: nature, in its wholeness, is beyond any termed description—any scientifically reproduced system of nature will always be incomplete.

BIONOMIC CENTRAL FORCE

For ages, thinkers have struggled to account for some general principal of natural motivation. Plato's desire to be eternal (*Symposium*) has been transformed into more modern notions such as dialectical force, will to power, libido, repetition compulsion, and desire for presence. Roughly, such theories suggest a drive towards a primordial oneness transcending time: time is a desire to go beyond time and be everything all at once. Such a motivation is evidenced in theories which try to account for everything, in the individual desires for property accumulation (the desire to own everything), and more globally in the tendency of discourses and institutions to attempt monopolization—it is as if each part struggled to be the whole (the singularity of which we see in each part). Rather than desire, this might also be passively and *technically* observed as temporal niche filling expansion—related "beings" seem *destined* to fall together into larger or more successful wholes. But a principle remains: there are *unities*, and proliferating unities expand and tend to survive in larger integrated aggregates—there is still a *force* here, but rather than "desire" it would be *time* coupled with primordial "building blocks," like subatomic particles, that "fit together" by their very unique "*design*" (either intelligent or random) and tend, through life, to survive and possibly expand as larger complex semi-autonomous structures. Also consider: the possibility that if time decelerates or accelerates with the expansion of the universe, it will eventually hit an (imperceptible?) limit of no time difference, or simply one "instant."

Nature, being the unified whole, already is what it wants to be; nature's laws do not change. Yet, as a part of nature, life struggles for unity: similar to physical gravity fusing parts together, biological life forms unities. Plants and animals, although ecologically interdependent, are naturally singular entities (our sense of objects having a unified singularity may be a projection, or extension of life's unification). This, which I call "bionomic central force," is akin to qualitative sentience, and is not reproducible (and hence is really un-nameable), for the unity involved is ultimately the oneness of the whole, which is beyond the plurality of a reproduction, but reverberates throughout it (cp. Moses' 2nd commandment, and Plato's aversion to mimesis). Life is the unifying bridge from the part to the whole; a division, or a fission, in the unity of life is death.

TECHNOLOGICAL REPRODUCTION

Technology, the de-centered itself, is death; and this death of the living creates the need for reproduction—repetition destroys unity, while maintaining life's struggle for it. Technology divides sentient qualitative nature from itself, and involves it in a temporal striving to be the technological unity which, temporally, nature already is. Life's unified desire, and deadly technological reproduction co-dependently originate—each arises with the other, and they are interlinked in the desire for reproduction.

More immediately, technology is any sort of tool, technique, knowledge, or any other stratification of parts that we might use, such as a sword, the dialectic method, or language itself—using this technology can multiply power, as with a lever, or with the knowledge gained by education, as mentioned above. Anything we can understand mechanistically is technological—to the extent that we understand nature mechanistically, it is technological; any quantifiable structure, anything reproducible in a consistent manner, even beyond causality and consistency, is technological. Thus, technology also operates through us, as with the desire for reproduction. Indeed, we are often unaware to what extent technologies, implicit in our actions, institutions, discourses, etc, predetermine our lives. Although our technological awareness continues to increase, complete technological control, technological perfection, is limited by death itself—to step out of *all* technology in order to use it (including the "mechanical" aspects of our brains) would be to step outside of death; it would entail being that which is un-reproducible, a transcendence beyond life and death.

ECONOMICAL "RESOURCE" TO PROPERTY CONVERSION

By the means and action of technology, the "resources" of nature are converted to power-property; with some knowledge or tool, given nature is transformed into a useable possession. (I put the word "resource" in quotes, for nature is seen as a resource only from an economical perspective.) This occurs with theories, as when science attempts to convert the implicitness of the unknown whole of nature into the explicitly reproducible verifiability of a relational system of parts.

Practically, we can see how clothing (a fig leaf perhaps), as a technological tool, works as a division which establishes a public space of individual privacies. Hence we have the romantic call, against the false pride of technological division—imagine the pride accompanying the discovery of fire (desire), or the wheel (of existence)—in favor of a poetical confession of natural nakedness.

Globally, we can see this conversion of nature or "resources" into property causing, through its biological consumption and digestion, increased non bio-degradable waste and pollution (as concrete and steel spread like a cancer of urban sprawl), the growing scarcity of natural resources, and the unequal distribution of wealth (technological advances, such as robotics, natural language processing, and artificial intelligence, may continue to increase this disparity). In response, some might prophetically call for a relinquishment of all private property; but it is already too late to give up all the dividing and mediating technologies such as clothing, shelter, language, and money. We must realign our technological knowledge with nature—technology, which converts nature into usable power, may also amplify this power enough that we may wisely use it to preserve nature (as with the development of recycling and biodegradable materials).

INSTITUTIONALLY INTERNALIZED TAXATION

It is therefore my suggestion that governments operate as regulators of this "resource"-property conversion pivot between nature and a society of privacies—a government must hold in check the "natural urge" to "violate" resources and possibly privacies, and accumulate private property (the urge to be a pre-technological nature which "owns" all of itself) which our human knowledge recognizes, while making sure, via regulation, that individual privacies and institutions do not corrupt the whole of nature. To respect private autonomy while maintaining the health of the whole, *taxation* might be institutionally internalized and the technological apparatus of institutional "resource"-property conversion might be regulated in such a way that the "resources" of these institutions, the life forms and materials of earth, are maintained (such, in the long run, would be in the best interest of all: sustainability). In this way, with the technical aspects of society taxed—skimming social / capital institutions' cash flows—and redirecting that power to preserve the ecological whole (including its unified parts: life-forms), technology itself might foot the bill of global regulation and maintenance, especially in a future augmented with ever more artificial intelligence and robotics.

Government, the institutional institution, might best be based on regulation, and not construction (e.g. communism: compare reigning in a live horse, with the compounding problems of building an automobile), for life's "desire" for unity cannot be constructed (again, any human construction is technologically de-centered); life's *force* can only be regulated, guided. Institutions technologically manipulate and profit from the life force of nature; products and services are not created or constructed, materials and labor are converted, reshaped, as life force is regulated. To try to engineer an entire society, to plan an economy from the ground up, is building something dead. The problem of exploitation could be avoided with employee-half-owned and operated businesses (with the other half offerred as public stock, for retirement investment diversification, etc). In turn, governments, as higher order institutions, should centralize regulation with a view towards the whole of life itself. We cannot let technology slip out of control (e.g. the atomic bomb); technological knowledges must be used to maintain that which they have threatened: life depends on it.

Institutionally internalized taxation would take the burden of taxation from the individual (the scope of governmental responsibility being beyond the capabilities of any single person), yet should aim to provide for individual health, as each individual is a part of the whole, the health of which should be maintained. Institutions must compensate the whole of ecology and society for their "resource"-property conversion and societal license. Yet, individuals should maintain the right of individual autonomy and privacy, the ability to dissociate from technological institutions (only their participation in institutions would indirectly tax their actions). Not only would such a scheme be more efficient, thus saving "everyone" money, it would also give governments more power to tax particular institutional aspects which are problematic—governments would thus be better steer the economy's ecological effect; and it would allow businesses more freedom in determining where to make cost cuts and increases for tax purposes.

We must take into account what Adam Smith's "invisible hand" and the Reagan era's deregulation in the United States failed to reckon with: the economic "resource"-property conversion's possible detriment to ecology (they might have the driver asleep at the wheel!—although the emergence of ecology advocates who are part of a de-centered public debate is a sure bet). Yet, the present scheme does not necessarily advocate Keynes' call for large government expenditure: a regulative government should try to stimulate and restrain such economic flows, rather than construct them (although some institutions may have to be commissioned to preserve the whole).

Popular representation would remain important too, as governments must be kept responsible to the whole (the mass media is important as a mediator here), and many borderline issues would continue to require the political play of conflict and compromise.

EQUALITY AND SAINTLY SHARING

This suggestion, to institutionally internalize taxation and dedicate revenues to global health and privacy regulation (government would not necessarily substantially provide, but would rather aim to guarantee, such), does not create fairness with competition, it only aims for a lowest common denominator: survival. Even if equal distribution of property were possible, some people's ability to use property would be greater than others. A balance in the distribution of property (power) and technology (tools, knowledges, and other power multipliers), beyond any balance between the whole and its parts, is an impossible dream. (Ironically, calls for equal distribution of property are often made by intellectuals whose sophisticated theories can only be "owned" by an elite few; shouldn't a government's tenets, e.g. privacy and global responsibility, concerning all its citizens, be understandable to as many as possible?) Moreover, if survival is eventually guaranteed, non-exploitative property acquisition becomes a game of surplus—materialism becomes an option, rather than a necessity.

The border law which divides private property can be suspended when one person shares personal space with another. Families and friends often share personal space—spouses usually share the same bed (where lovemaking can be a sharing, rather than a give and take). Also, a sacrifice of one's time and space can be quite saintly—to a greater or lesser extent, one can give away some of one's property, and devote some of one's time to helping others (there are saintly institutions as well—The Red Cross, Green Peace, the Peace Corps, United Way, etc). Such self "sacrifice," (extending self respect to respect for others as part of your / our world) goes beyond any type of governmental legislation (such as encouraging employee-owned businesses)—the flesh of morality can fill out the stick figure of justice.

LOVE'S MEDIAL SANITY: MURDER / APOCALYPSE

The notion of "mental health" (as distinct from biological dysfunctions), often equated with morality, but not morality itself, might be completely disregarded though, as long as compliance with the law remains. As most people have varying opinions as to what is real, or really important, one person believing only scientific proof is true while another believes only in the values of poetic revelation, no single perspective could be claimed as a basis of sanity—who would call an exploitative scientist more sane than a inspiring visionary poet? Although character and morality are important, these are often wrapped up in different traditions and individual preferences. As soon as the law is disregarded though, violation of either personal space or global security may occur—health (of one's body or its extensions, or of a part of the entire ecosystem of divisions and connections) is damaged. Breaking the law, damaging an individual or the whole of nature, would then be the only poor health action: mental illness (not physical brain damage) could only occur with breaking the law. Insanity is not a state of mind, it is criminal action; where justice ends, insanity begins.

Seeing that our actions have the potential of violating another's space (a degree of murder), or of damaging the ecological system (a contribution to apocalypse), we must attempt to healthily guide our each and every action between murder and apocalypse. Of course, avoiding murder can be easy, and the entire ecology is somewhat robust in nature—hence the way between murder and apocalypse is not narrow, although it might be refined indefinitely. Finding this way may be informed by heartfelt intuition as much as fretful awareness.

We may find that the dual aspect of the law—as the border dividing individualities, and as the ties that connect us and consistently intertwine us with our environment—this duality can be found in the singularity of love, where love includes a respect for other's differences, and the affinity of our interconnectedness. The singularity of love, while above the duality of the law, maintains both aspects. The "madness" of a singular love above the law, would also be, with its inclusion of respect and affinity, the foundation of sanity and law.

CONCLUSION: BORDERLINE CASES

There may be unclear cases concerning personal space, such as determining when seduction is an invitation to share—one has the right to prevent or stop sharing at any time, and say "No! get your un-coolness out of my space," "don't ruin my stuff!" or "I don't want to be associated with you." Establishing when and to what extent children have an autonomy distinct from their parents can also be difficult (e.g., when does a fetus gain its own privacy distinct from the privacy of the mother—and would a respect for the fetus' privacy mean that it would have to be kept in the mother, or should there simply be attempts to keep it alive, as with special machines or surrogate mothers?) Although the distinction between the whole and its parts, between connections and divisions, may help clarify issues and show where a balance is needed, issues, such as what would be included in "public health" (food, clothing, shelter, health-care, and education?), or whether preventative regulations (often infringing on private autonomy) or performative regulations (often coming too late) should be enforced, will most likely remain debatable as attitudes and technological resources shift.

It should be noted that our current legislation, which often diverges from the strict laws of personal privacy and global responsibility that I have described, may actually produce a disrespect for these laws. We have people calling the police "pigs" and "blue meanies" predominately because of sometimes questionable speeding laws and drug prohibitions—preventative measures, of which the first is a type of roadway safety contract, and the latter can be a blatant invasion of privacy (yet, how could we expect the government to pay for a recovering drug addict?) or drafts for questionable military actions (it seems any sort of mandatory draft would require the government to demonstrate that the draftee's personal privacy is already being threatened and give the option of national exile).

There are further questions concerning the multiplicity of governments—for true justice requires an attention to the entire world, while governments have local constituents. There is no need for a single world government, but there is a need for organizations like the United Nations to encourage a global perspective among diverse countries. These countries will continue to face tough global issues, such as how one nation ought to trade with another when the other's exports are produced under circumstances that would be deemed "unjust" and not allowed at home: why should we have lower standards for our competitors? (Maybe they need them to even compete.) And how frustrating would it be to provide population reduction incentives when a neighboring nation's population spills past its borders?

These questionable borderline cases and international issues demonstrate how difficult it can be to use the fundamental dialectical opposition to aid the balancing of laws. The purpose of this chapter has been to show how this fundamental duality in the structure of any knowledge or consistent understanding, between the interconnected whole and its individual parts, can be used to clarify political and social issues and identify imbalances (such as culture's "overtaking" of a nature which had threatened human survival in the past). Government is a type of structure (the U.S. Constitution is a textual mechanism, a technology), and philosophical political science can shed light on the effective engineering of such a fundamental structure, as well as the metaphysical ethical implications throughout its details.

No doubt, some might be skeptical with regard to any theoretical prescription for running society, possibly in favor of whatever political consensus dictates. Far from urging a radical enactment of "my" theoretical insights, however, I would suggest that our society has already been on the way towards respecting the dual law and the possibility of institutionally internalized taxation: programmatic theory, and pragmatic success may eventually coincide. I would be prophetically adamant about the dual law of love though: to me, the fine details of justice and the advocacy of moral saintliness remain in need of revolutionary activism which recognizes complex trade-offs. Until some sort of transcendental zillionaire declares heaven on earth, with the subsequent value increase of an earth included in the realty of a universal heaven, this dialectical insight may be the best we humans have. The redemption of a global community respecting and enacting the dual law of singular love may yet be on the horizon.

An autograph collection

Chapter 31

Epilogue: Zen & The Singularity

INDEFINABLES

Defining "terms" like "deconstruction" or "*différance*" can pose a challenge in that they are said to be pre-conceptual, being that which shapes the conceptual in itself, or defined, as with Saussure's structuralism, in a negative relation to related concepts. We *can* become familiar with these semi-concepts or "terms" (in qualifying quotation marks)—but in many ways they are undefinable, esp. with a Kantian-like recognition via the Understanding. In a way, these "terms" are "open-concepts"—open to a future of possible change, yet as stable or unstable as a dynamic democratic definition can be. One might claim that *all* words are actually this elusive... all words are defined by a context that can be expanded, changed, and never pinned down in its entirety. But there have also been traditions of "un-definable" terms, esp. in religious contexts (recall the "negative theology" slant of the present work)—much like the "good" and "one" being capstones to Plato's formal dialectic, we have "Brahman" from India, the "Tao" from China, and "Zen" from Japan:

Wikipedia defines "Brahman":

> "In Hinduism, Brahman [...] is the one supreme, universal Spirit that is the origin and support of the phenomenal uni-

verse. Brahman is sometimes referred to as the Absolute or Godhead which is the Divine Ground of all matter, energy, time, space, being, and everything in and beyond this universe. Brahman is conceived as personal ('with qualities'), impersonal ('without qualities') and / or supreme depending on the philosophical school.

The sages of the Upanishads teach that Brahman is the ultimate essence of material phenomena (including the original identity of the human self) that cannot be seen or heard but whose nature can be known through the development of self-knowledge (atma jnana). According to Advaita, a liberated human being (jivanmukta) has realised Brahman as his or her own true self" (<//en.wikipedia.org/wiki/Brahman>).

And from the *Upanishads*:

"7. Verily, at that time the world was undifferentiated. It became differentiated just by the name and the form [....]

10. Verily, in the beginning this world was Brahma.

It knew only itself (atamanam): 'I am Brahma!' Therefore it became the All" (Robert Ernest Hume (trans.), *The Thirteen Principal Upanishads*, Oxford University Press: Oxford (1968), pp. 82-83).

Wikipedia says of the "Tao":

"In Taoism, Chinese Buddhism and Confucianism, the object of spiritual practice is to 'become one with the tao' (Tao Te Ching) or to harmonise one's will with Nature (cf. Stoicism) in order to achieve 'effortless action' (Wu wei). This involves meditative and moral practices [....]

In all its uses, Tao is considered to have ineffable qualities that prevent it from being defined or expressed in words. It can, however, be known or experienced, and its principles (which can be discerned by observing Nature) can be followed or practiced" (<//en.wikipedia.org/wiki/Tao>).

And the *Tao Te Ching* opens with:

> "The way that can be spoken of
> Is not the constant way;
> The name that can be named
> Is not the constant name.
> The nameless was the beginning of heaven and earth;
> The named was the mother of the myriad creatures" (Lao Tsu, D.C. Lau (trans.), *Tao Te Ching*, Penguin Books: New York (1963), p. 57).

Wikipedia on Zen:

> "Zen emphasizes the attainment of enlightenment and the personal expression of direct insight in the Buddhist teachings. As such, it de-emphasizes mere knowledge of sutras and doctrine and favors direct understanding through zazen and interaction with an accomplished teacher" (<//en.wikipedia.org/wiki/Zen>).

And a verse from Mumon from the *Mumonkan*:

> "Before each step is taken, the goal is reached;
> Before the tongue is moved, the speech is finished.
> Though each move is ahead of the next,
> There is still a transcendent secret" (Katsuki Sekida (trans.), *Two Zen Classics: Mumonkan & Hekiganroku*, Weatherhill: New York (1977), p. 134).

So much is to suggest that there are aspects of our existence beyond words—possibly some undifferentiated "oneness" before names that pervades all the world, which we cannot learn like typical knowledge, but which we can come to "experience" or know in an unexplainable way.

OUTLIERS ON THE VERGE

As the heart of this book has been a history of philosophy, esp. as relevant to deconstruction and artificial intelligence, the present section will give a brief survey sample of some "other voices"—thinkers that I have found relevant to the intersection of AI and deconstruction—and who have overlapping interests connected to *technology*, *postmodernism*, *the ecological environment*, and *mythical-metaphorical interpretation*. Some of these "other voices" are, like me, outside the academic system, some within; some are more known than others, but most will have probably had more readers than this book. As an "intellectual outsider," these are thinkers that I have been inspired by.

In segueing from the previous comments on Zen, I should note that I was inspired by the reclusive author Robert Pirsig (1928 – present), and his book, *Zen and the Art of Motorcycle Maintenance* (William Morrow & Company: New York (1974)). Pirsig studied philosophy both in the US and India, had some military experience in the US Army, and was also diagnosed with schizophrenia at one point in his life—experiences which the present author can relate to. Pirsig's novel charts a 17 day motorcycle trip from Minnesota to California, with the narrator (Phædrus) and his son, Chris, accompanied partially by two other cyclists. Many philosophical discussions ensue, where Pirsig pursues "quality" as an unknowable—maybe like Zen enlightenment itself, but as contrasted to the technical knowhow of much Western thinking. There is a contrast between the romantic and the analytic—between being in the moment, with fresh experience, and more scientific reasoning towards a situated truth—and truth is not a one-size-fits-all affair. The format of *Zen and the Art of Motorcycle Maintenance* follows the form of this dichotomy between the rational and romantic, offering both a narrative of real life, and more abstract contemplation. Pirsig developed his idea of quality in his second book, *Lila*, noting a difference between the original indefinable dynamic quality, and the habitual customary patterns of static quality that we can define—and he questions the subject object dualism that has been a salient aspect of philosophy since at least Descartes. Pirsig's dichotomies have a clear relation to the un-definable terms of deconstruction and to the schisms between how we can understand AI technologically and how it may relate to conscious quality. Here Pirsig recounts a pursuit of Quality:

> "I want to talk now about Phædrus' exploration into the meaning of the term Quality, an exploration which he saw as a route through the mountains of the spirit. As best I can puzzle it out, there were two distinct phases.
>
> In the first phase he made no attempt at a rigid, systematic definition of what he was talking about. This was a happy, fulfilling and creative phase. It lasted most of the time he taught at the school back in the valley behind us.
>
> The second phase emerged as a result of normal intellectual criticism of his lack of definition of what he was talking about. In this phase he made systematic, rigid statements about what Quality is, and worked out an enormous hierarchic structure of thought to support them. He literally had to move heaven and earth to arrive at this systematic understanding and when he was done felt he'd achieved an explanation of existence and our consciousness of it better than any that had existed before.
>
> If it was truly a new route over the mountain it's certainly a needed one. For more than three centuries now the old routes common in this hemisphere have been undercut and almost washed out by the natural erosion and change of the shape of the mountain wrought by scientific truth. The early climbers established paths that were on firm ground with an accessibility that appealed to all, but today the Western routes are all but closed because of dogmatic inflexibility in the face of change. To doubt the literal meaning of the words of Jesus or Moses incurs hostility from most people, but it's just a fact that if Jesus or Moses were to appear to-

day, unidentified, with the same message he spoke many years ago, his mental stability would be challenged. This isn't because what Jesus or Moses said was untrue or because modern society is in error but simply because the route they chose to reveal to others has lost relevance and comprehensibility. 'Heaven above' fades from meaning when space-age consciousness asks, Where is 'above'? But the fact that the old routes have tended, because of language rigidity, to lose their everyday meaning and become almost closed doesn't mean that the mountain is no longer there. It's there and will be there as long as consciousness exists.

Phædrus' second metaphysical phase was a total disaster. Before the electrodes were attached to his head he'd lost everything tangible: money, property, children; even his rights as a citizen had been taken away from him by order of the court. All he had left was his one crazy lone dream of Quality, a map of a route across the mountain, for which he had sacrificed everything. Then, after the electrodes were attached, he lost that.

I will never know all that was in his head at that time, nor will anyone else. What's left now is just fragments: debris, scattered notes, which can be pieced together but which leave huge areas unexplained" (Robert Pirsig, *Zen and the Art of Motorcycle Maintenance*, Bantam Books: New York (1981), pp. 167-169).

Another thinker who has gone far beyond the subject-object dichotomy is W. John Coletta (1952 – present), who along with his colleges, has developed a cogent view that even rocks desire. Coletta, a professor of English focused on bio-semiotics, eco-hermeneutics, ecological and social justice, and "renewable historicism," introduced me to deconstruction when I was an undergraduate back in the late 1980's—and he plays as a professional flautist in the Celtic-fusion band Green Tea. In papers like "The Semiosis of Stone: A 'Rocky' Rereading of Samuel Taylor Coleridge through

Charles Sanders Peirce" (W. John Coletta, Dometa Wiegand, and Michael C. Haley, *Semiotica* 174-1/1 (2009), pp. 69-143), Coletta, et. al. develop the notion of an "outness of mind" where thought is characterized as an environmental phenomenon—so much so that rocks could be said to have intent. This is demonstrated in a couple of ways. Just as biological organisms are said to be "self-organizing"—the way in which they can be unified creatures with surfaces that distinguish them from a wider environment and other creatures—so too rocks can be seen as being grouped together due to geological and environmental factors, esp. the weather: with temperature variations and rain flows, etc, that form distinct phases between what is rocky and what is not. Moreover, with the example of the Rock Beauty, a fish that evolved a dark stripe to look like it is behind a rock, we see that living creatures have incorporated rocks—a predator may see a Rock Beauty, and be fooled as to the availability of the fish for a feast. In this way there is a co-determination between rocks and fish—as the rocks have shaped how the Rock Beauty has evolved; so to, if we are to see language as an exterior phenomenon—with signs being in the outer world, and not necessarily just "used by an ego"—we can see that Rock Beauties give false signals to predators: a predator misreads a Rock Beauty—interprets it as behind a rock—demonstrating that rocks are integral to an ecological linguistic network in themselves. In a reversal of thing and context, creature and environment, we can see that the context places the thing in any meaningful relationship, and that the environment shapes the evolution of the creature. Just as meaning is not contained in the nucleus of an atomic word, but rather is situated in contextual usage, intent is not contained in the interior of a creature, but is established through the entire stratification of an environment. Coletta, et. al, give many examples of language theory being instantiated in ecological domains—just as we noted that the Rock Beauty gives a false signification to an interpreting predator.

> "Our chief illustration in the deployment of our models of the signing action of nature (of the semiosphere) will be to

show how it is that even rocks may be said to have desire or intentionality. By 'desire' we mean how it is that rocks are not only involved in self-organization, that is, 'nonlinear, dissipative interaction among the small- and fast-scale constituents of a system giv[ing] rise to order at larger spatial and longer temporal scales' (see Kessler and Werner 2003 [...]), but also how rocks have succeeded in getting themselves written into biological, ecological, aesthetic, and even legal codes—a process whereby objects (rocks) become objectives (say, laws which protect rock formations or the fish that live within them)" (W. John Coletta, Dometa Wiegand and Michael C. Haley, "The semiosis of stone: A 'rocky' re-reading of Samuel Taylor Coleridge through Charles Sanders Peirce," *Semiotica* 174–1/4 (2009), p. 72; citing, M.A. Kessler & B.T. Werner. 2003. "Self-organization of sorted patterned ground," *Science* 299 (5605). pp. 380–383).

Moreover, though Coletta's et. al, "renewable historicism" we see how the ideas of past thinkers (e.g. Coleridge and Pierce) can be recycled due to the open-ended nature of language: old terms and ideas can mean something different in newer contexts. Meanings do not change radically though, but rather *stretch* to mean something else; much like a trait that evolved for one purpose can become valuable for some completely different purpose. Coletta and his colleges have thoroughly deconstructed the nature / culture opposition, at least in relation to language and signifying. Such is not to claim that the nature-nurture opposition has been eliminated, but rather that our use of language as humans has a vast history in the evolution of our eco-web as a whole—and that meaning and intent is fully embedded in the environmental context: rocks, as members of the environment have intent via their very situation in this milieu—and with consciousness as well possibly being on the interface between inside and outside (rather than experienced by a homunculus inside a creature), the very notion of conscious intent is flipped inside-out: we too intend only with regard to our situated place in the world as well. In this way, it may be suggested that artificially intelligent robots, as further members of our environment, could be said to have conscious intent: even already as they are *now*. This might be extended to the notion that computers have used us to come into existence: there is a symbiosis between technology and biology where the biology keeps technology "alive," just as technologies (like clothing) have kept us alive too.

Another ecological thinker, yet who has a distaste for both technology and deconstruction, is John Zerzan (1943 – present), an anarcho-primitivist. Zerzan has lived in my hometown (Eugene, Oregon), and although I thought I spotted him once at a local Taco Bell, I have never met him. He became more widely known when he befriended the Unibomber (Ted Kaczynski, who has his own critique of Zerzan), and with his intellectual exposition of various green anarchist activisms. According to Zerzan, ancient humans as hunter-gatherers were happy prior to being dominated by the technological division of labor, which came with domestication of animals and agriculture, and by the symbolic / reifying thinking that alienates people from nature such that they consequently dominate it and one another. Fundamentally, Zerzan sees technology as creating divisions that are hierarchal, and as an anarchist, he rejects any sort of technology that is not under immediate control but is rather more systematic in its direction of human behavior. Zerzan, who is well versed in contemporary theory, having been influenced by Adorno, but critical of Derrida, et. al, sees cultured institutions as intrinsically dominating, in that they are technologically and hence hierarchically divisive through and through; the very structures of institutions dominate human practices, and place people in a tool mediated relationship to the environment and one another, which again, is a relationship of domination. I disagree with Zerzan on theoretical grounds, in that via DNA, etc, we are always already technological through and through—in fact, we are unconsciously motivated in ways that we can only understand technically: a technological understanding can help free us, or at least make us more conscious of our biotechnological motivations. Much like Derrida's archi-writing operating as a technology that shapes our speech in a way that writing as a controlled tool could not: there is NO tool that we control that is not already wrapped up in technologies that control us to some extent. Yet, although I'm unwilling to forgo my fig leaf of technical clothing, I too have often envied monkeys pulling a banana off a tree—a life of leisure and *no social pressures* is enticing (although groups of monkeys seem to have their own soap-opera dramas going on). The stark contrast between the relaxed, bored monkey and a recently fired employee leaping to their death on the evening news gives one pause for thought. And it's not just a matter of attitude: we can't simply forage and hunt, given the strictures of modern society—hunting licenses, private property, etc. It is hard to tell, whether overpopulation is due to technology as well: city plumbing being a case and point—but it seems that a *return* to nature, the "Future Primitive," as it were, would require a drastic reduction in current world population. Unfortunately, global warming may take care of that!:

> "To 'define' a disalienated world would be impossible and even undesirable, but I think we can and should try to reveal the unworld of today and how it got this way. We have taken a monstrously wrong turn with symbolic culture and division of labor, from a place of enchantment, understanding and wholeness to the absence we find at the heart of the doctrine of progress. Empty and emptying, the logic of domestication with its demand to control everything now shows us the ruin of the civilization that ruins the rest. Assuming the inferiority of nature enables the domination of cultural systems that soon will make the very Earth uninhabitable" (John Zerzan, *Future Primitive Revisited*, Feral House: Port Townsend, WA (2012), p. 23).

Unlike the "celebrity" thinkers in the second half of this last epilogue chapter who have in a way been imposed on me by my cultural immersion, most of the thinkers in this first part are people I stumbled upon while pursing my various interests. One such serendipitous stumble came when I researched the notion of the "post-digital" which figured in the title of my poetry novella *Post-Digital*

Revelation. I happened across American electronic composer Kim Cascone's (1955 – present) paper "The Aesthetics of Failure: 'Post-Digital' Tendencies in Contemporary Computer Music" (*Computer Music Journal* 24:4, Winter 2000, pp. 12-18). In his article, Cascone, influenced by deconstructive theory, shows that the tools for making music have been used not only to convey notes, melodies, counterpoint, etc, but have been exploited in ways that they were not designed for. An example (not mentioned by Cascone) is the electronic music artist Burial's use of the sound of a phonograph stylus landing on a record—repeated as a source of rhythm in itself: beyond sampling the music, Burial has used the non-musical sounds of the technology itself to create new auditory experiences. In a similar way DJs had long been "scratching" vinyl, again drawing sounds out of the mechanism, snaps, crackles and pops that stretch the notion of playing an instrument. The "glitch" movement in music as well, via such techniques as CD-skipping, further brings these background accidents to the foreground—demonstrating how much creativity comes with taking advantage of whatever happens when one noodles or doodles. More broadly the "postdigital" movement appears to be a new Romanticism that strives to bring the human back to center stage—a sort of neo-humanism. Here in relation to Marshal McLuhan:

> "The medium is no longer the message in glitch music: the tool has become the message. The technique of exposing the minutiae of DSP [Digital Signal Processing] errors and artifacts for the own sonic value has helped further blur the boundaries of what is to be considered music, but it has also forced us to also [...] examine our preconceptions of failure and detritus more carefully" (Kim Cascone, "The Aesthetics of Failure: 'Post-Digital' Tendencies in Contemporary Computer Music," *Computer Music Journal* 24:4 (Winter 2000), pp. 17).

I came across "neo-humanist" John David Ebert (1968 – present) via his many YouTube videos explaining various deep thinkers, esp. from the continental philosophy tradition—his discussions of these thinkers are clear and profound, and marked by his use of extensive hand gesturing. Ebert operates as something of a "guerrilla scholar" with these videos, offering education for "free" outside the academic system. He is an historian of the present, a culture critic (esp. movies) with a background heavily invested in mythology (via Joseph Campbell), and a sort of post-postmodern philosopher. Ebert's views on technology can be gleaned from his book, *The New Media Invasion: Digital Technologies and the World They Unmake*, McFarland & Company, Inc., Publishers: Jefferson, NC (2011))—where he goes beyond the obvious benefits of contemporary technologies such as virtual internet content and various hardware techno-gadgets, and examines the other facets of life that these displace. He sees the virtual world of the internet and media in general as connecting with a Platonic mathematical transcendence—yet unlike the stable world of forms that Plato argued for, this virtual world is nearly as an unstable carrier of our cultural information as a crashing computer. In developing virtual selves, as on Facebook, we are following the paths blazed by media celebrities in the 20th century—creating ourselves as iconic images that bear only partial resemblance to real human beings (and we can see the devastating effects celebrity images had on the lives of Marilyn Monroe and James Dean). There seems to be a tendency towards moving out of the geographical public sphere and retreating to our homes and computers—as we buy more products online (hence less hanging out at the mall or in neighborhood meeting places), and spend more time communicating with our virtual friends as virtual selves. A danger in this is evidenced by the volatility of information found when moving away from concrete, real-world, instantiations of information, as with books, records, etc, towards more and more virtual environments. A case in point is Wikipedia, with its revisable articles and somewhat democratic approach to knowledge. Although I think Ebert underestimates the power and accuracy of distributed amateur intelligence (recall James Surowiecki's book, *The Wisdom of Crowds*), and that decentralization actually makes things *more* stable rather than less so (because they are tethered to so many more points of "authority"), I think being leery of a culture capable of being "unplugged" is a cogent position, esp. since contemporary techno-media devours and replicates almost all media of the past. Ebert implies that we need a middle-ground dose of reality, escaping the opposing dangers of a dehumanized insect-hive-mentality and finding ourselves in isolation from real human contact—there is a danger of our becoming radically inauthentic (non)humans. Ebert sees hope though, in a vein similar to Heidegger's opposition of poetry and technology, with Ebert's thinking that art can be a cure for our technical ailments, much as science fiction often is an allegory for coping with the present rather than a prophecy of the future: a sort of antidote to futureshock. Ebert on approaching an "extinction event":

> "When, however, worlds *close down* and visions *become extinct*, so too, their corresponding media disappear. Thus, Egyptian hieroglyphs vanish at the end of the fourth century and the vision that had animated Egyptian civilization for three thousand years had crumbled and fallen into desuetude. The cuneiform tablet, likewise, had by then already disappeared (end of the first century) along with the Mesopotamian way of life that Abraham was already paying farewell to in about 1800 B.C.
>
> Likewise, when the literate world configured in the fifteenth century by the advent of the printing press—a world structured by individuality, nationalism and the abstract, three-dimensional space—begins to disappear somewhere in the middle of the twentieth century in a haze of violent, earth-shaking battles and academic disputes in which their corresponding metanarratives are taken apart, dismantled and deconstructed—then it is only a matter of time before the media which made that world possible, which had fed and sustained it for centuries with printed books, magazines, newspapers and bookstores, also begin to vanish.
>
> Which brings us to the edges of history, where we, today, currently find ourselves gazing off into the abyss of a new

media extinction event in which absolutely *all* the media which have built and shaped a literate Gutenbergian cosmos for five centuries are, within the space of about a decade, *ceasing* to exist" (John David Ebert, *The New Media Invasion: Digital Technologies and the World They Unmake*, McFarland & Company, Inc., Publishers: Jefferson, NC (2011), p. 6).

Another thinker interested in technology and influenced by Derrida and deconstruction, is Bernard Stiegler (1952 – present), who became a philosopher while serving time in jail for an armed robbery. Stiegler is fairly well-known in contemporary continental philosophy, and has written, among others, three books in a series called *Technics and Time.* Stiegler theorizes that technological, human-made objects form a third type of "human" memory, alongside DNA and the brain. Far from being a mere object-tool for use, technological items, human-made objects, are actually what create the possibility of history (e.g. note how archeologists look at ancient artifacts to know how people lived in past times; as well as writing-as-tool passing down historical accounts). There seems to be a co-dependent symbiosis between humans and technology—the typical subject-object who-what dichotomy is overturned as we see that the technological *systems* shape who we are as humans (much like Derrida's archi-writing shapes a speaking subject who uses conventional writing; or Heidegger's technological enframing, with so many cybernetic information-flows, shapes humans who use tools). Moreover, technology may not only make history possible, the very nature of the technical may shape time itself, as repetition (so integral to time) is wrapped up in technology. Steigler traces the ever more prominent place of technology in culture—with eras such as the 18th century's industrial revolution demonstrating that tools are transforming to broader systems of technologies:

"Since the Industrial Revolution, 'technical becoming,' on the contrary, has compounded its systematic dimensions, becoming visible to the naked eye in various ways and sensible to the bodies and minds devastated by an entire universe of hellish machines, given that technical becoming operates through an evolutionary logic endemic to this dynamic system as a '*technoscientific* age' defining itself through a process of what would correctly be called 'technical *individuation*'—Simondon's word for the process of *concretization*, supplementing Leroi-Gourhan's 'technical tendencies' and Bertrand Gille's 'technical systems.' The concretization process, including the morphogenesis of industrial technical objects, controls not just the becoming of the object itself but also of technical assemblages, and thus, in the end, henceforth (as we have already seen), the global mnemotechnical system itself; i.e., the mind's retentional milieu.

The Industrial Revolution created a dynamic unique to the technical beings and to what I have called the reign of the organized inorganic. This dynamic has become possible at this particular historical moment for two reasons:

—on the one hand, archaeology and paleontology began to emphasize that the evolution of even the oldest artifacts can be traced through formal lineages that are comparable to those of the beings that are discernible in contemporaneous fossilized skeletons;

—on the other hand and most importantly, the complicity between technics and science during this technoscientific period, which has been produced by and given rise to industrial technology, has opened an era of permanent innovation in which the evolution of technical objects, now suddenly accelerating, becomes 'modernity' itself, foregrounding the problem of adoption of ever-new industrial products.

The co-operation of technics and science in complicity with industry, then, has become manifest, bodily and mentally, through the dynamics *inherent in* technology and technics" (Bernard Stiegler, Stephen Barker (trans.), *Technics and*

Time, 3: Cinematic Time and the Question of Malaise, Stanford University Press: Stanford, CA (2011), pp. 188-189).

As Stiegler has traced the history of technology and it's integral relationship to being a contemporary human—the technological inventor and futurist Ray Kurzweil (1948 – present) has projected this techno-evolution into the not-so-far future and claims we are heading towards a "technological singularity." That is, with computer power accelerating in both capacity and speed, Kurzweil predicts a point where artificially intelligent computers will design intelligences even surpassing their own, also at an accelerating rate: an event horizon beyond which we can only begin to imagine. Beyond seeing the human as shaped historically by the technical, Kurzweil predicts extensive human-machine interfacing in the literal sense, where our own memories will be augmented by computer memory, and possibly at some point we will "upload" our souls into a machine, and eventually there will be a human-machine continuum. A problem I see with Kurzweil's predictions is that although hardware capacity and speed have accelerated over the years, I do not think our understanding of intelligence has advanced quite as quickly. In fact, it is my contention that we are still in the process of inventing intelligence; and that intelligence, pragmatically understood, is not all that it's cracked up to be—yes, we may become better at augmenting our creativity with computers—decoding DNA, extending longevity, reducing needs for labor, etc; but so much of who we are has to do with politics, personality, etc: Would a super-intelligent computer have superior sense of humor? What sort of values would a computer have: liberal, conservative, etc? Moreover, technology seems to contribute to urban sprawl and population growth: a Malthusian ecology (e.g. with global warming) and determining who or what should regulate these super-computers politically on the way to the singularity could be a crisis in the making: would super-smart computers "educate" us too? Here is Kurzweil summarizing a book chapter on the year 2099:

> "There is strong trend toward a merger of human thinking with the world of machine intelligence that the human spe-

cies initially created. There is no longer any clear distinction between humans and computers. Most conscious entities do not have a permanent physical presence. Machine-based intelligences derived from the extended models of human intelligence claim to be human. Most of these intelligences are not tied to a specific computational processing unit. The number of software-based humans vastly exceeds those still using native neuron-cell-based computation. Even among those human intelligences still using carbon-based neurons, there is ubiquitous use of neural-implant technology that provides enormous augmentation of human perceptual and cognitive abilities. Humans who do not utilize such implants are unable to meaningfully participate in dialogues with those who do. Life expectancy is no longer a viable term in relation to intelligent beings" (Ray Kurzweil, *The Age of Spiritual Machines: When Computers Exceed Human Intelligence*, Viking: New York (1999), p. x).

Thinkers like John Zerzan would most likely take issue with this; and technophile, and LSD guru Timothy Leary noted in a presentation I attended years ago that even in a highly automated future, we'd still want a human to cut our hair:

> "It turns out, our HAL, paranoias are exaggerated.
>
> Computers will not replace real people. They will replace middle- and low-level bureaucrats. They will replace you only to the extent that you use artificial (rather than natural) intelligence in your life and work. If you think like a bureaucrat, a functionary, a manager, an unquestioning member of a large organization, or a chess player, beware: You may soon be out-thought!" (Timothy Leary, Michael Horowitz and Vicki Marshal (eds.), *Chaos & Cyber Culture*, Ronin Publishing Inc., Berkeley, CA (1994), pp. 32-33).

Speaking of LSD, our next thinker I stumbled upon may have taken a few trips: Michel Hoffman (~1966 – present) has theorized that we are by and large deluded into thinking that time and free-will exist and that religion is largely composed of metaphors for drug-induced enlightenment experiences where one becomes aware, through ego-death, that circumstances are beyond their control; he summarizes his stance at his Ego Death and Self-Control Cybernetics website—Ego Death Theory is:

> "a new systematic research framework and paradigm. The ego death theory holds that the essence and origin of religion is the use of visionary plants to routinely trigger the intense mystic altered state, producing loose cognitive-association binding, which then produces an experience of being controlled by frozen block-universe determinism with a single, pre-existing, ever-existing future.
>
> Experiencing this model of control and time initially destabilizes self-control power, and amounts to the death of the self that was conceived of as an autonomous control-agent. Self-control stability is restored upon transforming one's mental model to take into account the dependence of personal control on a hidden, separate thought-source, such as Necessity or a divine level that transcends Necessity.
>
> Myth describes this mystic-state experiential insight and transformation. Religious initiation teaches and causes this transformation of the self considered as a control-agent, through a series of visionary-plant sessions, interspersed with study of perennial philosophy. Most modern-era religion has been a distortion of this standard initiation system, reducing these concepts to a weak interpretation that is based in the ordinary state of consciousness. The ego death theory is, specifically, the Cybernetic Theory of Ego Transcendence, and it incorporates the entheogen theory of religion" (<//egodeath.com>).

Hoffman seems to take the notion of being an embodied technological agent quite literally: we already are robots to a certain extent—caught in a technological space-time continuum. Although sympathetic with some of Hoffman's views (I agree that a "dark night of the soul" can awaken one to a new perspective on non-ego "steering" of our habitual bodily styles and gestures)—I think the possibility of an a-technological perspective: quality on the surface, and a time conceived of through différance *prior* to space, a zero dimension rather than a fourth dimension, can radically rupture the space-time continuum, much like the concept of "subjectivity" (but not a subject / ego) disrupts objective accounts of reality. A grand active *performance*, after sublime passive *observation*.

Roger L. Satterlee (1950 – present), another thinker influenced by continental philosophy and interpreting it in a "mythical" framework—astrology—stumbled upon me: he saw my chapter on Hegel ("Zodiac Robot") on the internet and sent me an email inviting me to check out his work on "Horoscopic Expressionism." With copious anecdotal evidence (yet lacking statistical evidence), Satterlee has shown that with many artists the visual horoscope for their astrological birth chart maps to key works of visual self-expression (e.g. arrows or weapons will be oriented much like a martial Mars aspect, a crescent may be visible in parallel with a moon aspect, etc.) The evidence, time and again, is uncanny, startling, and bound to raise doubts about how it could be, or if there really is a systematic parallel between key works of self-expression and one's horoscope. In fact, in a reversal, Satterlee has tried to approximate birthdates based on works of art: the horoscope that would parallel the work could give a more exact time of birth for some historical artist—here at his "Pedantus Pontificus" blog:

> "I am always looking for examples of what I call, Horoscopic Expressionism—a work of art which at least seems to be an unconscious exposition of some connection between one's psyche and whatever astrology actually is. While browsing Google Images, I discovered the art of one Andy Goldsworthy.

[....In reference to a Goldsworthy artwork] I felt as though this man had projected a symbolism which has an astrological parallel. To me, it seems an expression of what astrology would call a conjunction of the natal Sun and Uranus. As well, the black circle part of the image is parallel to the idea of Sun trine Saturn—an angle of 120 degrees between these two astrological planets. Bear in mind that I cannot explain why any of this 'works' or why I am given to this kind of perception, but I try to either establish the facts of my suppositions, or maybe, with luck, make it all just go away...:) In searching Google for Goldsworthy's birth data I found literally hundreds of references to his birth year, 1956, and little else.

So, I looked at his birth year to see which day of the year the Sun was actually conjunct Uranus—in the same Zodical degree. It, this conjunction, falls on July 25, in 1956.

Thus, 'if' astrology works, then Andy Goldsworthy's art is pointing to his birthday. When I Googled, <'Goldsworthy was born' July 25>, I hit upon a web page citing Goldsworthy's birthday, a quote from an autobiographic text. It was posted by a college student who has since graduated, or just moved on, but that site no longer shows up on the web. <Updating here 11/21/08; Andy's birth date is suddenly a lot more accessible when Googled...and some are citing it as being the 26th of July...(which is my birthday of course...:)>

Anyway, I hit this 1 in 365 chance right on the nose...:) And, to sweeten the pot, the Sun is also trine (120 degrees of separation +/- 8 degrees) Saturn on that date. I don't think I'm just plain lucky enough or unlucky enough, to have this kind of 'guess' work out as it does. It's not at all like guessing coin tosses and such. I want to remain skeptical, but it's not easy...:)" (<//pedantus.blogspot.com>).

Now, art as confession of one's deepest self could evidence their astrological being. Where even astrologers might be skeptical is:

how could a person have an image of their chart in their unconscious? Why the wheel populated by aspects, etc? I don't think we NEED to find a scientific explication for such, but there are a couple of possible explanations, I think: 1) the illustrated horoscope corresponds to actual spatial arrangements of the planets, sun, moon and earth rotation (rising and orientation of houses) at birth, and thus could be as imprinted on the brain / mind as anything astrological could; or 2) just as regular astrology might be an unconscious cultural self-fulfilling prophecy, our artistic expressions could have that astrological projection as well. This second possibility seems at odds with children and other 'naïve' artists expressing their horoscopes—which brings innate ideas to mind: the very thought of innate knowledge of the zodiac, and possibly an intuitive insight into one's own individual chart—I think this reaches back to the origins of 'theoretical' knowledge itself: how we humans even began to articulate a self-understanding via our culturally refined biological intuition. Astrology was a very early "stab in the dark" at figuring out human psychology and a logic of temporal progression aligned with the clockwork of the stars.

Satterlee has also shown that the graphic orientation of theories (constellations of concepts) can be mapped onto the Zodiac, and a theorist's own personal horoscope. As with my comments in the chapter on Hegel, I think many philosophers, in deriding astrology, have denied, if not a foundation for knowledge, at least a stepping stone on the way towards interpreting culture and humanity via a coherent and stable paradigm—and the Zodiac as categorical paradigm is actually more diversified and comprehensive than many contemporary cultural interpretation schemes—astrology, besides being on a track towards astronomy, cosmology, and physics, was also on the track towards social hermeneutics: and those looking for a science of the cultural or personality, and moreover, a science of wisdom, might do well to look to the time before the art of mythology and science of history were split—a resurrection of astrology as a deeply intuitive answers to life's most fundamental questions: who are we, why are we here, where are we going?

FAN MAIL

Having covered some of my favorite contemporary scholars who are more or less outliers in the international intellectual scene, but ripe for moving towards the center of debate—I turn now to some thinkers who have risen to the level of "celebrity intellectual"—attention which is usually earned, in many ways, but is also a tribute, in some cases, to knowing how to promote one's work in the right circles, or being lucky enough to have one's interests and aptitudes match the interests and appetites of the public at large. The following is a sample of correspondence (or attempted correspondence) between me and a few of my intellectual heroes. That many took the time to respond, even if only to say "hello" or "thanks" shouldn't be taken for granted: I've written many less known scholars as well, and found replies less forthcoming from them! Possibly there is some connection between these busy scholars taking the time to read and respond to someone like me, and their having wide followings.

In 2003 I sent letters to the late Richard Rorty and Jacques Derrida, along with some of my art-work prints and poetry plus copies of symbols I'd found while investigating *Finnegans Wake*.

I asked Rorty this question:

> "Do you (or other pragmatists) think that the ecology has any value beyond what the human community has for it? Should our community include animals and plants?" (from a letter to Richard Rorty, May 26th, 2003).

To which Rorty responded:

> **"I wouldn't mind expanding the moral community to include plants /& animals, with the proviso that when things get tough, we humans come first**" (from a postcard response from **Richard Rorty**, June, 2003).

I asked Derrida this question:

> "I've found two themes that seem to be complimentary and which mark a difference between your earlier and later work. First, as in *A Question of Style,* you write about style as a movement, like differance, ever striving to close the veiled gap between itself and an ever deferred presence. While in works like *A Taste for the Secret* you speak of a being open to an incalculable future. Are these two approaches reconcilable? Are there any texts that you know of where you, or others, discuss the relation to these two?" (from a letter to Jacques Derrida, May 26th, 2003).

Derrida dodged the question, but was kind enough to respond:

> "**Pardonnez-moi de vous remercier si simplement pour le magnifique présent que vous le faites. J'y suis très sensible, et vous en remercie du fond du cœur**" (from **Jacques Derrida**'s response, July 22nd, 2003)

I translate this as:

> "Forgive me for so simply thanking you for the magnificent present that you made. I see you most sympathetically, and thank you for it from the bottom of the heart"

The Utilitarian ethicist Peter Singer was kind enough to send me his autograph when I sent him this diatribe argument:

> "a) We should save the starving children NOW.
>
> b) We should not focus on a population that grows due to under-developed economies (population grows slower where people need not have many children to ensure their security in old age) even though addressing economic problems would prevent even more suffering in the future (i.e. Dr. Singer goes for the image of immediate romantic liberal hero, rather than long-term conservative economic realist)

On a large scale your solution would probably be economically infeasible—yes people need aid, but there are always economic trade-offs; and if a richer economy gives more to a poorer economy too quickly, it will probably suffer and give less in the long run (consider the effect of rapid out-sourcing of jobs to poorer countries—what if we in wealthier countries reduced our economy to the bare necessities and gave all the excess to counties in need—I think it would mean an economic collapse, and no further aid). Inter-dependent economic growth that aspires to elevating everyone to self-sufficiency is a hope (and I believe that the U.S. economy, for instance, can grow, even with a trade-deficit through innovation and shifts away from material products to media and services).

But what about the individual?—possibly your shaming tactics are mostly motivators for generosity (maybe a good idea, yet, who wants to be shamed into generosity—what about positive tactics, like your leading by example with donations, donations that don't thumb their nose at those who don't donate 20% or so of their income—who feels ashamed when they hear of Mother Theresa? I feel inspired!) Now, as an individual, I am part of that larger scale local economy that needs my support, at least a little—and I personally feel responsibilities more for my own personal survival and well being, and the survival and well being of those close to me—this seems like an animal fact (soldiers in combat claim they fight for their fellow soldiers, not for some grander ambition). I am more integrated with my local community than with a global community—it shapes who I am, and who I care about more—it is more ME, and thus seems that MY resources would flow accordingly—Yes, I am a world citizen, but who I am is more determined by those I inter-relate to on a daily basis. Conversely, why should I expect others to support me and make me happy?

Each person has their own priorities, and for me, the crises of the world are one among many (like family, friends, philosophy (and Hegel), education, politics, art, science); many areas would be addressed by my resources which are not that critical (which really points out to me that they are my resources in question, usually my body's earned money, or even just my time and effort). Should my priorities be such that every spare moment and resource is committed to the cause? Is that all I am psychologically and humanly worth? Am I allowed to treat myself well, in Peter Singer's book? Why should my body's labor, sacrifices, efforts, time, etc. belong to the community—your solution to world poverty is a voluntary communism—but with a gun armed and aimed with shame. With your logical utilitarian equation, there is no moral choice to be made—the mathematics of suffering has decided it: no choice: no morality, just machines that suffer, and alleviate suffering.

Communism? This is what is implied by your solution to hunger, where all resources ought to belong to the community. Do you really think we can mechanistically construct social structures—I see capitalism as an animal that was discovered and whose taming is constantly being refined via democracy. There are so many variables in a society—government is a constant experiment where the terrain is always shifting—it would be impossible, I think, to try to plan so many variables in a totalitarian manner. I see shifting desires (not necessarily greed) necessary to get people motivated and working, and in need of democratic regulation. Again, how could so many possible variables like various desires be bureaucratically constructed in a planned economy? Tens of millions have died in communist countries in fell swoops due to poor planning. I digress here, though, for your main appeal is to an individual's generosity—but it stems from what seems like a logical argument that suggests going for a loin-cloth, a sleeping bag, and a post-office box, and three jobs so other people don't have shortened life spans—or else you're an apathetic barbarian who just doesn't care enough" (from a letter to Peter Singer, September 6th, 2004).

In retrospect, I think I may have suffered from some logorrhea in my emails and letters to my favorite "celebrity" intellectuals; here's a lengthy excerpt from an email to the deconstructive thinker, Mark C. Taylor:

> "It seems that just as I'm starting to get into a new realm of thinking (like deconstruction, or network theory and cybernetics) you've already written a book engaged with it. Most of my education revolves around the history of philosophy, deconstruction, and cognitive science. And like you, I am very interested in the newest thinking (which you are helping to define)—the leading edge of the Zeitgeist as it were; so I'm very curious as to what you're thinking about now.
>
> As to *Confidence Games,* despite my having to trudge through a snow blizzard of tormenting voices, I think I got some of what you were conveying; and I wonder how much I agree with you, and how much I may have some nuanced differences of opinion. A consistent theme of your work touches on how: 'As complexity increases, the rate of change accelerates and disruptions become more frequent. In other words, when webs and networks extend their reach, volatility and therefore instability increase.' (300) This seems connected to a running theme in *Confidence Games*, that elements of the economy separate from material things and events (like the gold standard), and fly off into the virtual (like the internet bubble rising above solid company performances, and revenue generating ability). Now, being a little simple minded, I see this concern with instability as connected with Derrida's thinking that language could be considered more unstable when the signifier is (always already) severed from the signified, in an interplay of differences between signifiers. However—and this is key to my

difference with you and Derrida, with regard to attitude, if not theory—isn't language *more* stable when it has many "ties that bind" (as in network connections, or *traces*) rather than some sort of "transcendental signified" that is itself not anchored anywhere? Or, imagine many floating signifier balloons tied to individual signified rocks, as compared to all those floating balloons tied to each other: which image seems more stable?

To bring this difference in relation to *Confidence Games*, although I agree that virtualized networks can accelerate changes that threaten equilibrium, would not counter-forces be accelerated as well? I'm not suggesting that there is a static equilibrium, but an evolution where cycles that feed back into themselves preserve a system that changes partially, and relatively slow on the whole relative to its parts. I think that our ecological and economic systems, since they are complex and interdependent are *robust* rather than unstable. That is, if you remove one element from an ecosystem, that may have a domino effect in many directions; but counter-forces will most definitely diffuse that effect and preserve the system, although it will be changed. (E.g., if you take out an abundant food source, some species may be reduced in population—but most likely will not be wiped out as they turn to other types of food source). I think we would agree that equilibrium is always in a state of flux. I would say the 'whole' is quite (but not absolutely) stable and secure, while the parts are radically insecure and ephemeral. I really don't foresee another Great Depression due to lessons learned in a trial and error mass-population self-correcting system. I agree that short term risk can't be calculated, but in the very long run Standard & Poors will most likely give an average of a ~10% return.

Another difference I may have with you revolves around the notion of efficiency. It seems to me that efficiency is usually measured around a specific task. Yet how efficient is a car when compared to a horse? Planned economies are more like a car, where free markets are more like a horse (that should be reined in a bit—regulated—in my opinion). It seems to me that in the over all eco-sphere a horse is probably more efficient than a car: efficiency in free markets would not be absolutely efficient in a particular measure, but the 'integrated' efficiency that arises out of co-dependent evolution and competitions seems to be better than what we could pre-program. Here complexity seems to be aligned with the free-market, where a planned economy would be like trying to program artificial intelligence with more rigid simple algorithms, rather than with flexible neural nets that would be vessels of adaptable common sense memes. So again, I see a whole / part contrast, where free markets are relatively efficient on the whole, while possibly less efficient in the parts. (The whole being simply our local earth: who knows what would happen if aliens make contact, and incorporate us within their economy (I imagine our art would be quite valuable)).

Some of my other current interests include Political Economics, especially Donald Wittman's book, *The Myth Of Democratic Failure—Why Political Institutions Are Efficient*; sustainability and ecological analogies found in free-market economics and democratic politics; the notion of *foci* in the absence of centers, and the relationship between Derridian différance, qualia, and neurological connections. Although a fan of complexity, I think simplicity can be a doorway to more complex thinking—you have helped bring together some really complex ideas and made them more coherent; I only wish I could pick up the ball and explain new complex ideas in an even clearer fashion, as to draw more people into ever-changing and complicated new worlds of thinking" (from an email to Mark C. Taylor, June 7th, 2005).

To which Taylor replied:

> **"I am, of course, delighted that you find my writings useful. The issues you mention and questions you raise are surely relevant to everything I write. Currently I am completing three books.**
>
> **After God - a book on religion and culture**
>
> **Mystic Bones - a book of original photographs and an essay**
>
> **About Nothing - collection of essays on literature and art**
>
> **I am sympathetic with your concern about ecological issues. Indeed, I think the analysis of complex systems provides a way to address these matters very effectively. That should become evident in the first two books listed above. Next year I am going to teach a course with a colleague in the Chemistry department titled What is Life?"** (email response from **Mark C. Taylor**, July 1st, 2005).

The literary critic and theorist Harold Bloom was kind enough thank me for an email I sent him:

> **"Thank you for your lucid and sensitive letter"** (a return email from **Harold Bloom**, July 28th, 2005).

My initial email contained these remarks on Bloom's work:

> "My intellectual background mostly concerns philosophy; specifically cognitive science and deconstruction—but also the history of philosophy and aesthetics. These interests have helped me to appreciate the work you are doing: preserving the canon, and reading it in a canonical manner—probing deeply into, and revealing the 'Truth' with an originality and clarity so rarely achieved by others.

Your essay, 'The Art of Reading Poetry,' from you recent book, *The Best Poems of the English Language*, evidences this clarity, artfully and engagingly. I like the way your essay has an underlying structural progression: (1) the foundations of poetry found in '*change*' (the transformative aspect of figurative language); (2) linguistic change reaching through *time* (historical etymology); (3) this history *remembered* through (re)cognition; (4) and used retrospectively through *allusion*; (5) next turning from these allusions to distinguish great *original poetic voices*; (6) poets' voices whose quality can be gauged by the '*inevitability*' of their phrasings; (7) this 'inevitability' approaching the *eternal, timeless*, aspect of the finest poetry; (8) which comes full circle back to change, with the authentically eternal revealing itself as the original and *strange*.

Your circle of progress loops from the changing to the timeless which itself may be change. Is there (and no doubt you have thought about this) a paradox, or tension, between good poetry being both 'inevitable' and the 'strange?' No doubt, good poetry may seem to write itself for the inspired poet, and find life with its inevitable niche located in the ecology of literary history, and the biology of humanness. Perhaps the evolution of poetry brings forth new strangeness which was always already predetermined by the seeds of the 'origin'; an origin time flees from with some anxiety and originality, reaching farther into the future from deep within the heart" (from email to Harold Bloom, July 28th, 2005).

I was also able to contact Noam Chomsky via the ZNet forum; below is the full interchange, as Chomsky responds to various questions I pose about legislating justice and language (Chomsky cited in bold):

"I've been giving talks abroad, and haven't been on e-mail. And since returning a few days ago have been working

through the rapidly accumulating chaos. Hence the delay.

Below, but can't comment on the references you cite; that would get out of hand and turn into a major essay. Only on your points here.

NC

From: 'J.D. Casten'
Human Rights in the UK: Judicial Interpretation or Legislative Refinement?

Noam—

First, some quotes (my question follows below):

Charles Moore—from the Telegraph ('Blair's Major moment: why Human Rights are like the ERM'—June 20th, 2006) claims:

> 'If you read the various universal declarations of human rights, they are, for the most part, impressive statements of things that any sensible society ought to have. They are a good basis upon which particular laws might be made. They are worse than useless, though, as a list of legal principles which judges can enforce. They transfer from elected people the duty to pass the right laws and give to judges the power to do politics' (<//www.telegraph.co.uk/opinion/main.jhtmlxml=/opinion/2006/05/20/do2002.xml&sSheet=/opinion/2006/05/20/ixopinion.html>).

Ronald Dworkin—from the Guardian ('It is absurd to calculate human rights according to a cost-benefit analysis'—May 23rd, 2006) notes:

> 'This is a dangerous time for freedom in Britain. The country's most powerful politicians have joined its irresponsible press in a shameful attack on the idea of human rights. Tony Blair says that the nation needs to re-examine what he calls the 'philosophy' behind the Human Rights Act so as to change the balance it strikes between individual freedom and the community's security. David Cameron says a Tory government would reform that act or repeal it entirely. The rightwing press goes still further: the Telegraph calls for Britain to renounce the European Convention on Human Rights that Britain sponsored (it was signed in London) in 1950' (<//politics.guardian.co.uk/print/0,,329487790-107865,00.html>).

Some Noam Chomsky comments on 'universality' and Human Rights can be found here:

<//www.zmag.org/content/showarticle.cfm?ItemID=8226>

Noam Chomsky—from ZNet Commentaries, May 20th, 2006 ('A Just War? Hardly') states:

> 'Let me state a couple of simple truths. The first is that actions are evaluated in terms of the range of likely consequences. A second is the principle of universality; we apply to ourselves the same standards we apply to others, if not more stringent ones.'

The above cited material often contrasts two areas of thought: high ideals vs. pragmatic laws, social security vs. individuals' liberties, theory vs. practice, and universal human rights vs. consequentialism /

utilitarianism (and also relevant are: principles vs. parameters, rules vs. exceptions, and majorities vs. minorities—these oppositions often being similar to a tree trunk and its selective branching out to the limits of its leaves). Pragmatism and utilitarianism go hand in hand (what 'works' is what we should go with), and since utilitarianism is designed to benefit the majority, it parallels direct democracy too. However, the enslavement or random sacrifice of the few might be justified on such simple grounds: minorities must be protected as well, and hence we also have the rule of law and universal rights to protect everyone equally on matters of personal integrity and survival. Rawls' A Theory of Justice includes the notion of a 'veil of ignorance' where we might make laws knowing not whether we would be in a majority or a minority, and thus might respect the rights of minorities from a sort of universal utilitarian perspective.

I doubt that Rawls would have been happy with that formulation. He was giving a critique of utilitarianism.

Maybe we could imagine the 'principle' of universality to vary via particular 'parameters' of utility / consequence? I think that brain structure is reflected in political ideologies (see links below) and am also interested in how the more rigid fundamental principled structures of language that have some 'play' or variation within limited possible parameters parallels fundamental government laws and legal interpretation within the bounds of these laws (laws being like principles, and 'precedents' being like parameters). I'm not sure if you'd care to comment on this parallel between the structures of language and social laws, if there is one, but my 'real world' question below touches on just this issue.

Rawls adopted linguistic models, explicitly. He saw himself as constructing something like a 'grammar of moral judgment,' and used as analogy the theories of generative grammar being developed in the sixties when he was writing his Theory of Justice. That move came under harsh attack from philosophers, and he put it to the side. A recent Phd dissertation by John Mikhail, soon to appear in expanded version as a book (Cambridge U press), reviews the critiques and, I think, adequately refutes them, then picks up the project and extends it, also to empirical investigations of moral judgments and a theory that might explain them. Primatologist / cognitive scientist Marc Hauser has a book coming out along similar lines, and others are working on it too. It's becoming an interesting domain of cognitive science and moral philosophy. The basic observations go back to Hume.

I doubt that parallels go very far; cognitive organs are different, just as the kidney and liver are. There are doubtless common properties, and they might turn out to be deep and significant, if enough were understood. That happens to be a personal interest of mine, particularly so in the past 20 years. And it runs through biology generally. For the moment, the analogies don't seem to me helpful, beyond the kinds of suggestions brought forth by Rawls and others since.

On the principle of universality—perhaps the most elementary moral principle—it's not clear to me what exceptions it might have in principle, though in practice it is almost universally ignored by systems of power and their acolytes. But maybe. Human life is too complex, and too poorly understood, for much confidence to be warranted. I'd like to see an example, though.

In light of Rawls' 'veil of ignorance' uniting universalism and utilitarianism, do you think that, actions 'evaluated in terms of a range of likely consequences' can still come into conflict with 'the principle of universality?' Criminals and enemy combatants loose certain rights to protect communities—but when something like justified deadly force, on a small or large scale becomes highly calculated, premeditated and for preemption and deterrence rather than immediate defense, the word 'murder' comes to mind. Criminals are something of the ultimate of minorities, and it is often up to the judicial branch to protect minorities from democratic and social desires by rule of law.

I don't see what this has to do with the principle of universality. You're bringing up some of the many factors that enter into moral judgment, but not giving any reason why, if we apply these standards to others, we should not apply them to ourselves. I don't see any conflict.

With regard to the is rather long-winded framing of my question: what do you think of (1) Tony Blair's wanting to deport the Afghan hijackers of 2000 to possible torture in their own country (and the related case of the prisoners at Guantánamo Bay), and (2) hints that he and others would like some Human Rights laws repealed to allow for more specific democratic-representative legislated laws rather than ('activist') judicial interpretation of broad inalienable human rights laws?

When human rights are called 'inalienable' it does not imply that there are no imaginable circumstances in which they might be violated. Life is too complex for that. Anyone who calls for violation of those rights must present a very strong case, and must also accept the principle of universality. I don't think Blair has presented much of a case, and he certainly rejects the principle of

universality. To mention only the most obvious illustration, I don't recall his offering himself for torture, though we can hardly doubt that he has been engaged in major crimes—including the 'supreme international crime,' but not only that.

Thank you,

J.D. Casten" (ZNet forum email exchange with **Noam Chomsky**, ~June 1st, 2006).

A couple other celebrity philosophers did not respond to my email letters; here's an excerpt from an email to Slavoj Žižek:

"'*Don't Rob Me of Coincidence*'

To engage your thinking a bit, I'll make a few remarks on your essay, 'Of Cells and Selves.'

If I may summarize your aims there, I would think that you were concerned with differentiating the 'personality' subject from being part of a deep ecological whole continuum on the one hand, or genetically determined (as in being identical to your clone) on the other. The self, rather, arises at the surface interface between the two (and moreover becomes more complex, singular, and possibly autonomous, due to a 'Self-Referential Loop' of internal cognition).

I'll set aside a debate on making sexual differences into metaphysical differences (are not sexual differences about as big or small as gendered-bodily differences?)—and possibly defend Habermas' contention that cloning is problematic and beyond.

Although there may someday be cloned humans, and these humans would have a right to their dignity—how boldly should we storm into creating such a psychological dilemma (and possible crazy crisis) for some self-searching teenager? No doubt, you would know that there is a difference between a random bird-dropping on your head, and some stranger spitting in your face. The event may be similar, but the psychology of it is radically different. I'd much rather deal with the bird shit, than the human psychology of 'who the hell do you think you are?'

So even if a person is not identical to their clone, due to the environmental interface, possible insults, such as, 'you're just like your father,' would take on a whole new twist. It just seems like a lot of psychological pressure to unduly put on some human being, which seems most likely an unnecessary exploration of curiosity.

Pragmatics aside, is there really a clear cut boundary between you and I. Yes we have separate bodies, but what about the similar DNA, and that same Earth environment. Maybe you've seen that psychological photocopy of a Dalmatian among dots—there's a sort of gestalt ability to, yes, *differentiate* the dog from the other black dots, but it's definitely not due to clear lines, as in a an Adami painting. It's the clusters of connections that are as important as the differentiating points. So, I would also say, that much of who we are, due to DNA and social phenomena such as memes, is part of that Gaia continuum (the connections). Lots of overlapping memes: Donald Davidson's 'principle of charity,' etc.

Also, what is 'autopoiesis?' I like the term when trying to think how the unity of subjective consciousness connects to the individual body, but shouldn't we try to explain as much as we can, such as about 'autonomous' individual bodies (and even the laws of the universe springing from a possible chaos of potential structure) via *natural selection?*

Plus, do you think that consciousness necessarily resides on the *inside* of an organism, or like me, do you think it may be right there on the surface, between inside and outside: a qualia laterally cut by differential structure (like so many

> living bodies inter-subjectively roaming about a 'nirvana' of quality)?
>
> I think *I'm* on the same pages of your essay in many places, I just wonder, if like so many contemporary thinkers, when rebelling against 'identity' with *difference*, you forget that you are emphasizing *differences* at the cost of *connections*. There is a political implication in this; one where I'd like to take some Taoist middle way, myself being a radical moderate (hence both appreciated and rejected by all sorts of sides, while being at the location of the 'real deal' where change materializes)." (from email to Slavoj Žižek, December 11th, 2006).

And from an email to Steven Pinker:

> "I don't think memes are incompatible with innate cognitive structures / functions. Differential thinking will help you overcome the notion that we can't learn identities from what is the associated in *different* circumstances. I don't think we learn that the word 'yellow' is associated with yellow only because we have an innate predisposition to know what yellow is, but also because the same word 'yellow' is used in various differing circumstances where yellow is the most salient common factor: Kant was a master of this notion (as well as the notion of what is *implied* by our phenomenal awareness). I bring up Kant, because I like his peacemaking between the empiricists (those 'blank slate' folks), and rationalists (like Chomsky, and possibly you). I too hope to bridge and connect the insights of those researching hard-wired cognitive abilities and those who might be called 'social constructivists'" (email to Steven Pinker, February 15th, 2007).

I emailed "Singularity" theorist and sci-fi author Vernor Vinge, to share my poetry novella *Post-Digital Revelation*, which I noted as:

> "the flip-side of my 'more serious' theoretical approach to AI (I'm a big fan of neural-nets, spreading activation, and 'cue-cluster-convergence,' as well as the work by Douglas Hofstadter & Daniel C. Dennett, Wittgenstein's 'Language Games,' and Foucault's 'Discourses'—although consciousness may be involved with Derrida's 'Deconstruction'—which is both of and against the mechanical—where the determination of the machine flips through the present and opens up to an impossibly and incalculably complex future... but I digress)" (from email to Vernor Vinge, June 21st, 2008).

Vinge replied:

> "**These are interesting times, for both upsides and down :-)**" (email excerpt from **Vernor Vinge**, July 10th, 2008).

Interesting times, indeed. And so this book comes to a close. It may have been a long journey, with many insights and challenges along the way, but this book veers far from exhausting the subject (we could have talked more about John Searle's engagement with Derrida and Strong AI, Rorty on cognitive science, Isaac Asimov's three robot laws, Jacques Lacan, Gilles Deleuze, David Marr and so much more; perhaps another book, perhaps another author).

> "Man makes man in his own image. This seems to be the echo or the prototype of the act of creation, by which God is supposed to have made man in His image. Can something similar occur in the less complicated (and perhaps more understandable) case of the nonliving systems that we call machines?
>
> What is the image of a machine? Can this image as embodied in one machine, bring a machine of a general sort, not yet committed to a particular specific identity, to reproduce the original machine, either absolutely or under some change that is construed as variation? Can the new and varied ma-

chine itself act as an archetype, even as to its own departures from its own archetypical pattern?" (Norbert Wiener, *God & Golem, Inc.: A Comment on Certain Points where Cybernetics Impinges on Religion*, The MIT Press: Cambridge, MA (1964), p. 29).

Bibliography

Adams, Hazard, and Leroy Searle (eds.), 1986, *Critical Theory Since 1965*. Gainesville: University Press of Florida.

Adorno, Theodor W., Shierry Weber Nicholsen and Samuel Weber (trans.), *Prisms*. Cambridge, MA: The MIT Press.

Adorno, Theodor W., E.B. Ashton (trans.), 1990, *Negative Dialectics*. London: Routledge.

Adorno, Theodor W., J.M. Bernstein (trans.), 2001, *The Culture Industry: Selected Essays on Mass Culture*. London: Routledge.

Adorno, Theodor W., E.F.N. Jephcott (trans.), 2006, *Minima Moralia: Reflections from Damaged Life*. New York: Verso.

Aesop, Handford, S.A. (trans.), 1984, *Fables of Aesop*. New York: Penguin Books.

Aiken, Henry D. (ed.), 1954, *The Age of Ideology: 19th Century Philosophers*. New York: New American Library

Alison, David B. (ed.), 1985, *New Nietzsche: Contemporary Styles of Interpretation*. Cambridge, Massachusetts: MIT Press.

Anderson, Perry, 1998, *The Origins of Postmodernity*. New York: Verso.

Aristotle, Jonathon Barnes (ed.), 1995, *The Complete Works of Aristotle*. Princeton: Princeton University Press.

Augustine, St., Vernon J. Bourke (ed.), 1983, *The Essential Augustine*. Indianapolis, IN: Hackett Publishing Company.

Augustine, St., 1991, *The Trinity*. New York: New City Press.

Augustine, St., 2006, *The City of God*. New York: Barnes & Noble.

Badiou, Alain, Louise Burchill (trans.), 1999, *Deleuze: The Clamor of Being*. Minneapolis: University of Minnesota Press.

Barabasi, Albert-Laszlo, 2003, *Linked: How Everything Is Connected to Everything Else and What it Means*. London: Penguin Books.

Barnes, Jonathan (ed.), 1995, *The Cambridge Companion to Aristotle*. Cambridge: Cambridge University Press.

Barrett, William, and Henry D. Aiken (eds.), 1962, *Philosophy in the Twentieth Century: An Anthology (Four Volumes)*. New York: Random House.

Barthes, Roland, Annette Lavers and Colin Smith (trans.), 1977, *Elements of Semiology*. New York: Hill and Wang.

Baudrillard, Jean, Phill Beitchman, Paul Foss and Paul Patton (trans.), 1983, *Simulations*. Los Angeles: Semiotext(e).

Beck, Lewis White, 1966, *18th – Century Philosophy*. New York: The Free Press.

Beckett, Samuel, et. al., 1972, *Our Exagmination Round His Factification For Incamination of Work in Progress*. New York: New Directions Books.

Benjafield, John G., Daniel Smilek and Alan Kingstone, 2010, *Cognition, 4th ed.* Oxford: Oxford University Press.

Benjamin, Walter, Hannah Arendt (ed.), Harry Zohn (trans.), 1969, *Illuminations: Essays and Reflections*. New York: Schocken.

Berkeley, George, David M. Armstrong (ed.), 1965, *Berkeley's Philosophical Writings*. New York: Macmillan Publishing Company.

Berlin, Isaiah (ed.), 1954, *The Age of Enlightenment: The 18th Century Philosophers*. New York: New American Library.

Bertalanffy, Ludwig von, 1969, *General System Theory: Foundations, Development, Applications*. New York: George Braziller Inc.

Best, Steven and Douglas Kellner, 1991, *Postmodern Theory*. New York: The Guilford Press.

Best, Steven and Douglas Kellner, 1997, *The Postmodern Turn*. New York: The Guilford Press.

Best, Steven and Douglas Kellner, 2001, *The Postmodern Adventure: Science, Technology, and Cultural Studies at the Third Millennium*. New York: The Guilford Press.

Blake, William, David V. Erdman (ed.), 1988, *The Complete Poetry & Prose of William Blake.* New York: Anchor Press Book.

Blonsky, Marshal (ed.), 1985, *On Signs.* Baltimore, MD: John Hopkins University Press.

Boden, Margaret A., 2006, *Mind as Machine: A History of Cognitive Science.* Oxford: Oxford University Press.

Brooke, C.F. Tucker and Nathaniel Burton Paradise (eds.), 1933, *English Drama 1580-1642.* Boston, MA: D.C. Heath and Company.

Brooks, Rodney, et. al, 2002, *Understanding Artificial Intelligence.* New York: Warner Books, Inc.

Brown, Stuart (ed.), 1998, *One Hundred Twentieth Century Philosophers.* London: Routledge.

Brun, Herbert, Stephen Sloan and Heinz von Foerster, 1995, *Cybernetics of Cybernetics.* Minneapolis, MN: Future Systems Inc.

Buchanan, Ian, 2000, *Deleuzism: A Metacommentary.* Durham, NC: Duke University Press.

Burgess, Anthony, 1966, *ReJoyce.* New York: Ballantine Books.

Bynum, Terrell Ward and James H. Moor (eds.), 1998, *The Digital Phoenix: How Computers are Changing Philosophy.* Malden, MA: Blackwell Publishers Inc.

Caesar, Michael, 1999, *Umberto Eco: Philosophy, Semiotics and the Work of Fiction.* Cambridge: Polity.

Cahoone, Lawrence E. (ed.), 2003, *From Modernism to Postmodernism: An Anthology.* Hoboken, NJ: Wiley-Blackwell.

Cain, Mark J., 2002, *Fodor: Language, Mind and Philosophy.* Malden, MA: Blackwell Publishers Ltd.

Campbell, James, 1996, *Understanding John Dewey.* Chicago: Open Court.

Campbell, Joseph, 1973, *A Hero with a Thousand Faces.* New Jersey: Princeton University Press.

Caputo, John D. (ed.), 1996, *Deconstruction in a Nutshell: A Conversation with Jacques Derrida.* New York: Fordham University Press.

Caputo, John D., 1997, *The Prayers and Tears of Jacques Derrida: Religion without Religion.* Indianapolis: Indiana University Press.

Carroll, Lewis, undated, *Alice's Adventures in Wonderland and Through the Looking Glass.* New York: J.J. Little and Ives Company.

Chomsky, Noam, James Peck (ed.), 1987, *The Chomsky Reader.* New York: Pantheon Books.

Chomsky, Noam, 1996, *Powers & Prospects: Reflections on Human Nature and the Social Order.* Boston: South End Press.

Chomsky, Noam, 2000, *New Horizons in the Study of Language and Mind.* Cambridge: Cambridge University Press.

Chomsky, Noam, 2002, *Media Control: The Spectacular Achievements of Propaganda, 2nd ed.* New York: Seven Stories Press.

Chomsky, Noam, Adriana Belletti anLuigi Rizzi (eds.), 2002, *On Nature and Language.* Cambridge: Cambridge University Press.

Chomsky, Noam and Michel Foucault, 2006, *The Chomsky-Foucault Debate: On Human Nature.* New York: The New Press.

Chomsky, Noam, 2007, *On Language – Chomsky's Classic Works: Language and Responsibility and Reflections on Language.* New York: The New Press.

Chomsky, Noam, James McGilvray (ed.), 2009, *Cartesian Linguistics: A Chapter in the History of Rationalist Thought, 3rd ed.* Cambridge: Cambridge University Press.

Cilliers, Paul, 1998, *Complexity and Postmodernism: Understanding Complex Systems.* New York: Routledge.

Clark, Andy, 1998, *Being There: Putting Brain, Body, and World Together Again.* Cambridge, MA: The MIT Press.

Clark, Andy, 2000, *Mindware: An Introduction to the Philosophy of Cognitive Science.* Oxford: Oxford University Press.

Clark, Andy, 2003, *Natural-Born Cyborgs: Minds, Technologies, and the Future of Human Intelligence.* Oxford: Oxford University Press.

Colebrook, Claire, 2001, *Gilles Deleuze.* London: Routledge.

Collins, John, 2008, *Chomsky: A Guide for the Perplexed.* New York: Continuum International Publishing Group.

Collonson, Diane, 1987, *Fifty Major Philosophers: A Reference Guide.* London: Routledge.

Copeland, Jack, 2012, *Artificial Intelligence: A Philosophical Introduction.* Malden, MA: Blackwell Publishers Inc.

Copleston, Frederick, 1985, *A History of Philosophy: Vols. I-III.* New York: Image.

Copleston, Frederick, 1985, *A History of Philosophy: Vols. IV-VI.* New York: Image.

Copleston, Frederick, 1985, *A History of Philosophy: Vols. VII-IX.* New York: Image.

Corrington, Robert S., 1993, *An Introduction to C.S. Peirce.* Lanham, MD: Rowan and Littlefield Publishers, Inc.

Cottingham, John, 1988, *A History of Western Philosophy 4: The Rationalists.* Oxford: Oxford University Press.

Crary, Alice Marguerite, and Rupert Read (eds.), 2000, *The New Wittgenstein.* London: Routledge.

Culler, Jonathan, 2000, *Literary Theory: A Very Short Introduction.* Oxford: University of Oxford Press.

Culler, Jonathan, 2002, *The Pursuit of Signs: Semiotics, Literature, Deconstruction.* New York: Cornell University Press.

Culler, Jonathan, 2008, *On Deconstruction: Theory and Criticism after Structuralism.* New York: Cornell University Press.

Cummins, Denise Dellarosa and Robert Cummins, 2000, *Minds, Brains, and Computers: An Historical Introduction to the Foundations of Cognitive Science.* Hoboken, NJ: Wiley-Blackwell.

Currie, Nicholas (aka Momus), 2009, *The Book of Jokes: a Novel by Momus.* London: Dalkey Archive Press.

Davidson, Donald, 2001, *Inquiries into Truth and Interpretation*. Oxford: Oxford University Press.

Davidson, Donald, 2001, *Subjective, Intersubjective, Objective*. Oxford: Oxford University Press.

Davis, Martin, 2000, *Engines of Logic: Mathematicians and the Origin of the Computer*. New York: W.W. Norton and Company.

Debord, Guy, Donald Nicholson-Smith (trans.), 2000, *Society of the Spectacle*. New York: Zone Books.

Deely, John, 2004, *Basics of Semiotics*. Houston: St. Augustine's Press.

Deleuze, Gilles, 1985, *Kant's Critical Philosophy: The Doctrine of the Faculties*. Minneapolis: University of Minnesota Press.

Deleuze, Gilles, Sean Hand (trans.), 1988, *Foucault*. Minneapolis: University of Minnesota Press.

Deleuze, Gilles, Paul Patton (trans.), 1995, *Difference and Repetition*. New York: Columbia University Press.

Deleuze, Gilles, Martin Joughin (trans.), 1997, *Negotiations 1972-1990*. New York: Columbia University Press.

Deleuze, Gilles, Michael Hardt (trans.), 2006, *Nietzsche and Philosophy*. New York: Columbia University Press.

Deleuze, Gilles and Felix Guattari, Robert Hurley, Mark Seem and Helen Lane (trans.), 2009, *Anti-Oedipus: Capitalism and Schizophrenia*. New York: Penguin.

De Man, Paul, 1982, *Allegories of Reading: Figural Language in Rousseau, Nietzsche, Rilke, and Proust*. New Haven, CT: Yale University Press.

De Man, Paul, 1983, *Blindness and Insight: Essays in the Rhetoric of Contemporary Criticism*. Minneapolis: University of Minnesota Press.

De Man, Paul, 1984, *The Rhetoric of Romanticism*. New York: Columbia University Press.

De Man, Paul, 1986, *Resistance to Theory*. Minneapolis: University of Minnesota Press.

De Man, Paul, 1996, *Aesthetic Ideology*. Minneapolis: University of Minnesota Press.

Dennett, Daniel C., 1991, *Consciousness Explained*. Boston: Little Brown and Company.

Dennett, Daniel C., 1996, *Darwin's Dangerous Idea: Evolution and the Meanings of Life*. New York: Touchstone.

Derrida, Jacques, David B. Allison (trans.), 1973, *Speech and Phenomena: And Other Essays on Husserl's Theory of Signs*. Evanston, IL: Northwestern University Press.

Derrida, Jacques, Alan Bass (trans.), 1978, *Writing and Difference*. Chicago: University of Chicago Press.

Derrida, Jacques, Barbara Harlow (trans.), 1981, *Spurs: Nietzsche's Styles / Eperons: Les Styles de Nietzsche*. Chicago: University of Chicago Press.

Derrida, Jacques, Alan Bass (trans.), 1982, *Positions*. Chicago: University of Chicago Press.

Derrida, Jacques, Barbara Johnson (trans.), 1983, *Dissemination*. Chicago: University of Chicago Press.

Derrida, Jacques, Alan Bass (trans.), 1985, *Margins of Philosophy*. Chicago: University of Chicago Press.

Derrida, Jacques, Alan Bass (trans.), 1987, *The Post Card: From Socrates to Freud and Beyond*. Chicago: University of Chicago Press.

Derrida, Jacques, Geoffrey Bennington and Ian McLeod (trans.), 1987, *The Truth in Painting*. Chicago: University of Chicago Press.

Derrida, Jacques, Jeffery Mehlman and Samuel Weber (trans.), 1988, *Limited Inc*. Evanston, IL: Northwestern University Press.

Derrida, Jacques, John P. Leavey Jr. (trans.), 1989, *Edmund Husserl's "Origin of Geometry": An Introduction*. Lincoln: University of Nebraska Press.

Derrida, Jacques, 1989, *Memoires for Paul de Man*. New York: Columbia University Press.

Derrida, Jacques, Peggy Kamuf (ed.), 1991, *A Derrida Reader: Between the Blinds*. New York: Columbia University Press.

Derrida, Jacques, Geoffrey Bennington and Rachel Bowlby (trans.), 1991, *Of Spirit: Heidegger and the Question*. Chicago: University of Chicago Press.

Derrida, Jacques, Gayatri Chakravorty Spivak (trans.), 1998, *Of Grammatology*. Baltimore, MD: Johns Hopkins University Press.

Derrida, Jacques, Peggy Kamuf (trans.), 2000, *Without Alibi*. Stanford, CA: Stanford University Press.

Derrida, Jacques and Maurizio Ferraris, 2001, *A Taste for the Secret*. London: Polity.

Derrida, Jacques, Rachel Bowlby (trans.), 2005, *Paper Machine*. Stanford, CA: Stanford University Press.

Descartes, René, Elizabeth S. Haldane and G.R.T. Ross (trans.), 1955, *Philosophical Works of Descartes, Vol. I*. New York: Dover Publications.

Descartes, René, Norman Kemp Smith (trans.), 1958, *Descartes Philosophical Writings*. New York: The Modern Library.

Descartes, René, Stephen Gaukgroger (trans.), 2004, *The World and Other Writings*. Cambridge: Cambridge University Press.

Descombes, Vincent, 1981, *Modern French Philosophy*. Cambridge: Cambridge University Press.

Dewey, John, 1958, *Experience and Nature*. New York: Dover Publications, Inc.

Dewey, John, 1967, *Reconstruction in Philosophy*. Boston: Beacon Press.

Dewey, John, John J. McDermott (ed.), 1981, *The Philosophy of John Dewey, Two Volumes in One: 1, The Structure of Experience; 2, The Lived Experience*. Chicago: University of Chicago Press.

Dewey, John, 1980, *Art as Experience*. New York: Perigee Books.

Dewey, John, 1980, *The Quest for Certainty*. New York: Perigee Books.

Docherty, Thomas, 1992, *Postmodernism: A Reader*. New York: Columbia University Press.

Dowling, John E., 2000, *Creating Mind: How the Brain Works.* New York: W.W. Norton and Company.

Dreyfus, Hubert L. and Stuart E. Dreyfus, 1988, *Mind over Machine.* New York: Free Press.

Dreyfus, Hubert L., 1992, *What Computers Still Can't Do: A Critique of Artificial Reason.* Cambridge, MA: The MIT Press.

Dummett, Michael, 1996, *Origins of Analytic Philosophy.* Cambridge, MA: Harvard University Press.

Dupuy, Jean-Pierre, M.B. DeBevoise (trans.), 2009, Cambridge, MA: The MIT Press.

Dyson, George B., 1997, *Darwin among the Machines: The Evolution of Global Intelligence.* Reading, MA: Perseus Books.

Eagleton, Terry, 1996, *The Illusions of Postmodernism.* Hoboken, NJ: Wiley-Blackwell.

Eagleton, Terry, 2008, *Literary Theory: An Introduction.* Minneapolis: University of Minnesota Press.

Ebert, John David, 2011, *The New Media Invasion: Digital Technologies and the World They Unmake.* Jefferson, NC: McFarland & Company, Inc.

Ebert, John David, 199, *Twilight of the Clockwork God: Conversations on Science and Spirituality at the End of an Age.* San Francisco, CA: Council Oak Books.

Eco, Umberto, 1978, *A Theory of Semiotics.* Bloomington: Indiana University Press.

Eco, Umberto, 1979, *The Role of the Reader.* Bloomington: Indiana University Press.

Eco, Umberto, 1986, *Semiotics and the Philosophy of Language.* Bloomington: Indiana University Press.

Eco, Umberto, 1991, *The Limits of Interpretation.* Bloomington: Indiana University Press.

Eco, Umberto, James Fentress (trans.), 1997, *The Search for the Perfect Language*. Cambridge, MA: Blackwell.

Eco, Umberto, 1999, *Serendipities: Language and Lunacy*. Boston: Mariner Books.

Eco, Umberto, 2000, *Kant and the Platypus: Essays on Language and Cognition*. Boston: Mariner Books.

Evnine, Simon, 1991, *Donald Davidson*. Palo Alto: Stanford University Press.

Fauconnier, Gilles and Mark Turner, 2003, *The Way We Think: Conceptual Blending and the Mind's Hidden Complexities*. New York: Basic Books.

Fenner, Peter, 1995, *Reasoning Into Reality: A System Cybernetics Model and Therapeutic Interpretation of Buddhist Middle Path Analysis*. Somerville, MA: Wisdom Publications.

Fodor, Jerry, 2000, *In Critical Condition: Polemical Essays on Cognitive Science and the Philosophy of Mind*. Cambridge, MA: The MIT Press.

Fodor, Jerry, 2001, *The Mind Doesn't Work That Way: The Scope and Limits of Computational Psychology*. Cambridge, MA: The MIT Press.

Forsythe, Robert Stanley, 1914, *Shirley's Plays and the Elizabethan Drama*. New York: Columbia University Press.

Foster, Hal (ed.), 2002, *The Anti-Aesthetic: Essays on Postmodern Culture*. New York: New Press.

Foucault, Michel, Richard Howard (trans.), 1988, *Madness and Civilization: A History of Insanity in the Age of Reason*. New York: Vintage Books.

Foucault, Michel, Robert Hurley (trans.), 1990, *The History of Sexuality, Vol. 1: An Introduction*. New York: Vintage Books.

Foucault, Michel, 1994, *The Order of Things: An Archaeology of the Human Sciences*. New York: Vintage Books.

Franklin, Stan, 1997, *Artificial Minds*. Cambridge, MA: The MIT Press.

Freeman, Walter J. and Rafael Nunez, 2000, *Reclaiming Cognition: The Primacy of Action, Intention and Emotion*. Bowling Green, OH: Imprint Academic.

Fremantle, Anne (ed.), 1954, *The Age of Belief: The Medieval Philosophers.* New York: New American Library.

Freud, Sigmund, A.A. Brill (trans.), 1938, *The Basic Writings of Sigmund Freud.* New York: Modern Library.

Freud, Sigmund, James Strachey (ed.), 1949, *An Outline of Psycho-Analysis.* New York: W.W. Norton and Company.

Freud, Sigmund, James Strachey (trans.), 1961, *Beyond the Pleasure Principle.* New York: W.W. Norton and Company.

Freud, Sigmund, Joan Riviere (trans.), 1969, *A General Introduction to Psychoanalysis.* New York: Pocket Books.

Freud, Sigmund, Peter Gay (ed.), 1989, *The Freud Reader.* New York: W.W. Norton and Company.

Freud, Sigmund, James Strachey (trans.), 2001, *The Standard Edition of Complete Psychological Works of Sigmund Freud, Volume I: Pre-Psychoanalytic Publications and Unpublished Drafts.* London: Vintage Books.

Friedenberg, Jay and Gordon Silverman, 2006, *Cognitive Science: An Introduction to the Study of Mind.* Thousand Oaks, CA: Sage Publications.

Frost, S.E., 1962, *Basic Teachings of the Great Philosophers.* New York: Anchor.

Gardiner, Patrick (ed.), 1969, *Nineteenth Century Philosophy.* New York: Free Press.

Gardner, Howard, 1987, *The Mind's New Science: A History of the Cognitive Revolution.* New York: Basic Books.

Garver, Newton and Seung-Chong Lee, 1995, *Derrida & Wittgenstein.* Philadelphia, PA: Temple University Press.

Gasché, Rodolphe, 1988, *The Tain of the Mirror: Derrida and the Philosophy of Reflection.* Cambridge, MA: The University of Harvard Press.

Gasché, Rodolphe, 1998, *Inventions of Difference: On Jacques Derrida.* Cambridge, MA: The University of Harvard Press.

Gasché, Rodolphe, 1998, *The Wild Card of Reading: On Paul de Man.* Cambridge, MA: The University of Harvard Press.

Gazzaniga, Michael S., Richard B. Ivry and George R. Mangun, 2002, *Cognitive Neuroscience: The Biology of the Mind*, 2nd ed. New York: W.W. Norton and Company.

Gladwell, Malcolm, 2002, *The Tipping Point: How Little Things Can Make a Big Difference*. New York: Back Bay Books.

Gladwell, Malcolm, 2005, *Blink: The Power of Thinking Without Thinking.* New York: Back Bay Books.

Gladwell, Malcolm, 2008, *Outliers: The Story of Success.* New York: Little Brown and Company.

Glass, Arnold Lewis and Keith James Holyoak, 1988, *Cognition, 2nd ed.* New York: Random House.

Globus, Gordon, 1995, *The Postmodern Brain.* Amsterdam: John Benjamins Publishing Company.

Gordon, John, 1986, *Finnegans Wake: a Plot Summary.* New York: Syracuse University Press.

Gray, Chris (ed.), 1995, *The Cyborg Handbook.* New York: Routledge.

Green, David W., et. al., 1996, *Cognitive Science: An Introduction.* Hoboken, NJ: Wiley-Blackwell.

Green, Karen, 2001, *Dummett: Philosophy of Language.* Cambridge: Polity.

Greenfield, Susan A., 1998, *The Human Brain: A Guided Tour.* New York : Basic Books.

Hahn, Lewis Edwin (ed.), 1999 , *The Philosophy of Donald Davidson.* Chicago: Open Court.

Hall, Clavin S., 1999, *A Primer of Freudian Psychology.* New York: Penguin.

Hampshire, Stuart (ed.), 1954, *The Age of Reason: The 17th Century Philosophers.* New York: New American Library.

Haraway, Donna J., 1990, *Simians, Cyborgs, and Women: The Reinvention of Nature.* New York: Routledge.

Harland, Richard, 1987, *Superstructuralism: The Philosophy of Structuralism and Post-Structuralism.* London: Routledge.

Haugeland, John, 1989, *Artificial Intelligence: The Very Idea*. Cambridge, MA: The MIT Press.

Haugeland, John, 1997, *Mind Design II: Philosophy, Psychology, and Artificial Intelligence*. Cambridge, MA: The MIT Press.

Hawkins, Jeff with Sandra Blakeslee, 2004, *On Intelligence: How a New Understanding of the Brain Will Lead to the Creation of Truly Intelligent Machines*. New York: Owl Books.

Hayles, N. Katherine, 1999, *How We Became Posthuman: Virtual Bodies in Cybernetics, Literature, and Informatics*. Chicago: University of Chicago Press.

Hegel, G.W.F., J. Glen Gray (ed.), 1970, *On Art, Religion, Philosophy: Introductory Lectures on the Realm of Absolute Spirit*. New York: Harper Torchbooks.

Hegel, G.W.F., J.B. Baillie (trans.), 1977, *The Phenomenology of Mind*. New York: Humanities Press Inc.

Hegel, G.W.F., Allen W. Wood (ed.), H.B. Nisbet (trans.), 1991, *Elements of the Philosophy of Right*. Cambridge: Cambridge University Press.

Heidegger, Martin, John Macquarrie and Edward Robinson (trans.), 1962, *Being and Time*. New York: Harper and Row.

Heidegger, Martin, William Lovitt (trans.), 1977, *The Question Concerning Technology*. New York: Harper and Row.

Heidegger, Martin, Albert Hofstadter (trans.), 1983, *Poetry, Language, Thought*. New York: Harper Colophon Books.

Heidegger, Martin, David Farell Krell (ed.), 1993, *Basic Writings*. San Francisco: Harper.

Held, David, 1980, *Introduction to Critical Theory: Horkheimer to Habermas*. Berkeley: University of California Press.

Herman, Edward S., and Noam Chomsky, 2002, *Manufacturing Consent: The Political Economy of the Mass Media*. New York: Pantheon Books.

Hofstadter, Douglas R., 1999, *Gödel, Escher, Bach: An Eternal Golden Braid*. New York: Basic Books.

Hofstadter, Douglas R. and Daniel C. Dennett, 2001, *The Mind's I: Fantasies and Reflections on Self & Soul.* New York: Basic Books.

Hofstadter, Douglas R., 2007, *I am a Strange Loop.* New York: Basic Books.

Holderlin, Friedrich, Willard R. Trask (trans.), 1965, *Hyperion.* New York: The New American Library.

Honderich, Ted (ed.), 2005, *The Oxford Guide to Philosophy.* Oxford: Oxford University Press.

Hookway, Christopher, 1988, *Quine: Language, Experience, and Reality.* Palo Alto: Stanford University Press.

Horkheimer, Max and Theodor W. Adorno, Edmund Jephcott (trans.), 2007, *Dialectic of Enlightenment.* Palo Alto: Stanford University Press.

Hospers, John, 1964, *An Introduction to Philosophical Analysis.* Upper Saddle River, New Jersey: Prentice Hall.

Howells, Christina, 1998, *Derrida: Deconstruction from Phenomenology to Ethics.* London: Polity.

Hume, David, V.C. Chappell (ed.), 1963, *The Philosophy of David Hume.* New York: The Modern Library.

Hume, Robert Ernest (trans.), 1968, *The Thirteen Principal Upanishads.* Oxford: Oxford University Press.

Husserl, Edmund, W.R. Boyce Gibson (trans.), 2012, *Ideas: General Introduction to Pure Phenomenology.* New York: Routledge.

Inwood, Brad and Lloyd P. Gerson (trans.), 2008, *The Stoics Reader: Selected Writings and Testimonia.* Indianapolis, IN: Hackett Publishing Company, Inc.

Jahnke, John C. and Ronald H. Nowaczyk, 1997, *Cognition.* Upper Saddle River, NJ: Prentice Hall.

Jameson, Fredric, 1990, *Postmodernism, or, The Cultural Logic of Late Capitalism.* Durham, NC: Duke University Press.

Jarvis, Simon, 1998, *Adorno: A Critical Introduction.* London: Polity Press.

Jay, Martin, 1973, *The Dialectical Imagination: History of the Frankfurt School and the Institute of Social Research 1923-1950.* Boston: Little Brown and Company.

Jay, Martin, 1984, *Adorno.* Cambridge, MA: Harvard University Press.

Johnson, Christopher, 1999, *Derrida: The Great Philosophers.* London: Routledge.

Johnson, Michael L., 1988, *Mind, Language, Machine: Artificial Intelligence in the Poststructuralist Age.* New York: St. Martin's Press.

Johnson, Steven, 2001, *Emergence: The Connected Lives of Ants, Brains, Cities, and Software.* New York: Scribner.

Joyce, James, 1968, *Finnegans Wake.* New York: The Viking Press.

Joyce, James, Richard Ellman (ed.), 1975, *Selected Joyce Letters.* New York: Viking Press.

Joyce, Stanislaus, 1969, *My Brother's Keeper.* New York: Viking Press.

Jung, Carl, R.F.C. Hull (trans.), 1971, *The Portable Jung.* New York: The Viking Press.

Jung, Carl, Mark Kyburz, John Peck and Sonu Shamdasani (trans.), 2009, *The Red Book.* New York: W.W. Norton and Company.

Kahneman, Daniel, 2011, *Thinking, Fast and Slow.* New York: Farrar, Straus and Giroux.

Kant, Immanuel, Lewis White Beck (ed.), Paul Carus (trans.), 1950, *Prolegomena to Any Future Metaphysics.* New York: Macmillan Publishing Company.

Kant, Immanuel, Norman Kemp Smith (trans.), 1965, *Immanuel Kant's Critique of Pure Reason – Unabridged Edition.* New York: St. Martin's Press.

Kant, Immanuel, Mary J. Gregor (trans.), 1974, *Anthropology from a Pragmatic Point of View.* The Hague: Martenus Nijhoff.

Kant, Immanuel, Werner S. Pluhar (trans.), 1987, *Critique of Judgment.* Indianapolis: Hackett Publishing Company.

Kant, Immanuel, Abbot Thomas Kingsmill (trans.), 2004, *Critique of Practical Reason*. New York: Dover Publications.

Kates, Joshua, 2005, *Essential History: Jacques Derrida and the Development of Deconstruction*. Evanston: IL, Northwestern University Press.

Kellogg, Ronald T., 2007, *Fundamentals of Cognitive Psychology*. Thousand Oaks, CA: Sage Publications.

Kelly, Kevin, 1994, *Out of Control: The Rise of Neo-Biological Civilization*. Reading, MA: A William Patrick Book.

Kierkegaard, Søren, Edna H. Hong and Howard V. Hong (trans.), 1983, *Fear and Trembling / Repetition: Kierkegaard's Writings, Vol. 6*. Princeton: Princeton University Press.

Kierkegaard, Søren, Edna H. Hong and Howard V. Hong (trans.), 1992, *Concluding Unscientific Postscript to Philosophical Fragments 1: Kierkegaard's Writings, Vol. 12.1)*. Princeton: Princeton University Press.

Kornfield, Jack, Gil Fronsdal (eds.), 1993, *The Teachings of the Buddha*. Boston, MA: Shambhala.

Krell, David Farrell, 1990, *Of Memory, Reminiscence, and Writing: On the Verge*. Bloomington, IN: Indiana University Press.

Kripke, Saul A., 1984, *Wittgenstein on Rules and Private Language: An Elementary Exposition*. Cambridge, Massachusetts: Harvard University Press.

Kristeva, Julia, Margaret Waller (trans.), 1984, *Revolution in Poetic Language*. New York: Columbia University Press.

Kristeva, Julia, Kelly Oliver (ed.), 1997, *The Portable Kristeva*. New York: Columbia University Press.

Kurzweil, Ray, 1999, *The Age of Spiritual Machines: When Computers Exceed Human Intelligence*. New York: Viking.

Lacan, Jacques, 1977, *Écrits: A Selection*. New York: W.W. Norton.

Laclau, Ernesto and Chantal Mouffe, 2001, *Hegemony and Socialist Strategy: Towards a Radical Democratic Politics*. New York: Verso.

Lakoff, George and Mark Johnson, 1980, *Metaphors We Live By*. Chicago: University of Chicago Press.

Lakoff, George and Mark Johnson, 1999, *Philosophy in the Flesh: The Embodied Mind and Its Challenge to Western Thought.* New York: Basic Books.

Laszlo, Ervin, 1972, *Introduction to Systems Philosophy.* New York: Harper Torchbooks.

Laszlo, Ervin, 1996, *The Systems View of the World: A Holistic Vision for Our Time.* Cresskill, NJ: Hampton Press, Inc.

Leary, Timothy, Michael Horowitz and Vicki Marshal (eds.), 1994, *Chaos & Cyber Culture.* Berkeley, CA: Ronin.

Leary, Timothy, 2008, *Cyber Punks, Cyber Freedom: Change Reality Screens.* Berkeley, CA: Ronin.

Lechte, John (ed.), 1994, *Fifty Key Contemporary Thinkers: From Structuralism to Postmodernity.* London: Routledge.

Leibniz, G.W., G.H.R. Parkinson (ed.), 1973, *Leibniz: Philosophical Writings.* London: J.M. Dent & Sons.

Leibniz, G.W., Roger Ariew And Daniel Garber (trans.), 1989, *Philosophical Essays.* Indianapolis: Hackett Publishing Company.

Leitch, Vincent B., 1982, *Deconstructive Criticism: An Advanced Introduction.* New York: Columbia University Press.

Lemaire, Anika, 1996, *Jacques Lacan.* London: Routledge.

Lepore, Ernest and Zenon Pylyshyn (eds.), 1999, *What is Cognitive Science.* Malden, MA: Blackwell Publishers Inc.

Levitt, Steven D. and Stephen J. Dubner, 2005, *Freakonomics.* New York: William Morrow.

Lindberg, Jordan J., 2000, *Analytical Philosophy: Beginnings to the Present.* New York : McGraw Hill.

Liu, Catherine, 2000, *Copying Machines: Taking Notes for the Automaton.* Minneapolis, MN: University of Minnesota Press.

Locke, John, Peter H. Nidditich (ed.), 1979, *An Essay Concerning Human Understanding.* Oxford: Oxford University Press.

Lowe, E.J., 1995, *Locke: on Human Understanding.* London: Routledge.

Lyotard, Jean François, Geoff Bennington (trans.), 1984, *The Postmodern Condition: A Report on Knowledge.* Minneapolis: University of Minnesota Press.

Macy, Joanna, 1991, *Mutual Causality in Buddhism and General Systems Theory: The Dharma of Natural Systems.* Albany, NY: State University of New York Press.

Magill, Frank N. (ed.), 1990, *Masterpieces of World Philosophy.* New York: HarperCollins, Publishers.

Malone, John C., 2009, *Psychology: Pythagoras to Present.* Cambridge, MA: The MIT Press.

Marcuse, Herbert, 1987, *Eros and Civilization: A Philosophical Inquiry into Freud.* London: Routledge.

Marcuse, Herbert, 1991, *One-Dimensional Man.* Boston: Beacon Press.

Marx, Karl, Lawrence H. Simon (ed.), 1994, *Selected Writings.* Indianapolis: Hackett Publishing Company.

Matt, Daniel C. (trans.), *Zohar: Annotated & Explained.* Woodstock, VT: Skylight Paths Publishing.

Maturana, Humberto R. and Francisco Verela, 1992, *The Tree of Knowledge: The Biological Roots of Human Understanding.* Boston: Shambhala Publications, Inc.

McGilvray, James, 1999, *Chomsky: Language, Mind, and Politics.* Malden, MA: Blackwell Publishers Inc.

McGreal, Ian P. (ed.), 1995, *Great Thinkers of the Eastern World: The Major Thinkers and the Philosophical and Religious Classics of China, India, Japan, Korea, and the world of Islam.* New York: HarperCollins, Publishers.

McLuhan, Marshal, 1964, *Understanding Media: The Extensions of Man.* New York: Mentor.

McLuhan, Marshal and Quentin Fiore, 1968, *The Medium is the Massage: An Inventory of Effects.* New York: Bantam Books.

McQuillan, Martin (ed.), 2001, *Deconstruction: A Reader*. London: Routledge.

McQuillan, Martin, 2001, *Paul de Man*. London: Routledge.

Megill, Allan, 1987, *Prophets of Extremity: Nietzsche, Heidegger, Foucault, Derrida*. Berkeley: University of California Press.

Merrell, Floyd, 1985, *Deconstruction Reframed*. West Lafayette, IN: Purdue University Press.

Minsky, Marvin and Seymour A. Papert, 1987, *Perceptrons: An Introduction to Computational Geometry, Expanded Edition*. Cambridge, MA: The MIT Press.

Minsky, Marvin, 1988, *The Society of Mind*. New York: Simon and Schuster, Inc.

Minsky, Marvin, 2007, *The Emotion Machine: Commonsense Thinking, Artificial Intelligence, and the Future of the Human Mind*. New York: Simon and Schuster, Inc.

Mitchell, Melanie, 2009, *Complexity: A Guided Tour*. Oxford: Oxford University Press.

Mitchell, W.J.T., 1985 , *Against Theory: Literary Studies and the New Pragmatism*. Chicago: University of Chicago Press.

Moriarty, Michael, 1991, *Roland Barthes*. Palo Alto: Stanford University Press.

Mouffe, Chantal, 1996, *Deconstruction and Pragmatism: Simon Critchley, Jacques Derrida, Ernesto Laclau & Richard Rorty*. New York: Routledge.

Nancy, Jean-Luc, Jason E. Smith and Steven Miller (trans.), 2002, *Hegel: The Restlessness of the Negative*. Minneapolis: University of Minnesota Press.

Natanson, Maruice, 1974, *Edmund Husserl: Philosopher of Infinite Tasks*. Evanston, Illinois: Northwestern University Press.

Nelson, Lynn Hankinson and Jack Nelson, 1999, *On Quine*. Stamford: Connecticut: Wadsworth Publishing.

Neumann, John von, 2000, *The Computer and the Brain: 2nd ed.* New Haven, CT: Yale University Press.

Newton, Isaac, I. Bernard Cohen and Richard S. Westfall (eds.), 1995, *Newton: Texts, Backgrounds, Commentaries*. New York: W.W. Norton and Company.

Nietzsche, Friedrich, Walter Kaufmann and R.J. Hollingdale (trans), 1968, *The Will to Power*. New York: Vintage Books.

Nietzsche, Friedrich, Walter Kaufmann (trans.), 1954, *The Portable Nietzsche*. New York: Viking Penguin Inc.

Nietzsche, Friedrich, Walter Kaufmann (trans.), 1992, *Basic Writings*. New York: Modern Library.

Nilsson, Nils J., 2010, *The Quest for Artificial Intelligence: A History of Ideas and Achievements*. Cambridge: Cambridge University Press.

Noonan, Harold, 2007, *Hume*. Oxford: Oneworld.

Norris, Christopher, 1988, *Derrida*. Cambridge, MA: Harvard University Press.

Norris, Christopher, 2002, *Hilary Putnam: Realism, Reason and the Uses of Uncertainty*. Manchester: Manchester University Press.

Olkowski, Dorothea, 1999, *Gilles Deleuze and the Ruin of Representation*. Berkeley: University of California Press.

Papadakis, Andreas, Catherine Cooke and Andrew Benjamin, 1989, *Deconstruction: Omnibus Volume*. Milan, Italy: Rizzoli.

Parmenides of Elea, David Galop (ed., trans.), 2000, *Fragments*. Toronto: University of Toronto Press.

Pears, David, 1986, *Ludwig Wittgenstein*. Cambridge, Massachusetts, Harvard University Press.

Peirce, Charles Sanders, Morris R. Cohen, Kenneth Laine Ketner (eds.), 1998, *Chance, Love, and Logic: Philosophical Essays*. Lincoln: University Of Nebraska Press.

Penrose, Roger, 2002, *The Emperor's New Mind: Concerning Computers, Minds, and the Laws of Physics*. Oxford: Oxford University Press.

Perkins, Franklin, 2007, *Leibniz: A Guide for the Perplexed*. London: Continuum.

Philo, C.D. Younge (trans.), 1993, *The Works of Philo: Complete and Unabridged.* Peabody, MA: Hendrickson Publishers, Inc.

Pinker, Steven, 1995, *The Language Instinct: How the Mind Creates Language.* New York: HaprerCollins Publishers Inc.

Pinker, Steven, 1997, *How the Mind Works.* New York: W.W. Norton and Company.

Pinker, Steven, 2000, *Words and Rules: The Ingredients of Language.* New York: HaprerCollins Publishers Inc.

Pinker, Steven, 2002, *The Blank Slate: The Modern Denial of Human Nature.* New York: Penguin Books.

Pinker, Steven, 2008, *The Stuff of Thought: Language as a Window into Human Nature.* New York: Penguin Books.

Plato, Edith Hamilton and Huntington Cairns (eds.), 1989, *The Collected Dialogues of Plato.* Princeton: Princeton University Press.

Popkin, Richard H. (ed.), 1966, *The Philosophy of the 16th and 17th Centuries.* New York: The Free Press.

Putnam, Hilary, 1981, *Reason, Truth and History.* Cambridge: Cambridge University Press.

Putnam, Hilary, 1988, *The Many Faces of Realism.* Chicago: Open Court.

Putnam, Hilary, 1994, *Representation and Reality.* Cambridge, MA: A Bradford Book.

Putnam, Hilary, 1993, *Pragmatism: An Open Question.* Hoboken, NJ: Wiley-Blackwell

Putnam, Hilary, 2004, *The Collapse of the Fact/Value Dichotomy and Other Essays.* Cambridge, MA: Harvard University Press.

Quine, W.V., 1964, *Word and Object.* Cambridge, MA: MIT Press.

Quine, W.V., 1980, *From a Logical Point of View: Nine Logico-Philosophical Essays, 2nd Revised Ed.* Cambridge, MA: Harvard University Press.

Rajchman, John and Cornel West, 1985 , *Post-Analytic Philosophy.* New York: Columbia University Press.

Rajchman, John, 2000, *The Deleuze Connections*. Cambridge, MA: MIT Press.

Rapaport, Herman, 2002, *Later Derrida: Reading the Recent Work*. London: Routledge.

Rasch, William and Carry Wolfe (eds.), *Observing Complexity: Systems Theory and Postmodernity*. Minneapolis: University of Minnesota Press.

Reijen, Willem Van, 1991, *Adorno: An Introduction*. Philadelphia, PA: Pennbridge Books.

Robinson, James Mansley, 1968, *An Introduction to Early Greek Philosophy: The Chief Fragments and Ancient Testimony With Connecting Commentary*. Boston: Houghton Mifflin School.

Roediger, Henry L. III, Elizabeth D. Capaldi, Scott G. Paris, Janet Polivy and C. Peter Herman, 1997, *Psychology*. Stamford, CT: Wadsworth Publishing.

Rorty, Richard, 1981, *Philosophy and the Mirror of Nature*. Princeton: Princeton University Press.

Rorty, Richard, 1989, *Contingency, Irony and Solidarity*. Cambridge: Cambridge University Press.

Rorty, Richard, 1990, *Objectivity, Relativism, and Truth: Philosophical Papers, Vol. 1*. Cambridge: Cambridge University Press.

Rorty, Richard, 1991, *Essays on Heidegger and Others: Philosophical Papers, Vol. 2*. Cambridge: Cambridge University Press.

Rorty, Richard, 1998, *Truth and Progress: Philosophical Papers, Vol. 3*. Cambridge: Cambridge University Press.

Rorty, Richard, 2000, *Philosophy and Social Hope*. New York: Penguin Books.

Rorty, Richard, 2007, *Philosophy as Cultural Politics: Philosophical Papers, Vol. 4*. Cambridge: Cambridge University Press.

Ross, Sir David, 1923, *Aristotle*. London: Routledge.

Royle, Nicholas, 2000, *Deconstruction: A User's Guide*. Basingstoke, UK: Palgrave Macmillan.

Russell, Bertrand, 1967, *A History of Western Philosophy*. New York: Simon & Schuster.

Russel, Stuart J. and Peter Norvig, 1995, *Artificial Intelligence: A Modern Approach*. Upper Saddle River, NJ: Prentice Hall.

Sagan, Carl, 1979, *Broca's Brain: Reflections on the Romance of Science*. New York: Random House.

Santillana, Giorgiode (ed.), 1954, *The Age of Adventure: The Renaissance Philosophers*. New York: New American Library.

Sarup, Madan, 1993, *An Introductory Guide to Post-Structuralism and Postmodernism*. Athens, GA: University of Georgia Press.

Saunders, Jason L. (ed.), 1966, *Greek and Roman Philosophy after Aristotle*. New York: The Free Press.

Saussure, Ferdinand de, 1997, *Course in General Linguistics*. Chicago: Open Court.

Schmitt, Richard, 1997, *Introduction to Marx and Engels: A Critical Introduction*. Boulder, Colorado: Westview Press.

Searle, John, 1986, *Minds, Brains and Science*. Cambridge, MA: Harvard University Press.

Sekida, Katsuki (trans.), 1977, *Two Zen Classics: Mumonkan and Hekiganroku*. New York: Weatherhill.

Shakespeare, William, G. Blakemore Evans (ed.), 1974, *The Riverside Shakespeare*. Boston, MA: Houghton Mifflin Company.

Shelley, Mary, 1963, *Frankenstein, or, The Modern Prometheus*. New York: Signet Classic.

Shelley, Mary, 2010, *The Last Man*. Mineola, NY: Dover Publications, Inc.

Silver, Nate, 2012, *The Signal and the Noise: Why So Many Predictions Fail – but Some Don't*. New York: Penguin Press.

Sim, Stuart, 1999, *A Routledge Critical Dictionary of Postmodern Thought*. London: Routledge.

Singer, Peter, 2001, *Hegel: A Very Short Introduction*. Oxford: Oxford University Press.

Singer, Peter, 2001, *Marx: A Very Short Introduction*. Oxford: Oxford University Press.

Smith, Neilson Voyne, 2004, *Chomsky: Ideas and Ideals*. Cambridge: Cambridge University Press.

Smith, T.V. (ed.), 1956, *Philosophers Speak for Themselves: From Aristotle to Plotinus*. Chicago: University of Chicago Press.

Solomon, Robert C. and Kathleen M. Higgins, 1996, *A Short History of Philosophy*. Oxford: Oxford University Press.

Spinoza, Benedict, Edwin Curley (trans.), 1996, *Ethics*. New York: Penguin Books.

Staten, Henry, 1986, *Wittgenstein and Derrida*. Lincoln: University of Nebraska Press.

Sternberg, Robert J., 2006, *Cognitive Psychology, 4th ed.* Belmont, CA: Thomson Wadsworth.

Stiegler, Bernard, Richard Beardsworth & George Collins (trans.), 1998, *Technics and Time, 1: The Fault of Epimetheus*. Stanford, CA: Stanford University Press.

Stiegler, Bernard, Stephen Barker (trans.), 2009, *Technics and Time, 2: Disorientation*. Stanford, CA: Stanford University Press.

Stiegler, Bernard, Stephen Barker (trans.), 2011, *Technics and Time, 3: Cinematic Time and the Question of Malaise*. Stanford, CA: Stanford University Press.

Stroll, Avrum, 2001, *Twentieth-Century Analytic Philosophy*. New York: Columbia University Press.

Stumpf, Samuel Enoch, 1994, *Philosophy: History and Problems*. New York: McGraw-Hill, Inc.

Sturrock, John, 1981, *Structuralism and Since: From Lévi Strauss to Derrida*. Oxford: Oxford University Press.

Surowiecki, James, 2005, *The Wisdom of Crowds*. New York: Anchor Books.

Taleb, Nassim Nicholas, 2007, *The Black Swan: The Impact of the Highly Improbable*. New York: Penguin Books.

Tarnas, Richard, 1991, *The Passion of the Western Mind: Understanding the Ideas That Have Shaped Our World View*. New York: Ballantine Books.

Taylor, Charles, 1977, *Hegel*. Cambridge: Cambridge University Press.

Taylor, Mark C., 1987, *Alterity*. Chicago: University of Chicago Press.

Taylor, Mark C., 1987, *Erring: A Postmodern A/theology*. Chicago: University of Chicago Press.

Taylor, Mark C., 2003, *The Moment of Complexity: Emerging Network Culture*. Chicago: University of Chicago Press.

Taylor, Mark C., 2007, *After God*. Chicago: University of Chicago Press.

Taylor, Mark C., 2008, *Confidence Games: Money and Markets in a World without Redemption*. Chicago: University of Chicago Press.

Thagard, Paul, 1996, *Mind: Introduction to Cognitive Science*. Cambridge, MA: The MIT Press.

Theodoridis, Sergios and Konstantinos Koutroumbas, 1999, *Pattern Recognition*. San Diego, CA: Academic Press.

Thomson, Alex, 2006, *Adorno: A Guide for the Perplexed*. New York: Continuum.

Tindall, William York, 1959, *A Reader's Guide to James Joyce*. New York: The Noonday Press.

Trupp, Michael, 2000, *On Freud*. Stamford, Connecticut: Wadsworth Publishing.

Turing, Alan M., B. Jack Copland (ed.), 2004, *The Essential Turing: Seminal Writings on Computing, Logic, Philosophy, Artificial Intelligence, and Artificial Life plus the Secrets of Enigma*. Oxford: Oxford University Press.

Twain, Mark, 1990, *The Wit and Wisdom of Mark Twain*. Philadelphia, PA: Running Press.

Tzu, Lao, D.C. Lau (trans.), 1963, *Tao Te Ching*. New York: Penguin Books.

Ulmer, Gregory L., 1984, *Applied Grammatology: Post(e)-Pedagogy from Jacques Derrida to Joseph Beuys*. Baltimore, MD: Johns Hopkins University Press.

Van Doren, Charles, 1991, *A History of Knowledge: Past, Present, Future*. New York: Ballantine Books.

Ward, Glenn, 2003, *Teach Yourself: Postmodernism*. Chicago: McGraw-Hill.

Watson, Peter, 2001, *The Modern Mind: An Intellectual History of the 20th Century*. New York: HarperCollins, Publishers.

Watson, Peter, 2005, *Ideas: A History of Thought and Invention, from Fire to Freud*. New York: Harper Collins, Publishers.

Watts, Alan, 1989, *The Way of Zen*. New York: Vintage Books.

Watts, Duncan J., 2004, *Six Degrees: The Science of a Connected Age*. New York: W.W. Norton and Company.

Weiten, Wayne, 2006, *Psychology: Themes and Variations*. Stamford, Ct: Wadsworth Publishing.

Weitz, Morris, et. al. (eds.), 1967, *Twentieth Century Philosophy: The Analytic Tradition*. New York: Free Press.

Wheeler, Samuel, III, 2000, *Deconstruction as Analytic Philosophy*. Palo Alto: Stanford University Press.

White, Morton (ed.), 1954, *The Age of Analysis: 20th Century Philosophers*. New York: New American Library.

Wiener, Norbert, 1965, *Cybernetics: or the Control and Communication in the Animal and the Machine, 2nd ed.* Cambridge, MA: The MIT Press.

Wiener, Norbert, 1966, *God and Golem, Inc.: A Comment on Certain Points where Cybernetics Impinges on Religion*. Cambridge, MA: The MIT Press.

Winograd, Terry and Fernando Flores, 1987, *Understanding Computers and Cognition: A New Foundation for Design*. Boston: Addison-Wesley.

Winston, Patrick Henry, 1992, *Artificial Intelligence, 3rd ed.* Boston: Addison-Wesley.

Wippel, John F. and Allan B. Wolter (eds.), 1969, *Medieval Philosophy: From St. Augustine to Nicholas of Cusa.* New York: The Free Press.

Wittgenstein, Ludwig, 1965, *The Blue and Brown Books.* New York: Harper and Row.

Wittgenstein, Ludwig, Cyril Barrett (ed.), 1966, *Lectures and Conversations on Aesthetics, Psychology and Religious Belief.* Oxford: Blackwell.

Wittgenstein, Ludwig, 1972, *On Certainty.* New York: Harper Pernnial.

Wittgenstein, Ludwig, G.E.M. Anscombe (trans.), 1973, *Philosophical Investigations, 3rd Ed.* London: Pearson.

Wittgenstein, Ludwig, Peter Winch (trans.), 1984, *Culture and Value.* Chicago: University of Chicago Press.

Wittgenstein, Ludwig, D.F. Pears and B.F. McGuinness (trans.), 1988, *Tractatus Logico-Philosophicus.* London: Routledge.

Woolhouse, R.S., 1988, *A History of Western Philosophy 5: The Empiricists.* Oxford: Oxford University Press.

Zerzan, John, 2012, *Future Primitive Revisited.* Port Townsend, WA: Feral House.

Žižek, Slavoj, 2006, *The Parallax View.* Cambridge, MA: The MIT Press.

Made in United States
Cleveland, OH
26 September 2026

44836272R00423